Phosphorus Management in Crop Production

Phosphorus Management in Crop Production

Nand Kumar Fageria
National Rice and Bean Research Center of EMBRAPA, Brazil

Zhenli L. He
University of Florida, Institute of Food and Agricultural Science,
Indian River Research and Education Center, USA

Virupax C. Baligar
USDA-ARS – Beltsville Agricultural Research Center, USA

CRC Press is an imprint of the
Taylor & Francis Group, an **informa** business

CRC Press
Taylor & Francis Group
6000 Broken Sound Parkway NW, Suite 300
Boca Raton, FL 33487-2742

First issued in paperback 2021

ISBN 13: 978-1-03-209713-8 (pbk)
ISBN 13: 978-1-4987-0586-8 (hbk)

Library of Congress Cataloging-in-Publication Data

Names: Fageria, N. K., 1942- author. | He, Zhenli, author. | Baligar, V. C., author.
Title: Phosphorus management in crop production/authors: Nand Kumar Fageria, Zhenli L. He, Virupax C. Baligar.
Description: Boca Raton, FL: CRC Press, 2017. | Includes bibliographical references and index.
Identifiers: LCCN 2016036026 | ISBN 9781498705868 (hardback: alk. paper)
Subjects: LCSH: Phosphorus in agriculture.
Classification: LCC S587.5.P56 F34 2017 | DDC 631.8/5--dc23
LC record available at https://lccn.loc.gov/2016036026

Visit the Taylor & Francis Web site at
http://www.taylorandfrancis.com

and the CRC Press Web site at
http://www.crcpress.com

Contents

Preface

World population is projected to be 9.1 billion by 2050 and most of this increase will occur in the developing countries. In addition, living standards are steadily improved worldwide. This will enhance the demand for food and fiber in the future. It is projected that global food supply will have to be increased by 60%–70% by 2050 to feed the increasing world population. There are three possible alternatives to increase yield of food and fiber crops: (1) improving yield per unit land area, (2) expanding land area for crop production, and (3) increasing cropping intensity. All these will require higher inputs of resources such as water, nutrients, and other chemicals such as fungicides, insecticides, and herbicides. Since arable land area in the world is limited, further yield increase in food, fiber, and energy will have to rely on yield increases per unit land area, especially in Asia, Europe, and North America.

Breeding and improved management practices have been and will continue to have a major role in future world food security. Improved management practices include use of essential plant nutrients in adequate amounts, proper balance, appropriate timing of application, effective methods, and sources of application. Nitrogen (N), phosphorus (P), and potassium (K) are the three major essential nutrients that are critical in determining crop yields. After N, P is the next most important nutrient that affects sustainable crop production. Most of the available lands for increasing world food production are in Africa and South America where soils are acidic in reaction and low in nutrient availability. Phosphorus deficiency is one of the most important yield-limiting factors on acid soils in the tropical and subtropical regions, which is related to low bioavailability of P attributed to chemical fixation by sesquioxides of iron and aluminum.

Application of P fertilizers is essential to sustain crop production in these regions. However, many water bodies are sensitive to P input; loss of P from agricultural practices has been blamed to trigger water eutrophication and algal bloom. Moreover, phosphate reserves are limited in the world. It is projected that phosphorus production will peak by 2034 and decline thereafter. Therefore, optimized management of phosphorus fertilization is imperative to both sustainable agriculture and environmental quality.

The proposed book will be a timely publication that discusses efficient management of phosphorus in sustainable crop production. The last book of this nature was published by the American Society of Agronomy, the Crop Science Society of America, and the Soil Science Society of America in 2005. Several articles are available on this topic, but scattered in various journals and book chapters. Hence, compiling all the information in a single book will be useful for multiple disciplines of agriculture and environmental sciences, including soil science, agronomy, horticulture, plant breeding, plant pathology, entomology, and environmental science. It can be a good reference book for students, professors, research scientists, extension specialists, private consultants, and government agents. A large number of tables and figures are included to make the book more valuable and informative to the readers.

Dr. Nanda Kumar Fageria completed the major part of the book write-up before his untimely death on July 6, 2014, at his home in Goiania, Goias, Brazil.

Dr. Fageria was appreciative of the National Rice and Bean Research Center, Empresa Brasileira de Pesquisa Agropecuaria (EMBRAPA), Santo Antonio de Goias, Brazil, for providing the facilities necessary to write this book, and, furthermore, he was grateful to all the staff members of National Rice and Bean Research Center of EMBRAPA for the cooperation and friendship. The financial support provided to Dr. Fageria by the Brazilian Scientific and Technological Research Council (CNPq) facilitated his accomplishments of many research projects since 1989. Some of the research results are included in this book.

Dr. Fageria always took pride in expressing his deep appreciation to his beloved wife, Shanti Fageria; his daughter, Savita; his sons, Rajesh and Satya Pal; his daughter-in-law, Neera; and his three grandchildren, Anjit, Maia, and Sofia, for their love, encouragement, and understanding.

Junior authors of this book have more than 20 years of collaborative research association and friendship with Dr. Fageria. They are honored to be coauthors with Dr. Fageria to complete and publish this book. They greatly appreciate Dr. Peter J Stoffella and Xiaoping Xin for their assistance with reading and revision of the manuscripts and thank Randy Brehm, Kate Gallo, and other staff at the Taylor & Francis Group, CRC Press, for their excellent advice and efficient assistance in publishing this book.

Authors

Nand Kumar Fageria received his PhD in agronomy, and had been senior research soil scientist at the National Rice and Bean Research Center, Empresa Brasileira de Pesquisa Agropecuária, since 1975. Dr. Fageria was a nationally and internationally recognized expert in the area of mineral nutrition of crop plants and a research fellow and an ad hoc consultant for the Brazilian Scientific and Technological Research Council (CNPq) since 1989. Dr. Fageria was the first to identify zinc deficiency in upland rice grown on Brazilian Oxisols in 1975. He had developed crop genotype screening techniques for aluminum and salinity tolerance and nitrogen, phosphorus, potassium, and zinc use efficiency. Dr. Fageria also established adequate soil acidity indices like pH, base saturation, Al saturation, and Ca, Mg, and K saturation for dry bean grown on Brazilian Oxisols in conservation or no-tillage system. He also determined adequate and toxic levels of micronutrients in soil and plant tissues of upland rice, corn, soybean, dry bean, and wheat grown on Brazilian Oxisols. Dr. Fageria determined adequate rate of N, P, and K for lowland and upland rice grown on Brazilian lowland soils, locally known as "Varzea" and Oxisols of "Cerrado" region, respectively. He also screened a large number of tropical legume cover crops for acidity tolerance and N, P, and micronutrient use efficiency. Dr. Fageria characterized chemical and physical properties of Varzea soils of several states of Brazil, which can be helpful in better fertility management of these soils for sustainable crop production. Dr. Fageria also determined adequate rate and sources of P and acidity indices for soybean grown on Brazilian Oxisols. The results of all these studies have been published in scientific papers, technical bulletins, book chapters, and congress or symposium proceedings.

Dr. Fageria was the author/coauthor of twelve books and more than 320 scientific journal articles, book chapters, review articles, and technical bulletins. His four books, individually titled *The Use of Nutrients in Crop Plants* published in 2009; *Growth and Mineral Nutrition of Field Crops*, third edition, published in 2011; *The Role of Plant Roots in Crop Production* published in 2013; and *Mineral Nutrition of Rice* published in 2014, are among the best-sellers of CRC Press. Dr. Fageria had been invited several times by the editor of *Advances in Agronomy*, which is a well-established and highly regarded serial publication, to write review articles on nutrient management, enhancing nutrient use efficiency in crop plants, ameliorating soil acidity by liming on tropical acid soils for sustainable crop production, and the role of mineral nutrition on root growth of crop plants. He had been an invited speaker to several national and international congresses, symposiums, and workshops. He was a member of the editorial board of the *Journal of Plant Nutrition* and *Brazilian Journal of Plant Physiology* and since 1990 he was member of the international steering committee of symposiums on plant–soil interactions at low pH. He was an active member of the American Society of Agronomy and Soil Science Society of America.

Zhenli He received his PhD in soil environmental chemistry from Zhejiang University of China and is currently research foundation professor of soil and water sciences at the University of Florida, Institute of Food and Agricultural Sciences, Indian River Research and Education Center, Fort Pierce, Florida. Dr. He teaches *soil quality* and *nanotechnology application in food, agriculture, and environment* courses to graduate and undergraduate students. He has trained over 60 graduate students, postdoctoral fellows, and visiting scientists. His research interests include biogeochemical processes of nutrients and contaminants in soil and the environment, soil/water quality and remediation, bioavailability and utilization efficiency of nutrients and fertilizers, plant nutrition, and nanotechnology application. Dr. He studied soil chemistry of phosphorus since 1985 and published more than 60 papers related to biogeochemistry, bioavailability, and cycling of phosphorus in agroecosystems of tropical and subtropical regions. He is the author/coauthor of two books, 28 book chapters, and more than 200 refereed academic journal articles. He serves as an editor for the

Journal of Soils and Sediments, and on the editorial board for several other journals. Dr. He is a fellow of the American Society of Agronomy and Soil Science Society of America.

Virupax C. Baligar earned his PhD in agronomy and physiology and currently is a senior research scientist with the United States Department of Agriculture, Agricultural Research Service, at Beltsville Agricultural Research Center, Beltsville, Maryland. He is author, coauthor, and coeditor of several books and more than 290 scientific journal articles, book chapters, and review papers and is an elected fellow of the American Society of Agronomy and Soil Science Society of America. Dr. Baligar has served as advisor, consultant, and collaborator on international research programs in Brazil, Peru, China, Chile, Greece, New Zealand, Germany, the Netherlands, and Serbia, Croatia, Bosnia, Slovenia, Montenegro, and Macedonia (former Yugoslavia). He served as advisor to the World Bank, OAS, and Empresa Brasileira de Pesquisa Agropecuária programs in Brazil and served as advisor and committee member for MS and PhD students at Virginia Tech, Blacksburg, Virginia; West Virginia University, Morgantown, West Virginia; Beijing Agricultural University, Beijing, China; National Agricultural University, La Molina, Lima, Peru; University of Florida, IRREC Fort Pierce, TREC Homestead, Florida; and California State University, Dominguez, Carson, California.

1 World Phosphate Situation and Factors Affecting Phosphorus Availability to Plants in Soil

1.1 INTRODUCTION

World food and fiber requirements are increasing rapidly due to increasing world population and aspiration for better quality of life, especially in developing countries. Phosphorus (P) is one of the most important essential plant nutrients in crop production. Ozanne (1980) reported that P is indispensable for all forms of life because of its genetic role in ribonucleic acid and function in energy transfers via adenosine triphosphate. After nitrogen (N), P has more widespread influence on both natural and agricultural ecosystems than any other essential plant elements (Brady and Weil, 2002; Fageria, 2009). Vassilev et al. (2006) and Abbasi et al. (2013) reported that P is the second most limiting nutrient after N in the majority of soils throughout the world and is unavailable to plants under most soil conditions. Without P in the environment, no living organisms could exist. It is an essential nutrient for both plants and animals. It is necessary for such life processes as photosynthesis, the synthesis and breakdown of carbohydrates, and the transfer of energy within plants. In addition, P does not occur as abundantly in soil as other major nutrients such as N and potassium (K). Farming systems are changing, and farmers are placing ever greater emphasis on judicious use of fertilizers to increase yield, lowering cost of crop production and reducing environmental pollution. In addition, most of the cereals and legumes translocate a large part of the absorbed P in the grains. P supplement to the soil is an essential component of modern crop production systems. There is no substitute for P in the production of crops and animals for food, fiber, and other essential needs.

It has been estimated that 5.7 billion ha of land worldwide contains minimal available P for sustainable crop production (Batjes, 1997; Hinsinger, 2001). Phosphorus deficiency is very common in crop plants, especially in highly weathered acid soils of tropical as well as temperate climates (Fageria, 1989, 2009; Fageria and Baligar, 2008). P deficiency on such soils is related to a low natural level as well as high immobilization capacity of P of these soils. The authors studied the influence of essential micro- and macronutrients on the growth of upland rice (*Oryza sativa*) grown on a Brazilian Oxisol (Figure 1.1). Phosphorus was the most growth-limiting nutrient in upland rice as compared to other essential plant nutrients. However, deficiencies of calcium (Ca), magnesium (Mg), boron (B), and zinc (Zn) were very acute. Due to a low natural P level and high immobilization capacity, a heavy dose of P is needed to achieve high crop production on these soils (Fageria et al., 1982). Fageria and Barbosa Filho (1987) studied P fixation capacity of Brazilian Oxisol (Table 1.1) using triple superphosphate as a source of P fertilization. The P recovery varied from 10% to 23% depending on fertilizer rate and reaction time. These results indicated that a large part of applied P in soluble fertilizers is immobilized in the highly weathered Oxisols.

Phosphorus deficiency has been identified as one of the major limiting factors for crop production in highly weathered Oxisols and Ultisols in many parts of the world (Sanchez and Salinas, 1981; Fageria et al., 1982; Haynes, 1984; Fageria, 2001). For example, in Brazil upland rice is mostly

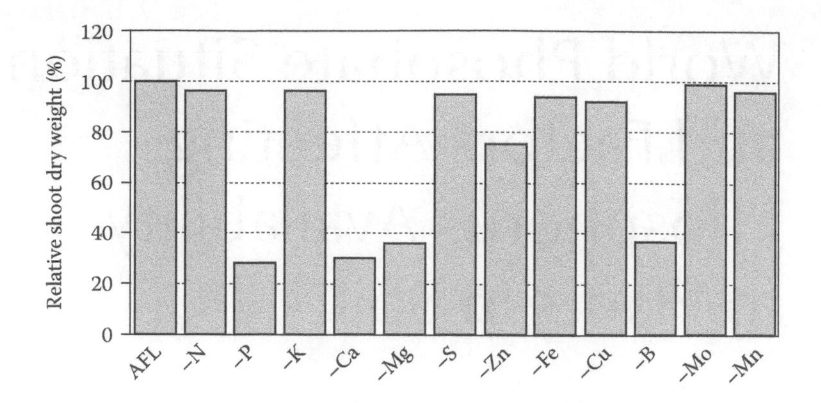

FIGURE 1.1 Relative shoot dry weight of upland rice grown on a Brazilian Oxisol under different nutrient treatments. AFL, adequate fertility level. (From Fageria, N.K. and Baligar, V.C., *J. Plant Nutr.*, 20, 1279, 1997a.)

TABLE 1.1
Phosphorus Recovery (%) in the Oxisol as a Function of P Rate and Reaction Time

	Reaction Time (Days)						
P Rate (mg kg^{-1})	0	17	31	45	60	80	Mean
0	—	—	—	—	—	—	—
25	13	7	7	6	12	13	10
50	17	16	10	16	12	13	14
75	16	16	19	14	13	14	15
100	15	17	22	21	13	20	18
125	20	17	20	18	17	18	18
150	21	22	18	19	21	15	19
175	23	20	20	19	19	17	20
200	29	20	19	23	24	23	23
LSD (0.05)	5.82	4.38	5.43	4.32	5.39	4.73	17

Source: Fageria, N.K., *Tropical Soils and Physiological Aspects of Crops*, EMBRAPA, Brasilia, Brazil, 1989.

grown in the central part locally known as "Cerrado" region. This area represents about 22% of the area of the country. Most of the soils in the Cerrado region are Oxisols and Ultisols. These soils are acidic in reactions and are characterized by low fertility, including P (Fageria and Baligar, 2008; Fageria, 2013, 2014). Similarly, in Brazil there are 35 million hectares of lowlands, known locally as "Varzea." These areas represent one of the largest lowlands of the world, which can be brought under agricultural production. At present less than 2% of the Varzea soils are under crop production. Lowland rice is the main crop grown during the rainy season. But during the dry season, other crops can be grown. These soils are acidic in reactions, and P fixation is one of the main problems (Fageria et al., 1991, 1997; Fageria and Baligar, 1996). Application of adequate rate of P is an important factor for rice production on these soils. Applied P has an influence on straw yield, grain yield, and panicle density of upland rice grown on an Oxisol of central Brazil (Table 1.2). Straw yield, grain yield, as well as panicle density were significantly increased with P fertilizers. Similar responses occurred for upland rice to P fertilization (Figure 1.2). Upland rice growth was

TABLE 1.2
Response of Upland Rice to P Application on a Brazilian Oxisol

P Level (mg kg^{-1})	Straw Yield (g/4 Plants)	Grain Yield (g/4 Plants)	Panicle Number (per 4 Plants)
0	48.7	16.7	21
50	54.2	53.9	23
100	57.6	7.9	24
150	58.2	56.2	24
200	62.0	57.1	25
400	83.5	58.9	34
R^2	0.99**	0.90**	0.99**

Source: Fageria, N.K., *Commun. Soil Sci. Plant Anal.*, 32, 2603, 2001.
**Significant at the 1% probability level.

FIGURE 1.2 Upland rice growth at three P levels grown on a Brazilian Oxisol. (From Fageria, N.K., *Mineral Nutrition of Rice*, CRC Press, Boca Raton, FL, 2014.)

significantly increased with the addition of 100 and 200 mg P kg^{-1} soil as compared to control (without P fertilizer). In the control, P deficiency was so severe that plants did not produce tillering and no panicles. Fageria and Baligar (2004) also studied influence of N, P, and K fertilizer on the growth and yield of upland rice and dry bean (Fabaceae) grown on soils of termite mound in Oxisols of the Cerrado region (Table 1.3). Phosphorus was the most yield-limiting nutrient in these upland rice and dry bean production (Table 1.3).

Fageria et al. (2013a) studied the influence of P on the growth of 14 tropical legume cover crops (Table 1.4). The P × cover crops interaction for shoot dry weight was highly significant (Table 1.4).

TABLE 1.3
Response of Upland Rice and Dry Bean to N, P, and K Fertilization on a Termite Mound Soil of Cerrado Region of Brazil

NPK Treatments[a]	Shoot Dry Wt. (g/Pot)	Grain Yield (g/Pot)
Upland rice		
Control	38.33b	18.40b
NPK	01.95a	52.93a
PK	48.43b	29.93b
NK	57.85b	25.53b
NP	91.10a	50.56a
Dry bean		
Control	13.98b	15.15b
NPK	51.20a	29.50a
PK	25.53b	16.15b
NK	16.20b	15.20b
NP	49.50a	19.48b

Source: Fageria, N.K. and Baligar, V.C., *Commun. Soil Sci. Plant Anal.*, 35, 2097, 2004.

[a] Means in the same column followed by the same letter are not significantly different at the 5% probability level by Tukey's test. NPK treatment received 200 mg N, 200 mg P, and 200 mg K per kg soil, without N (PK), without P (NK), and without K (NP). With the exception of the control and without N (PK) treatments, all the other treatment received 180 mg N as topdressing. Topdressing of N for rice was done 45 days after sowing and that for bean was done 35 days after sowing.

Therefore, the response of cover crops to P varied with the variation in P levels. Screening for P use efficiency should be performed at specific P levels. Shoot dry weight varied from 0.13 g plant^{-1} produced by *Crotalaria breviflora* to 5.81 g plant^{-1} produced by *Canavalia ensiformis*, with a mean value of 1.31 g plant^{-1} at a low (0 mg kg^{-1}) P level. At a medium P level (100 mg kg^{-1}), shoot dry weight varied from 0.54 g plant^{-1} produced by *C. breviflora* to 8.76 g plant^{-1} produced by *C. ensiformis*, with a mean value of 2.89 g plant^{-1}. At a higher P level (200 mg kg^{-1}), the shoot dry weight varied from 0.26 to 9.28 g plant^{-1}, with a mean value of 3.50 g plant^{-1}. Across three P levels, the shoot dry weight varied from 0.46 to 7.95 g plant^{-1}. White jack bean species produced the highest shoot dry weight at the three P application rates.

Overall, the shoot dry weight also increased (1.31–3.50 g plant^{-1}) with increasing P levels from 0 to 200 mg kg^{-1}. Interspecies variability in shoot dry weight of tropical legume cover crops has been widely reported (Fageria et al., 2005; Baligar et al. 2006; Baligar and Fageria, 2007; Fageria, 2009). Inter- and intraspecies variations for plant growth are known to be genetically and physiologically controlled and are modified by plant interactions with environmental variables (Fageria, 1992, 2009; Baligar et al., 2001). Growth of Sunn hemp (*Crotalaria juncea*) and lablab (*Dolichos lablab* L.) cover crops increased with increasing P levels from 0 to 200 mg kg^{-1} soil (Figures 1.3 and 1.4).

Fageria et al. (2012b) also studied response of dry bean to P fertilization (Table 1.5). Phosphorus × genotype interactions for shoot dry weight and grain yield were significant (Table 1.5), indicating that genotypes differently responded to variation in P levels. Shoot dry weight varied from 1.36 to 1.98 g plant^{-1} at the low P level (25 mg kg^{-1}), with a mean value of 1.71 g plant^{-1}. Similarly, at a high P level (200 mg P kg^{-1}), shoot dry weight varied from 4.51 to 10.79 g plant^{-1}, with a mean value of 7.23 g plant^{-1}. Genotype CNFC 10470 produced lower shoot dry weight at the lower P level and did not produce lower shoot dry weight at the higher P level. Overall, an increase in shoot dry

TABLE 1.4
Shoot Dry Weight of 14 Legume Cover Crops as Influenced by P Levels

	Shoot Dry Weight (g Plant^{-1})			
Cover Crops	**0**	**100**	**200 mg P kg^{-1}**	**Mean**
Crotalaria	0.13f	0.54f	0.70fg	0.46i
Sunn hemp	1.19de	3.38d	4.49d	3.02e
Crotalaria	0.20f	0.77ef	0.74fg	0.57hi
Crotalaria	0.31f	1.01ef	1.37efg	0.89fghi
Crotalaria	0.30f	1.38ef	2.37e	1.35fg
Calopogonium	0.26f	0.93ef	0.26g	0.48i
Pueraria	0.17f	0.74ef	1.03efg	0.64ghi
Pigeon pea (black)	0.63ef	1.90e	1.97ef	1.50f
Pigeon pea (mixed color)	0.39cf	1.57cf	1.76efg	1.24fgh
Lablab	0.92def	4.03cd	5.60bcd	3.51de
Mucuna bean ana	2.26c	4.54bcd	5.35cd	4.05cd
Black mucuna bean	1.61cd	5.71b	6.92bc	4.75c
Gray mucuna bean	4.21b	5.27bc	7.14b	5.54b
White Jack bean	5.81a	8.76a	9.28a	7.95a
Mean	1.31	2.89	3.50	2.57
F-test				
P	**			
Cover crops (C)	**			
P × C	**			
CV(%)	16.05			

Source: Fageria, N.K. et al., *Commun. Soil Sci. Plant Anal.*, 44, 3340, 2013a.
**Significant at the 1% probability level. Means followed by the same letter in the same column are significantly not different by Tukey's test at the 5% probability level.

FIGURE 1.3 Growth of Sunn hemp at three P levels. (From Fageria, N.K. et al., *Commun. Soil Sci. Plant Anal.*, 44, 3340, 2013a.)

FIGURE 1.4 Growth of lablab at three P levels. (From authors' unpublished figure.)

TABLE 1.5
Shoot Dry Weight and Grain Yield of Dry Bean Genotypes at Two P Levels (mg kg^{-1})

	Shoot Dry Weight (g Plant^{-1})		Grain Yield (g Plant^{-1})	
Genotype	**25**	**200**	**25**	**200**
CNFP 10103	1.83a	5.96b	0.63a	7.89ab
CNFP 10104	1.89a	6.78ab	2.74a	10.18a
CNFC 10429	1.61a	4.51b	1.37a	9.92a
CNFC 10431	1.98a	7.10ab	1.79a	7.74ab
CNFP 10120	1.58a	10.79a	1.34a	3.67c
CNFC 10470	1.36a	8.24ab	2.22a	5.89bc
Mean	1.71	7.23	1.68	7.55
F-test				
P level (P)	**		**	
Genotype (G)	**		**	
P × G	**		**	

Source: Fageria, N.K. et al., *Commun. Soil Sci. Plant Anal.*, 43, 2289, 2012b.
**Significant at the 1% probability level. Means followed by the same letter within the same column are not significantly different at the 5% probability level by Tukey's test.

weight was 323% at the high P level as compared to the low P level. Grain yield varied from 0.63 to 2.74 g plant^{-1} at the low P level, with a mean value of 1.68 g plant^{-1}. Similarly, at the high P level, the variation in grain yield was 3.67 to 10.18 g plant^{-1}, with a mean value of 7.55 g plant^{-1}. Grain yield increased with the application of 200 mg P kg^{-1}, which was 349%, as compared to the low P level (25 mg P kg^{-1}). Fageria (2009) and Fageria et al. (2011) reported significant variations in shoot dry weight and grain yield among dry bean genotypes grown on Brazilian Oxisols. Dry bean growth at

FIGURE 1.5 Response of dry bean genotype CNFC 10467 to P fertilization grown on a Brazilian Oxisol. (From Fageria, N.K. et al., *Commun. Soil Sci. Plant Anal.*, 43, 1, 2012a.)

zero mg P kg^{-1} was significantly reduced, and P deficiency symptoms were visible as compared to 200 mg P kg^{-1} level (Figure 1.5).

Approximately 80% of the area devoted to grain crops and 83% of the total grain production in the United States are in 15 Midwestern states (Hanway and Olson, 1980). Evidence has been accumulated to suggest that P deficiency is common in these grain-producing areas. Although P availability in the soils varies widely, general use of P fertilizers is essential for the high yields of crops. Approximately 61% of the total P fertilizers marketed in the United States are used in 15 major grain-producing states (Hanway and Olson, 1980).

Importance of P in crop production is enormous, mainly because of limited P supply in most soils. Phosphate rock (PR), is a nonrenewable natural resource of P; thus, the authors believe that this introductory chapter is needed to provide information on world phosphate production, reserves, and its transformation in soil and P availability to plants.

1.2 HISTORICAL OVERVIEW

Manures, plant materials, and bones have been used by man for stimulating plant growth since the beginning of agriculture; however, it was not until 1849 that Liebig, the German chemist, suggested that dissolve bones in sulfuric acid made the better P fertilizer to plants (Cathcart, 1980). The era of field experimentation for evaluation of fertilizers began in 1834, when J.B. Boussingault, a French chemist, established the first field experiments at Bechelbonn, Alsace (France), which was conducted with a modern scientific method by Liebig's report in 1840 (Collis-George and Davey, 1960; Fageria, 2007b). The first field experiments conducted in the method used today were established by Lawes and Gilbert at Rothamsted in 1843 (Williams, 1993). Since then, the field experiments have sought for and have confirmed the importance of essential elements in improving the production of field crops. However, evidence for the discovery of the essentiality of nutrients has been in laboratory experiments with nutrient solution, not from field experiments (Collis-George and Davey, 1960; Fageria, 2007b).

Limited supply of bones prompted the development in the utilization of rock phosphates where Lawes obtained the first patent concerning the utilization of acid-treated rock phosphate in 1842 (Williams, 1993). In 1842, patents were granted in the United Kingdom to both John Bennet Lawes and James Murray for the manufacture of P fertilizer by the process of acidulation. Although others, including Justus Von Liebig, had been studying the process, Lawes and Murray have been credited as the laymen who placed the idea into permanent commercial practice (Van Kauwenbergh et al., 2013). Practically all P fertilizers today are made by this wet process of treating RP with acid, like sulfuric, nitric, or phosphoric acid, to produce phosphoric acid or triple superphosphate. Phosphoric acid is then used to produce both granular and fluid P fertilizers. The first commercial production of rock phosphate began in Suffolk, U.K., in 1847. Mining phosphate in the United States began in 1867 in South Carolina, although the deposits were known as early as 1837. Thus, began the P fertilizer industry (Sanchez, 2007). The first recorded production of P fertilizers in that year was about 6.2 mt (Cathcart, 1980). Phosphate deposits were discovered in Florida in 1888, in Tennessee in 1894, and in the western United States in 1906. Deposits in North Africa (Algeria and Tunisia) were discovered in 1873, mining began in 1889 (Cathcart, 1980), and production from large Moroccan deposits began in 1921, although they were known as early as 1914 (Cathcart, 1980).

Early progress in the understanding of soil fertility and plant nutrition concepts was slow, although the Greeks and Romans made significant contributions in the years 800 to 200 BC (Westerman and Tucker, 1987). Marschner (1983) states that it was mainly to the credit of Justus von Liebig (1803–1873) that the scattered information concerning the importance of mineral elements for plant growth was collected and summarized and that mineral nutrition of plants was established as a scientific discipline.

In 1840 Liebig published results from his studies on the chemical analysis of plants and the mineral contribution of soils. These studies initiated modern research on plant nutrition and highlighted the importance of individual minerals in stimulating plant growth. From these studies evolved the concept that individual minerals were limiting factors on the growth potential of plants (Sinclair and Park, 1993). These findings led to a rapid increase in the use of chemical fertilizers. By the end of the nineteenth century, large amounts of potash, superphosphate, and, later, inorganic N were used in cropping systems to improve plant growth, especially in Europe (Marschner, 1995). In the twentieth century, great progress has been made in developing extracting solution for P and its relationship with plant growth. Early contributions of Dyer (1894), Truog (1930), Morgan (1941), and Bray and Kurtz (1945) are noteworthy (Sanchez, 2007). Importance of P for plant growth and development was discovered by Posternak in 1903 (Fageria et al., 2011). However, it was not until the twentieth century that the list of 16 essential elements was completed and the fundamental concepts of plant nutrition were developed. The quest for an understanding of plant nutrition is not complete, however (Glass, 1989).

1.3 WORLD PHOSPHATE ROCK RESERVES AND RESOURCES

Practically all of the inorganic P fertilizers are produced from PR. PR is an imprecise term that describes naturally occurring geological minerals that contain a relatively high concentration of P. The term PR is often used to include nonbeneficiate phosphate ores and concentrated products. There is no accepted worldwide system for classifying PR reserves and resources (Zapata and Roy, 2004). The U.S. Geological Survey (USGS) defines resources as a concentration of naturally occurring solid, liquid, or gaseous material in or on the Earth's crust in such form and amount that economic extraction of a commodity from the concentration is currently or potentially feasible (U.S. Geological Survey, 2011). Reserve base is defined as the part of an identified resource that meets specified minimum physical and chemical criteria related to current mining and production practice, including those for grade, quality, thickness, and depth (Heffer and Prudhomme, 2013).

International Fertilizer Development Center (IFDC) defines PR reserves as the amount that can be produced with current technology at current prices and current costs and PR resources as naturally occurring phosphate material in such a form or amount that economic extraction of product is currently or potentially feasible (Heffer and Prudhomme, 2013). However, Van Kauwenbergh (2010) defined reserves of PR as that can be economically produced at the time of the determination using existing technology. He also defined resources as the PR of any grade, including reserves that may be produced at some time in the future.

Currently, PR is the only economical source of P for the production of phosphate fertilizers and phosphate chemicals. Phosphate of almost all minable deposits is one of the minerals of the apatite group, that is, $Ca_{10}(PO_4,CO_3)_6(F,OH)_{2-3}$ (Cathcart, 1980). A small percentage, however, is mined from secondary [illegible] phosphate deposits, in which the phosphate mineral was derived from apatite by weathering.

PR occurs in both sedimentary and igneous deposits across the world (Van Kauwenbergh et al., 2013). Most (80%–90%) of the PRs used to produce fertilizer are sedimentary in origin and were deposited in ancient marine continental shelf environments. Sedimentary deposits, sometimes called phosphorites, occur throughout geological time (Van Kauwenbergh et al., 2013). Most phosphate deposits contain silica in the form of quartz; other common diluting materials include calcite, dolomite, Fe oxide minerals, and clay minerals. Some deposits contain diluting materials such as zeolites derived from the alteration of volcanic ash, glauconite, cristobalite, pyrite, and so on. Apatite must be separated from the gangue minerals, and methods of beneficiation have to be tailored that will suite to the minerals present in the PR. It is essential to determine the mineralogy as a first step in evaluating the economics of the deposit (Cathcart, 1980).

Mining method used to extract the rock will depend on the physical character of the rock and its geologic setting. If the rock is unconsolidated and flat lying, open pit mining methods can be used, whereas if the rock is consolidated and steeply dipping, some methods of underground mining will have to be used (Cathcart, 1980). Most PR is mined by an open pit method, but a significant amount of deposits in China, Russia, and other countries are extracted by underground mining (Van Kauwenbergh et al., 2013).

In 2010, the worldwide phosphate (PR) mine production was estimated to be 176 million mt, and world reserves were estimated to be 65 billion mt (Table 1.6) (U.S. Geological Survey, 2011).

TABLE 1.6
World Mine Production and Reserve of Phosphate Rock (Data Are in Thousand Metric Tons)

Country	Mine Production in 2010	Reserves	% of Total Reserves
United States	26,100	1,400,000	2.00
Algeria	2,000	2,200,000	3.00
Australia	2,800	82,000	0.13
Brazil	5,500	340,000	0.52
Canada	700	5,000	0.01
China	65,000	3,700,000	6.00
Egypt	5,000	100,000	0.15
Israel	3,000	180,000	0.28
Jordan	6,000	1,500,000	2.00
Morocco and Western Sahara	26,000	50,000,000	77.00
Russia	10,000	1,300,000	2.00
Senegal	650	180,000	0.28
South Africa	2,300	1,500,000	2.00
Syria	2,800	1,800,000	3.00
Togo	800	60,000	0.01
Tunisia	7,600	100,000	0.15
Other countries	9,500	620,000	0.95
World total (rounded)	176,000	65,000,000	100

Source: U.S. Geological Survey, Mineral commodity summaries 2011, accessed December 2013, at http://minerals.usgs.gov/minerals/pubs/mcs/2011/mcs2011.pdf.

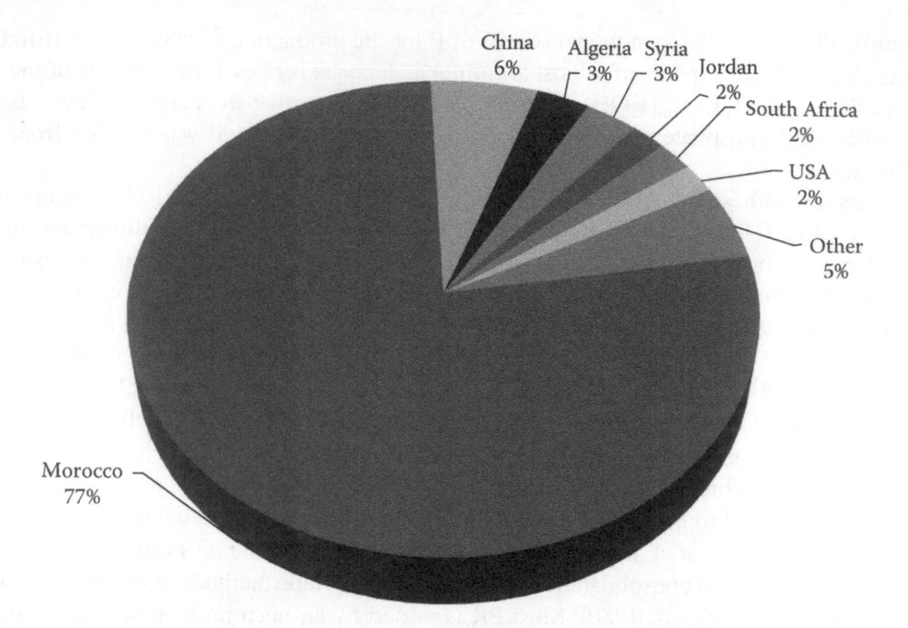

FIGURE 1.6 World phosphate rock reserve in 2011 in various countries, estimated at 65 billion metric tons. (Data from U.S. Geological Survey, Mineral commodity summaries 2011, accessed December 2013, http://minerals.usgs.gov/minerals/pubs/mcs/2011/mcs2011.pdf.)

Phosphate rock is produced in some 33 countries, but 16 countries are reported to be major PR producing, and among these countries, Morocco, China, Algeria, Syria, Jordan, South Africa, and the United States have reserve of 95% of the total world PR (Figure 1.6). Quantity of PR reserves and resources worldwide has been an issue of speculation over the last 20 years. It has been hypothesized that PR production would peak in 2033–2034, with production unavoidably decreasing as the reserves are depleted (Cordel et al., 2009). Assuming current rate of production, IFDC estimates that there are sufficient PR concentrate reserves to produce fertilizer with current technology at current prices and current costs for the next 300–400 years. However, in the authors' opinion, PR for P fertilizer production will last more than 500 years because new reserves will be discovered and new modern technology will be available in the future to extract this natural resource. However, PR is a nonrenewable natural resource, and its judicious and efficient use in agriculture is fundamental to improving crop yields, reducing cost of production, and curtailing environmental pollution.

Phosphoric fertilizer production primarily involves the mining and the conversion of PR to more soluble P compounds, which can be effectively utilized by plants. Certain chemical processes, such as the solubilization of PR in acids, have been basic to the phosphate industry (Phillips and Webb, 1971). PR from various sources varies widely in chemical composition, but those rich in carbonate apatite are the most commonly used fertilizer materials (Stevenson, 1986). Inorganic or soluble phosphate fertilizers are produced by treating PR with sulfuric and phosphoric acids. Superphosphate is a product resulting from the mixing of approximately equal quantities of 60%–70% H_2SO_4 and PR. Overall reaction to produce simple superphosphate or ordinary superphosphate is as follows (Stevenson, 1986):

$$3\left[Ca_3\left(PO_4\right)_2\right]\cdot CaF_2 + 9H_2SO_4 + H_2O \rightarrow 3Ca\left(H_2PO_4\right)_2 \cdot H_2O + 9CaSO_4 + 2HF$$

In this reaction Ca dihydrogen phosphate $3Ca(H_2PO_4)_2 \cdot (H_2O)$ and gypsum ($7CaSO_4$) are produced. When PR is treated with excess H_2SO_4 and removal of much of the gypsum gives concentrated or triple superphosphate, a product containing about 20% P. The main reaction leading to the production of concentrated P is shown in the following equation (Stevenson, 1986):

$$3\left[Ca_3(PO_4)_2\right]\cdot CaF_2 + 10H_2SO_4 + xH_2O \rightarrow 10CaSO_4\cdot xH_2O + 6H_3PO_4 + 2HF$$

Stevenson (1986) reported that three environmental problems are encountered with the mining of phosphate ores and the production of P fertilizers. These problems are emission of fluorine, disposal of gypsum, and accumulation of Cd and other heavy metals in soils and possibly in plants.

1.4 MINERALOGY OF PHOSPHATE ROCKS

Knowledge of PR mineralogy is important to the understanding of their chemical composition. Mineralogical analysis is the estimation or determination of the kinds or amounts of minerals present in a rock. Mineralogy of phosphate apatites has been determined by techniques including chemical analysis, X-ray powder diffraction, petrographic microscopy, infrared spectroscopy, and electron microscopy. In general, the apatites are in the form of carbonate apatite (francolite) with varying degrees of isomorphic substitution of carbonate for phosphate (Hammond et al., 1986). Degree of isomorphic substitution in the apatite structure is the key factor in determining the chemical reactivity of PR containing carbonate apatite (Hammond et al., 1986).

Phosphatic raw materials are composed of several minerals grouped under the generic heading of *PR* or phosphorite. PR is a trade name that covers a wide variety of rock types that have widely different textures and mineral compositions (McClellan and Gremillion, 1980). PR is a sedimentary rock composed principally of phosphate minerals. Grade of commercial PR is expressed in terms of tricalcium phosphate [$(Ca_3(PO_4)_2$], known in the trade as bone phosphate of lime (BPL). According to McClellan and Gremillion (1980), BPL was originated when tricalcium phosphate was thought to be the chief constituent of bone and PR. It is now known that both bone and PR are apatites and not tricalcium phosphate. Because of the wide use of the term BPL and the present trend toward elemental notation, the conversion factors are (McClellan and Gremillion, 1980) as follows:

$$\%P_2O_5 = 0.4576 \times \%BPL$$

$$\%P = 0.1997 \times \%BPL$$

Commercial PR varies in grade from about 83% BPL to about 60% BPL (17% to 12% P). About 85% of the world's annual PR production is processed to yield P and phosphoric acid, which are converted into a wide variety of fertilizer materials ((McClellan and Gremillion, 1980). Phosphate deposits fall into three classes (Fe–Al phosphate, Ca–Fe–Al phosphates, and Ca phosphates) based upon their mineral composition. These three classes form a natural weathering sequence in which the stable Fe–Al phosphates represent the final stage of weathering (McClellan and Gremillion, 1980).

1.5 GLOSSARY OF PHOSPHORUS-RELATED TERMS

Many P fertilizer-associated terms are generally used in the literature. Defining or explaining these terms is important to the understanding and application in agricultural science of mineral nutrition. These terms are compiled from many references, and readers may refer these references for further clarification (Soil Science Society of America, 2008; Fageria, 2009). In addition, some references are also cited in the terminology of the terms:

1. **Phosphate rock (PR):** If an apatite-bearing rock is high enough in P content to be used directly to make fertilizer or as a furnace charge to make elemental P, it is called PR. The term is also used to designate a beneficiated apatite concentrate (Cathcart, 1980).

2. **Phosphorite:** It is a rock term for sediment in which a phosphate mineral is a major constituent.
3. **Ore or matrix:** It can be defined as material in the ground that can be mined and processed at a profit. Matrix is a term used in synonymous with ore.
4. **Absolute citrate solubility:** AOAC citrate soluble P_2O_5%/theoretical P_2O_5% concentration of apatite, where AOAC refers to Association of Official Analytical Chemists method.

   As the use of other solvents became more common, the absolute solubility was expressed in more general terms to include any type of solvent (McClellan and Gremillion, 1980). The term absolute solubility index (ASI) then was defined:

$$\text{ASI} = \text{Solvent} - \text{soluble } P_2O_5, \%/\text{theoretical } P_2O_5 \text{ concentration of apatite}, \%$$

   Solvents most commonly used throughout the world in making solubility tests are neutral ammonium citrate, 2% citric acid, and 2% formic acid (McClellan and Gremillion, 1980).
5. **Acidulation processes:** Acidulation involves treating phosphatic raw materials with mineral acids to prepare water and citrate-soluble P compounds. Most widely used processes involve treatment of apatitic PR with sulfuric acid to prepare normal superphosphate or phosphoric acid; triple or concentrated superphosphate is prepared from PR and phosphoric acid.
6. **Phosphorus immobilization:** Soluble P compounds, when added to soil, become chemically or biologically attached to the solid phase of soil so as not to be recovered by extracting the soil with specific extracting solution under specific conditions. Immobilized P is generally not absorbed by plants during the first cropping cycle.
7. **Chemisorbed phosphorus:** P adsorbed or precipitated on the surface of clay minerals or other crystalline materials as a result of the attractive forces between the phosphate ion and constituents in the surface of the solid phase.
8. **Water-soluble phosphorus:** Water-soluble, citrate-soluble, citrate-insoluble, available, and total P are the terms frequently used in describing phosphates that are present in fertilizers. Tisdale et al. (1985) described all these terms as follows: (1) *Water soluble:* A small sample of the material to be analyzed is first extracted with water for a prescribed period of time. The slurry is then filtered, and the amount of P contained in the filtrate is determined. Expressed by a percentage (by weight) of the sample, it represents the fraction of the sample that is *water soluble.* (2) *Citrate soluble*: The residue from the leaching process is added to a solution of neutral 1N ammonium citrate. It is extracted for a prescribed period of time by shaking, and the suspension is filtered. P content of the filtrate is determined, and the amount determined is expressed as a percentage of the total weight of the sample and is termed as the citrate soluble P. (3) *Citrate insoluble*: The residue remaining from the water and citrate extraction is analyzed. The amount of P found is referred to as citrate insoluble. (4) *Available phosphorus:* The sum of water-soluble and citrate-soluble P is termed available P. (5) *Total phosphorus*: The sum of available and citrate-insoluble P is termed as total P.
9. **Labile phosphorus:** From the plant nutrition point of view, three soil phosphate fractions are important. These soil fractions are known as solution P, labile P, and nonlabile P. Phosphorus dissolved in the soil solution is known as solution P. Fraction of P, which is held on the solid surface and is in rapid equilibrium with soil solution phosphate, is referred to as labile P. It can be determined by means of isotopic exchange. The third fraction is the insoluble phosphate. The phosphate in this fraction can be released only very slowly into the labile pool and is known as nonlabile P. Isotopically exchangeable

P, sampled by a growing plant over the span of a growing season, is called the L-value. It is customary to describe soil P in terms of the following relationship:

$$\text{Soil solution P} \Leftrightarrow \text{Labile soil P} \Leftrightarrow \text{Nonlabile P}$$

where equilibrium is rapidly established between labile soil solution P, whereas true equilibrium is seldom, if ever, established between the labile and nonlabile pools of soil P (Olsen and Khasawneh, 1980).

10. **Phosphorus buffer power of soils:** Buffer power can be defined as the total amount of diffusible ion (solution plus sorbed) per unit of volume of soil required to increase the solution concentration by one unit. Buffer power, however, has also been described as the relationship between the concentration of ions adsorbed on the solid phase and ions in solution.
11. **Phosphate beneficiation:** Almost all PR is mined by strip mining. It usually contains about 6.55% P (15% P_2O_5) and must be upgraded for use as fertilizers. Upgrading removes much of the clay and other impurities. This process is called *beneficiation*. It raises P from 13.1% to 15.3% (30% to 35% P_2O_5). Following beneficiation, the PR is finely ground and treated to make the P more soluble. Fertilizer phosphates are classified as either *acid treated or thermal processed*. Acid-treated P is by far the most important. Sulfuric and phosphoric acids are basic in producing acid-treated phosphate fertilizers. Sulfuric acid is produced from elemental S or from sulfur dioxide. More than 60% of this industrial acid is used to produce fertilizers. Treating PR with sulfuric acid produces a mixture of phosphoric acid and gypsum. Filtration removes the gypsum to leave green or wet process phosphoric acid.
12. **Polyphosphates:** Most liquid P sources start with wet process phosphoric acid. But wet process acid can be further concentrated to form superphosphoric acid. In this process water is driven off and molecules with two or more P atoms are formed. Such molecules are called *polyphosphates*.
13. **Phosphorus longevity:** Longevity is defined as the time when the P concentration in the center of the applied band is five times the original water-soluble soil P concentration.
14. **Phosphorus-efficient plant:** Plants able to absorb, translocate, and utilize P effectively in growth.
15. **Phosphoric acid:** In commercial fertilizer manufacturing, phosphoric acid is used to designate orthophosphoric acid, H_3PO_4. In fertilizer labeling, it is the common term used to represent the phosphate concentration in terms of available P, expressed as percent P_2O_5 (Soil Science Society of America, 2008).
16. **Phosphogypsum:** Phosphogypsum is the term used for the gypsum by-product of wet acid production of phosphoric acid from PR. It is essentially hydrated $CaSO_4 \cdot 2H_2O$ with small proportions of P, F, Si, Fe, and Al, several minor elements, heavy metals, and radionuclides as impurities (Alcordo and Rechcigl, 1993).
17. **Apatite:** A mineral containing mainly calcium and phosphate ions, $Ca_5(PO_4)_3(OH, Cl, F)$.
18. **Band application:** A method of fertilizer or other agrichemical applications above, below, or alongside the planted seed row.
19. **Broadcast application:** Application of fertilizer on the surface of the soil. Usually applied prior to planting and normally incorporated with tillage, but may be unincorporated in no-till production systems.
20. **Bulk density:** The mass of dry soil per unit bulk volume. Bulk density is expressed as mega gram per cubic meter ($Mg\ m^{-3}$).
21. **Bulk fertilizer:** Solid or liquid fertilizer in a nonpackaged form.

22. **Calcareous soils:** Soil containing sufficient free $CaCO_3$ and other carbonates to effervesce visibility or audibility when treated with 0.1 M HCl. These soils usually contain from 10 to almost 1000 g kg^{-1} $CaCO_3$ equivalent.
23. **Cation exchange capacity (CEC):** The sum of exchangeable bases plus total soil acidity at a specific pH value, usually 7.0 or 8.0. When acidity is expressed as salt extractable acidity, the CEC is called the effective cation exchange capacity because this is considered to be the CEC of the exchanger at the native pH value. It is usually expressed in centimoles of charge per kilogram of exchanger ($cmol_c$ kg^{-1}) or millimoles of charge per kilogram of exchanger (mmol kg^{-1}).
24. **Citrate-soluble phosphorus:** Fraction of total P in fertilizer that is insoluble in water but soluble in neutral 0.33 M ammonium citrate. Together with water-soluble phosphate, this represents the readily available P content of the fertilizer.
25. **Critical nutrient concentration:** Nutrient concentration in the plant, or specific plant part, above which additional plant growth response slows. Crop yield, quality, or performance is less than optimum when the concentration is less than the critical nutrient concentration.
26. **Critical soil test concentration:** That concentration at which 95% of maximum relative yield is achieved.
27. **Desorption:** Mitigation of adsorbed entities of the adsorption sites. Desorption is the inverse of adsorption.
28. **Diffusion coefficient:** Proportionality constant that indicates the ability of a material to allow gases and ions to flow under a partial pressure or concentration gradient.
29. **Economic rate of phosphorus:** Application rate of P fertilizer that provides the highest economic returns for the crop produced.
30. **Eutrophication:** Condition in an aquatic ecosystem where excessive nutrient concentrations result in high biological productivity, typically associated with algae blooms, that causes sufficient oxygen depletion to be detrimental to other organisms.
31. **Superphosphate:** A product obtained when PR is treated with H_2SO_4, H_3PO_4, or a mixture of those acids.
32. **Ammoniated phosphate:** A product obtained when superphosphate is treated with NH_3 and/or other NH_4–N–containing compounds.
33. **Concentrated phosphate:** Also called triple or treble superphosphate, made with phosphoric acid and usually containing 45% P_2O_5.
34. **Enriched phosphate:** Superphosphate derived from a mixture of sulfuric acid and phosphoric acid. This includes any grade between 10% and 19% P (22% and 44% P_2O_5), commonly 11% to 13% P (25% to 30% P_2O_5).
35. **Normal phosphate:** Also called ordinary or single superphosphate. Superphosphate made by reaction of PR with sulfuric acid, usually containing 7% to 10 %P (16% to 22% P_2O_5).
36. **Superphosphoric acid:** Acid form of polyphosphates, consisting of a mixture of orthophosphoric and polyphosphoric acids. Species distribution varies with concentration, which is typically 30% to 36% P (68% to 83% P_2O_5).
37. **Agronomic potential of phosphate rock:** Agronomic potential refers to the inherent capability of the P-containing rock to supply plant-available P under a specific set of conditions. It is determined primarily by the chemical solubility of the rock and differs in meaning from the term agronomic effectiveness in that the latter refers to the actual performance of a given PR source as influenced by both agronomic potential and external conditions under which it was used (Hammond et al., 1986).
38. **PR reactivity:** PR reactivity is the combination of PR properties that determine the rate of dissolution of the PR in a given soil under given field conditions (Rajan et al., 1996).
39. **Phosphorus sorption:** Phosphorus sorption includes adsorption and precipitation reactions of P ions in the soil.

1.6 CYCLE IN SOIL–PLANT SYSTEM

Phosphorus is the 10th most abundant element on earth and essential nutrient for all organisms (Tamburini et al., 2014). Phosphorus is fundamental to many biological processes since it is involved in energy transfer and is the constituent of a number of organic molecules (Westheimer, 1987). When in excess in the environment, however, P can become a pollutant, causing eutrophication of water bodies (Sutton et al., 2013) and eventually be attributed to important shifts in ecosystems. For all these reasons, P chemistry and biochemistry; its cycle in marine, aquatic, and terrestrial environments; and its transfer from source to sinks have been extensively studied (Ruttenberg, 2003; Frossard et al., 2011; Paytan and McLaughlin, 2011). Along with advances in technology, new analytical tools have provided deeper insights into P forms, pool sizes and fluxes, and processes affecting P cycling (Frossard et al., 2012).

Knowledge of P cycle is an important aspect in the efficient management of a nutrient, including P for efficient and sustainable crop production. Iyamuremye and Dick (1996) reported that P cycle can be characterized as the flow of P between plants, animals, microorganisms, and solid phases of the soil. Fageria (2009) reported that P cycle refers to addition, transformation, uptake, and loss of P from soil–plant system. Hence, P cycle in soil–plant system is very complex since soil, climate, plant factors, and their interactions are involved. Losses of soil P occur through leaching and erosion. In addition, soils comprise a multiple-phase system consisting of numerous solid phases (about 50%), a liquid phase (about 25%), and a gas phase (about 25%) (Lindsay, 1979). These soil phases significantly influence solubilization, immobilization, availability, and loss of P from soil–plant system. Compared with the other major nutrients, P is by far the least mobile and available to plants in diverse soil conditions (Hinsinger, 2001).

Of the P in the plant–animal system, commonly over 90% is derived from the soil. Of this less than 10% enters the plant–animal life cycle (Ozanne, 1980). A simplified diagram of P cycle in soil–plant system is depicted (Figure 1.7). Main features of P cycle involve its addition in the soil–plant systems, transformation, losses, and uptake by plants (Figure 1.7). Solubilization and immobilization are the main transformation processes of P in soil–plant systems that control its availability to plants and potential losses. Immobilization or fixation is defined as the strong adsorption or precipitation of P ions on Al and Fe hydroxides. In most soils, soil solution P ranges between <0.01 and 1 mg L^{-1}, and a value of 0.2 mg P L^{-1} is commonly accepted as the solution P concentration required to meet the plant nutritional needs of most agronomic crops (Wood, 1998). However, Brady and Weil (2002) reported that the concentration of P in the soil solution is very low, generally ranging from 0.001 mg L^{-1} in very infertile soils to 1 mg L^{-1} in heavily fertilized soils. Soil erosion, surface and subsurface runoff, leaching, and uptake by plants are the main processes of P losses from soil–plant systems (Figure 1.7). With no volatilization and with usually very low leaching losses, erosion and runoff are by far the most important sources of the nutrient carried in inorganic and organic particulates by streams to the ocean. Mean lithospheric content of 0.1% P and a mean global denudation rate of around 750 kg ha^{-1} (Froelich et al., 1982) would release about 10 mt P annually from P-bearing rocks (Smil, 2000).

Phosphorus cycle is similar to N in many aspects. However, N is less spectacular in that no valence charges occur during assimilation of organic phosphate by living organisms or during breakdown of organic P compounds by microorganisms. Furthermore, the difference of P cycle to N is that P cycle does not have a gaseous component. Next to N, P is the most abundant nutrient contained in microbial tissue, comprising of up as much as 2% of the dry weight. Partly for this reason, P is the second most abundant nutrient in soil organic matter (Stevenson, 1986). In natural ecosystems, the P cycle is virtually closed, and most plant P is recycled by microbial breakdown of litter and organic debris. For example, in Brazilian rainforest, most of the P are in the living and dead organic matter, and where the underlying soil contains such a low level of P, that optimum crop growth is achieved only when P fertilizer is added to the newly cleared forest land. Much of the P in the grassland soils resides in the biomass (Stevenson, 1986).

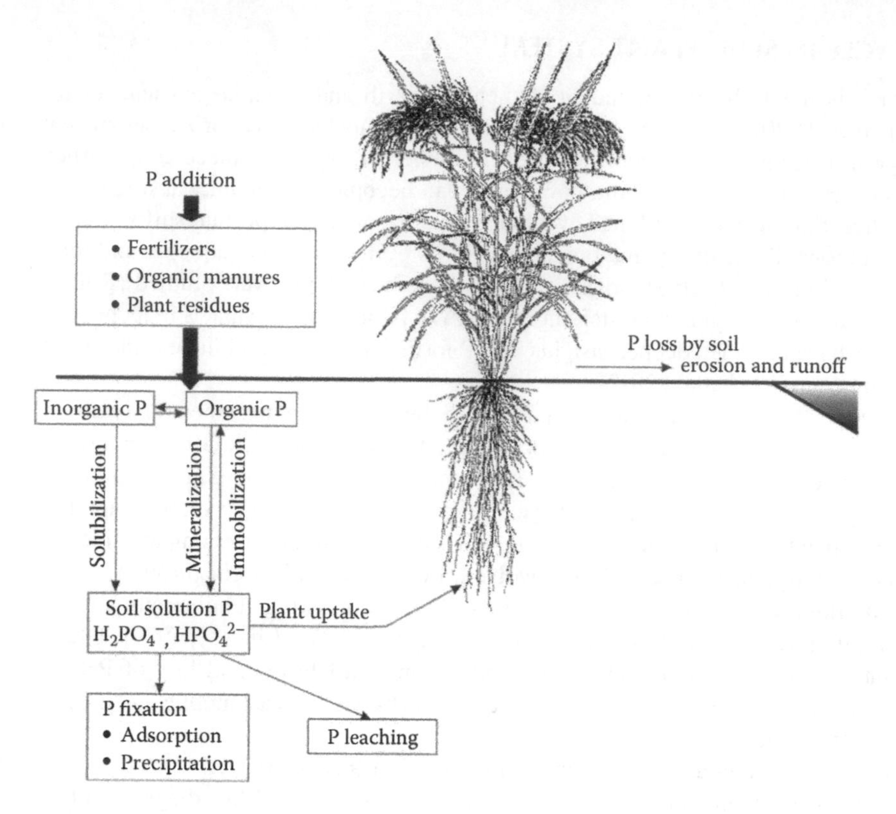

FIGURE 1.7 Simplified version of P cycle in soil–plant system. (From Fageria, N.K., *The Use of Nutrients in Crop Production*, CRC Press, Boca Raton, FL, 2009.)

Recovery efficiency of applied soluble P fertilizers by annual crops during their growth cycle is less than 20% in most of the acid soils (Baligar et al., 2001; Fageria et al., 2003). Ionic species of phosphate in the soil solution is dependent on pH. In dilute solution, orthophosphoric acid dissociates into three forms according to the following equation (Lindsay, 1979: Fageria, 1992):

$$H_3PO_4^0 \leftrightarrow H_2PO_4^- + H^+$$

$$H_2PO_4 \leftrightarrow HPO_4^{2-} + H^+$$

$$HPO_4^{2-} \leftrightarrow PO_4^{3-} + H^+$$

Phosphorus uptake by plants mainly occurs in the form of $H_2PO_4^-$ ion in acid soils and in the form of HPO_4^{2-} ion in basic or alkaline soils. Proportion of these two ions in the soil solution is governed by pH. At pH 5, most P are in the form of $H_2PO_4^-$, and at pH 7 both of these ions are present more or less in equal amounts (Mengel et al., 2001). Foth and Ellis (1988) reported that some studies with excised roots suggested that plants preferred the $H_2PO_4^-$ ion over HPO_4^{2-} ion by about 10 to 1. But since conversion between the two species in solution is very rapid, this preference is probably of minimal importance for soils in the pH range of 4–8 (Foth and Ellis, 1988). Overall, maximum uptake of P by crop plants occurs at pH value of 6.0–7.0 (Stevenson, 1986). Uptake of P by plants is governed by the ability of a soil to supply P to plant roots and by the desorption characteristics of the soil (Fageria et al., 2003). In acid soils, P is mainly immobilized or fixed by Al and Fe ions by following reactions (Fageria, 2009):

$$Al^{3+} + H_2PO_4^{-}(\text{soluble}) + 2H_2O \Leftrightarrow Al(OH)_2H_2PO_4\ (\text{insoluble}) + 2H^{+}$$

$$Fe^{3+} + H_2PO_4^{-}(\text{soluble}) + 2H_2O \Leftrightarrow Fe(OH)_2H_2PO_4\ (\text{insoluble}) + 2H^{+}$$

In basic or alkaline soils, P immobilization or fixation occurs by the following reaction:

$$Ca(H_2PO_4)_2\ (\text{soluble}) + 2Ca^{2+} \Leftrightarrow Ca_3(PO_4)_2\ (\text{insoluble}) + 4H^{+}$$

Phosphorus immobilization is high in soils containing higher amount of amorphous Fe and Al hydroxides and allophane where adsorption or retention reduces phosphate mobility and renders a large proportion of the total inorganic P insoluble and unavailable to plants. Thus, P acquisition is not an issue of total supply but of being unavailable caused by the extreme insolubility of P at both acidic and alkaline pH. As a result concentrations of P in soil solution are often low for adequate plant nutrition (Jayachandran et al., 1989). Reaction of P with soil may involve both adsorption and precipitation, which are thought to result from the same chemical force (Lin et al., 1983; Bolan et al. 1999). Many believe that adsorption mechanisms prevail at low concentrations and precipitation mechanisms dominate at high P concentrations (Lin et al., 1983).

Van Riemsdijk and Haan (1981) reported that the reaction of phosphate with soils, which are initially free of sorbed P, and with metal oxides/hydroxides is a very fast reaction at the beginning, slowing down substantially during the course of reaction. Phosphorus recovery increased with increasing levels of P. Fageria and Barbosa Filho (1987) reported that P immobilization capacity of Brazilian Oxisol varied from 77% to 90%, depending on the level of soluble P applied. High P fixation capacity of Brazilian Oxisol is related to low pH and high iron oxide contents, Al saturation, and clay fractions that are commonly composed primarily of kaolinite, gibbsite, and iron oxides (Smyth and Sanchez, 1982). Phosphorus immobilization is higher in soils containing high clay content as compared with coarse-textured soils.

Fageria and Gheyi (1999) reported that in Oxisols and Ultisols of tropics, P immobilization capacity is high. They recommended that in these soils, soluble P should not be applied prior to sowing crops. Since most crops need P throughout their growth cycle, if P is applied in advance, a large amount may be fixed initially and crops may subsequently suffer from P deficiency.

Phosphorus fixation is related to well-drained soils. However, a large area around the world is planted under flooded condition. Sah and Mikkelsen (1986) reported that flooding and subsequent draining of soil affected P transformations, increased amorphous Fe levels and P sorption, and induced P deficiency in crops grown after flooded rice (*O. sativa*). In California, wheat (*Triticum* spp.), safflower (*Carthamus tinctorius* L.), corn (*Zea mays*), and sorghum (*Sorghum bicolor*) have shown P deficiency following flooded rice (Brandon and Mikkelsen, 1979). Similar observations were also reported from Australia (Willet and Higgins, 1980). Process leading to increased immobilization of P in flooded-drained soils perhaps starts during the flooding periods of soil. Decrease in redox potential of flooded soils causes transformation of several chemical species. Reversal of these processes after soil drainage leads to increased chemical reactivity of soil minerals with P (Sah and Mikkelsen, 1986), which may immobilize P for several years (Willet and Higgins, 1980). A flooding period as short as 2–4 days increased P sorptivity in flooded-drained soils (Willet, 1982). Factors such as supplemented organic matter and favorable soil temperatures that accelerate soil anaerobiosis may also enhance P sorption. Under aerobic soil conditions, addition of organic matter has been reported to decrease P sorption and increase P desorption (Singh and Jones, 1976; Kuo, 1983). Application of organic matter to a flooded soil intensifies the soil reduction processes, increases the transformation of soil Fe and P minerals, and leads to high P sorption in flooded-drained soil (Sah and Mikkelsen, 1989; Sah et al. 1989a,b) reported that increased soil P sorption in flooded-drained soil is related to an increase in amorphous Fe oxides in these environments, and they concluded that the sorptivity in flooded-drained soils is correlated with Fe transformation.

1.7 ORGANIC PHOSPHORUS IN SOILS

Phosphorus is present in both organic and inorganic forms in soils. Organic form of P is bonded to C in some way (Condron et al., 2005; Pierzynski et al., 2005). While the percentage of each can vary widely, substantial amounts of both forms generally are present (Daroub and Snyder, 2007). Most of the P in the mineral soils is in the organic form. Kamprath and Foy (1971) reviewed the literature on lime–fertilizer interactions in acid soils and reported that one-half to two-third of the total P in mineral soils is present in the organic form. They further reported that extreme values range from 4% for a podzol to 90% for an alpine humus. Stevenson (1986) reported that 15%–80% of the P in the soils occurs in organic forms, the exact amount being dependent on the nature of the soil and its composition. Higher percentages are typical of peats and uncultivated forest soils, although much of the P in tropical soils and certain prairie grassland soils may occur in organic forms. In fertilized temperate zone soils, however, the contribution of organic forms is likely to be rather small relative to inorganic P (Stevenson, 1986).

Soil organic P compounds can be classified into three groups (Anderson, 1980): (1) inositol phosphate, which composes up to 605 of soil organic P (Tate, 1984), (2) nucleic acids, and (3) phospholipids. Another important dynamic organic P pool is the biomass P, which is 1%–2% of the total soil P (Stevenson, 1986). Isotopic double-labeling techniques have shown that recently added organic residue P apparently is an important component of the microbial biomass P. A field experiment indicated that 22%–28% of the ^{33}P applied in medic plant residues was recovered in the microbial biomass (McLaughlin et al., 1988a). Furthermore, there appears to be rapid transformation of plant P to organic P fractions in soils. For example, McLaughlin et al. (1988b) reported that after 7 days, 40% of the plant residue ^{33}P was incorporated into organic P fractions of soil.

For uptake to occur, P in soil organic matter must be converted to orthophosphate anions by soil microorganisms, a process known as mineralization (Daroub and Snyder, 2007). Release of P from organic matter, like that of N, depends on parent material, cultivation, depth of the soil, soil temperature, moisture, O_2, and pH. When soils are first placed under cultivation, the content of organic C and N usually declines. A similar pattern is also followed for organic P (Stevenson, 1986). In acid soils the very insoluble Al and Fe phytates are believed to be the most abundant organic P compounds. Increasing soil pH generally causes mineralization of phytate P and, therefore, increases its availability to plants (Cosgrove, 1967). Thompson et al. (1954) reported that mineralization of P was positively correlated with soil pH.

Most naturally occurring organic forms of P are esters of orthophosphoric acid, and numerous mono- and diesters have been characterized (Anderson, 1980). Some phosphate esters, such as nucleic acids, nucleotides, and sugar phosphates, are essential to life and occur in all living cells, though the proportion of the total P may vary widely. Others, for example, teichoic acids and inositol phosphates, have been detected in some organisms only and only in selected tissues of these organisms (Anderson, 1980). Since a large part of P in the soils are present in organic forms, the availability of P to plants depends on its mineralization. Olsen and Watanabe (1966) reported that only a fraction of the total amount of the organic P mineralized to inorganic P will be taken up by the roots, since the inorganic P will be distributed in a volume of soil, only a part of which contribute to the uptake of P by plants. In addition, the inorganic P released will be subjected to reactions that fix P similar to those which occur with fertilizer P (Olsen and Flowerday, 1971).

Extraction and ignition are the two methods of determining organic P in the soils. In the extraction method, P is extracted with acid and base. Organic P is converted to orthophosphate, and the content of organic P is determined from the increase in inorganic phosphate as compared to a dilute acid extract of the original soil. The following equation is used to calculate organic P (Stevenson, 1986):

$$\text{Organic P} = \text{Total P in alkaline extract} - \text{Inorganic P in acid extract}$$

In the ignition method, organic P is converted to inorganic P by ignition of the soil at elevated temperatures and is calculated as the difference between inorganic P in acid extracts of ignited and nonignited soil by using the following equation (Stevenson, 1986):

$$\text{Organic P} = \text{Inorganic P of ignited soil} - \text{Inorganic P of untreated soil}$$

1.8 INORGANIC PHOSPHORUS IN SOILS

Phosphorus in soil can be divided into organic and inorganic P (Condron et al., 2005). Major part of P uptake by plants is in the form of inorganic. Inorganic P is divided into orthophosphate, pyrophosphate, and polyphosphate. At the pH of most soils, orthophosphate occurs as $H_2PO_4^-$ or HPO_4^{2-}. Polyphosphates are chains of orthophosphate, ranging in length from two orthophosphate groups (pyrophosphate) to >100 (Cade-Menum and Liu, 2014). The inorganic forms of P occur in numerous combinations with iron, aluminum, calcium, fluorine, and other elements. Inorganic forms are $H_2PO_4^-$ and HPO_4^{2-}. Concentration of the inorganic forms is the most important single factor governing the availability of P to crop plants. Uptake of these ions of P depends on pH and their concentrations in the soil solution. However, uptake of $H_2PO_4^-$ is more rapid than HPO_4^{2-} (Tisdale et al. 1985). Primary inorganic P-containing compounds are presented (Table 1.7). These organic P compounds like calcium phosphate have high solubility at lower pH, and Al containing P compounds are less soluble in acid soils. Organic source of P in the soil remains for a longer time and reduces its solubility (Brady and Weil, 2002).

1.9 REACTION OF PHOSPHORUS FERTILIZERS IN SOIL

When P fertilizers are applied to the soil, they are dissolved in soil solution and react with soil constituents. However, the reactions of applied P fertilizer in soils depend on the pH. Hence, P fertilizer supplied to a soil has different reactions in acid and alkaline soils. When inorganic fertilizers are applied to acid soils, they react with oxides of Al and Fe, and subsequently the P ions are unavailable to plants.

TABLE 1.7
Major Inorganic P-Containing Compounds in Soils

Calcium Compounds	Aluminum and Iron Compounds
Fluorapatite [{$3Ca_3(PO_4)_2$} · CaF_2]	Variscite ($AlPO_4 \cdot 2H_2O$)
Carbonate apatite [{$3Ca_3(PO_4)_2$} · $CaCO_3$]	Berlinite ($AlPO_4$)
Hydroxyapatite [{$3Ca_3(PO_4)_2$} · $Ca(OH)_2$]	Strengite ($FePO_4 \cdot 2H_2O$)
Oxyapatite [{$3Ca_3(PO_4)_2$} · CaO]	Vivianite {$Fe_3(PO4)_2 \cdot 8H_2O$}
Tricalcium phosphate $Ca_3(PO_4)_2$	
Octacalcium phosphate $Ca8H_2(PO_4)_6 \cdot 5H_2O$	
Dicalcium phosphate $CaHPO_4 \cdot 2H_2O$	
Monocalcium phosphate $Ca(H_2PO_4)_2 \cdot H_2O$	

Decrease in availability of P by these compounds is known as P immobilization (precipitation and fixation). This phenomenon of P fixation or retention has been known for well over the century (Way, 1850) and has been studied more than any other aspects of soil–fertilizer–plant interactions (Sample et al., 1980). Sample et al. (1980) reviewed the literature on P retention by soil constituents and concluded that hydrous oxides of Fe and Al and calcium carbonate played key roles in P immobilization. They suggested either that the P was precipitated as Fe, Al, or Ca phosphate or that P was chemically bonded to these cations at the surface of the soil minerals. From the early days, the choices of mechanisms used to explain P immobilization involved precipitation or adsorption. More studies have proposed several mechanisms through which P may be retained by soils. These include physical adsorption, chemisorption, anion exchange, surface precipitation, and precipitation of separate solid phases (Sample et al., 1980).

Chemical mechanisms involved in the adsorption of P have been reviewed by Mattingly (1975). At present it is sufficient to note that the phosphate is closely and chemically bonded to the surface of Fe and Al oxides by chemical bonds. The reaction may be regarded as partly a displacement of water molecules and partly displacement of hydroxyls, so that the negative charge conveyed to the surface is usually lower than the charge on the anion (Barrow, 1980). Phosphorus immobilization reactions by Fe and Al oxides or hydroxides occurred in the heading P Cycle in Soil–Plant System.

In calcareous soils (higher pH), P fertilizer reacts with Ca and makes P unavailable to plants. These reactions are discussed under the heading P Cycle in Soil–Plant System. Conversion of soluble fertilizer P to extremely insoluble calcium phosphate forms mostly occurs in calcareous soils of low rainfall regions. Iron and aluminum impurities in calcite particles may also adsorb considerable amounts of phosphate in these soils (Brady and Weil, 2002). Because of the various reactions with $CaCO_3$, P availability tends to be nearly low in Aridisols, Inceptisols, and Mollisols of arid regions as in the highly acid Spodosols, Oxisols and Ultisols of humid regions, where iron, aluminum, and manganese limit P availability (Brady and Weil, 2002). The recovery efficiency of P is less than 20% in most agroecosystems. Fertilizer P is most available to plants immediately after soil application. Phosphorus availability from inorganic P fertilizer becomes less available with increasing time application. Therefore, P fertilizers should be applied to crops at the time of sowing or transplanting.

1.10 EVALUATION AND UTILIZATION OF RESIDUAL PHOSPHORUS IN SOILS

When optimum amounts of fertilizer P are used for intensive cropping, most soils tend to accumulate residual P. Residual P in soils consists of adsorbed and firmly held phosphate, phosphate held as insoluble precipitates, and organic P (Haynes, 1984). When soluble P fertilizer is applied to soils deficient in P, its availability decreases with the time of application. This decrease in availability may be due to changes occurring in the zone of precipitation and slow movement of P from this zone. One of the methods to measure the availability of residual phosphate is by the net uptake of phosphate by successive crops in comparison with that from unfertilized soils (Haynes, 1984). Although such studies do not differentiate between uptake of fertilizer or native phosphate from inorganic or organic origins, they have demonstrated the importance of residual phosphate in crop production under diverse soil conditions (Haynes, 1984). Spratt et al. (1980) have emphasized the importance of residual phosphate in wheat production on Chenozemic soils. Their results indicated that an initial application of 100 kg P ha^{-1} is enough to sustain 8 years of wheat flax (*Phormium tenax*) rotational cropping. Sanchez and Uehara (1980) noted that in high phosphate fixing Ultisols and Oxisols, an initial application of at least 175 kg P ha^{-1} appears necessary for a residual effect to last about 10 years. Where repeated annual applications of phosphate are applied, the residual pool can become more important than the currently applied phosphate in determining yield responses (Haynes, 1984).

Another method of determining the residual effect of P is comparing the results of crop responses to newly added and applied P earlier. In a containerized trial, Devine et al. (1968) reported that after 1-, 2-, and 3-year contact, the effectiveness of powdered single superphosphate on four soils averaged 58%, 38%, and 20% of that of fresh superphosphate, respectively. These percentages are accorded to those reported in several field trials even though some of the other factors would have also operated (Barrow, 1980). Barrow (1980) concluded that published results appear to be consistent with the idea that the effectiveness of soluble P fertilizers decreases with time in a similar manner so that initial effectiveness is followed by a markedly decreased effectiveness. Experimental data reported since the review of Khasawneh and Doll (1978) as well as some earlier data (Rajan et al. 1996) show that residual effects of PRs can exceed those of soluble P fertilizers and are greatly influenced by the PR dissolution rate and the rate of loss of P from the plant-available P pool in the soil.

There was a higher yield of rice straw and grain occurring in plots, which had residual fertilizer treatments, as compared to control treatment (Table 1.8). However, maximum straw and grain yield

TABLE 1.8
Straw and Grain Yield of Lowland Rice under Different Fertilizer Treatments

Fertilizer Treatment	Straw Yield (kg ha^{-1})	Grain Yield (kg ha^{-1})
T_1	4688ab	4050c
T_2	5646ab	4972abc
T_3	4297b	4141c
T_4	6107a	5342ab
T_5	4915ab	4016c
T_6	5507ab	4762abc
T_7	6160a	5598a
T_8	5706ab	4562bc
F-test	**	**
CV(%)	16	11

Source: Fageria, N.K. et al., *Rev. Bras. Eng. Agric. Amb.*, 4, 177, 2000.
(Values are means of 3-year field experimentation.)
**Significant at the 1% probability level. Means followed by the same letter within the same column are not significantly different at the 5% probability level by Tukey's test. T_1 = Control; T_2 = Control + 40 kg N ha^{-1} in topdressing 45 days after sowing; T_3 = Residual effect of medium level of soil fertility; T_4 = Residual effect of medium level of soil fertility + 50 kg N ha^{-1} + 60 kg P_2O_5 ha^{-1} + 40 kg K_2O ha^{-1}; T_5 = Residual effect of high level of soil fertility; T_6 = Residual effect of high level of soil fertility + 25 kg N ha^{-1} + 30 kg P_2O_5 ha^{-1} + 20 kg K_2O ha^{-1}; T_7 = Residual effect of medium soil fertility + residual effect of green manure + 75 kg N ha^{-1} + 90 kg P_2O_5 ha^{-1} + 60 kg K_2O ha^{-1}; and T_8 = Residual effect of medium level of fertility + residual effect of green manure + 100 kg N ha^{-1} + 120 kg P_2O_5 ha^{-1} + 80 kg K_2O ha^{-1}. Before imposing the fertilizer treatments, rice and bean were cultivated in rotation for 3 years. Rice received the following fertility treatment: (1) control (low soil fertility), (2) 100 kg N ha^{-1} + 100 kg P_2O_5 ha^{-1} + 60 kg K_2O ha^{-1} + 40 kg FTE-BR-12 ha^{-1} as micronutrients (medium soil fertility), (3) 200 kg N ha^{-1} + 200 kg P_2O_5 ha^{-1} + 120 kg K_2O ha^{-1} (high soil fertility level), and (4) 28 Mg ha^{-1} green or fresh weight of pigeon pea (*Cajanus cajan* L.) as green manure. Similarly, dry bean crop received the following soil fertility levels: (1) control (low soil fertility), (2) 35 kg N ha^{-1} + 120 kg P_2O_5 ha^{-1} + 60 kg K_2O ha^{-1} + 40 kg ha^{-1} FTE-Br 12 (medium soil fertility level), (3) 70 kg N ha^{-1} + 240 kg P_2O_5 ha^{-1} + 120 kg K_2O ha^{-1}+ 80 kg FTE- BR-12 ha^{-1} (high soil fertility level), and (4) medium soil fertility level + green manure (residual effect of rice green manure).

was obtained in the plots with residual effect of fertility + residual effect of green manure + 75 kg N ha^{-1} + 90 kg P_2O_5 ha^{-1} + 60 kg K_2O ha^{-1}. Residual effect of fertilizer remains for crop production, but it may not be sufficient to achieve maximum economic yield of crops without fresh fertilizer application.

1.11 FACTORS INFLUENCING PHOSPHORUS AVAILABILITY TO PLANTS

There are several climatic, soil, and plant factors that influence the availability of P to plants. Principal climatic factors that influence P uptake are temperature and precipitation. Soil factors are aeration and compaction, concentration of P in the soil, soil pH, presence of Al and Fe oxides, fertilizer granules, and solubility and exudation of organic acids/anions in the rhizosphere. Important plant factors are crop species and genotypes within species. In addition presence of microorganisms in the rhizosphere may also influence availability of P to plants. Therefore, many factors are responsible for P uptake to crop plants. The discussion of these factors is given in this section.

1.11.1 Climatic Factors

Main climatic factors that affect availability of P to plants are temperature and precipitation. These two factors are responsible for many reactions of P in the soil, including solubilization and transport of P ions in soil–plant system.

1.11.1.1 Precipitation

Adequate precipitation during crop growth is essential to maintain optimum soil moisture in the soil–plant system. Ochsner et al. (2013) reported that soil moisture is an essential climate variable influencing land–atmosphere interactions, an essential hydrologic variable impacting rainfall–runoff processes, an essential ecological variable regulating net ecosystem exchange, and an essential agricultural variable constraining food security. Soil moisture affects soil reactions governing the release and diffusion of P in the soil solution and ultimately the positional availability of P relative to root growth. Generally, maximum availability of P for most crops has been associated with a soil water tension of about 1/3 bar ($0.33 \times 0.1 = 0.033$ megapascal, MPa) (Sanchez, 2007).

Soil water deficit is one of the major abiotic stresses that adversely affect crop growth and yield (Hsiao et al., 2007; Saseendran et al., 2014). This adverse effect is brought about in two major ways. Lack of adequate soil and water supply and reduced plant water uptake reduce cell division of leaf elongation and root enlargement, which lead to a decline in leaf area for photosynthesis and nutrient ion transport to the root surface in the soil (Saseendran et al. 2014). Water stress also directly affects many biochemical reactions and physiological growth processes, such as photosynthesis, C allocation and partitioning, phasic development rates, and phenology of crop plants (Chen and Reynolds, 1997; Chaves et al., 2002; Cakir, 2004; Saseendran et al., 2014).

Under anaerobic conditions (saturation with water), flooded rice, for example, the reduction of ferric phosphate to ferrous phosphate due to increase in soil pH of acid soils, might result in additional increased P solubility (Ponnamperuma, 1972; Holford and Patric, 1979; Fageria et al., 2011; Fageria, 2013, 2014). Similarly, P uptake in flooded alkaline soils also improves because of the liberation of P from Ca and calcium carbonate resulting from the decrease in pH. Formation of insoluble tricalcium phosphate is favored at a high pH (Fageria et al., 2011). Nevertheless, it is the general view that with the exception of aquatic crops, excessive water resulting in poor aeration would actually restrict P uptake by crops in spite of this enhanced solubility (Sanchez, 2007).

Diffusion of P through the soil to the root is the dominant mechanism governing the supply of P to roots growing in all except soils extremely high in P. Furthermore, diffusion coefficient of P is an important factor in availability of P to plants. If soil moisture content is near field capacity, diffusion coefficient varies with the variation of soil moisture. Lowering soil moisture decreases diffusion coefficient of P and subsequently its availability. Diffusion coefficient of $H_2PO_4^-$ in water at 25°C is 0.89×10^{-5} cm^2 s^{-1} (Parsons, 1959). At pH between 4.0 and 6.5, most of the inorganic P is in $H_2PO_4^-$ form. Assuming that P diffuses mainly through solution, Barber (1980) proposed the following equation for calculating the effective diffusion coefficient:

$$De = D\theta f \frac{dCl}{dC}$$

where

D is the diffusion coefficient in water
θ is the volumetric moisture content of the soil
f is a tortuosity factor
dCl/dC is the inverse of the differential buffer capacity of the soil

Therefore, θ, f, and dCl/dc are the three principal factors that influence the size of De.

In addition, major part of P remains in the upper soil layer, and during a dry season, plant roots take water from the lower soil layer, and subsequently P deficiency occurs in dry agriculture (Hanway and Olson, 1980). Increasing soil moisture to an optimum level increases P availability of all carriers, particularly with the more water-soluble sources (Beaton and Cough, 1962; Beaton and Read, 1963). Matocha et al. (1970), studying the residual effects of P fertilizers, reported that greater amounts of applied P were required for maximum yields in years of drought stress. They suggested that the lack of water severely reduced P availability in the soil.

Hammond et al. (1986) reviewed the 1950s work in Africa (Senegal) and reported that the fertilizer efficiency of PR increased with increasing rainfall. In a series of trials over a range of mean annual rainfall between 500 and 1300 mm, the yield increases of peanuts (*Arachis hypogaea*) over control had a highly significant linear correlation with the mean annual rainfall for the first 2 years following basal fertilization (Hammond et al., 1986). Hammond et al. (1986) also reported that in Francophone West Africa, PR was not recommended for either annual or basal fertilization where the rainfall was below 700 mm. All these results showed importance of water in P availability to crop plants. Precipitation during the year in the central part of Brazil, locally known as "Cerrado" region , with a large land area having high potential for crop production, is presented (Table 1.9).

TABLE 1.9
Precipitation in mm in the Cerrado Region Central Part of Brazil

Month	Year 2012	Year 2013
January	338.0	380.0
February	310.8	229.6
March	163.0	257.0
April	54.0	112.2
May	15.6	74.0
June	13.2	20.4
July	5.2	0.0
August	0.0	3.2
September	109.6	2.4
October	51.6	92.0
November	183.4	218.6
December	212.6	354.4
Total	1457.0	1743.8

Source: Data were taken from the record of the meteorological Station of National Rice and Bean Research Center, EMBRAPA, Santo Antônio de Goias, Brazil.

1.11.1.2 Temperature

Temperature is one of the important factors affecting distribution of plants on the earth. Root zone temperature has a dominant influence on seed germination, vegetative and reproductive growth stages in crop plants. As with the shoot, temperature affects both the expansion of the root system through effects on development and growth and a range of metabolic processes that determine the activity of the root system (Gregory, 2006). Root zone temperature affects root function and metabolism (Cooper, 1973; Bland, 1993; McMichael and Burk, 1998). At optimum temperature, cell division is more rapid but of shorter duration than at lower temperatures. At cooler temperatures, roots are usually whiter, thicker in diameter, and less branched than at warmer temperatures (Ketellapper, 1960; Nielsen and Cunningham, 1964; Garwood, 1968), although there are exceptions (Bowen, 1970). Soil temperature regimes are defined in Soil Taxonomy using measurements taken at a depth of 50 cm (Rodriguez et al., 2010). Root temperature is generally lower than that of air, but seasonal fluctuations can occur with depth depending on soil and above ground factors (McMichael and Burk, 1998).

Impact of soil temperature on the function of the root systems has been documented in a number of species (McMichael and Burk, 1998). At low temperatures water and nutrient uptake (including P) by root systems may be reduced (Nielsen and Humphries, 1966; Nielsen, 1974). In general, root growth tends to increase with increasing temperature until an optimum is reached above which root growth is reduced (Glinski and Lipiec, 1990; Fageria, 2013). Higher root temperature can affect overall enzymatic activity of root systems (Nielsen, 1974). Efficiency of extracellular enzymes increases with temperature, and microbes in warmer soils may invest fewer resources in their production in order to incur less of a metabolic cost (Allison, 2005; Bell et al., 2010). Alternatively, warming can decrease the soil water content during the growing season, which can limit enzyme and substrate diffusion (Allison, 2005).Warming during the plant growth season can decrease microbial biomass, possibly due to decreased soil moisture or increased predation (Cole et al., 2002; Rinnan et al., 2007). Warming can also shift the balance toward higher fungal dominance over bacteria in microbial communities (Zhang et al., 2005), although the direct effects of warming on soil microbes are potentially confounded by the indirect effects of warming on plant productivity and species composition (Jonasson et al., 1999) or microbial consumers (Rinnan et al. 2008; Bell et al. 2010).

McMichael and Burk (1994) reported that root metabolism may become more temperature sensitive as the mobilization of reserves in the cotyledons declines during early seedling growth, indicating that temperature dependency is developmentally regulated. McMichael and Quisenberry (1993) observed that the optimum temperature for root growth in cotton (*Gossypium*) was between 28°C and 35°C vs. between 23°C and 25°C for sunflower (*Helianthus annuus*). Genetic variability exists for root growth in response to changes in temperature both between and within species (McMichael and Burk, 1998). McMicheal and Quisenberry (1991) also reported that the genetic variability observed in root development in cotton genotypes was somewhat independent of their variability in shoot development, suggesting that breeding for more favorable root traits (i.e., deeper, more branched roots, less sensitive to low temperature) may be possible.

Crop species respond differently to temperature throughout their life cycles. Each species has a defined range of maximum and minimum temperatures within which growth occurs and an optimum temperature at which plant growth progresses at its fastest rate (Hatfield et al., 2011). Vegetative development usually has a high optimum temperature than reproductive development. Progression of a crop through phonological phases is accelerated by increasing temperatures up to the species dependence on optimum temperature. Exposure to higher temperatures causes faster development in food crops, which does not translate into an optimum for maximum production because the shorter life cycle means shorter reproductive period and shorter radiation interception period (Hatfield et al., 2011). Optimum temperature for maximum yield and root growth of several crop species are presented (Tables 1.10 and 1.11).

TABLE 1.10
Optimum Soil Temperature for Maximum Yield of Important Field Crops

Crop	Optimum Temperature (°C)	Reference
Barley (*Hordeum vulgare* L.)	18	Power et al. (1970)
Oats (*Avena sativa* L.)	15–20	Case et al. (1964)
Wheat (*Triticum aestivum* L.)	20	Whitfield and Smika (1971)
Corn (*Zea mays* L.)	25–30	Dormaar and Ketcheson (1960)
Cotton (*Gossypium hirsutum* L.)	28–30	Pearson et al. (1970)
Potato (*Solanum tuberosum* L.)	20–23	Epstein (1966)
Rice (*Oryza sativa* L.)	25–30	Owen (1971)
Bean (*Phaseolus vulgaris* L.)	28	Mack et al. (1964)
Soybean (*Glycine max* L. Merr.)	30	Voorhees et al. (1981)
Sugar beet (*Beta vulgaris* L.)	24	Radke and Bauer (1969)
Sugarcane (*Saccharum officinarum* L.)	25–30	Hartt (1965)
Alfalfa (*Medicago sativa* L.)	28	Heinrichs and Nielsen (1966)

Sources: Compiled from Case, V.W. et al., *Soil Sci. Soc. Am. Proc.*, 28, 409, 1964; Dormaar, J.F. and Ketcheson, J.W., *Can. J. Soil Sci.*, 40, 177, 1960; Epstein, E., *Agron. J.*, 58, 169, 1966; Hartt, C.E., *Plant Physiol.*, 40, 74, 1965; Heinrichs, D.H. and Nielsen, K.F., *Can. J. Plant Sci.*, 46, 291, 1966; Mack, H.J. et al., *Proc. Am. Soc. Hortic. Sci.*, 84, 332, 1964; Owen, P.C., *Field Crop Abstr.*, 24, 1, 1971; Pearson, R.W. et al., *Agron. J.*, 62, 243, 1970; Power, J.F. et al., *Agron. J.*, 62, 567, 1970; Radke, J.F. and Bauer, R.E., *Agron. J.*, 61, 860, 1969; Voorhees, W.B. et al., Alleviating temperature stress, in: *Modifying the Root Environment to Reduce Crop Stress*, eds. G.F. Arkin and H. Taylor, American Society of Agricultural Engineering, St. Joseph, MI, 1981, pp. 217–266; Whitfield, C.J. and Smika, D.E., *Agron. J.*, 63, 297, 1971.

TABLE 1.11
Optimum Temperature for Root Growth of Some Crop Species

Crop Species	Parameters	Temperature (°C)
Sunflower (*Helianthus annuus*)	Root elongation rate	20
Tomato (*Solanum lycopersicum*)	Root elongation rate	30
Corn (*Zea mays*)	Root elongation rate	30
Corn	Root mass	26
Oats (*Avena fatua*)	Root mass	5
Cotton (*Gossypium*)	Root elongation rate	33
Soybean (*Glycine max*)	Taproot extension rate	25
Grape (*Vitis vinifera*)	Root extension	23
Rice (*Oryza sativa*)	Root growth	25–37
Lolium perenne	Root mass	17

Sources: McMichael, B.L. and Burk, J.J., *Environ. Exp. Bot.*, 31, 461, 1994; Fageria, N.K., *The Role of Plant Roots in Crop Production*, CRC Press, Boca Raton, FL, 2013.

In addition to influence of temperature on root growth, soil temperature affects reactions that govern the dissolution, adsorption, and diffusion of P. Although sorption and desorption generally occur concurrently, an increase in soil temperature increases kinetics of reaction and enables more rapid equilibrium among nonlabile, labile, and solution P pools, resulting in more rapid replenishment of solution P as P is taken up by crops (Gardner and Jones, 1973; Sanchez, 2007). Sutton (1969) reported that most of the effects of temperature on available P were due to inorganic reactions,

TABLE 1.12
Minimum and Maximum Temperatures (°C) in the Cerrado Region Central Part of Brazil

	Year 2012		Year 2013	
Month	Minimum	Maximum	Minimum	Maximum
January	18.7	26.4	19.6	27.7
February	18.3	28.2	18.9	29.4
March	18.8	29.4	19.9	28.8
April	18.9	29.6	18.3	28.0
May	16.4	27.2	16.7	28.4
June	16.7	28.1	16.9	27.6
July	15.0	28.3	14.6	27.9
August	16.2	28.4	14.5	28.6
September	18.9	32.0	18.1	29.8
October	19.9	32.5	19.4	29.0
November	19.8	27.6	19.0	28.5
December	19.7	29.6	18.9	26.9

Source: Data were taken from the record of meteorological Station of National Rice and Bean Research Center, EMBRAPA, Santo Antônio de Goiás, Brazil.

since the effect occurred too rapidly to be explained by microbial mineralization. Olsen and Flowerday (1971) reported that increasing soil temperature, within limits, effects a more rapid mineralization of organic P, may enhance microbial activity, and may increase soil chemical activity, all of which may differentially influence the plant availability of P fertilizers.

Soil temperature affects physical, chemical, and biological processes in the rhizosphere and nutrient availability. Lower as well as higher temperatures are detrimental for rhizosphere environmental changes. Crops originated in the tropical climate such as corn grow well in the temperature range of 25°C–30°C, whereas crops that originated in the temperate climate, such as rye (*Elymus* sp.), grow well in the temperature range of 12°C–18°C (Brady and Weil, 2002). Although there are exceptions, C_4 plants are generally more tolerant of high temperature than C_3 plants (Edwards et al., 1983). Many C_4 plants like sugarcane and corn are better able to grow under high temperatures than are C_3 plants such as wheat and barley (*Hordeum vulgare*) (Fageria et al., 2011). Data related to mean minimum and maximum temperature in the central part of Brazil from 2012 to 2014 are presented (Table 1.12).

1.11.2 Soil Factors

Soil is defined as the unconsolidated mineral or organic materials on the immediate surface of the earth that serves as a natural medium for the growth of terrestrial plants (Soil Science Society of America, 2008). There are several soil factors that influence uptake of nutrients, including P. These factors are soil aeration and compaction, level and forms of soil P, soil pH, presence of aluminum and iron hydroxides, fertilizer granule and solubility, and exudation of organic acid/anions in the rhizosphere.

1.11.2.1 Aeration and Compaction

Soil aeration is defined as the condition and sum of all processes affecting soil pore space gaseous composition, particularly with respect to the amount and availability of oxygen for use by soil biota and/or soil chemical oxidation reactions (Soil Science Society of America, 2008). Similarly, soil compaction is defined as increasing the soil bulk density, and concomitantly decreasing the soil porosity, by the application of mechanical forces to the soil (Soil Science Society of America, 2008). Soil compaction is a major problem caused by the use of heavy machinery to cultivate the soil for plant production ([illegible] 2013). Compression of a soil beyond the limit of purely elastic deformation results in a persistent reduction of pore space and pore connectivity, which by decreasing soil water drainage can lead to severe waterlogging and impair soil aeration (Soane and Van Ouwerkerk, 1995). Sensitivity toward deformation can vary considerably among different structures of soil pore space (Blackwell et al., 1990; Richard et al., 2001). P uptake is higher in well-aerated and uncompacted soils as compared to soils that are compacted and have low aeration.

1.11.2.2 Concentration and Forms of Soil Phosphorus

Concentration and forms of P in the soil are important factors in determining its availability. Phosphorus uptake by plant roots is usually in direct proportion to root surface concentration of P. Olsen et al. (1961) reported with corn seedlings that uptake of P at a constant moisture level on soils differing in texture and soluble P was a linear function of solution P concentration. Olsen and Flowerday (1971) reported that the intensity factor, the concentration of P in the soil solution, is important since the difference between the initial concentration of the soil solution P and the P concentration at the root surface mainly controls the rate of P uptake by roots. Maximum yield or P uptake was obtained at variable P concentrations, ranging from 2 to 24 μM (Table 1.13). However, Foth and Ellis (1988) reported that the concentration of P found in soil solution may range from <0.01 to 8 mg L^{-1} (0.32 to 256 μM). Similarly, Australian researchers reported that concentration of 0.2–0.3 mg L^{-1} (6.4–9.6 μM) is adequate for a variety of crops (Tisdale et al. 1985). This variability is partly due to the differences among crop species and partly an indication that while intensity of P is the main factor that determines uptake, additional factors come into play in soil systems (Olsen and Khasawneh, 1980). These other factors were listed as quantity and buffering capacity by Khasawneh (1971) and as capacity and diffusion by Olsen and Flowerday (1971). Tisdale et al. (1985) reported that optimum solution concentrations of P are probably not constant for a specific crop. Stage of growth and the occurrence of stress caused by disease and adverse climatic conditions are factors that are expected to substantially modify the desirable amounts of soil solution P.

Phosphorus is mainly absorbed by plants in the form of inorganic or orthophosphate ions ($H_2PO_4^-$ and HPO_4^{2-}), which are present in the soil solution. Amount of each P form present depends on pH. At pH below 7.0, major ionic form of P is $H_2PO_4^-$ in many agricultural soils. At pH 7.0 both are present in equal amount. At pH higher than 7.0, HPO_4^{2-} is the dominant form. Plant uptake of divalent ion (HPO_4^{2-}) is reported to be slower as compared to monovalent ionic form ($H_2PO_4^-$) (Tisdale et al., 1985). Most of the P in soils present is in the organic form. Therefore, organic forms

TABLE 1.13
Minimum Concentration of P in Soil Solution or Soil Extract Needed to Produce Maximum Growth of Important Field Crops

Crop Species	Critical P Level (μM)	Reference
Barley (*Hordeum vulgare*)	24	Olsen and Watanabe (1970)
Cotton (*Gossypium*)	2	Khasawneh and Copeland (1973)
Millet (*Pennisetum glaucum*)	6	Fox and Kamprath (1970)
Oats (*Avena fatua*)	10	Ozanne and Shaw (1968)
Sorghum Sudan grass	12	Soltanpour et al. (1974)
Rice (*Oryza sativa*)	3	Hossner et al. (1973)
Most crop plants	6	Beckwith (1965)

of P should be transformed into inorganic forms by mineralization. In addition, use of inorganic P fertilizers is required to produce desired economic yield of crops in soils that are deficient in P.

1.11.2.3 Soil pH

Soil pH is one of the important factors affecting availability of P to plants. In acid soils, P availability is limited by Al and Fe oxides because these oxides immobilize P. Similarly at higher pH (>7.0), P is fixed by Ca and its availability to plants is decreased. Brady and Weil (2002) reported that as a general rule in mineral soils, P immobilization is at its lowest and plant availability is at its highest when soil pH is maintained in the range of 6.0–7.0. Foth and Ellis (1988) also reported that maximum level of P in solution will be in soils with a pH value of 6.5–8.0. However, in our opinion (authors of this book), maximum availability of P occurs in soils within a pH range of 6.0–7.5.

Dry bean shoot dry weight, grain yield, and its components were significantly improved with the improvement in soil pH on a Brazilian Oxisol (Figures 1.8 and 1.9). Adequate pH for shoot, grain,

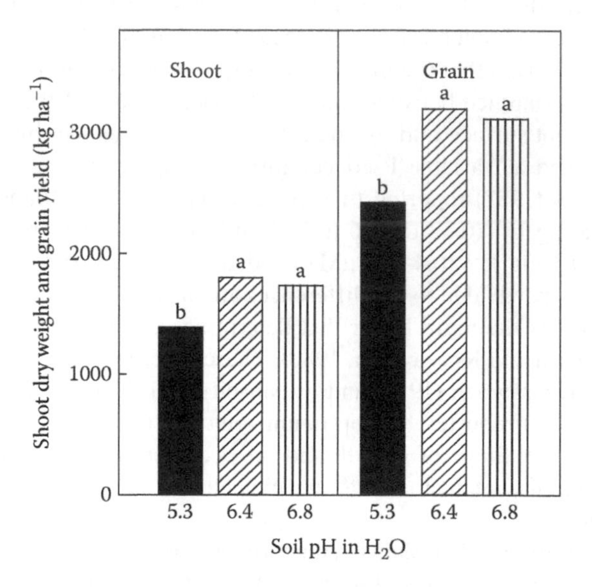

FIGURE 1.8 Influence of soil pH on shoot dry weight and grain yield of dry bean. (From Fageria, N.K. and Barbosa Filho, M.P., *Commun. Soil Sci. Plant Anal.*, 39, 1016, 2008.)

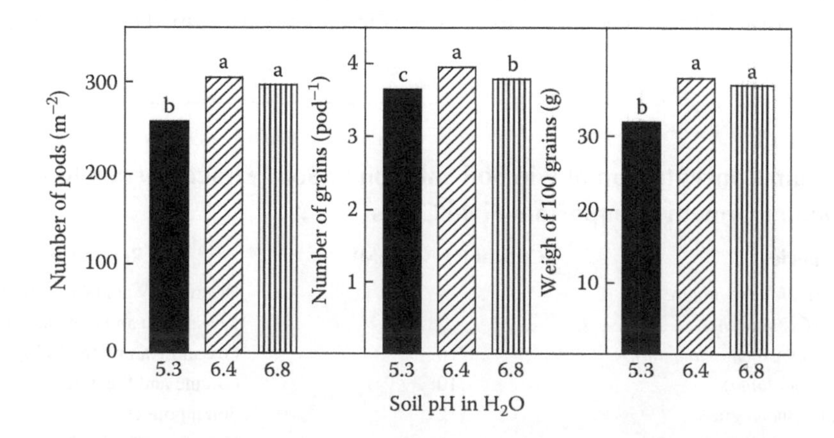

FIGURE 1.9 Influence of soil pH on yield components of dry bean. (From Fageria, N.K. and Barbosa Filho, M.P., *Commun. Soil Sci. Plant Anal.*, 39, 1016, 2008.)

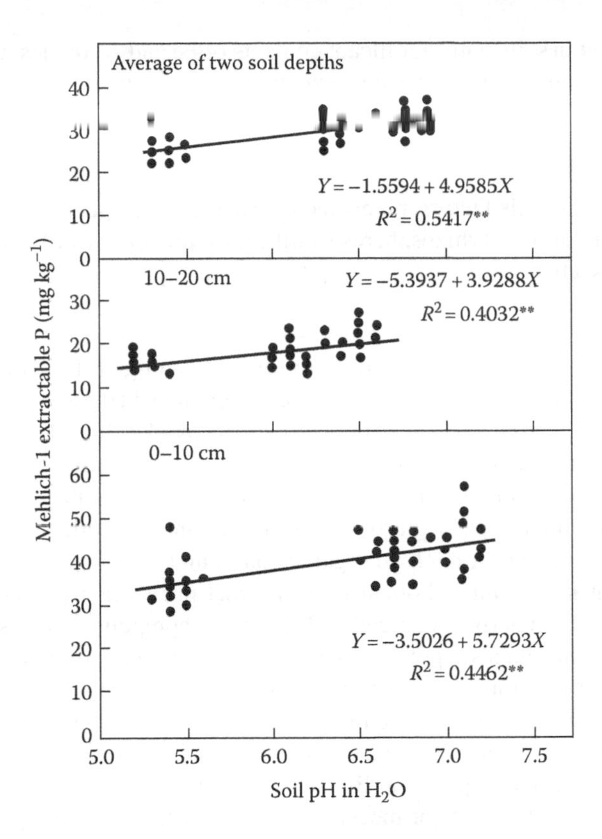

FIGURE 1.10 Relationship between soil pH and Mehlich 1 extractable soil phosphorus. (From Fageria, N.K. and Barbosa Filho, M.P., *Commun. Soil Sci. Plant Anal.*, 39, 1016, 2008.)

and yield components was 6.4. An increase in grain yield with increasing soil pH was associated with availability of nutrients, especially N, P, Ca, and Mg and reduction of Al^{3+} toxicity (Fageria and Baligar, 2003; Menzies, 2003). Foy (1984) reported that with increasing H^+ activity in the soil solution, uptake of P, Ca, and Mg was reduced and Al^{3+} activity increased.

Phosphorus deficiency is one of the most yield-limiting factors in the Cerrado soils of Brazil. In this regard, effect of soil pH was evaluated on P availability (Figure 1.10). Phosphorus extracted by Mehlich 1 method increased linearly in the pH range of 5.3–7.3 in the 0–10 cm soil depth. Mehlich 1 extractable soil P increased from 34 mg kg^{-1} at pH 5.3 to 45 mg kg^{-1} at pH 7.3 in the 0–10 cm soil depth. In the 10–20 cm soil depth, extractable soil P was 15 mg kg^{-1} at pH 5.3 and increased to 20 mg kg^{-1} at pH 6.6. Mean soil P at two soil depths was 25 mg kg^{-1} at pH 5.3 and 32 mg kg^{-1} at pH 6.9. An increase in extractable P in acid soils as a function of liming is justified by neutralization of positive charges of Fe and Al oxides and hydroxides, thereby reducing P immobilization (Haynes, 1982). According to Lindsay (1979), diminishing soil acidity by liming occurs due to hydrolysis of Fe and Al compounds and thereby liberation of P in the soil solution.

1.11.2.4 Presence of Aluminum and Iron Oxides

Presence of Al and Fe hydrous metal oxides in the soils significantly influences P availability. These substances have very high capacity to sorb a large amount of P. Although these hydroxides are present in most soils, they are most abundant in highly weathered soils like Oxisols and Ultisols. Aluminum and Fe oxides and their hydrous oxides can occur as discrete particles in soils or as coating or films on other soil particles. They also exist as amorphous Al hydroxyl compounds between the layers of expandable Al silicates (Tisdale et al., 1985). Tisdale et al. (1985) reported that it is generally accepted that in soils with significant contents of Fe and Al oxides, the less crystalline the oxides are, the larger their phosphate fixing capacity because of greater surface areas.

1.11.3 Plant Factors

In addition to climate and soils factors, plants factors such as crop species/genotypes within species and microorganisms in the plant rhizospheres significantly influence availability of P. These factors are discussed in this section.

1.11.3.1 Release of Organic and Inorganic Compounds in the Rhizosphere

Plant roots not only absorb water and nutrients to support plant growth but also release in the rhizosphere organic and inorganic compounds. These compounds bring several chemical changes in the root environment affecting microbial population and availability of nutrients (Neumann and Romheld, 2001; Fageria and Stone, 2006). Release may occur as an active exudation, a passive leaking, the production of mucilage, or with the death and sloughing of root cells. Releases increase under a variety of conditions, particularly under abiotic and/or biotic stress (Marschner, 1995). Rovira et al. (1979) classified root-released organic compounds as (1) exudates—compounds of low molecular weight that leak nonmetabolically from intact plant cells; (2) secretions—compounds metabolically released from active plant cells; (3) lysates—compounds released from the autolysis of older cells; (4) plant mucilages—polysaccharides from the root cap, root cap cells, primary cell wall, and other cells; and (5) mucilage—gelatinous material of plant and microbial origin.

Terms exudates and exudation are sometimes used collectively and perhaps incorrectly to include all of the organic compounds released from the roots and most if not all the mechanisms involved in the release of organic compounds (Pepper and Bezdicek, 1990). Important organic compounds released by roots are present (Table 1.14). Major mechanisms of releasing these compounds are leakage and secretion. Leakage involves simple diffusion of these compounds due to the higher concentrations of compounds within the roots as compared to the soil (Pepper and Bezdicek, 1990). Secretion, however, requires metabolic energy because it occurs against concentration gradients. Sugars and amino acids

TABLE 1.14
Organic and Inorganic Compounds Released by the Roots in the Rhizosphere

Root Exudates	Compounds
Diffusitives	Sugars and polysaccharides (arabinose, fructose, galactose, glucose, maltose, mannose, oligosaccharides, ribose, sucrose, xylose)
	Organic acids (acetic, butyric, citric, oxalic, tartaric, succinic, propionic, malic, glycolic, benzoic)
	Amino acids (glutamine, glycine, serine, tryptophan, aspartic, cystine, cystathionine, α-alanine, β-alanine, γ-aminobutyric)
	Inorganic ions, oxygen
Secretives	Mucilage, protons, electrons, enzymes (amylase, invertase, peroxidase, phenolase, phosphatases, adenine, uridine/cytidine, nucleotides)
Excretives	CO_2, HCO_3, protons, electrons, ethylene
Root debris	Rot cap cells, cell content

Sources: Compiled from Bertin, C. et al., *Plant Soil*, 256, 67, 2003; Dakora, F.D. and Phillips, D.A., *Plant Soil*, 245, 35, 2002; Fageria, N.K. and Stone, L.F., *J. Plant Nutr.*, 29, 1327, 2006; Neumann, G. and Romheld, V., The release of root exudates as affected by the plants physiological status, in: *The Rhizosphere: Biochemistry and Organic Substances at the Soil–Plant Interface*, eds. R. Pinto, Z. Varanini, and P. Nannipieri, Marcel Dekker, New York, 2001, pp. 41–93; Uren, N.C., Types, amounts, and possible functions of compounds released into the rhizosphere by soil-grown plants, in: *The Rhizosphere: Biochemistry and Organic Substances at the Soil–Plant Interface,* eds. R. Pinto, Z. Varanini, and P. Nannipieri, Marcel Dekker, New York, 2001, pp. 19–40.

provide energy for microorganisms in the rhizosphere, which mineralize or solubilize many nutrients. Similarly, acids reduce the pH, and the availability of many micro- and macronutrients improved. Release of mucilages protects the root tips from injury and desiccation as well as plays a role in nutrient uptake through its pH-dependent cation exchange capacity (Jenny and Grossenbacher, 1963). Quality and quantity of organic compounds release are determined by plant species and genotypes within species, plant age, soil type, soil physical properties, and presence of microorganisms.

1.11.3.2 Crop Species/Genotypes within Species

Genetic variability in plant species and genotypes within species in P uptake and use efficiency is widely reported in the literature (Devine et al., 1990; Duncan and Baligar, 1990; Fageria et al., 2008). Variation within a species is readily capitalized upon by breeders in developing new cultivars. Variability between species requires more intricate and complex operations for utility, particularly in the legume family, where barriers to sexual hybridization between species are formidable (Devine et al., 1990). However, biotechnology can overcome these barriers in the future in these species. In breeding for any trait, it is important to assess the genetic variability available for use. If sufficient genetic variability is not available in agronomic cultivars, a search for the desired genetic variability in new germplasm sources such as plant introductions, undomesticated forms, and other species may be undertaken (Devine et al., 1990).

To exploit the genetic variability for nutrient use efficiency in crop species, four basic criteria should be adopted (Mahon, 1986; Duncan and Baligar, 1990). These criteria are as follows: (1) genetic variability exhibiting a range of expression is needed to assess the trait; (2) improvement strategies rely on detailed information about the genetic systems (genotypic performance stability (broad and heritability) over a range of spatial and temporal environments is necessary for useful selection); (3) improvement in nutritional and physiological traits must be related to some feature of agronomic importance, such as yield stability, improved quality of harvestable product, or reduced production costs; and (4) practical exploitation of the traits will be difficult unless phenotypic expression can be expressed in large-scale trials, in large germplasm collections, or in segregating populations.

Chisholm and Blair (1988) compared the P efficiency of two pasture legume species, Ladino (*Trifolium repens*) and Verano (*Stylosanthes hamata*). They reported that the amount of dry matter accumulated per unit of P indicated that white clover (*T. repens*) was more P efficient than Verano in the early stage of growth. Caradus (1983) reported significant differences among white clover populations for P uptake per unit of root length in solution culture following growth at high P but not following growth at low P. Populations originating from low P soils had lower rates of P uptake per unit of root length than those from high P soils. Caradus and Snaydon (1986) reported that white clover populations collected from low P soils had a higher percentage survival and higher yields than populations collected from high P soils. Populations from low P soils had finer root systems than populations from high P soils when grown in pure sands.

Fageria et al. (2013c) studied P use efficiency in shoot and grain of upland rice, dry bean, corn, and soybean (*Glycine max*). They calculated by using the following equation:

$$\text{P-Use efficiency}\left(\text{kg kg}^{-1}\right)=\frac{\text{Dry weight of grain or shoot in kg}}{\text{P-uptake in grain or shoot in kg}}$$

Phosphorus use efficiency of these four important food crops is presented (Figures 1.11 and 1.12). Phosphorus use efficiency in shoot as well as grain varied in four crop species. Phosphorus use efficiency in shoot as well as grain followed the pattern of corn > upland rice > soybean > dry bean. Higher P use efficiency in cereals as compared to legumes may be related to higher yield of corn and upland rice as compared to dry bean (*Phaseolus vulgaris*) and soybean (*G. max*). Phosphorus use efficiency in crop plants expressed in terms of dry matter production per unit of P uptake is maximum as compared to N and K (Fageria et al. 2006). This is in contrast with P recovery efficiency

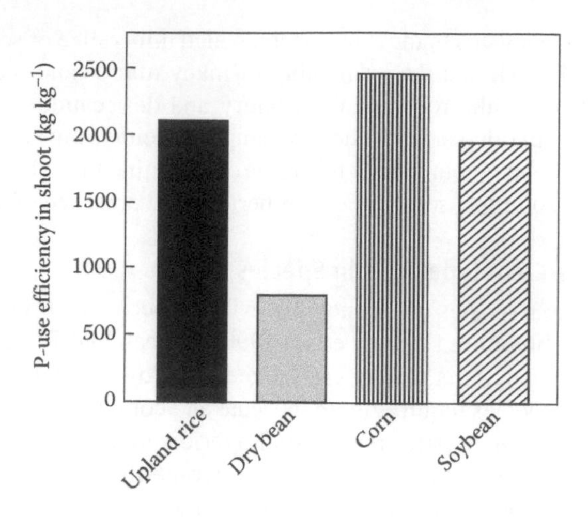

FIGURE 1.11 Phosphorus use efficiency in the shoot of upland rice, dry bean, corn, and soybean. Values are means of 2-year field experimentation for each crop. (From Fageria, N.K. et al., *J. Plant Nutr.*, 36, 2013, 2013c.)

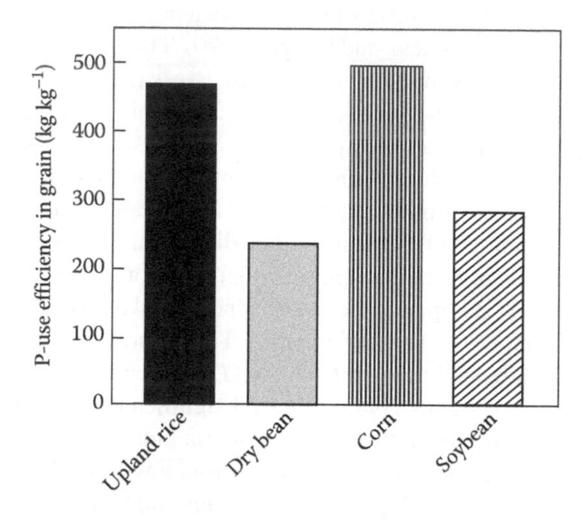

FIGURE 1.12 Phosphorus use efficiency in the grain of upland rice, dry bean, corn, and soybean. Values are means of 2-year field experimentation for each crop. (From Fageria, N.K. et al., *J. Plant Nutr.*, 36, 2013, 2013c.)

of applied fertilizer in the soils. The P recovery efficiency of applied fertilizers in the soil by crop plants is less than 20%, N recovery efficiency is 50% or less, and K recovery efficiency is near 40% (Baligar et al., 2001).

Fageria et al. (2013b) also evaluated upland rice genotypes for P use efficiency. P level × genotype interaction for grain yield was significant, indicating genotype responses differently varied according to P level (Table 1.15). At low P rates, grain yield varied from 7.58 to 17.59 g plant^{-1}, with a mean yield of 10.91 g plant^{-1}. The lowest grain yield producing genotype was BRA032051, and maximum grain-producing genotype was BRA052015. Difference in grain yield between lowest and highest yield–producing genotypes was 132% at the low P level. At higher P levels, grain yield varied from 10.49 g plant^{-1} produced by genotype BRA032039 to 21.35 g plant^{-1} produced by genotype BRS Primavera. Grain yield difference between lowest and highest grain yield–producing genotypes was 104% at high P levels. Overall, grain yield increase at the high P level was 52% as compared to the low P level, indicating high importance of P fertilization for upland rice in the Brazilian Oxisols.

TABLE 1.15
Grain Yield of 20 Upland Rice Genotypes as Influenced by P Levels (mg kg^{-1}) and Grain Yield Efficiency Index

	Grain Yield (g plant^{-1})		
Genotype	**Low P (25)**	**High P (200)**	**Grain Yield Efficiency Index**
BRA01506	12.93abcde	17.63abcde	1.27abc
BRA01596	12.16bcdef	15.53cdefg	1.05bcde
BRA01600	8.31ef	15.39cdefg	0.71cde
BRA02535	9.10def	15.35cdefg	0.76cde
BRA02601	11.96bcdef	12.20fg	0.81cde
BRA032033	11.88bcdef	16.24abcdef	1.07bcde
BRA032039	9.42cdef	10.49g	0.55e
BRA032048	8.50ef	15.20cdefg	0.71cde
BRA032051	7.58f	14.99cdefg	0.64de
BRA042094	9.58cdef	16.81abcdef	0.91cde
BRA042156	10.37bcdef	16.75abcdef	0.94cde
BRA042160	8.86def	13.47efg	0.67de
BRA052015	17.59a	16.70abcdef	1.63ab
BRA052023	14.95ab	19.49abcd	1.62ab
BRA052033	8.81def	14.81defg	0.72cde
BRA052034	9.55cdef	19.40abcd	1.03cde
BRA052045	10.14bcdef	21.12ab	1.21abcd
BRA052053	14.67abc	20.31abc	1.68a
BRS Primavera	14.04abcd	21.35a	1.67a
BRS Sertaneja	7.74ef	15.90bcdef	0.69cde
Mean	10.91	16.46	1.02
F-test			
P level (P)	**		
Genotype (G)	**		**
P × G	**		

Source: Fageria, N.K. et al., *J. Plant Nutr*., 36, 1868, 2013b.
**Significant at the 1% probability level. Means in the same column followed by the same letter are not significantly different at the 5% probability level by the Tukey's test.

Response of upland rice to P fertilization in Brazilian Oxisol has been reported by Fageria et al. (1982), Fageria and Baligar (1997a), and Fageria and Baligar (2001). The response of annual crops to P fertilization is due to the low level of this element in the Oxisols and high P immobilization capacity (Fageria, 1989).

Fageria et al. (2013b) classified upland rice genotypes into efficient and inefficient groups in P use based on grain yield efficiency index (GYEI). The GYEI was calculated by using the following equation:

$$\text{GYEI} = \frac{\text{GY at low P levels}}{\text{AGY of 20genotypes at low P levels}} \times \frac{\text{GY at high P levels}}{\text{AGY of 20genotypes at high P levels}}$$

where

GY is the grain yield
AGY is the mean grain yield

The GYEI is useful in separating high-yield, stable, P-efficient genotypes from low-yield, unstable P-inefficient genotypes (Fageria et al., 1988a). Genotypes having GYEI higher than 1 were considered P efficient, inefficient genotypes were in the range of 0–0.5 P efficiency index, and genotypes in between these two limits were considered intermediate in P use efficiency. These ratings, although selected arbitrarily, are supported when genotype grain yield means were calculated by Tukey's test (Table 1.15). In addition, GYEI had a significant quadratic response with grain yield ($Y = 5.28 + 10.39X - 1.82X^2$, $R^2 = 0.93^{**}$), which further emphasizes usefulness of this index in classifying rice genotypes for P use efficiency. There was 93% variation in grain yield of genotypes due to GYEI, indicating a close relationship between these two variables. Based on GYEI, efficient genotypes in P use efficiency were BRA052053, BRS primavera, BRA052015, BRA052023, BRA01506, BRA052045, BRA032033, BRA01596, and BRA052034. Moderately efficient genotypes were BRA032039, BRA 032051, BRA042160, BRS Sertaneja, BRA032048, BRA01600, BRA052033, BRA02535, BRA02601, BRA042094, and BRA042156. None of the genotypes fall into inefficient group. Variation in P use efficiency among upland rice genotypes has been reported by Fageria et al. (1988a,b) and Fageria et al. (2011), Fageria and Baligar (1997b), and Fageria (2009). Difference in P use efficiency of rice genotypes may be associated with their different ability in P uptake and utilization (Fageria et al., 2006; Fageria, 2007a).

One of the mechanisms that are responsible for the difference in P use efficiency is the root system (Fageria, 2009, 2013; Fageria et al., 2011). Crop cultivars with improved root systems, that are able to unlock and absorb P from soil-bound P resources may be of additional value for increasing the efficiency of P fertilizers (Abelson, 1999; Gahoonia and Nielsen, 2004). Although the varietal differences in P uptake and their link to the size of root systems were previously reported (Smith, 1934; Lyness, 1936), breeding for efficient root systems has received minimal attention (Clark, 1990; Fageria, 2013). This is attributed to wide variation that has indeed been reported in the ability of crop genotypes to perform and produce economic yields under P-limited soil conditions (Clark, 1990). Gahoonia and Nielsen (2004) reviewed the literature and synthesized existing information about intraspecific genetic variation in root traits and also discussed the possibility and limitations regarding accumulated information that can be used as a tool for selecting and breeding P-efficient crop cultivars.

Fageria et al. (2013b) evaluated the root systems of this upland rice and reported significant differences in root dry weight and maximum root length at low and high P levels (Table 1.16). Root dry weight of 20 upland rice genotypes at low P levels varied from 2.00 to 5.68 g plant^{-1}, with a mean value of 3.41 g plant^{-1}. At high P levels, root dry weight varied from 2.43 to 8.55 g plant^{-1}, with a mean value of 4.01 g plant^{-1}. Increase in dry weight at the high P level was about 18% as compared to the low P level. Root length varied from 23.00 to 38.33 cm, with a mean value of 30.9 cm at the low P level. At the high P level, root length varied from 23.67 to 34.33 cm, with a mean value of 28.20 cm. There was a 10% decrease in root length at the high P level as compared to the low P level. Fageria and Baligar (1997b) and Baligar et al. (1998) have reported an increase in root dry weight of upland rice with the addition of P in Brazilian Oxisols. Similarly, these authors also reported a decrease in root length at higher P levels as compared to the lower P levels. At the higher P level, roots had more fine hairs as compared to the lower P level. Therefore, roots at the higher P level had the capacity for uptake of nutrients and water, as compared to the lower P level.

There are several references in the literature that report that larger root systems provide greater root–soil contact, which is particularly important for the uptake of P (Gahoonia and Nielsen, 2004; Fageria, 2009, 2013). Mobile nutrients like N can be depleted at low rooting density, while for less mobile ions like P, uptake is often closely related to root length (Atkinson, 1991). Leon and Schwang (1992) reported that yield stability of oats (*Avena fatua*) and barley cultivars was related to their total root length. Barraclough (1984) reported that total root length of winter wheat was positively correlated to grain yield. Large differences in root morphology and distribution exist between genotypes of plant species (O'Toole and Bland, 1987; Romer et al., 1988; Atkinson, 1991;

TABLE 1.16
Root Dry Weight and Root Length of 20 Upland Rice Genotypes as Influenced by P Levels (mg kg^{-1})

	Root Dry Weight (g plant^{-1})		Root Length (cm)	
	Low P	High P	Low P	High P
Genotype	(25)	(200)	(25)	(200)
BRA01506	3.92ab	3.22c	26.00ab	26.67a
BRA01596	2.78ab	2.73c	35.67ab	28.00a
BRA01600	2.81ab	3.03c	36.00ab	29.00a
BRA02535	3.12ab	4.30c	28.67ab	33.33a
BRA02601	4.42ab	3.20c	31.33ab	27.00a
BRA032033	3.70ab	3.62c	23.00b	29.67a
BRA032039	2.91ab	4.36c	27.67ab	27.00a
BRA032048	3.96ab	3.91c	37.00a	33.33a
BRA032051	2.00b	2.58c	36.00ab	30.67a
BRA042094	2.82ab	3.92c	30.00ab	27.33a
BRA042156	2.50b	2.91c	29.00ab	27.00a
BRA042160	5.68a	8.32ab	32.67ab	33.00a
BRA052015	3.91ab	2.98c	27.00ab	27.00a
BRA052023	4.69ab	8.55a	29.67ab	34.33a
BRA052033	2.23b	2.43c	27.00ab	29.00a
BRA052034	3.18ab	3.99c	31.00ab	24.67a
BRA052045	3.07ab	3.08c	38.33a	24.67a
BRA052053	2.57ab	3.87c	28.33ab	23.67a
BRS Primavera	3.56ab	5.21bc	29.67ab	25.00a
BRS Sertaneja	4.36ab	3.92c	34.00ab	23.67a
Mean	3.41	4.01	30.9	28.20
F-test				
P level (P)	NS		*	
Genotype (G)	**		**	
P × G	*		**	

Source: Fageria, N.K. et al., *J. Plant Nutr.*, 36, 1868, 2013b.

*,**, and NS Significant at the 5% and 1% probability level and nonsignificant, respectively. Means within the same column followed by the same letter are not significantly different at the 5% probability level by the Tukey's test.

Fageria, 2013). In addition to variation in root system, Yadav et al. (1997), by studying a double haploid population of 105 lines derived from a cross between *indica* and *japonica* rice, reported that the main quantitative trait loci were common for root thickness and maximum root length. They suggested that there was a possibility of modifying several aspects of root morphology simultaneously. The reported heritability of root length is 0.14–0.51 in wheat (Gahoonia and Nielsen, 2004), 0.83 in oats (Barbour and Murphy, 1984), and 0.35 in rice (Ekanayake et al., 1985), which are sufficiently high to consider worthy for manipulation of root length (Gahoonia and Nielsen, 2004).

1.11.3.3 Colonization of Rhizosphere by Plant Growth–Promoting Rhizobacteria

Colonization of rhizosphere by microorganisms results in modifications in plant growth and development. These microorganisms are designated as "plant growth–promoting rhizobacteria" (PGPR)

(Fageria and Stone, 2006). The PGPR have been divided into two classes according to whether they can affect plant growth either directly or indirectly (Bashan and Holguin, 1998). Direct influence is related to increased solubilization and uptake of nutrients and production of phytohormones, whereas indirect effect is associated with pathogen suppression, production of iron-chelating siderophores and antibiotics, and induction of plant resistance mechanisms (Persello-Cartieaux et al., 2003). Biosynthesized plant growth regulators include auxins, gibberellins, cytokinins, ethylene, and abscisic acid (Arshad and Frankenberger, 1998). Detailed discussion regarding microorganisms involved and quantity of growth-promoting hormones produced and functions are reported by Arshad and Frankenberger (1998); Persello-Cartieaux et al. (2003). Properties of PGPR offer a great promise for agronomic applications, but interactions with other bacteria and environmental factors are still a problem for sustainable application of the PGPR (Persello-Cartieaux et al., 2003; Fageria and Stone, 2006).

Tinker (1980) summarizes the mechanisms by which microorganisms could potentially alter P uptake rates, which seem to be related to the following: (1) alteration of root morphology, in particular root hair length and density, or change in active root length; (2) change of mean absorbing power of the root over all or part of its surface; (3) displacement of sorption equilibria to produce higher local P concentrations in the soil solution, thereby allowing a higher flux toward the root surface and a higher uptake rate; and (4) facilitated transport of P to the root, again allowing a larger uptake rate, and possibly from a larger soil volume.

1.11.3.4 Mycorrhiza Fungi Symbiosis by Plant Roots

Mycorrhizal fungi are one of the most important groups of soil microorganisms and vary widely in structure and function. Mycorrhizae associated with crop plants are primarily arbuscular mycorrhizal fungi (AMF). The AMF association represents an ancient symbiosis with 80% of all terrestrial plants forming this type of association (Harrier and Watson, 2003). In this mutual symbiosis, the fungi receive a carbon (C) source from the host, and in exchange, it supplies minerals to the host. Development of AMF association with host plant is a complex process and is characterized by distinct development stages. These stages are spore germination, hyphal differentiation, aspersorium formation, root penetration, intercellular growth, intracellular arbuscular formation, and nutrient exchange (Harrier and Watson, 2003). The AMF fungi have not been cultured in the absence of the host plant, and this has hampered their mass production and utilization in cropping systems (Jarstfer and Sylvia, 1992).

The AMF form beneficial symbioses with roots to facilitate plants to grow and develop considerably better than would be expected under relatively harsh mineral stress conditions (Fageria et al., 2002). The AMF improve host plant nutrition by improving the acquisition of P and other minerals, especially the low mobile micronutrients Zn, Fe, and Cu (Marschner, 1995). Low availability of P in bulk soil limits plant uptake. For this reason, AMF fungi are key for P acquisition since fungal hyphae greatly increase the volume of bulk soil that the plant roots can explore. Quantitatively, P is the most important nutrient taken up by the extraradical hyphae, and influx of P in roots colonized by AMF fungi can be three to five times higher than in a nonmycorrhizal root (Harrier and Watson, 2003). The AMF accomplish this primarily by extension of root geometry. That is, AMF hyphae are smaller (mean diameter = 3–4 μm) than roots and/or root hairs (diameter = >10 μm) and can make contact with soil particles and/or explore pores/cavities that roots would not otherwise contact (Clark and Zeto, 2000).

Hyphae also extend away from roots and explore greater volume of soil than roots themselves. Nutrients are then transported via the hyphal network to the plant root whereby they are passed to the plant in exchange for C. The AMF may also protect plants from excessive uptake of some toxic minerals (Clark and Zeto, 2000; Brady and Weil, 2002). Root colonization with AMF can decrease the risk of plants to Mn, Fe, B, and Al toxicity in acid soils (Clark and Zeto, 2000; Fageria et al., 2002). Toxicity factors may be reduced by inhibiting the acquisition of toxic minerals and/or from root/hyphae exudations to decrease reactions in the rhizosphere like Mn reduction (Marschner, 1995).

Mycorrhizae are also involved in the biological control of root pathogens and nutrient cycling (solubilization, mineralization) (Marschner, 1995).

1.11.3.5 Phosphatase Activity in the Rhizosphere

Increased phosphatase enzyme activity in the rhizosphere has been reported to increase the hydrolysis of soil phosphate esters and increase P availability to plants (Bieleski and Johnson, 1972; Tarafder and Junk, 1987). Phosphatase activity in the rhizosphere can improve the solubility of organic P and consequently soil P availability to plants (Asmar et al., 1995).

1.12 CONCLUSIONS

Phosphorus is an essential nutrient for plants and animals, and its use in adequate amount is important for increasing food, feed, fiber, and fuel to support an expanding world population. Most of the P fertilizers are produced from PR, a nonrenewable natural resource. World PR reserves are estimated to be around 65 billion mt. More reserves may be discovered in the future with the innovation of new technology. Hence, there will not be P shortage at least in the coming 500 years. However, a judicious use in agriculture is fundamental, not only to increase crop yields but also to reduce costs of crop production and environmental pollution. PR is a trade name that covers a wide variety of rock types that have widely different textures and mineral compositions.

Phosphorus cycle in soil–plant system is very dynamic and complex due to the involvement of climate, soil, plant, and microorganisms and their interactions. Major processes include addition, solubilization, mineralization by microorganisms, immobilization at clay and oxides surfaces, and uptake by plants. In acid soils, P is mainly immobilized by fixation on Fe and Al oxides, and in calcareous soils it is mainly immobilized by the formation of insoluble Ca phosphates. Soluble P fertilizers are most available to plants immediately after soil application. P availability from soluble inorganic P fertilizer becomes less available with increasing time after application. Therefore, P fertilizers should be applied to crops at the time of sowing or transplanting. When optimum amounts of fertilizer P are used for intensive cropping, most soils tend to accumulate residual P. Residual effects of PRs can exceed those of soluble P fertilizers and are greatly influenced by the PR dissolution rate and the rate of loss of P from the plant-available P pool in the soil.

Phosphate availability to plants is influenced by climatic, soil, and plant factors and their interactions. Important climatic factors that influence P uptake are temperature and availability of water. Both of these factors influence solubility and transport of P to plant roots and hence its availability. Soil factors that influence P availability are presence of Al and Fe oxides and hydroxides in acid soils and higher amounts of Ca in alkaline soils. Presence of these elements in the soil–plant system immobilizes P (precipitation and adsorption) and reduces its availability to plants. Plant factor that influences P availability is the presence of microorganisms in the rhizosphere including beneficial microorganisms and fungi, which changes the root structure in favor of higher uptake of P. In addition, crop species and genotypes within species also differ in P uptake and utilization. Planting P-efficient crop species or genotypes within species is an important strategy from the economical and environmental point of view.

REFERENCES

Abbasi, M. K., S. Mansha, N. Rahim, and A. Ali. 2013. Agronomic effectiveness and phosphorus utilization efficiency of rock phosphate applied to winter wheat. *Agron. J.* 105:1606–1612.

Abelson, P. H. 1999. A potential phosphate crisis. *Science* 283:2015.

Alcordo, I. S. and J. E. Rechcigl. 1993. Phosphogypsum in agriculture: A review. *Adv. Agron.* 49:55–118.

Allison, S. D. 2005. Cheaters, diffusion and nutrients constrain decomposition by microbial enzymes in spatially structural environment. *Ecol. Lett.* 8:626–635.

Anderson, G. 1980. Assessing organic phosphorus in soils. In: *The Role of Phosphorus in Agriculture*, ed. R. C. Dinauer, pp. 411–431. Madison, WI: ASA, SCSA, and SSSA.

Arshad, M. and W. T. Frankenberger Jr. 1998. Plant growth-regulating substances in the rhizosphere: Microbial production and functions. *Adv. Agron.* 62:45–151.

Asmar, F., T. S. Gahoonia, and N. E. Nielsen. 1995. Barley genotypes differ in extracellular phosphatase activity and depletion of organic phosphorus from rhizosphere. *Plant Soil* 172:117–122.

Atkinson, D. 1991. Influence of root system morphology and development on the need for fertilizers and the efficiency of use. In: *Plant Roots: The Hidden Half*, eds. Y. Waisel, A. Eshel, and U. Kalkaki, pp. 411–451. New York: Marcel Dekker.

Baligar, V. C. and N. K. Fageria. 2007. Agronomy and physiology of tropical cover crops. *J. Plant Nutr.* 30:1287–1339.

Baligar, V. C., N. K. Fageria, and M. Elrashidi. 1998. Toxicity and nutrient constraints on root growth. *Hortic. Sci.* 33:960–965.

Baligar, V. C., N. K. Fageria, and Z. L. He. 2001. Nutrient use efficiency in plants. *Commun. Soil Sci. Plant Anal.* 32:921–950.

Baligar, V. C., N. K. Fageria, A. Q. Paiva, A. Silveira, A. W. V. Pomella, and R. C. R. Machado. 2006. Light intensity effects on growth and micronutrient uptake by tropical legume cover crops. *J. Plant Nutr.* 29:1959–1974.

Barber, S. A. 1980. Soil–plant interactions in the phosphorus nutrition of plants. In: *The Role of Phosphorus in Agriculture*, ed. R. C. Dinauer, pp. 591–615. Madison, WI: ASA, CSSA and SSSA.

Barbour, W. and C. F. Murphy. 1984. Field evaluation of seedling root length selection in oats. *Crop Sci.* 24:165–169.

Barraclough, P. B. 1984. The growth and activity of winter wheat roots in the field: Root growth of high yielding crops in relation to shoot growth. *J. Agric. Sci.* 103:439–442.

Barrow, N. J. 1980. Evaluation and utilization of residual phosphorus in soils. In: *The Role of Phosphorus in Agriculture*, ed. R. C. Dinauer, pp. 333–359. Madison, WI: ASA, CSSA and SSSA.

Bashan, Y. and G. Holguin. 1998. Proposal for the division of plant growth-promoting rhizobacteria into two classifications:biocontrol-PGPB (plant growth promoting bacteria) and PGPB. *Soil Biol. Biochem.* 30:1225–1228.

Batjes, N. H. 1997. A world data set of derived soil properties by FAO-UNISCO soil unit for global modelling. *Soil Use Manage.* 13:9–16.

Beaton, J. D. and N. A. Cough. 1962. The influence of soil moisture regime and phosphorus source on the response of alfalfa to phosphorus. *Soil Sci. Soc. Am. Proc.* 26:265–270.

Beaton, J. D. and D. W. L. Read. 1963. Effects of temperature and moisture on phosphorus uptake from a calcareous Saskatchewan soil treated with several pelleted sources of phosphorus. *Soil Sci. Soc. Am. Proc.* 27:61–65.

Beckwith, R. S. 1965. Sorbed phosphate at standard supernatant concentration as an estimate of the phosphate needs of soils. *Aust. J. Exp. Agric. Anim. Husb.* 5:52–58.

Bell, T. H., J. N. Klironomos, and H. A. L. Henry. 2010. Seasonal responses of extracellular enzyme activity and microbial biomass and nitrogen addition. *Soil Sci. Soc. Am. J.* 74:820–828.

Bertin, C., X. Yang, and L. A. Weston. 2003. The role of root exudates and allelochemicals in the rhizosphere. *Plant Soil* 256:67–83.

Bieleski, R. L. and P. N. Johnson. 1972. The external location of phosphatase activity in phosphorus deficient *Spirodela oligorrhiza*. *Aust. J. Biol. Sci.* 25:707–720.

Blackwell, P. S., T. W. Green, and W. K. Mason. 1990. Responses of biopore channels from roots to compression by vertical stresses. *Soil Sci. Soc. Am. J.* 54:1988–1091.

Bland, W. L. 1993. Cotton and soybean root system growth in three soil temperature regimes. *Agron. J.* 85:906–911.

Bolan, N. S., R. Naidu, J. K. Syers, and R. W. Tillman. 1999. Surface charge and solute interactions in soils. *Adv. Agron.* 67:88–141.

Bowen, G. D. 1970. Effects of soil temperature on root growth and on phosphate uptake along *Pinus radiata* roots. *Aust. J. Soil Res.* 8:31–42.

Brady, N. C. and R. R. Weil. 2002. *The Nature and Properties of Soils*, 13th edn. Upper Saddle River, NJ: Prentice Hall.

Brandon, D. M. and D. S. Mikkelsen. 1979. Phosphorus transformations in alternately flood California soils I. Cause of plant phosphorus deficiency in rice rotation crops and correctional methods. *Soil Sci. Soc. Am. J.* 43:989–994.

Bray, R. H. and L. T. Kurtz. 1945. Determination of total, organic, and available forms of phosphorus in soils. *Soil Sci.* 59:39–45.

Cade-Menum, B. C. and C. W. Liu. 2014. Solution phosphorus-31 nuclear magnetic resonance spectroscopy of soils from 2005 to 2013. A review of sample preparation and experimental parameters. *Soil Sci. Soc.* [illegible]

Cakir, R. 2004. Effect of water stress at different development stages on vegetative and reproductive growth of corn. *Field Crops Res.* 89:106.

Caradus, J. R. 1983. Genetic differences in phosphorus absorption among white clover populations. *Plant Soil* 72:379–383.

Caradus, J. R. and R. W. Snaydon. 1986. Response to phosphorus of populations of white clover I. Field studies. *N. Z. J. Agric. Res.* 29:155–162.

Case, V. W., N. C. Brady, and D. J. Lathwell. 1964. The influence of soil temperature and phosphorus fertilizers of different water solubilities on the yield and phosphorus content of oats. *Soil Sci. Soc. Am. Proc.* 28:409–412.

Cathcart, J. B. 1980. World phosphate reserves and resources. In: *The Role of Phosphorus in Agriculture*, ed. R. C. Dinauer, pp. 1–18. Madison, WI: ASA, CSSA and SSSA.

Chaves, M. M., J. S. Pereira, J. Maroco, M. L. Rodrigues, C. P. P. Richard, and M. L. Osorio. 2002. How plants cope with water stress in the field: Photosynthesis and growth. *Ann. Bot.* 89:907–916.

Chen, J. L. and J. F. Reynolds. 1997. A coordination model of whole plant carbon allocation in relation to water stress. *Ann. Bot.* 80:45–55.

Chisholm, R. H. and G. J. Blair. 1988. Phosphorus efficiency in pasture species I. Measures based on total dry weight and P content. *Aust. J. Agric. Res.* 39:807–816.

Clark, R. B. 1990. Physiology of cereals for mineral nutrient uptake, use and efficiency. In: *Crops as Enhancers of Nutrient Use*, eds. R. R. Duncan and V. C. Baligar, pp. 131–183. New York: Academic Press.

Clark, R. B. and S. K. Zeto. 2000. Mineral acquisition by arbuscular mycorrhizal plants. *J. Plant Nutr.* 23:867–902.

Cole, L., R. D. Bardgett, P. Ineson, and P. J. Hobbs. 2002. Enchytraeid worm (Oligochaeta) influences on microbial community structure, nutrient dynamics and plant growth in blamket peat subjected to warming. *Soil Biol. Biochem.* 34:83–92.

Collis-George, N. and B. G. Davey. 1960. The doubtful utility of present day field experimentation and other determinations involving soil–plant interactions. *Soil Fertiliz.* 23:307–310.

Condron, L. M., B. L. Turner, and B. J. Cade-Menun. 2005. Chemistry and dynamics of soil organic phosphorus. In: *Phosphorus: Agriculture and the Environment*, eds. J. T. Sims and A. N. Sharpley, pp. 87–121. Madison, WI: ASA, CSSA, and SSSA.

Cooper, A. J. 1973. *Root Temperature and Plant Growth*. Slough, U.K.: Commonwealth Agricultural Bureau.

Cordel, D., J. O. Drangert, and S. White. 2009. The story of phosphorus: Global food security and food for thought. *Glob. Environ. Change* 19:292–305.

Cosgrove, D. J. 1967. Metabolism of organic phosphates in soil. In: *Soil Biochemistry*, eds. A. D. McLaren and G. H. Peterson, pp. 216–226. New York: Marcel Dekker.

Dakora, F. D. and D. A. Phillips. 2002. Root exudates as mediators of mineral acquisitional in low-nutrient environments. *Plant Soil* 245:35–47.

Daroub, S. H. and G. H. Snyder. 2007. The chemistry of plant nutrients in soil. In: *Mineral Nutrition and Plant Disease*, eds. L. E. Datnoff, W. H. Elmer, and D. M. Huber, pp. 1–7. St. Paul, MI: The American Phytopathological Society.

Devine, J. R., D. Gunary, and S. Larsen. 1968. Availability of phosphate as affected by duration of fertilizer contact with soil. *J. Agric. Sci.* 71:359–364.

Devine, T. E., J. H. Bouton, and T. Mabrahtu. 1990. Legume genetics and breeding for stress tolerance and nutrient efficiency. In: *Crops as Enhancers of Nutrient Use*, eds. R. R. Duncan and V. C. Baligar, pp. 211–252. New York: Academic Press.

Dormaar, J. F. and J. W. Ketcheson. 1960. The effect of nitrogen form and soil temperature on the growth and phosphorus uptake of corn grown in the greenhouse. *Can. J. Soil Sci.* 40:177–184.

Duncan, R. R. and V. C. Baligar. 1990. Genetics, breeding, and physiological mechanisms of nutrient uptake and use efficiency: An overview. In: *Crops as Enhancers of Nutrient Use*, eds. R. R. Duncan and V. C. Baligar, pp. 3–35. New York: Academic Press.

Dyer, B. 1894. Analytical determination of probably available mineral plant food in soils. *Trans. Chem. Soc.* 65:115–167.

Edwards, G. E., S. B. Ku, and J. G. Foster. 1983. Physiological constraints to maximum yield potential. In: *Challenging Problems in Plant Health*, eds. T. Kommedahl and P. H. Williams, pp. 105–119. St. Paul, MI: American Phytopatholgy Society.

Ekanayake, I. J., J. C. O'Toole, J. P. Garrity, and T. M. Masajo. 1985. Inheritance of root characters and their relations to drought resistance in rice. *Crop Sci.* 25:927–933.
Epstein, E. 1966. Effect of soil temperature at different growth stages on growth and development of potato plants. *Agron. J.* 58:169–171.
Fageria, N. K. 1989. *Tropical Soils and Physiological Aspects of Crops*. Brasilia, Brazil: EMBRAPA.
Fageria, N. K. 1992. *Maximizing Crop Yields*. New York: Marcel Dekker.
Fageria, N. K. 2001. Nutrient management for improving upland rice productivity and sustainability. *Commun. Soil Sci. Plant Anal.* 32:2603–2629.
Fageria, N. K. 2007a. Yield physiology of rice. *Plant Nutr.* 30:843–879.
Fageria, N. K. 2007b. Soil fertility and plant nutrition research under field conditions: Basic principles and methodology. *J. Plant Nutr.* 30:203–223.
Fageria, N. K. 2009. *The Use of Nutrients in Crop Plants*. Boca Raton, FL: CRC Press.
Fageria, N. K. 2013. *The Role of Plant Roots in Crop Production*. Boca Raton, FL: CRC Press.
Fageria, N. K. 2014. *Mineral Nutrition of Rice*. Boca Raton, FL: CRC Press.
Fageria, N. K. and V. C. Baligar. 1996. Response of lowland rice and common bean grown in rotation to soil fertility levels on a varzea soil. *Fertiliz Res.* 45:13–20.
Fageria, N. K. and V. C. Baligar. 1997a. Response of common bean, upland rice, corn, wheat, and soybean to soil fertility of an Oxisol. *J. Plant Nutr.* 20:1279–1289.
Fageria, N. K. and V. C. Baligar. 1997b. Upland rice genotypes evaluation for phosphorus use efficiency. *J. Plant Nutr.* 20:499–509.
Fageria, N. K. and V. C. Baligar. 2001. Improving nutrient use efficiency of annual crops in Brazilian acid soils for sustainable crop production. *Commun. Soil Sci. Plant Anal.* 32:1303–1319.
Fageria, N. K. and V. C. Baligar. 2003. Fertility management of tropical acid soils for sustainable crop production. In: *Handbook of Soil Acidity*, ed. Z. Rengel, pp. 359–385. New York: Marcel Dekker.
Fageria, N. K. and V. C. Baligar. 2004. Properties of termite mound soils and response of rice and bean to nitrogen, phosphorus, and potassium fertilization on such soil. *Commun. Soil Sci. Plant Anal.* 35:2097–2109.
Fageria, N. K. and V. C. Baligar. 2008. Ameliorating soil acidity of tropical Oxisols by liming for sustainable crop production. *Adv. Agron.* 99:345–399.
Fageria, N. K., V. C. Baligar, and B. A. Bailey. 2005. Role of cover crops in improving soil and row crop productivity. *Commun Soil Sci. Plant Ana.* 36:2733–2757.
Fageria, N. K., V. C. Baligar, and R. B. Clark. 2002. Micronutrients in crop production. *Adv. Agron.* 77:185–268.
Fageria, N. K., V. C. Baligar, and R. B. Clark. 2006. *Physiology of Crop Production*. New York: The Haworth Press.
Fageria, N. K., V. C. Baligar, and C. A. Jones. 2011a. *Growth and Mineral Nutrition of Field Crops*, 3rd edn. Boca Raton, FL: CRC Press.
Fageria, N. K., V. C. Baligar, and Y. C. Li. 2008. The role of nutrient efficient plants in improving crop yields in the twenty first century. *J. Plant Nutr.* 32:1121–1157.
Fageria, N. K., V. C. Baligar, A. Moreira, and L. A. C. Moraes. 2013a. Soil phosphorus influence on growth and nutrition of tropical legume cover crops in acid soil. *Commun. Soil Sci. Plant Anal.* 44:3340–3364.
Fageria, N. K. and M. P. Barbosa Filho. 1987. Phosphorus fixation in Oxisolof Central Brazil. *Fertiliz. Agric.* 94:33–37.
Fageria, N. K. and M. P. Barbosa Filho. 2008. Influence of pH on productivity, nutrient use efficiency by dry bean, and soil phosphorus availability in a no-tillage system. *Commun. Soil Sci. Plant Anal.* 39:1016–1025.
Fageria, N. K., M. P. Barbosa Filho, and J. R. P. Carvalho. 1982. Response of upland rice to phosphorus fertilization on an Oxisol. *Agron. J.* 74:51–56.
Fageria, N. K., G. D. Carvalho, A. B. Santos, E. P. B. Ferreira, and A. M. Knupp. 2011b. Chemistry of lowland rice soils and nutrient availability. *Commun. Soil Sci. Plant Anal.* 42:1913–1933.
Fageria, N. K. and H. R. Gheyi. 1999. *Efficient Crop Production*. Paraiba, Brazil: Federal University of Paraiba.
Fageria, N. K., L. C. Melo, J. P. Oliveira, and A. M. Coelho. 2012a. Yield and yield components of dry bean genotypes as influenced by phosphorus fertilization. *Commun. Soil Sci. Plant Anal.* 43:1–15.
Fageria, N. K., O. P. Moraes, and M. J. Vasconcelos. 2013b. Upland rice genotypes evaluation for phosphorus use efficiency. *J. Plant Nutr.* 36:1868–1880.
Fageria, N. K., O. P. Morais, V. C. Baligar, and R. J. Wright. 1988a. Response of rice cultivars to phosphorus supply on an Oxisol. *Fertiliz. Res.* 16:195–206.
Fageria, N. K., A. Moreira, and A. M. Coelho. 2012b. Nutrient uptake in dry bean genotypes. *Commun. Soil Sci. Plant Anal.* 43:2289–2302.
Fageria, N. K., A. Moreira, and A. B. Santos. 2013c. Phosphorus uptake and use efficiency in field crops. *J. Plant Nutr.* 36:2013–2022.

Fageria, N. K., A. B. Santos, and V. C. Baligar. 1997. Phosphorus soil test calibration for lowland rice on an Inceptisol. *Agron. J.* 89:737–742.

Fageria, N. K., A. B. Santos, and F. J. P. Zimmermann. 2000. Response of irrigated rice to residual fertilization and to applied levels of fertilizers in lowland soil. *Rev. Bras. Eng. Agric. Amb.* 4:177–182.

Fageria, N. K., N. A. Slaton, and V. C. Baligar. 2003. Nutrient management for improving lowland rice productivity and sustainability. *Adv. Agron.* 80:63–152.

Fageria, N. K. and L. F. Stone. 2006. Physical, chemical and biological changes in the rhizosphere and nutrient abvailability. *J. Plant Nutr.* 29:1327–1356.

Fageria, N. K., R. J. Wright, and V. C. Baligar. 1988b. Rice cultivar evaluation for phosphorus use efficiency. *Plant Soil* 111:105–109.

Fageria, N. K., R. J. Wright, V. C. Baligar, and C. M. R. Sousa. 1991. Characterization of physical and chemical properties of varzea soils of Goias state of Brazil. *Commun. Soil Sci. Plant Anal.* 22:1631–1646.

Foth, H. D. and B. G. Ellis. 1988. *Soil Fertility*. New York: John Wiley & Sons.

Fox, R. L. and E. J. Kamprath. 1970. Phosphate sorption isotherms for evaluating the phosphate requirements of soils. *Soil Sci. Soc. Am. Proc.* 34:902–907.

Foy, C. D. 1984. Physiological effects of hydrogen, aluminum, and manganese toxicities in acid soil. In: *Soil Acidity and Liming*, 2nd edn., ed. F. Adams, pp. 57–97. Madison, WI: ASA, CSSA and SSSA.

Froelich, P. N., M. L. Bender, N. A. Luedtke, G. R. Heath, and T. D. Vries. 1982. The marine phosphorus cycle. *Am. J. Sci.* 282:474–511.

Frossard, E., D. L. Achar, S. M. Bernasconi, E. K. Bunemann, J. C. Fardeau, and J. Jansa 2011. The use of tracers ti investigate phosphate cycling in soil-plant systems. In: *Phosphorus in Action*, ed. E.K. Bunemann, pp. 59–91. Berlin, Germany: Springer Verlag.

Frossard, E., E. K. Bunemann, A. Obserson, J. Jansa, and M. A. Kertesz. 2012. Phosphorus and sulfur in soil. In: *Handbook of Soil Sciences: Properties and Processes, Part IV. Soil Biology and Biochemistry*, P. M. Huang, Y. Li, and M. E. Summer (Ed.), 26, pp. 2–15. Boca Raton, FL: CRC Press.

Gahoonia, T. S. and N. E. Nielsen. 2004. Root traits as tools for creating phosphorus efficient crop varieties. *Plant Soil* 26:47–57.

Gardner, B. R. and J. P. Jones. 1973. Effects of temperature on phosphate sorption isotherms and phosphate desorption. *Commun. Soil Sci. Plant Anal.* 4:83–93.

Garwood, E. A. 1968. Some effects of soil water conditions and temperature on the roots of grasses and clovers II. Effects of variation in the soil water content and soil temperature on root growth. *J. Brit. Grassland Soc.* 23:117–127.

Glass, A. D. M. 1989. *Plant Nutrition: An Introduction to Current Concepts*. Boston, MA: Jones and Bartlett Publishers.

Glinski, J. and J. Lipiec. 1990. *Soil Conditions and Plant Roots*. Boca Raton, FL: CRC Press.

Gregory, P. 2006. *Plant Roots: Growth, Activity and Interactions with Soils*. Oxford, U.K.: Blackwell Publishing.

Hammond, L. L., S. H. Chien, and A. U. Mokwunye. 1986. Agronomic value of unacidulated and partially acidulated PRs indigenous to the tropics. *Adv. Agron.* 40:89–140.

Hanway, J. J. and R. A. Olson. 1980. Phosphate nutrition of corn, sorghum, soybeans, and small grains. In: *The Role of Phosphorus in Agriculture*, ed. R. C. Dinauer, pp. 681–692. Madison, WI: ASA, CSSA and SSSA.

Harrier, L. A. and C. A. Watson. 2003. The role of arbuscular mycorrhiza fungi in sustainable cropping systems. *Adv. Agron.* 20:185–225.

Hartt, C. E. 1965. The effects of temperature upon translocation of C^{14} in sugarcane. *Plant Physiol.* 40:74–81.

Hatfield, J. L., K. J. Boote, B. A. Kimball, L. H. Zisks, R. C. Izaurralde, D. Ort, A. M. Thompson, and D. Wolfe. 2011. Climate impacts on agriculture: Implications for crop production. *Agron. J.* 103:351–370.

Haynes, R. J. 1982. Effects of liming on phosphate availability in acid soils: A critical review. *Plant Soil* 68:289–308.

Haynes, R. J. 1984. Lime and phosphate in the soil-plant system. *Adv. Agron.* 37:249–315.

Heffer, P. and M. Prudhomme. 2013. Nutrients as limited resources: Global trends in fertilizer production and use. In: *Improving Water and Nutrient Use Efficiency in Food Production Systems*, 1st edn., ed. Z. Rengel, pp. 79–91. Ames, IA: John Willey & Sons.

Heinrichs, D. H. and K. F. Nielsen. 1966. Growth response of alfalfa varieties of diverse genetic origin to different root zone temperatures. *Can. J. Plant Sci.* 46:291–298.

Hinsinger, P. 2001. Bioavailability of soil inorganic P in the rhizosphere as affected by root induced chemical changes: A review. *Plant Soil* 237:173–195.

Holford, I. C. R. and W. H. Patric Jr. 1979. Effect of reduction and pH changes on phosphate sorption and mobility in an acid soil. *Soil Sci. Soc. Am. J.* 43:292–297.

Horn, R., H. Domzal, A. Slowinska-Jukiewicz, and C. Van Ouwerkerk. 1995. Soil compaction processes and their effects on the structure of arable soils and the environment. *Soil Tillage Res.* 35:23–36.

Hossner, L. R., J. A. Freeouf, and B. L. Folsom. 1973. Solution phosphorus concentration and growth of rice in flooded soils. *Soil Sci. Soc. Am. Proc.* 27:405–408.

Hsiao, T. C., P. Steduto, and E. Fereres. 2007. A systematic and quantative approach to improve water use efficiency in agriculture. *Irrig. Sci.* 25:209–231.

Iyamuremye, F. and R. P. Dick. 1996. Organic amendments and phosphorus sorption by soils. *Adv. Agron.* 56:139–185.

Jarstfer, A. G. and D. M. Sylvia. 1992. Inoculum production and inoculation technologies of vesicular arbuscular mycorrhizal fungi. In: *Soil Technologies: Applications in Agriculture, Forestry and Environmental Management*, ed. B. Metting, pp. 349–377. New York: Marcel Dekker.

Jayachandran, K., A. P. Schwab, and B. A. D. Hetrick. 1989. Micorrhizal mediation of phosphorus availability: Synthetic iron chelates effects on phosphorus solubilization. *Soil Sci. Soc. Am. J.* 53:1701–1706.

Jenny, H. and K. Grossenbacher. 1963. Root-soil boundary zones as seen in the electron microscope. *Soil Sci. Soc. Am. Proc.* 27:273–277.

Jonasson, S., A. Michel, I. K. Schmidt, and E. V. Nielsen. 1999. Responses in microbes and plants to changed temperature, nutrient, and light regimes in the arctic. *Ecology* 80:1828–1843.

Kamprath, E. J. and C. D. Foy. 1971. Lie–fertilizer–plant interactions in acid soils. In: *Fertilizer Technology and Use*, ed. R. C. Dinauer, pp. 105–151. Madison, WI: SSSA.

Ketellapper, H. J. 1960. The effect of soil temperature on the growth of *Phalaris tuberso* L. *Physiol. Plant* 13:641–647.

Khasawneh, F. E. 1971. Solution ion activity and plant growth. *Soil Sci. Soc. Am. Proc.* 35:426–436.

Khasawneh, F. E. and J. P. Copeland. 1973. Cotton root growth and uptake of nutrients: Relation of phosphorus uptake to quantity, intensity and buffering capacity. *Soil Sci. Soc. Am. Proc.* 37:250–254.

Khasawneh, F. E. and E. C. Doll. 1978. The use of phosphate rock for direct application to soils. *Adv. Agron.* 30:159–203.

Kuo, S. 1983. *Effects of Organic Residues and Nitrogen Transformations on Phosphorus Sorption Desorption by Soil.* Madison, WI: Agronomy abstracts, ASA.

Leon, J. and K. U. Schwang. 1992. Description and application of a screening method to determine root morphology traits of cereals cultivars. *Z. Acker. Pflanzenbau.* 169:128–134.

Lin, C., W. J. Busscher, and L. A. Douglas. 1983. Multifactor kinetics of phosphate reactions with minerals in acidic soils. I. Modeling and simulation. *Soil Sci. Soc. Am. J.* 47:1097–1103.

Lindsay, W. L. 1979. *Chemical Equilibrium in Soils.* New York: John Wiley & Sons.

Lyness, A. S. 1936. Varietal differences in phosphorus feeding capacity of plants. *Plant Physiol.* 11:665–683.

Mack, H. J., S. C. Fang, and S. B. Butts. 1964. Effects of soil temperature and phosphorus fertilization on snap beans and peas. *Proc. Am. Soc. Hortic. Sci.* 84:332–338.

Mahon, J. D. 1986. Limitations to the use of physiological variability in plant breeding. *Can. J. Plant Sci.* 63:11–21.

Marschner, H. 1983. General introduction to the mineral nutrition of plants. In: *Encyclopedia of Plant Physiology, New Ser.*, Vol. 15A, eds. A. Lauchli and R. L. Bieleski, pp. 5–60. New York: Springer-Verlag.

Marschner, H. 1995. *Mineral Nutrition of Higher Plants*, 2nd edn. New York: Academic Press.

Matocha, J. E., B. E. Conard, L. Reyes, and C. W. Thomas. 1970. Residual value of phosphorus fertilizer on a calcareous soil. *Agron. J.* 62:572–575.

Mattingly, G. E. G. 1975. Labile phosphate in soils. *Soil Sci.* 119:369–375.

McClellan, G. H. and L. R. Gremillion. 1980. Evaluation of phosphate raw materials. In: *The Role of Phosphorus in Agriculture*, ed. R. C. Dinauer, pp. 43–128. Madison, WI: ASA, CSSA and SSSA.

McLaughlin, M. J., A. M. Alston, and J. K. Martin. 1988a. Phosphorus cycling in wheat-pasture rotations I. The source of phosphorus taken up by wheat. *Aust. J. Soil Res.* 26:323–331.

McLaughlin, M. J., A. M. Alston, and J. K. Martin. 1988b. Phosphorus cycling in wheat-pasture rotations II. The role of the microbial biomass in phosphorus cycling. *Aust. J. Soil Res.* 26:333–342.

McMichael, B. L. and J. J. Burk. 1994. Metabolic activity of cotton roots in response to temperature. *Environ. Exp. Bot.* 31:461–470.

McMichael, B. L. and J. J. Burk. 1998. Soil temperature and root growth. *Hortscience* 33:947–951.

McMichael, B. L. and J. E. Quisenberry. 1991. Genetic variation for root-shoot relationships among cotton germplasm. *Environ. Exp. Bot.* 31:461–470.

McMichael, B. L. and J. E. Quisenberry. 1993. The impact of the soil environment on the growth of root systems. *Environ. Exp. Bot.* 33:53–61.

Mengel, K., E. A. Kirkby, H. Kosegarten, and T. Appel. 2001. *Principles of Plant Nutrition*, 5th edn. Dordrecht the Netherlands: Kluwer Academic Publishers.

Menzies, N. W. 2003. Toxic elements in acid soils: Chemistry and management. In: *Handbook of Soil Acidity*, ed. Z. Rengel, pp. 267–296. New York: Marcel Dekker.

Morgan, M. F. 1941. Chemical soil diagnosis by the universal soil testing system. Connecticut Agricultural Experimental Station Bulletin 45. Rome, Italy: FAO.

Neumann, G. and V. Romheld. 2001. The release of root exudates as affected by the plants physiological status. In: *The Rhizosphere: Biochemistry and Organic Substances at the Soil-Plant Interface*, eds. R. Pinto, Z. Varanini, and P. Nannipieri, pp. 41–93. New York: Marcel Dekker.

Nielsen, K. F. 1974. Roots and root temperatures. In: *The Plant Root and Its Environment*, ed. E. W. Carson, pp. 293–333. Charlottesville, VA: University Press of Virginia.

Nielsen, K. F. and R. K. Cunningham. 1964. The effects of soil temperature, form and level of N on growth and chemical composition of Italian ryegrass. *Soil Sci. Soc. Am. Proc.* 28:213–218.

Nielsen, K. F. and E. C. Humphries. 1966. Effects of root temperature on plant growth. *Soil Fertiliz.* 29:1–7.

O'Toole, J. C. and W. L. Bland. 1987. Genotypic variation in crop plant root systems. *Adv. Agron.* 41:91–145.

Ochsner, T. E., M. H. Cosh, R. H. Cuenca, W. A. Dorigo, C. S. Draper, Y. Hagimoto, Y. H. Kerr, K. M. Larson, E. G. Njoku, E. E. Small, and M. Zreda. 2013. State of the art in large-scale soil moisture monitoring. *Soil Sci. Soc. Am. J.* 77:1888–1919.

Olsen, S. R. and A. D. Flowerday. 1971. Fertilizer phosphorus interactions in alkaline soils. In: *Fertilizer Technology and Use*, ed. R. C. Dinauer, pp. 153–185. Madison, WI: SSSA.

Olsen, S. R. and F. E. Khasawneh. 1980. Use and limitations of physical-chemical criteria for assessing the status of phosphorus in soils. In: *The Role of Phosphorus in Agriculture*, ed. R. C. Dinauer, pp. 361–410. Madison, WI: ASA, CSSA and SSSA.

Olsen, S. R. and F. S. Watanabe. 1966. Effective volume of soil around roots determined from phosphorus diffusion. *Soil Sci. Soc. Am. Proc.* 30:598–602.

Olsen, S. R. and F. S. Watanabe. 1970. Diffusive supply of phosphorus in relation to soil textural variations. *Soil Sci.* 110:318–327.

Olsen, S. R., F. S. Watanabe, and R. E. Danielson. 1961. Phosphorus absorption by corn roots as affected by moisture and phosphorus concentration. *Soil Sci. Soc. Am. Proc.* 25:289–294.

Owen, P. C. 1971. The effects of temperature on the growth and development of rice. *Field Crop Abstr.* 24:1–8.

Ozanne, P. G. 1980. Phosphate nutrition of plants: A general treatise. In: *The Role of Phosphorus in Agriculture*, ed. R. C. Dinauer, pp. 559–589. Madison, WI: ASA, CSSA and SSSA.

Ozanne, P. G. and T. C. Shaw. 1968. Advantages of the recently developed phosphate sorption test over the older extracting methods for soil phosphate. *Int. Cong. Soil Sci. Trans. 9th (Adelaide, Australia)* 2:273–280.

Parsons, R. 1959. *Handbook of Electrochemical Constants*. New York: Academic Press.

Paytan, A. and K. McLaughlin. 2011. Tracing the source and biogeochemical cycling of phosphorus in aquatic systems using isotopes of oxygen in phosphorus. In: *Handbook of Environmental Isotope Geochemistry*, ed. M. Baskarn, pp. 419–436. Berlin, Germany: Springer Verlag.

Pearson, R. W., L. F. Ratliff, and H. M. Taylor. 1970. Effect of soil temperature, strength, and pH on cotton seedling root elongation. *Agron. J.* 62:243–246.

Pepper, I. L. and D. F. Bezdicek. 1990. Root microbial interactions and rhizosphere nutrient dynamics. In: *Crops as Enhancer of Nutrient Use*, eds. V. C. Baligar and R. R. Duncan, pp. 375–410. San Diego, CA: Academic Press.

Persello-Cartieaux, E., L. Nussaume, and C. Robaglla. 2003. Tales from the underground: molecular plant-rhizobacteria interactions. *Plant Cell Environ.* 26:189–199.

Phillips, A. B. and J. R. Webb. 1971. Production, marketing, and use of phosphorus fertilizers. In: *Fertilizer Technology and Use*, ed. R. C. Dinauer, pp. 271–301. Madison, WI: SSSA.

Pierzynski, G. M., R. W. McDowell, and J. T. Sims. 2005. Chemistry, cycling and potential movement of inorganic phosphoruMs in soils. In: *Phosphorus: Agriculture and the Environment*, eds. J. T. Sims and A. N. Sharpley, pp. 53–86. Madison, WI: ASA, CSSA, and SSSA.

Ponnamperuma, F. N. 1972. The chemistry of submerged soils. *Adv. Agron.* 24:29–96.

Power, J. F., D. L. Grunes, G. A. Reichman, and W. O. Willis. 1970. Effect of soil temperature on rate of barley development and nutrition. *Agron. J.* 62:567–571.

Radke, J. F. and R. E. Bauer. 1969. Growth of sugarbeets as affected by root temperatures Part I. Greenhouse studies. *Agron. J.* 61:860–863.

Rajan, S. S. S., J. H. Watkinson, and A. G. Sinclair. 1996. Phosphate rocks for direct application in soils. *Adv. Agron.* 57:77–159.

Richard, G., I. Cousin, J. F. Sillon, A. Bruand, and J. Guerif. 2001. Effect of compaction on the porosity of a silty soil: Influence on unsaturated hydraulic properties. *Eur. J. Soil Sci.* 52:49–58.

Rinnan, R., A. Michelsen, E. Baath, and S. Jonasson. 2008. Fifteen years of climate change manipulations alter soil microbial communities in a subarctic heath ecosystems. *Glob. Change Biol.* 13:28–39.

Rinnan, R., A. Michelsen, and S. Jonasson. 2007. Effects of litter addition and warming on soil carbon, nutrient pools and microbial communities in a subarctic heath ecosystem. *Appl. Soil Ecol.* 39:271–281.

Rodriguez, M., J. Neris, M. Tejedor, and C. Jimenez. 2010. Soil temperature regimes from different latitudes on a subtropical island (Tenerife, Spain). *Soil Sci. Soc. Am. J.* 74:1662–1669.

Romer, W. J., J. Augustin, and G. Schilling. 1988. The relationship between phosphate absorption and root-length in nine wheat cultivars. *Plant Soil* 111:199–201.

Rovira, A. D., R. C. Foster, and J. K. Martin. 1979. Note on terminology: Origin, nature and nomenclature of the organic materials in the rhizosphere. In: *The Soil–Root Interface*, eds. J. L. Harley and R. Scott Russell, pp. 1–4. London, U.K.: Academic Press.

Ruttenberg, K. 2003. The global phosphate cycle. In: *Treatise on Geochemistry*, ed. W.H. Schlesinger, pp. 585–643. Oxford, U.K.: Pergamon Press.

Sah, R. N. and D. S. Mikkelsen. 1986. Transformations of inorganic phosphorus during flooding and draining periods of soil. *Soil Sci. Soc. Am. J.* 50:62–67.

Sah, R. N. and D. S. Mikkelsen. 1989. Phosphorus behavior in flooded-drained soils I. Effects on phosphorus sorption. *Soil Sci. Soc. Am. J.* 53:1718–1722.

Sah, R. N., D. S. Mikkelsen, and A. A. Hafez. 1989a. Phosphorus behavior in flooded-drained soils. I. Iron transformation and phosphorus sorption. *Soil Sci. Soc. Am. J.* 53:1723–1729.

Sah, R. N., D. S. Mikkelsen, and A. A. Hafez. 1989b. Phosphorus behavior in flooded-drained soils III. Phosphorus desorption and availability. *Soil Sci. Soc. Am. J.* 53:1729–1732.

Sample, E. C., R. J. Soper, and G. J. Racz. 1980. Reactions of phosphate fertilizers in soils. In: *The Role of Phosphorus in Agriculture*, ed. R. C. Dinauer, pp. 263–310. Madison, WI: ASA, SCSA, and SSSA.

Sanchez, C. A. 2007. Phosphorus. In: *Handbook of Plant Nutrition*, ed. A. V. Barker, D. J. Pilbeam, pp. 51–90. Boca Raton, FL: CRC Press.

Sanchez, P. A. and J. G. Salinas. 1981. Low input technology for managing Oxisols and Ultisols in tropical America. *Adv. Agron.* 34:279–398.

Sanchez, P. A. and G. Uehara. 1980. Management considerations for acid soils with high phosphorus fixation capacity. In: *The Role of Phosphorus in Agriculture*, ed. R.C. Dinauer, pp. 471–514. Madison, WI: ASA, CSSA and SSSA.

Saseendran, S. A., L. R. Ahuja, L. Ma, D. C. Nielsen, T. J. Trout, A. A. Andales, J. L. Chavez, and J. Ham. 2014. Enhancing the water stress factors for simulation of corn in RZWQM2. *Agron. J.* 106:81–94.

Schaffer, B., R. Schulin, and P. Boivin. 2013. Shrinkage properties of repacked soil at different states of uniaxial compression. *Soil Sci. Soc. Am. J.* 77:1930–1943.

Sinclair, T. R. and W. I. Park. 1993. Inadequacy of the Liebig limiting-factor paradigm for explaining varying crop yields. *Agron. J.* 85:742–746.

Singh, B. B. and J. P. Jones. 1976. Phosphorus sorption and desorption characteristics of soil as affected by organic residues. *Soil Sci. Soc. Am. J.* 40:389–394.

Smil, V. 2000. Phosphorus in the environment: Natural flows and human interferences. *Annu. Rev. Energy Environ.* 25:53–88.

Smith, F. 1934. Response of inbred lines and crosses in maize to variation of nitrogen and phosphorus supplies as nutrients. *J. Am. Soc. Agron.* 26:785–804.

Smyth, T. T. and P. A. Sanchez. 1982. Phosphate rock dissolution and availability in cerrado soils as affected by phosphorus sorption capacity. *Soil Sci. Soc. Am. J.* 46:339–345.

Soane, B. D. and C. Van Ouwerkerk. 1995. Implications of soil compaction in crop production for the quality of the environment. *Soil Tillage Res.* 35:5–22.

Soil Science Society of America. 2008. *Glossary of Soil Science Terms.* Madison, WI: Soil Science Society of America.

Soltanpour, P. N., F. Adams, and A. C. Bennett. 1974. Soil phosphorus availability as measured by displaced soil solutions, calcium-chloride extracts, dilute-acid extracts and labile phosphorus. *Soil Sci. Soc. Am. Proc.* 38:225–228.

Spratt, E. D., F. G. Warder, L. D. Bailey, and D. W. L. Read. 1980. Measurement of fertilizer phosphorus residues and its utilization. *Soil Sci. Soc. Am. J.* 44:1200–1204.

Stevenson, F. J. 1986. *Cycles of Soil Carbon, Nitrogen, Phosphorus, Sulfur and Micronutrients*. New York: John Wiley & Sons.

[illegible] 1969. Effect of low temperature on phosphate nutrition of plants: A review. *J. Sci. Food Agric.* 20:1–3.

Sutton, M. A., A. Bleeker, C. M. Howard, M. Bekunda, B. Grizzetti, and W. Varies. 2013. *Our Nutrient World: The Challenge to Produce More Food and Energy with Less Pollution*. Edinburgh, U.K.: Center for Ecology and Hydrology.

Tamburini, F., V. Pfahler, C. V. Sperber, E. Frossard, and S. M. Bernasconi. 2014. Oxygen isotope for unraveling phosphorus transformations in the soil-plant systems: A review. *Soil Sci. Soc. Am. J.* 78:38–46.

Tarafder, J. C. and A. Junk. 1987. Phosphatase activity in the rhizosphere and its relation to the depletion of organic phosphorus. *Biol. Fertiliz. Soils* 3:199–204.

Tate, K. R. 1984. The biological transformation of phosphorus in soil. *Plant Soil* 76:245–256.

Thompson, L. M., C. A. Black, and J. A. Zoellner. 1954. Occurrence and mineralization of organic phosphorus in soils, with particular reference to associations of nitrogen, carbon, and pH. *Soil Sci.* 77:185–196.

Tinker, P. B. 1980. Role of rhizosphere microorganisms in phosphorus uptake by plants. In: *The Role of Phosphorus in Agriculture*, ed. R. C. Dinauer, pp. 617–654. Madison, WI: ASA, CSSA and SSSA.

Tisdale, S. L., W. L. Nelson, and J. D. Beaton. 1985. *Soil Fertility and Fertilizer*, 4th edn. New York: Macmillan Publishing Company.

Truog, E. 1930. Determination of the readily available phosphorus of soils. *Agron. J.* 22:874–882.

Uren, N. C. 2001. Types, amounts, and possible functions of compounds released into the rhizosphere by soil-grown plants. In: *The Rhizosphere: Biochemistry and Organic Substances at the Soil–Plant Interface*, eds. R. Pinto, Z. Varanini, and P. Nannipieri, pp. 19–40. New York: Marcel Dekker.

U.S. Geological Survey. 2011. Mineral commodity summaries 2011. Accessed December 2013, at http://minerals.usgs.gov/minerals/pubs/mcs/2011/mcs2011.pdf.

Van Kauwenbergh, S. J. 2010. *World Phosphate Rock Reserves and Resources*. Muscle Shoals, AL: IFDC. www.ifde.org.

Van Kauwenbergh, S. J., M. Stewart, and R. Mikkelsen. 2013. World reserves of phosphate rock : A dynamic and unfolding story. *Better Crops Plant Food* 97:18–20.

Van Riemsdijk, W. H. and D. Haan. 1981. Reaction of orthphosphate with a sandy soil at constant supersaturation. *Soil Sci Soc. Am. J.* 45:261–266.

Vassilev, N., A. Medina, R. Azcon, and M. Vassileva. 2006. Microbial solubilization of rock phosphate on media containing agro-industrial wastes and effect of the resulting products on plant growth and P uptake. *Plant Soil* 287:77–84.

Voorhees, W. B., R. R. Allmaras, and C. E. Johnson. 1981. Alleviating temperature stress. In: *Modifying the Root Environment to Reduce Crop Stress*, eds. G. F. Arkin and H. Taylor, pp. 217–266. St. Joseph, MI: American Society of Agricultural Engineering.

Way, J. T. 1850. On the power of soils to absorb manure. *J. Roy. Agric. Soc. Engl.* 11:313–379.

Westerman, R. L. and T. C. Tucker. 1987. Soil fertility concepts: Past, present and future. In: *Future Developments in Soil Science Research*, ed. L. L. Boersma, pp. 171–179. Madison, WI: Soil Science Society of America.

Westheimer, F. H. 1987. Why nature chose phosphates. *Science* 235:1173–1178.

Whitfield, C. J. and D. E. Smika. 1971. Soil temperature and residue effects on growth components and nutrient uptake of four wheat varieties. *Agron. J.* 63:297–300.

Willet, I. R. 1982. Phosphorus availability in soils subject to short period of flooding and drying. *Aust. J. Soil Res.* 20:131–138.

Willet, I. R. and M. L. Higgins. 1980. Phosphorus sorption and axtractable iron in soils during irrigated rice-upland crop rotation. *Aust. J. Exp. Agric. Anim. Husb.* 20:346–353.

Williams, I. H. 1993. Rothamsted: Cradle of agricultural and apicultural research: 1. *Orig. Dev. Bee World* 74:20–26.

Wood, C. W. 1998. Agricultural phosphorus and water quality: An overview. In: *Soil Testing for Phosphorus: Environmental Uses and Implications*, ed. J. Thomas, pp. 5–12. Newark, DE: University of Delaware.

Yadav, R. L., R. V. Singh, R. Singh, and V. K. Srivastava. 1997. Effect of planting geometry and fertilizer N rates on nitrate leaching, nitrogen use efficiency, and sugarcane yield. *Trop. Agric.* 74:115–120.

Zapata, E. and N. R. Roy. 2004. Use of phosphate rocks for sustainable agriculture. Fertilizer Plant Nutrients Bulletin 13. Rome, Italy: FAO.

Zhang, W., K. M. Parker, Y. Luo, S. Wan, L. L. Wallace, and S. Hu. 2005. Soil microbial response to experimental warming and clipping in a tallgrass prairie. *Glob. Change Biol.* 11:266–277.

2 Functions of Phosphorus in Crop Plants

2.1 INTRODUCTION

The 17 elements considered essential for plant growth are carbon (C), hydrogen (H), oxygen (O), nitrogen (N), phosphorus (P), potassium (K), calcium (Ca), magnesium (Mg), sulfur (S), iron (Fe), zinc (Zn), copper (Cu), boron (B), manganese (Mn), molybdenum (Mo), chlorine (Cl), and nickel (Ni) (Fageria, 2014). Sodium (Na), silicon (Si), selenium (Se), vanadium (V), and cobalt (Co) are beneficial for some plants but have not been established as essential elements for all higher plants (Mengel et al., 2001; Fageria et al., 2011). Among essential plant nutrients, C, H, O, and N are major constituents of the principal classes of compounds that make up about 95% of the fresh weight of most living plants (Epstein and Bloom, 2005). Essential plant nutrients can also be classified as metals or nonmetals. Metals include K, Ca, Mg, Fe, Zn, Mn, Cu, Ni, and Mo. Nonmetals include C, H, O, N, P, S, B, and Cl (Bennett, 1993).

According to Mengel et al. (2001), classification of plant nutrients based on their biochemical behavior and their physiological functions appears to be more appropriate. Based on such a physiological approach, plant nutrients may be divided into the following four groups:

1. *Group 1*: C, H, O, N, and S. These nutrients are major constituents of organic material, involved in enzymic processes and oxidation–reduction reactions
2. *Group 2*: P and B. These elements are involved in energy transfer reactions and esterification with native alcohol groups in plants
3. *Group 3*: K, Ca, Mg, Mn, and Cl. This group plays osmotic and ion balance roles, plus more specific functions in enzyme conformation and catalysis
4. *Group 4*: Fe, Cu, Zn, and Mo. These elements present as structural chelates or metalloproteins and enable electron transport by valence change.

Even though some of these elements have been known since ancient times, their essentiality has been established only within the last century (Marschner, 1995; Fageria et al., 2011; Fageria, 2014). The discovery of essential nutrients, their chemical symbols, and their principal forms for uptake are presented in Table 2.1. Now the question arises: "What are the criteria of essentiality of nutrients for plant growth?" Long ago, Arnon and Stout (1939) proposed certain criteria of essentiality of mineral nutrients, and these criteria are still valid. According to these researchers, the essentiality of a nutrient is based on the following criteria: (1) omission of the element results in abnormal growth, failure to complete the life cycle (i.e., from seed germination through the production of viable seeds), or premature death of the plant, (2) the element forms part of a molecule or constituent of the plant that is itself essential in the plant (examples are N in protein and Mg in chlorophyll), and (3) the element must be directly involved in plant metabolisms and not by some indirect effects such as antagonism of another element present at a toxic level.

Among these essential nutrients, the first three elements (C, H, and O) are supplied to plants by air and water and need not to be applied to soils. The remaining 14 nutrients need to be applied to soils by organic and/or inorganic fertilizers if deficient for normal plant growth and development.

TABLE 2.1
Essential Nutrients for Plant Growth, Their Principal Forms for Uptake, and Discovery

Nutrient	Chemical Symbol	Principal Forms of Uptake	Year of Discovery	Discovered Essentiality to Plants by
Carbon	C	CO_2	1882	J. Sachs
Hydrogen	H	H_2O	1882	J. Sachs
Oxygen	O	H_2O, O_2	1804	T. De Saussure
Nitrogen	N	NH_4^+, NO_3^-	1872	G. K. Rutherford
Phosphorus	P	$H_2PO_4^-$, HPO_4^{2-}	1903	Posternak
Potassium	K	K^+	1890	A. F. Z. Schimper
Calcium	Ca	Ca^{2+}	1856	F. Salm-Horstmar
Magnesium	Mg	Mg^{2+}	1906	Willstatter
Sulfur	S	SO_4^{2-}, SO_2	1911	Peterson
Iron	Fe	Fe^{2+}, Fe^{3+}	1860	J. Sachs
Manganese	Mn	Mn^{2+}	1922	J. S. McHargue
Boron	B	H_3BO_3	1923	K. Warington
Zinc	Zn	Zn^{2+}	1926	A. L. Sommer and C. B. Lipman
Copper	Cu	Cu^{2+}	1931	C. B. Lipman and G. MacKinney
Molybdenum	Mo	MoO_4^{2-}	1938	D. I. Arnon and P. R. Stout
Chlorine	Cl	Cl^-	1954	T. C. Broyer et al.
Nickel	Ni	Ni^{2+}	1980	Welch and Eskew

Source: Fageria, N.K. et al., *Growth and Mineral Nutrition of Field Crops*, 3rd edn., CRC Press, Boca Raton, FL, 2011.

Among these nutrients, N, P, K, Ca, Mg, and S are known as major or macronutrients and Fe, Zn, Cu, B, Mn, Mo, Cl, and Ni are known as minor or micronutrients (Fageria et al., 2002). All the essential nutrients are equally important for the growth and development of plants. Division between macro and micro is only based on the amount required by the plants. Macronutrients are required in higher amounts compared to micronutrients. Generally, uptake of macronutrients is in several kg ha^{-1} by most crop plants, whereas micronutrients are less than 1 kg ha^{-1}. There is an exception to this general rule: for example, Cl is considered a micronutrient but its uptake is equal to that of macronutrients (Fageria et al., 2002). Among the essential plant nutrients, P plays an important role in the growth and development of crop plants. In addition, P is the theme of this book. Hence, functions of P in plants are discussed in the succeeding section.

2.2 FUNCTIONS

P is essential for the growth of plants and animals. Plants need P for growth, utilization of sugar and starch, photosynthesis, nucleus formation, and cell division. The role of P in energy transfer in plants is well known. It is an essential component of the organic compound often called the energy currency of the living cell, that is, adenosine triphosphate (ATP) (Brady and Weil, 2002). Energy from photosynthesis and the metabolism of carbohydrates is stored in phosphate compounds for later use in growth and reproduction. In addition to influencing plant growth and yield, P affects crop maturity, plants' ability to deal with environmental stresses, such as drought and disease, and it interacts with other nutrients, especially N and K, to increase grain yields (Bundy et al., 2005). Grant et al. (2001) reported that P plays an important role in plant metabolisms, such as cellular energy transfer, and is a structural component of the nucleic acids of plant genes and chromosomes and of many coenzymes, phosphoproteins, and phospholipids. They note that an adequate supply of P is needed especially throughout the early stages of plant development to optimize growth (Bundy et al., 2005). Functions of P in the growth and development may be divided in morphological, physiological, and/or biochemical. It promotes early maturity of cereals and legumes and also improves seed quality. P has been described as ubiquitous in plants, being involved in nearly all metabolic processes (Bennett, 1993).

2.2.1 Morphological Functions

Adequate P nutrition enhances many morphological traits such as plant height, tillering, dry matter of shoot, leaf area, panicle density, and panicle or ear length in cereals. In legumes, P at an adequate level improves pod number, pod weight, or seed weight. It also improves root growth in both cereals and legumes (Fageria et al., 2006; Fageria and Moreira, 2011; Fageria, 2013). Since most morphological characteristics have positive association with grain yield (Figure 2.1), discussion on influence of P on these traits is pertinent and essential. Plant height, shoot dry weight, panicle number, and panicle length had significant quadratic association with grain yield of upland rice (Figure 2.1). Hence, detailed discussion of this topic is worthwhile to understand the role of P in improving these characteristics and consequently crop yields.

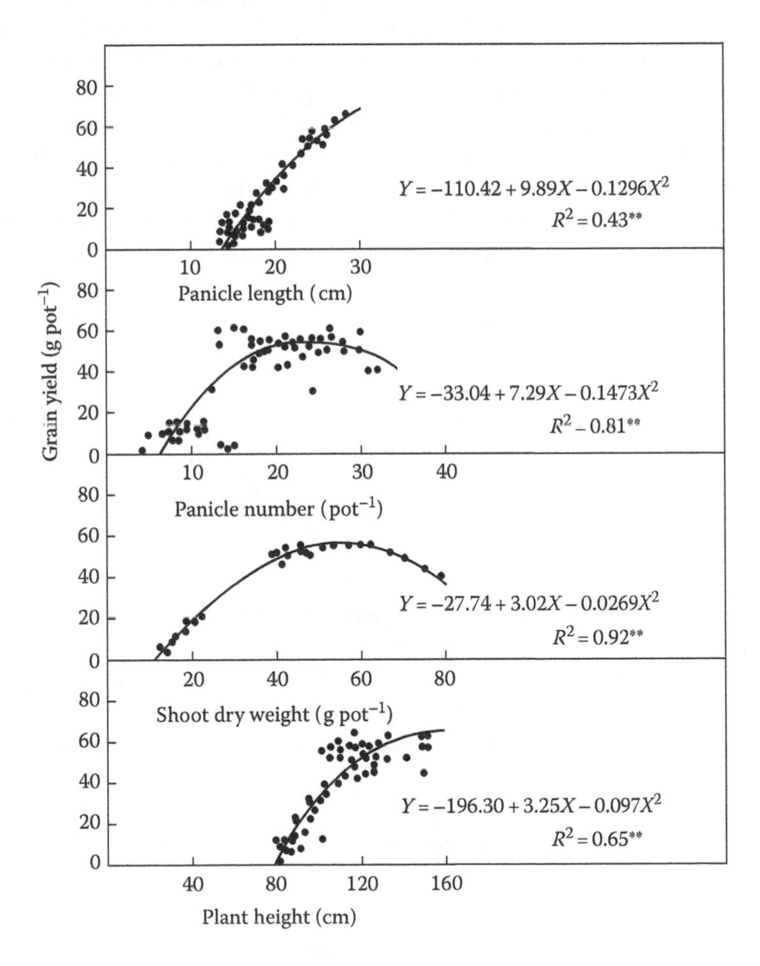

FIGURE 2.1 Relationship between morphological traits and yield of upland rice. (From Fageria, N.K. et al., *J. Plant Nutr.*, 33, 1696, 2010.)

2.2.1.1 Plant Height

Plant height is a major morphological trait in both cereals and legumes. Adequate level of P applied to a P-deficient soil significantly affects plant height of cereals as well as legumes. In cereals such as rice (*Oryza sativa*), wheat (*Triticum* spp.), barley (*Hordeum vulgare*), and oats (*Avena fatua*) for seedlings or juvenile plants, plant height is the distance from ground level to the tip of the tallest leaf. For mature plants, it is the distance from ground level to the tip of the tallest panicle or ear (Fageria, 2007). Plant height is an important trait because it is associated with plant lodging. Old cultivars of cereals (rice and wheat) were taller as compared to modern cultivars and were also more susceptible to lodging when soil fertility is high, especially N. This scenario changed during the latter half of the twentieth century with a dramatic increase in grain yield of cereals (rice, wheat, and corn (*Zea mays*)), and the term "Green Revolution" was used for this change (Fageria, 2007). In addition, plant height has a positive relationship with grain yield in cereals (Fageria, 2007). Lowland rice yield increased linearly with increasing plant height (Figure 2.2). Similarly, yield increase of upland rice was linear with increasing plant height (Figure 2.3). Fageria et al. (2013a) reported that plant height was having a significant quadratic association with grain yield of lowland rice. Fageria et al. (2010) also reported that plant height was having a significant quadratic influence on the grain yield of upland rice.

Fageria et al. (2013a) studied the influence of P on plant height of 20 upland rice genotypes (Table 2.2). The P level × genotype interaction was significant for this growth trait. Hence, values of this growth variable at the two P levels (25 and 200 mg kg^{-1}) are presented in Table 2.2. Plant height at the low P level varied from 94 to 127.7 cm, with a mean value of 112.7 cm. Minimum plant height at the low P level was produced by genotype BRA02601, and maximum plant height was produced by genotype BRA042156. At the high P level, plant height varied from 101 cm produced by genotype BRA02601 to 137 cm produced by genotype BRA052053, with a mean value of 116.3 cm. Overall, a 3% increase in plant height occurred at the high P level as compared to the low P level. Fageria and Baligar (1997a,b) have reported the improvement in plant height of upland rice genotypes with the addition of P in Brazilian Oxisol. Similarly, variability in plant

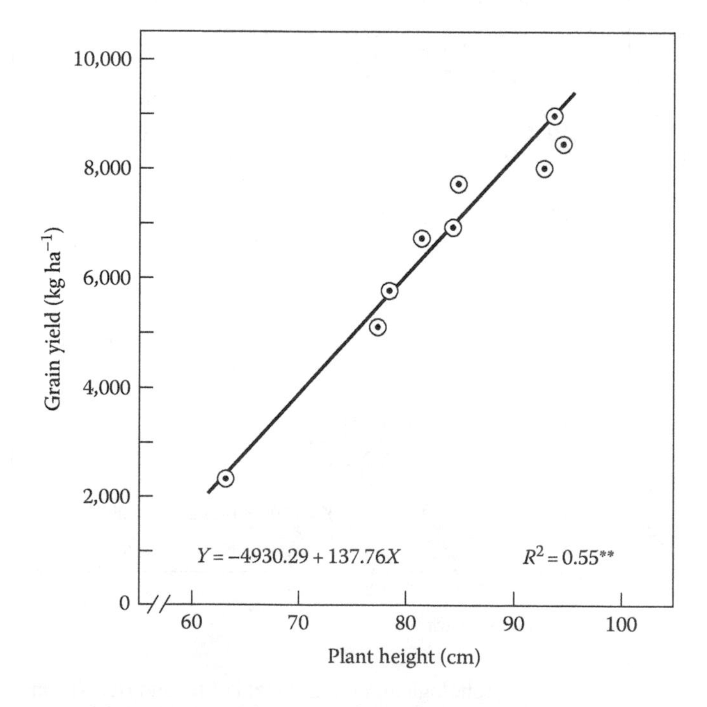

FIGURE 2.2 Relationship between plant height and grain yield of lowland rice.

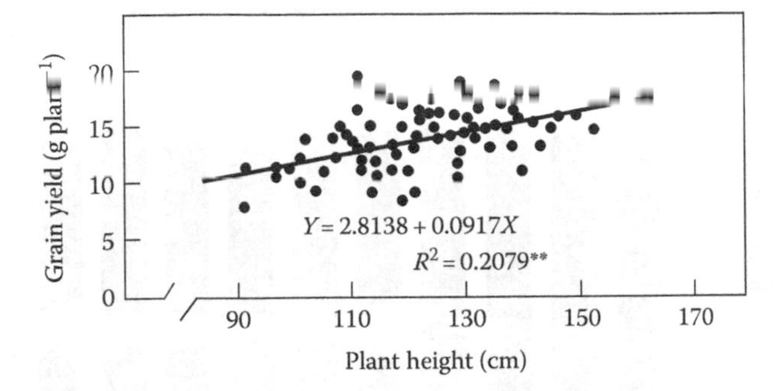

FIGURE 2.3 Relationship between plant height and grain yield. (From Fageria, N.K. et al., *Commun. Soil Sci. Plant Anal.*, 44, 2656, 2013b.)

TABLE 2.2
Influence of P Level (mg kg^{-1}) on Plant Height (cm) of 20 Upland Rice Genotypes

Genotype	Low P Level (25)	High P Level (200)
BRA01506	114.58abc	120.00abcde
BRA01596	109.00abcd	107.00ef
BRA01600	112.33abcd	122.33abcde
BRA02535	118.67abc	119.67abcdef
BRA02601	94.00d	101.00f
BRA032033	109.00abcd	103.67ef
BRA032039	104.50cd	117.67cdef
BRA032048	103.25cd	109.00def
BRA032051	119.33abc	136.67ab
BRA042094	121.58abc	118.00bcdef
BRA042156	127.67a	127.33abcd
BRA042160	112.33abcd	112.33cdef
BRA052015	115.67abc	119.67abcdef
BRA052023	106.33bcd	104.67ef
BRA052033	106.67bcd	109.00def
BRA052034	107.67bcd	107.67ef
BRA052045	125.00ab	106.00ef
BRA052053	121.33abc	137.00a
BRS Primavera	115.33abc	130.00bc
BRS Sertaneja	110.33abcd	117.00cdef
Average	112.73	116.28
F-test		
P level (P)	NS	
Genotype (G)	**	
P × G	**	

Source: Fageria, N.K. et al., *J. Plant Nutr.*, 36, 1868, 2013a.

** and NS Significant at the 5% and 1% probability level and nonsignificant, respectively. Means within the same column followed by the same letter are not significantly different at the 5% probability level by Tukey's test.

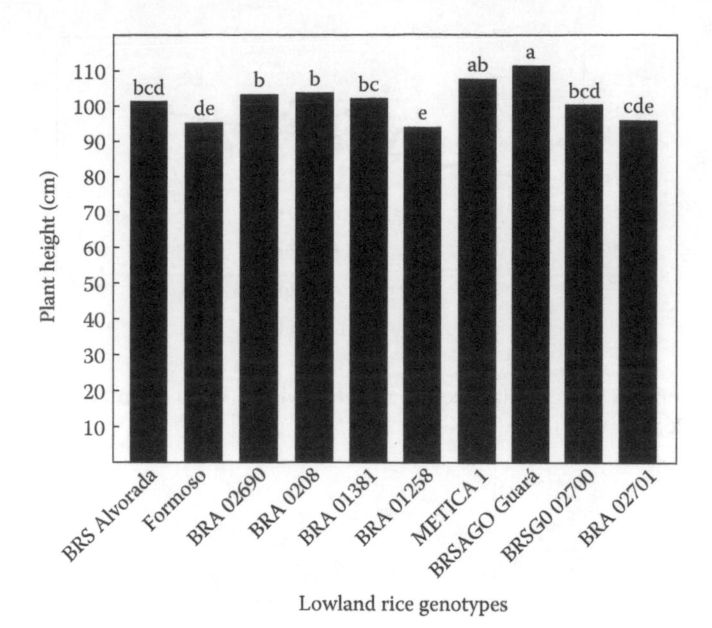

FIGURE 2.4 Plant height of 19 lowland rice genotypes. (From Fageria, N.K., *J. Plant Nutr.*, 30, 843, 2007.)

height among upland rice genotypes has been reported by Fageria et al. (2004). Although plant height is influenced by environmental factors, it is a genetically controlled plant trait (Figure 2.4) and the heritability of dwarfism is high and easy to identify, select, and recombine with other traits (Jennings et al., 1979). These authors also reported that dwarf segregates have a fairly narrow range in height, presumably from minor gene action. Although a few are so short that they are undesirable, the great majority fall within the useful range of from 80 to 100 cm with some reaching 120 cm under certain conditions (Jennings et al., 1979). During the 1960s, rice breeders made excellent progresses in the development of dwarf cultivars that responded to heavy applications of N (Jennings et al., 1979; Fageria, 2009).

2.2.1.2 Tillering

Tillering is an important morphological trait in cereals because it is related to the number of panicles or ear produced in the final stand. Mineral nutrition significantly influences tillering in cereals. Tillering followed a quadratic increase in cereals (rice, wheat, barley, and oats) with the advancement of plant age (Fageria, 2007) and significantly influenced by N and P fertilization (Fageria et al., 2003; Fageria, 2005). The period in which the increase of tiller number per unit length of time is great is defined as the active tillering stage (Fageria, 2007). The stage in which the number of tillers reaches maximum is known as maximum tiller number stage. Tillers that do not produce panicles degenerate, and their number decreases until they become equal to the number of panicles. Growth juncture of this period is called the ineffective tillering stage (Murayama, 1995).

P fertilization significantly increases tillering in rice (Fageria et al., 2003). Murata and Matsushima (1975) reported that P concentration is also correlated with tillering and a P concentration of >2.5 g kg^{-1} (0.25%) in the mother stem is necessary for tillering. Grain yield in cereals is highly dependent upon the number of spikelet-bearing tillers produced by each plant (Power and Alessi, 1978; Nerson, 1980). The number of productive tillers depends on environmental conditions during tiller bud initiation and subsequent developmental stages (Handa, 1995). Tiller appearance, abortion, or both are affected by environmental conditions, especially nutrient deficiencies (Black and Siddoway, 1977; Power and Alessi, 1978; Masle, 1985).

TABLE 2.3
Correlation Coefficients (r) between Lowland Rice Grain Yield and Tiller Number at Different Growth Stages

Parameter	First Year	Second Year	Third Year
Tiller number m^{-2} at IT	0.59**	0.41*	0.23^{NS}
Tiller number m^{-2} at AT	0.69**	0.43*	0.34*
Tiller number m^{-2} at IP	0.79**	0.59**	0.68**
Tiller number m^{-2} at B	0.67**	0.52**	0.46**
Tiller number m^{-2} at F	0.70**	0.37*	0.52**
Tiller number at PM	0.77**	0.48**	0.44*

Source: Fageria, N.K. and Baligar, V.C., *Commun. Soil Sci. Plant Anal.*, 32, 1405, 2001a.
IT, initiation of tillering; AT, active tillering; IP, initiation of panicle; B, booting; F, flowering; PM, physiological maturity.
*,**, and NS Significant at the 5% and 1% probability level and nonsignificant, respectively.

Tiller number decrease was attributed to the death of some of the last tillers as a result of their failure in competition for light and nutrients (Fageria et al., 2011). Another explanation is that during the period of growth beginning with panicle development, competition for assimilates exists between developing panicles and young tillers. Eventually, growth of many young tillers is suppressed, and they may senesce without producing seed (Dofing and Karlsson, 1993). Tillering was significantly correlated to grain yield at all the growth stages; however, highest correlation in all the 3 years of experimentation was obtained at the initiation of panicle growth stage (Table 2.3). This means that the number of tillers developed at this growth stage had more significance than that at any other growth stages in lowland rice.

Tiller number is quantitatively inherited. Its heritability is low to intermediate depending on the cultural practices used and the uniformity of the soil. Although often associated with early vigor in short statured materials, tiller number is inherited independently of all other major characters. In many crosses, tiller erectness or compactness is recessive to a spreading culm arrangement (Jennings et al., 1979). Developing good plant types with high tillering capacity is rather simple. Many sources of heavy tillering are available in traditional tropical rice cultivars. When their culms are shortened, their tillering ability generally does not decrease and may increase (Jennings et al., 1979).

2.2.1.3 Shoot Dry Weight

Shoot dry weight is an important morphological trait in determining the yield of annual crops. Generally, shoot dry weight has positive significant relationship with grain yield in crop plants (Fageria, 2013). Adequate P nutrition improves shoot dry weight of crop plants. Increase in shoot weight is mainly associated with an increase in leaf and culm weights during growth cycle of crop plants. Example of increasing shoot dry weight of upland rice during growth cycle is presented in Figure 2.5. Shoot dry weight increased significantly in the vegetative as well as reproductive growth stages. Upland rice shoot weight decreased from flowering to physiological maturity (Figure 2.5). Dry matter loss from the vegetative tissues during the interval from flowering to maturity was 35%, suggesting active transport of assimilates to the panicles, which resulted in a grain yield of 3811 kg ha^{-1}. Fageria et al. (1997) and Fageria et al. (2006) reported more or less similar reduction in shoot dry weight of upland rice from flowering to physiological maturity.

Fageria and Santos (2008b) also studied shoot dry weight of dry bean (Fabaceae) during growth cycles (Figure 2.6). Shoot dry weight of dry bean increased with increasing plant age in an exponential quadratic fashion, and maximum dry weight was produced 98 days after sowing. From 18- to 98-day growth periods, shoot dry weight was almost linear. This was the period in which canopy development was maximum. Shoot growth includes stem, branching, trifoliate, flowers, and pods

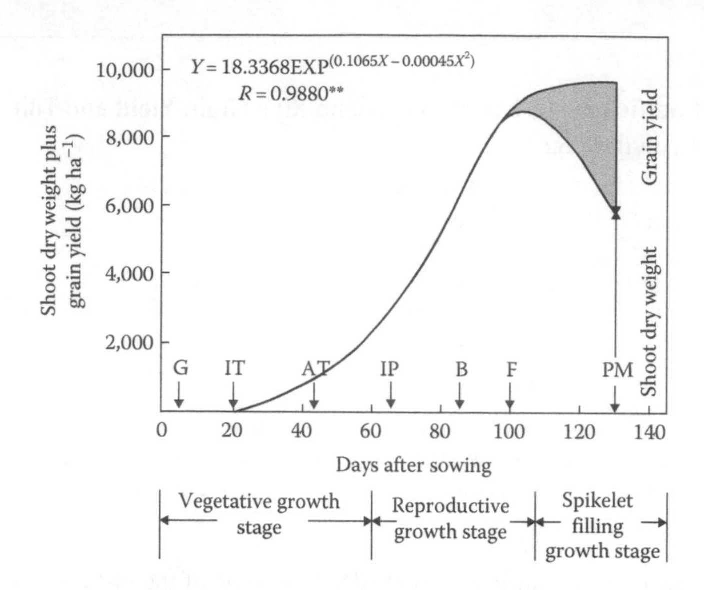

FIGURE 2.5 Shoot dry weight accumulation and grain yield of upland rice during the growth cycle of the crop in central Brazil. G, germination; It, initiation of tillering; AT, active tillering; IP, initiation of panicle primordia; B, booting; F, flowering; PM, physiological maturity. (From Fageria, N.K., *J. Plant Nutr.*, 30, 843, 2007.)

in this case. Trifoliate leaves number was increased in an exponential quadratic manner from 15 to 80 days of growth period (Figure 2.6). Wallace et al. (1972) reported that leaf area is a major physiological component of yield and relative growth rate.

Fageria et al. (2013a) studied the influence of P on the shoot (straw) weight of 20 upland rice genotypes (Table 2.4). Shoot dry weight varied from 5.84 to 20.91 g plant^{-1} at the low P level (25 mg kg^{-1}), with a mean value of 10.9 g plant^{-1}. At the high P level (200 mg kg^{-1}), shoot dry weight varied from 11 to 23.4 g plant^{-1}, with a mean value of 15.5 g plant^{-1}. Overall, the increase in shoot dry weight at the high P level was 43% as compared to the low P level. Fageria et al. (1982) and Fageria and Baligar (1997b) also reported improvement in shoot dry weight with the addition of P in Brazilian Oxisols. Shoot weight is characteristics of genotypes and is also influenced by environmental factors. Grain yield differences occur among plants or genotypes having the same amount of dry matter, since there exist differences in the utilization of photosynthates among them (Hayashi, 1995). N, P, and K fertilization influences shoot dry weight (Fageria et al., 1997; Fageria and Baligar, 2005). Growth of three upland rice genotypes at the two P levels is presented (Figure 2.7). Growth of genotypes at the level of 200 mg P kg^{-1} was more vigorous, as compared to 25 mg P kg^{-1}.

Fageria et al. (2012) studied the influence of P on shoot (straw) weight of 30 dry bean genotypes (Table 2.5). Dry matter yield of straw was significantly influenced by P as well as genotype treatments (Table 2.5). The P × genotype interaction was also significant for this trait, indicating different responses of genotypes at the two P levels. Straw yield varied from 1.03 g plant^{-1} produced by genotype BRS Valente to 2.13 g plant^{-1} produced by genotype BRS Pontal, with a mean value of 1.58 g plant^{-1} at 0 mg P kg^{-1} P level. At the high P level, straw yield varied from 1.94 to 5.09 g plant^{-1} produced by genotype Corrente, with a mean value of 3.68 g plant^{-1}. Straw yield differences between the lowest and highest straw yield–producing genotypes were 2-fold at the low P level and 2.6-fold at the high P level. Straw yield increase was 133% at the high P level as compared to the low P level. Straw yield production was reported to differ among dry bean genotypes by Fageria (2002). Similarly, Fageria (1989a,b, 1998) and Fageria and Santos (1998) also reported straw yield differences among dry bean genotypes at the low and high P level grown on a Brazilian Inceptisol. Fageria and Baligar (1989) also reported dry matter yield differences in dry bean grown in nutrient

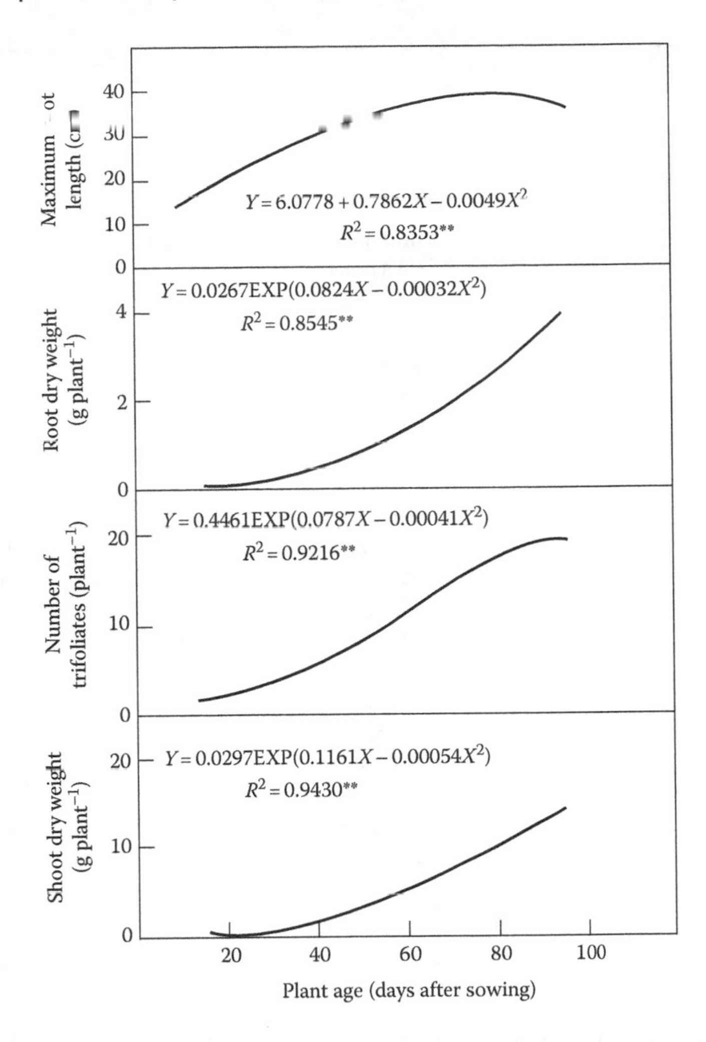

FIGURE 2.6 Relationship between plant age and dry bean growth parameters. (From Fageria, N.K. and Santos, A.B., *J. Plant Nutr.*, 31, 983, 2008a.)

solution at different P levels. P deficiency was very severe at 0 mg P kg^{-1} treatment, and dry bean plants were very healthy at 200 mg P kg^{-1} soil treatment (Figure 2.8).

Fageria et al. (2013c) studied the influence of P on the shoot dry weight of 14 tropical legume cover crops (Table 2.6). P × cover crop interaction for shoot dry weight was significant (Table 2.6). It suggests that the response of cover crops to P varied with the variation in P levels and screening for P use efficiency should be performed at various P levels. Shoot dry weight varied from 0.13 g plant^{-1} produced by *Crotalaria breviflora* to 5.81 g plant^{-1} produced by *Canavalia ensiformis*, with a mean value of 1.31 g plant^{-1} at low (0 mg kg^{-1}) P level. At medium P level (100 mg kg^{-1}), shoot dry weight varied from 0.54 g plant^{-1} produced by *C. breviflora* to 8.76 g plant^{-1} produced by *C. ensiformis*, with a mean value of 2.89 g plant^{-1}. At the higher P level (200 mg kg^{-1}), the shoot dry weight varied from 0.26 to 9.28 g plant^{-1}, with a mean value of 3.50 g plant^{-1}. Across three P levels, the shoot dry weight varied from 0.46 to 7.95 g plant^{-1}. White jack bean species produced highest shoot dry weight at the three P levels.

Shoot dry weight was also increased (1.31–3.50 g plant^{-1}) with increasing P level from 0 to 200 mg kg^{-1}. Interspecies variability in shoot dry weight of tropical legume cover crops has been widely reported (Fageria et al., 2005, 2009; Baligar et al., 2006; Baligar and Fageria, 2007; Fageria, 2009).

TABLE 2.4
Influence of Phosphorus Level (mg kg^{-1}) on Shoot Dry Weight (g Plant^{-1}) of 20 Upland Rice Genotypes

Genotype	Low P Level (25)	High P Level (200)
BRA01506	10.42bcde	12.53def
BRA01596	9.84bcde	11.00f
BRA01600	7.81cde	12.77def
BRA02535	12.69bc	15.87bcdef
BRA02601	12.36bc	12.83def
BRA032033	11.89bcd	15.84bcdef
BRA032039	9.79bcde	15.50bcdef
BRA032048	6.25de	11.58ef
BRA032051	5.84e	12.36def
BRA042094	10.15bcde	17.79abcd
BRA042156	9.46bcde	15.09cdef
BRA042160	20.91a	20.82ab
BRA052015	14.70b	14.53cdef
BRA052023	14.96b	23.44a
BRA052033	7.75cde	13.10cdef
BRA052034	8.65cde	16.20bcdef
BRA052045	9.38bcde	15.67bcdef
BRA052053	12.06bc	16.74bcde
BRS Primavera	11.33bcde	18.70abc
BRS Sertaneja	11.18bcde	17.41bcd
Average	10.87	15.49
F-test		
P level (P)	*	
Genotype (G)	**	
P × G	**	

Source: Fageria, N.K. et al., *J. Plant Nutr.*, 36, 1868, 2013a.
*,**Significant at the 5% and 1% probability level, respectively. Means within the same column followed by the same letter are not significantly different at the 5% probability level by Tukey's test.

FIGURE 2.7 Growth of three upland rice genotypes at the two P levels.

TABLE 2.5
Shoot (Straw) Yield (g Plant^{-1}) of 30 Dry Bean Genotypes as Influenced by P Fertilization Rate

	P Fertilization Rate (mg kg^{-1})	
Genotypes	**0**	**200**
1. Aporé	1.09hi	2.39m
2. Pérola	1.71a–g	4.25b–e
3. BRSMG Talisma	1.64a–h	3.19f–l
4. BRS Requinte	1.72a–g	3.03h–l
5. BRS Pontal	2.13a	3.58d–i
6. BRS 9435 Cometa	1.87a–c	3.32d–j
7. BRS Estilo	1.76a–f	3.38d–j
8. BRSMG Majestoso	1.81a–e	3.42d–j
9. CNFC 10429	1.64a–h	2.45m
10. CNFC 10408	1.70a–g	2.17m
11. CNFC 10467	1.92a–c	4.87a–c
12. CNFC 10470	2.07ab	4.34b–d
13. Diamante Negro	1.77a–f	4.16b–f
14. Corrente	1.45c–i	1.94m
15. BRS Valente	1.03i	5.61a
16. BRS Grafite	1.19f–i	4.12b–g
17. BRS Campeiro	1.63a–h	2.61im
18. BRS 7762 Supermo	1.52b–i	3.33d–j
19. BRS Esplendor	1.61a–h	4.01c–h
20. CNFP 10104	1.65a–h	4.20b–f
21. Bambuí	1.53b–i	2.94im
22. BRS Marfim	1.45c–i	3.10g–l
23. BRS Agreste	1.89a–c	5.05a–c
24. BRS Pitamda	1.15g–i	4.11b–g
25. BRS Verede	1.24e–i	4.36b–d
26. EMGOPA Ouro	1.28d–i	2.46m
27. BRS Radiante	1.41c–i	3.23e–j
28. Jalo Precoce	1.58a–i	5.09ab
29. BRS Executivo	1.84a–d	4.26b–e
30. BRS Embaixador	1.11hi	3.69a
Average	1.58b	3.68a
F-test		
P levels (P)	**	
Genotype (G)	**	
P × G	**	
CVP (%)	11.03	
CVG (%)	9.97	

Source: Fageria, N.K. et al., *Commun. Soil Sci. Plant Anal.*, 43, 2752, 2012.
**Significant at the 1% probability level. Means followed by the same letter within the same column or same line (P levels) are not significantly different at the 5% probability level by Tukey's test.

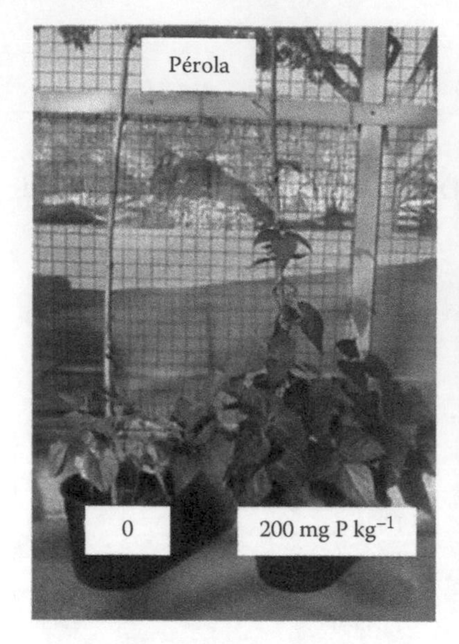

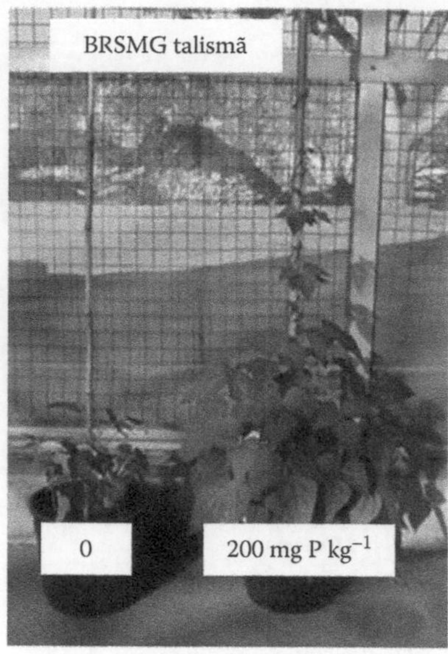

FIGURE 2.8 Growth of two dry bean genotypes at the two P levels.

TABLE 2.6
Shoot Dry Weight of 14 Legume Cover Crops as Influenced by P Levels (mg kg^{-1})

	Shoot Dry Weight (g Plant^{-1})			
Cover Crops	**0**	**100**	**200**	**Mean**
Crotalaria	0.13f	0.54f	0.70fg	0.46i
Sunn hemp	1.19de	3.38d	4.49d	3.02e
Crotalaria	0.20f	0.77ef	0.74fg	0.57hi
Crotalaria	0.31f	1.01ef	1.37efg	0.89fghi
Crotalaria	0.30f	1.38ef	2.37e	1.35fg
Calopogonium	0.26f	0.93ef	0.26g	0.48i
Pueraria	0.17f	0.74ef	1.03efg	0.64ghi
Pigeon pea (black)	0.63ef	1.90e	1.97ef	1.50f
Pigeon pea (mixed color)	0.39ef	1.57ef	1.76efg	1.24fgh
Lablab	0.92def	4.03cd	5.60bcd	3.51de
Mucuna bean ana	2.26c	4.54bcd	5.35cd	4.05cd
Black mucuna bean	1.61cd	5.71b	6.92bc	4.75c
Gray mucuna bean	4.21b	5.27bc	7.14b	5.54b
White Jack bean	5.81a	8.76[a]	9.28a	7.95a
Average	1.31	2.89	3.50	2.57
F-test				
P	**			
Cover crops (C)	**			
P × C	**			
CV (%)	16.05			

Source: Fageria, N.K. et al., *Commun. Soil Sci. Plant Anal.*, 44, 3340, 2013c.

**Significant at the 1% probability level. Means followed by the same letter within the same column are significantly not different by Tukey's test at the 5% probability level.

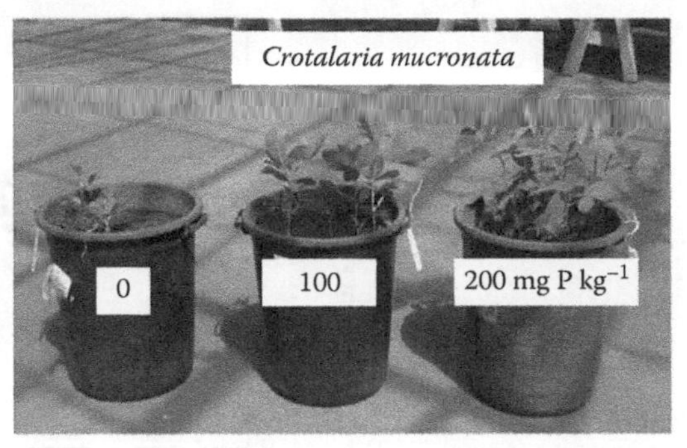

FIGURE 2.9 Growth of two tropical legume cover crops at three P levels.

Inter- and intraspecific variations for plant growth are known to be genetically and physiologically controlled and are modified by plant interactions with environmental variables (Fageria, 1992, 2009; Baligar et al., 2001). Growth of two tropical cover crops was significantly reduced at the 0 mg P kg^{-1} level, as compared to 100 and 200 mg P kg^{-1} treatment (Figure 2.9).

An increase in shoot weight is important since it is significantly associated with grain yield (Table 2.7; Figure 2.10). Fageria et al. (2013b) reported that shoot dry weight had a significant

TABLE 2.7
Correlation Coefficients (r) between Lowland Rice Grain Yield and Shoot Dry Matter Production at the Different Growth Stages

Parameters	First Year	Second Year	Third Year
Dry matter yield at IT	0.36*	0.37*	0.29^{NS}
Dry matter yield at AT	0.71**	0.55**	0.42*
Dry matter yield at IP	0.63**	0.51**	0.63**
Dry matter yield at B	0.72**	0.81**	0.61**
Dry matter yield at F	0.81**	0.80**	0.57**
Dry matter yield at PM	0.78**	0.80**	0.53**

Source: Fageria, N.K. and Baligar, V.C., *Commun. Soil Sci. Plant Anal.*, 32, 1405, 2001a.
IT, initiation of tillering; AT, active tillering; IP, initiation of panicle; B, booting; F, flowering; PM, physiological maturity.
*,**, and NS Significant at the 0.05 and 0.01 probability level and nonsignificant, respectively.

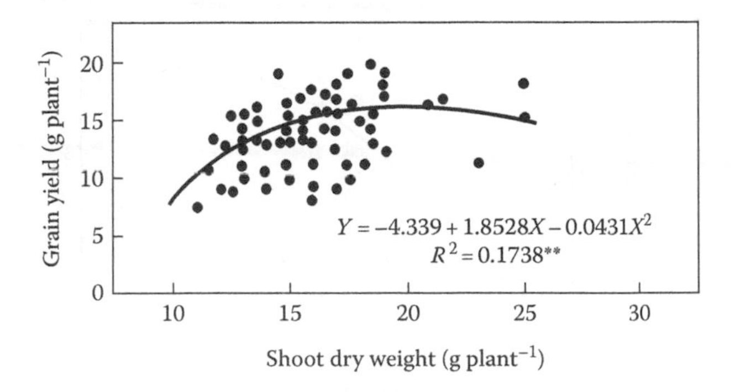

FIGURE 2.10 Relationship between shoot dry weight and grain yield of upland rice. (From Fageria, N.K. et al., *Commun. Soil Sci. Plant Anal.*, 44, 2656, 2013b.)

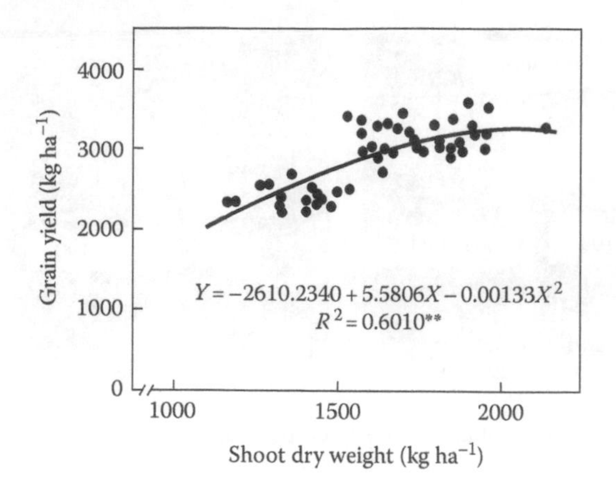

FIGURE 2.11 Relationship between shoot dry weight of dry bean and grain yield. (From Fageria, N.K. and Santos, A.B., *J. Plant Nutr.*, 31, 983, 2008a.)

quadratic relationship with grain yield of upland rice ($Y = -3.97 + 2.30X - 0.07X^2$, $R^2 = 0.26^{**}$). This means that improvement in shoot dry weight improved grain yield of upland rice. Fageria et al. (2004) reported a significant positive relationship between shoot dry weight of upland rice and grain yield. Fageria and Santos (2008a) reported that shoot dry weight of dry bean had a significant quadratic relationship with grain yield (Figure 2.11). Maximum grain yield of 3244 kg ha^{-1} was obtained with shoot dry weight of 2098 kg ha^{-1}. Variation of 60% in grain yield of dry bean was attributed to shoot dry weight (Figure 2.11). Peet et al. (1977) reported a positive association between shoot dry weight and grain yield of dry bean. Wallace et al. (1972) reported that genetic improvement in economic yield of several crops derives in part from higher percentage of biological yield being portioned to the plant organs constituting economic yield. Similarly, relationships occurred between shoot dry weight and grain yield of lowland rice (Figure 2.12). Rice yield was significantly and quadratically increased with increasing shoot dry weight.

2.2.1.4 Leaf Area Index

Leaf area index (LAI) is the plant growth index that measures leaf area per unit of land area. Wells and Norman (1991) defined LAI as the total leaf area per unit ground area, and LAI is commonly used to quantify vegetative canopy structure. Similarly, Narayanan et al. (2014) defined LAI as a widely used physiological parameter to quantify the vegetative structure of crops. LAI influences photon capture, photosynthesis, assimilate partitioning, growth, and yield formation (Rajcan and

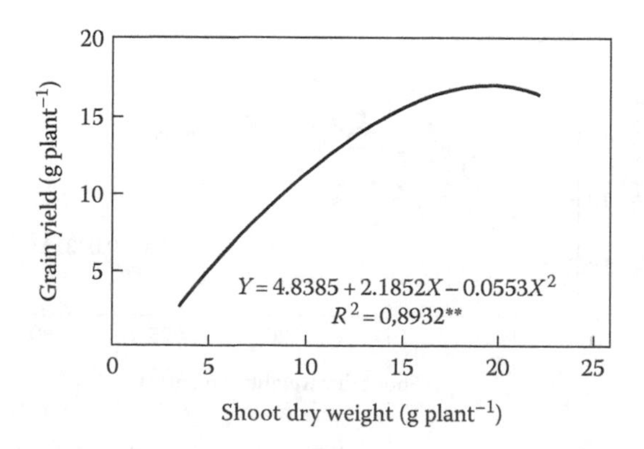

FIGURE 2.12 Relationship between shoot dry weight and grain yield of lowland rice.

Tollenaar, 1999; Yin et al., 2000; Tsialtas and Maslaris, 2008). Thus, accurate quantification of LAI is important for crop growth and development models. Watson (1958) formulated this index and reported that the crop growth rate (CGR) was a product of LAI and net assimilation rate (NAR). The equation is as follows:

$$\text{CGR} = \text{LAI} \times \text{NAR}$$

NAR is the dry matter accumulation per unit of leaf area and is expressed as g (m of leaf area)$^{-2}$ day^{-1} (Brown, 1984). It can be computed with the following equation (Fageria, 1992):

$$\text{NAR}\left(\text{g m}^{-2}\ \text{day}^{-1}\right) = \frac{1}{A}\frac{dW}{dt}$$

where

A is the leaf area

dW/dt is the change in plant dry matter per unit time

NAR determines the efficiency of plant leaves in dry matter production (Fageria, 1992). NAR decreases with increasing crop growth due to mutual shading of leaves and reduced photosynthetic efficiency of older leaves.

LAI in cereals can be calculated by the following equation:

$$\text{LAI} = \frac{\text{Leaf area of one tiller}\left(\text{cm}^2\right) \times \text{Number of tillers}\left(\text{m}^{-2}\right)}{10{,}000}$$

For legumes, tiller can be replaced by plant, and LAI can be calculated by the equation described earlier.

LAI is influenced by climatic, soil, and plant factors. Among soil factors, mineral nutrition is one of the important factors affecting this index. N and P significantly influence leaf area of the crop plants (Fageria et al., 2006). Fageria et al. (1982) studied the influence of P on LAI of upland rice throughout the crop growth cycle. LAI increased with the advancement of plant age and P levels. At 80 days of growth, the LAI significantly increased with P application up to 44 kg P ha^{-1} as compared to 0 and 22 kg P ha^{-1} treatments. At 100 days of growth in the first year, LAI values were 1.14, 2.21, 2.26, and 2.85 for 0, 22, 44, and 66 kg P ha^{-1} treatments, respectively. In the second year at 80 days of growth, values were 0.97, 1.59, 2.16, and 2.15 for the same treatments, respectively. Optimum LAI occurred between 2 and 3 at 85–100 days after sowing upland rice, but LAI varied with crop species and environmental conditions (Fageria, 1992). LAI values were reported to be 3.2, 5, and 6, respectively, for soybean, corn, and wheat (Yoshida, 1972). Critical LAI values for the large leaf tropical legumes generally fall in the range of 3–4 (Muchow, 1985), but they can exceed 5 for small leaf pigeon pea (*Cajanus cajan*) (Rowden et al., 1981).

2.2.1.5 Root Growth

Roots are important organ of the plants because they absorb water and nutrients, which are important components of growth and development. In addition, roots also provide mechanical support to the plants. Growth hormones are supplied by roots for optimum growth and development. Roots improve the organic matter content of the soil, which is responsible for improving the soil, that is, the physical, chemical, and biological properties favored for higher crop yields (Fageria, 2013). Root growth (dry weight and length) is positively related to crop yields (Fageria, 2013). Grain yield of lowland rice was significantly related to dry weight of roots (Figure 2.13). Root dry weight was significantly and positively related to shoot dry weight of upland rice, dry bean, corn, wheat, and soybean (Table 2.8). Variation in shoot dry weight due to root dry weight was 23%, 76%, 45%,

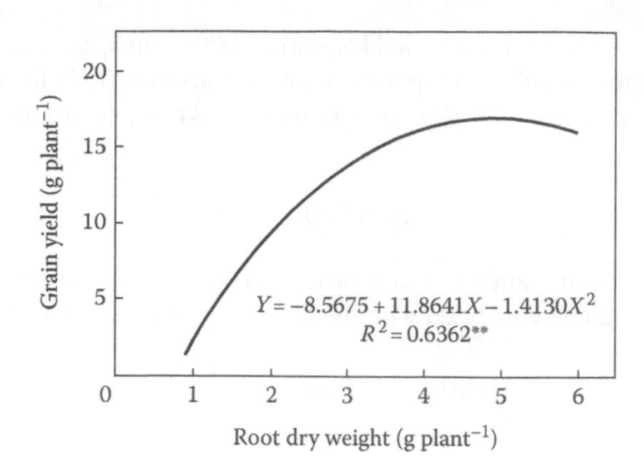

FIGURE 2.13 Relationship between root dry weight and grain yield of lowland rice.

TABLE 2.8
Relationship between Root Dry Weight and Shoot Dry Weight of Upland Rice, Dry Bean, Corn, Wheat, and Soybean

Variables	Regression Equation	R^2
RDW of rice vs. SDW of rice	$Y = 3.50 + 0.78X$	0.23*
RDW of dry bean vs. SDW of dry bean	$Y = 0.79 + 20.13X - 6.75X^2$	0.76**
RDW of corn vs. SDW of corn	$Y = -119.65 + 10.85X - 0.22X^2$	0.45**
RDW of wheat vs. SDW of wheat	$Y = 0.15 + 4.86X$	0.55**
RDW of soybean vs. SDW of soybean	$Y = -0.10 + 2.06X - 1.42X^2$	0.83**

RDW, root dry wt; SDW, shoot dry wt.
*,**Significant at the 5% and 1% probability level, respectively.

55%, and 83%, respectively, for upland rice, dry bean, corn, wheat, and soybean. These indicated the importance of roots in the production of shoots of many food crops.

Roots can sequestrate C, which is a major issue in the twenty-first century due to climate change and having adverse consequences on the environment and consequently human and animal health. Soil organic C is considered a key component in removing CO_2 from the atmosphere to decrease greenhouse gas emissions and mitigate global climate change (Christopher et al., 2009). Hinsinger et al. (2009) reported that roots of higher plants anchor the aboveground diversity of terrestrial ecosystems and provide much of the C to power the soil ecosystem. Production of ramified root system is important, especially under environmental stresses like drought and mineral nutrition (Fageria et al., 2006; Jaleel et al., 2009). Early growth of vigorous roots has been proven to be a major factor to increase N uptake in wheat (Liao et al., 2004; Noulas et al., 2010). Larger root systems provide greater root–soil contact, which is particularly important for the uptake of P (Gahoonia and Nielsen, 2004). Mobile nutrients like nitrate can be depleted in rhizosphere at low rooting density, while less mobile nutrients like P and K are often closely related to root length (Atkinson, 1991).

Root growth is genetically controlled and also influenced by environmental factors (Fageria and Moreira, 2011; Fageria, 2013). Partitioning of dry matter into the root relative to the shoot is high in the seedling stages of growth and steadily declines throughout the growth cycle of a crop plant (Evans and Wardlaw, 1976). Among the environmental factors, adequate supply of P is important to the growth and development of field crops. Fageria et al. (2012) studied the influence of P on the root length and root dry weight of 30 dry bean genotypes (Table 2.9).

TABLE 2.9
Maximum Root Length and Root Dry Weight of 30 Dry Bean Genotypes at the Two P Levels (mg kg⁻¹)

Genotype	Maximum Root Length (cm)		Root Dry Weight (g plant⁻¹)	
	0	200	0	200
1. Aporé	21.00c–f	22.00b–h	0.33ab	1.47a–d
2. Pérola	22.33b–d	25.00a–g	0.44ab	1.02a–d
3. BRSMG Talisma	20.33c–f	30.67a	0.32ab	1.29a–d
4. BRS Requinte	28.00ab	23.66a–h	0.51ab	1.53a0d
5. BRS Pontal	20.67c–f	19.00e–h	0.54a	1.57a–d
6. BRS 9435 Cometa	19.67c–g	26.33a–d	0.43ab	1.35a–d
7. BRS Estilo	22.00b–e	21.33c–h	0.43ab	1.00a0d
8. BRSMG Majestoso	23.33a–d	30.67a	0.26ab	1.33a–d
9. CNFC 10429	29.67a	29.00ab	0.36ab	1.12a–d
10. CNFC 10408	19.00c–h	20.00d–h	0.27ab	0.88b–d
11. CNFC 10467	19.33c–g	28.00a–c	0.39ab	1.49a–d
12. CNFC 10470	20.67c–f	26.00a–e	0.36ab	1.45a–d
13. Diamante Negro	18.67d–h	23.67a–h	0.42ab	1.87ab
14. Corrente	15.67e–i	17.00h	0.30ab	0.74d
15. BRS Valente	25.33a–c	25.67a–f	0.47ab	1.97a
16. BRS Grafite	15.00f–i	20.33d–h	0.42ab	1.42a–d
17. BRS Campeiro	20.33c–f	20.33d–h	0.38ab	1.17a–d
18. BRS 7762 Supermo	13.33g–j	18.67f–h	0.43ab	1.31a–d
19. BRS Esplendor	21.67b–e	24.00a–h	0.52ab	1.06a–d
20. CNFP 10104	23.67a–d	21.67c–h	0.49ab	1.49a–d
21. Bambuí	22.33b–d	19.33d–h	0.24ab	1.16a–d
22. BRS Marfim	13.67g–j	18.33gh	0.40ab	0.86b–d
23. BRS Agreste	13.67g–j	19.33d–h	0.35ab	1.81a–c
24. BRS Pitamda	15.00f–i	24.00a–h	0.21b	1.05a–d
25. BRS Verede	23.00b–d	25.67a–f	0.41ab	1.96a
26. EMGOPA Ouro	8.00j	22.00b–h	0.26ab	0.79cd
27. BRS Radiante	10.67ij	18.67f–h	0.23ab	0.60d
28. Jalo Precoce	13.33g–j	21.00c–h	0.25ab	1.03a–d
29. BRS Executivo	12.67h–j	21.00c–h	0.38ab	1.20a–d
30. BRS Embaixador	13.67g–j	25.33a–g	0.30ab	1.13a–d
Average	18.86b	22.65a	0.38b	1.27a
F-test				
P levels (P)	**		**	
Genotype (G)	**		**	
P × G	**		**	
CVP (%)	8.07		72.39	
CVG(%)	12.78		26.28	

Source: Fageria, N.K. et al., *Commun. Soil Sci. Plant Anal.*, 43, 2752, 2012.

**Significant at the 1% probability level. Means followed by the same letter within the same column or same line (P levels) are not significantly different at the 5% probability level by Tukey's test.

Maximum root length and root dry weight were significantly influenced by P level as well as genotype treatments (Table 2.9). Maximum root length varied from 8.00 to 29.7 cm, with a mean value of 18.9 cm at the low P level (0 mg kg^{-1}). At the high P level (200 mg kg^{-1}), maximum root length varied from 17.0 to 30.7 cm, with a mean value of 22.65 cm. A 20% increase in maximum root length occurred at the high P level as compared to the low P level. Root dry weight varied from 0.21 to 0.54 g $plant^{-1}$ at the low P level. Similarly, at the high P level, root dry weight varied from 0.60 to 1.97 g $plant^{-1}$, with a mean value of 1.27 g $plant^{-1}$. Mean increase in root weight with the addition of P was 234% as compared to the control treatment. Root growth of three dry bean genotypes was more vigorous at the high P level as compared to the low P level (Figure 2.14). Improvement in root dry weight with the addition of P in dry bean is reported by Fageria (2009). Root growths of dry bean, upland rice, and soybean (*Glycine max*) at different P levels are presented (Figures 2.15 through 2.17).

Fageria et al. (2013c) also studied the influence of P on root growth of tropical legume cover crops (Tables 2.10 and 2.11). Root dry weight was significantly influenced by P, crop species, and P × crop species interaction. P × C interaction indicated significant variation in shoot dry

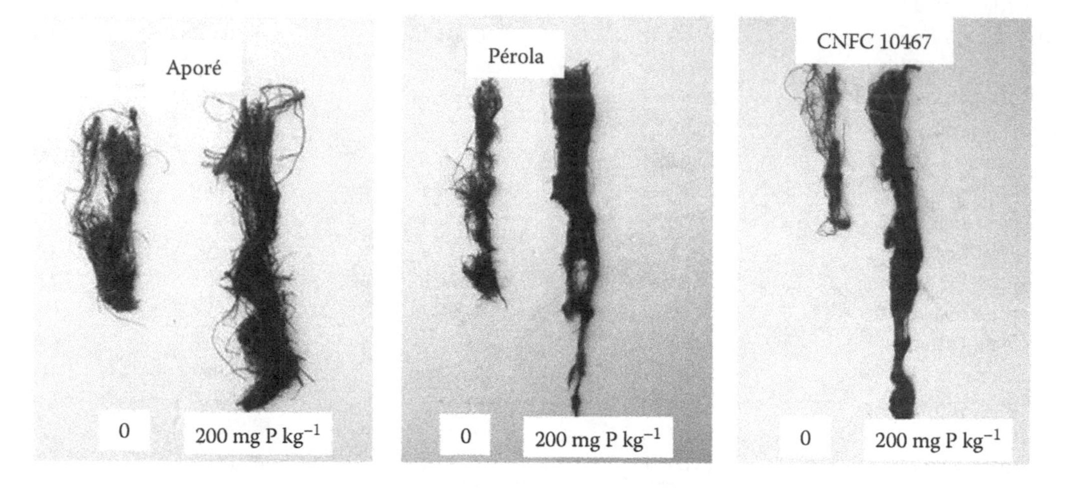

FIGURE 2.14 Root growth of three dry bean genotypes at the two P levels.

FIGURE 2.15 Dry bean root growth at different P levels.

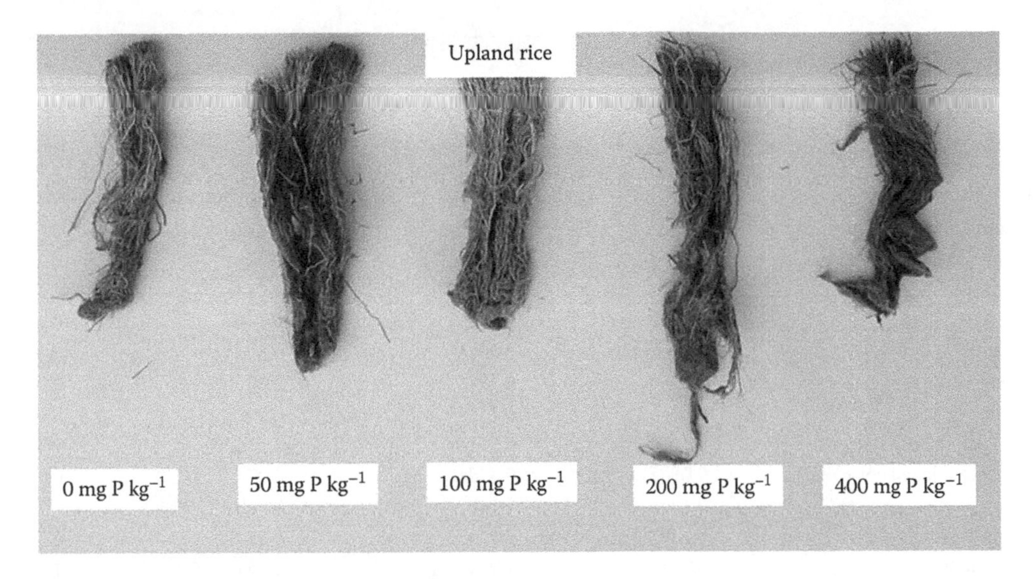

FIGURE 2.16 Upland rice root growth at different P levels.

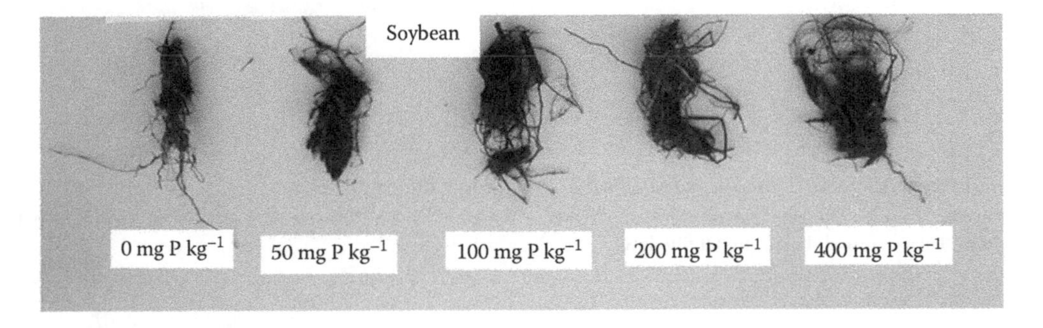

FIGURE 2.17 Soybean root growth at different P levels.

weight with P levels. At the low P level (0 mg kg^{-1}), maximum root dry weight of 0.77 g $plant^{-1}$ was produced by white jack bean and minimum root dry weight of 0.01 g $plant^{-1}$ was produced by Crotalaria (*Crotalaria mucronata*) and Pueraria (*Pueraria phaseoloides*). At the medium P level (100 mg P kg^{-1}), maximum root dry weight of 1.91 g $plant^{-1}$ was produced by black mucuna bean and minimum root dry weight of 0.07 was produced by *Crotalaria breviflora*, with a mean value of 0.63 g $plant^{-1}$. At the high P level (200 mg P), maximum root dry weight of 1.42 g $plant^{-1}$ was produced by gray mucuna bean and minimum root dry weight of 0.09 g $plant^{-1}$ was produced by Calopogonium and Pueraria, with a mean value of 0.55 g $plant^{-1}$. Across the three P levels, maximum root dry weight was produced by black mucuna bean and minimum by Calopogonium and Pueraria. Variation in root dry weight is genetically controlled and also influenced by environmental variables, such as supply of mineral nutrition (Caradus, 1990; Baligar et al., 2001; Fageria et al., 2006; Fageria and Moreira, 2011).

Maximum root length varied from 15.5 to 36 cm at the low P, from 20.5 to 50.3 cm at the medium P, and 18.3 to 53.0 cm at the high P level (Table 2.11). Across the three P levels, maximum root length of 46.2 cm was produced by white jack bean and minimum root length of 20.7 cm was produced by *Crotalaria*. Overall, root length also increased with increasing P levels.

TABLE 2.10
Root Dry Weight of 14 Legume Cover Crops as Influenced by P Levels

	Root Dry Weight (g Plant^{-1})			
Cover Crops	**0**	**100**	**200 mg P kg^{-1}**	**Average**
Crotalaria	0.03de	0.07f	0.18de	0.09g
Sunn hemp	0.17cd	0.83bc	0.64c	0.54d
Crotalaria	0.01e	0.18ef	0.18de	0.12fg
Crotalaria	0.04de	0.33ef	0.13de	0.17efg
Crotalaria	0.02e	0.40ef	0.50c	0.30e
Calopogonium	0.03de	0.13f	0.09e	0.08g
Pueraria	0.01e	0.14f	0.09e	0.08g
Pigeon pea (black)	0.16cd	0.51cde	0.13de	0.27ef
Pigeon pea (mixed color)	0.08de	0.43def	0.27d	0.26ef
Lablab	0.13cde	1.14b	0.96b	0.74c
Mucuna bean ana	0.54b	0.79bcd	0.82	0.72c
Black mucuna bean	0.26c	1.91a	1.36a	1.17a
Gray mucuna bean	0.53b	0.83bc	1.42a	0.93b
White Jack bean	0.77a	1.12b	0.93b	0.94b
Average	0.20	0.63	0.55	
F-test				
P levels (P)	**			
Cover crops (C)	**			
P × C	**			
CV (%)	17.16			

Source: Fageria, N.K. et al., *Commun. Soil Sci. Plant Anal.*, 44, 3340, 2013c.
**Significant at the 1% probability level. Means followed by the same letter within the same column are significantly not different by Tukey's test at the 5% probability level.

The improvement in root length by improved P nutrition has been reported by Fageria (2009) and Fageria and Moreira (2011) in various crop species, including cover crops. Barber (1995), Marschner (1995), Mengel et al. (2001), and Fageria et al. (2006) reported that mineral nutrition has tremendous effects on root growth, development, and function and subsequently the ability of roots to absorb and translocate nutrients. They reported that mineral deficiency induces considerable variations in growth and morphology of roots and such variations are strongly influenced by plant species and genotypes.

Root dry weight has a highly significant positive quadratic relationship with shoot dry weight ($Y = 0.34 \exp(5.52X - 2.72X^2$, $R^2 = 0.97^{**}$). Variations (97%) in shoot dry weight were associated with root dry weight. Similarly, root length also improved shoot dry weight in a quadratic relationship ($Y = 2.98 - 0.26X + 0.0076X^2$, $R^2 = 0.88^{**}$). Improvement in shoot dry weight with increasing root dry weight and root length may be associated with more absorption of water and nutrients (Fageria et al., 2009). Variations in shoot dry weight were about 88% due to root length and 97% due to root dry weight. Hence, it can be concluded that root dry weight is a better indication in determining shoot dry weight as compared to maximum root length. This suggests that root growth (maximum length and dry weight) was significantly affected by P nutrition. In addition, root growth varied among crop species and genotypes of same species. Heritability of root parameters is relatively high and a genetically controlled system as compared to those of shoot parameters (O'Toole and Bland, 1987).

TABLE 2.11
Maximum Root Length of 14 Legume Cover Crops as Influenced by P Levels

	Maximum Root Length (cm)			
Cover Crops	**0**	**100**	**200 mg P kg^{-1}**	**Average**
Crotalaria	21.0cde	20.5f	28.50c	23.33def
Sunn hemp	31.5ab	24.5def	53.00ab	36.33c
Crotalaria	20.0de	24.5def	22.00cd	22.17cf
Crotalaria	20.0de	23.67ef	18.33d	20.67f
Crotalaria	21.0cde	31.0cde	24.0cd	25.56de
Calopogonium	23.5cd	30.67cde	21.00cd	25.06de
Pueraria	15.5e	27.5def	23.50cd	22.17ef
Pigeon pea (black)	21.33cde	30.00cde	23.00cd	24.78de
Pigeon pea (mixed color)	23.0cd	26.50def	28.50c	26.00de
Lablab	19.67de	32.00bcd	29.00c	26.89d
Mucuna bean ana	24.33cd	47.00a	53.00ab	41.44b
Black mucuna bean	27.5bc	37.50bc	60.50a	41.83b
Gray mucuna bean	31.5ab	39.00b	51.00b	40.50b
White Jack bean	36.0a	50.33a	52.33ab	46.22[a]
Average	23.99	31.76	34.88	30.21
F-test				
P	**			
Cover crops (C)	**			
P × C	**			
CV(%)	8.31			

Source: Fageria, N.K. et al., *Commun. Soil Sci. Plant Anal.*, 44, 3340, 2013c.
**Significant at the 1% probability level. Means followed by the same letter within the same column are significantly not different by Tukey's test at the 5% probability level.

2.2.1.6 Yield and Yield Components and Related Traits

Grain yield of cereals and legumes is formulated by yield components, and formulation of yield components occurs during the growth cycle of crop plants. For example, in cereals (rice as a case study) grain yield can be expressed in the form of an equation by taking into account the yield components (Fageria, 2007, 2009):

$$\text{GY (Mg ha}^{-1}) = \text{NP m}^{-2} \times \text{spikelet per panicle} \times \%\text{ filled spikelets} \times 1000\text{ GW (g)} \times 10^{-5}$$

where
GY is grain yield
NP is number of panicles

This equation indicates that grain yield of rice is a combination of various yield components. For example, to produce a grain yield of 6 Mg ha^{-1}, a combination of the following yield components is required:

"400 panicles m^{-2}, 80 spikelets per panicle, 85% filled spikelets, and 22 g weight of 1000 grains." If we incorporate these values into the earlier equation,

$$\text{GY (Mg ha}^{-1}) = 400 \times 80 \times 0.85 \times 22 \times 10^{-5} = 6.0$$

Similarly in legumes (cowpea as a case study), grain yield can be expressed in the form of the following equation (Fageria, 1989b):

$$\text{GY (Mg ha}^{-1}) = \text{NP m}^{-2} \times \text{NS per pod} \times \text{weight of 1000 seeds (G)} \times 10^{-5}$$

where
GY is the grain yield
NP is the number of pods
NS is the number of seeds

For example, to obtain a seed yield of 1.5 Mg ha^{-1}, the following yield components are required:

"155 pods m^{-2}, 7 seeds per pod, 140 g weight of 1000 seeds,
and GY (Mg ha^{-1}) = $155 \times 7 \times 140 \times 10^{-5} = 1.5$"

Panicle number, spikelet sterility, 1000 grain or seed weight, pod number, and seeds per pod are important yield components in determining yield in cereals and legumes. Improvement in these traits contributes significantly to the yield of cereals and legumes (Fageria, 2007; Fageria and Santos, 2008a). Fageria (2007) and Gravois and Helms (1992) reported that optimum rice yield could not be attained without optimum panicle density of uniform maturity. Similarly, Ottis and Talbert (2005) reported a high correlation ($R^2 > 0.85$) between yield and panicle density of rice. P is an important element in improving these yield components. Fageria and Santos (2008a) reported that number of pods per unit area, grain per pod, and weight of 100 grain should be at an appropriate balance to achieve maximum economic yield of legumes. Adequate concentration of P in the soil solution is important to obtain maximum economic yield of crop plants under all agroecological conditions.

2.2.1.6.1 Yield

Yield is defined as the amount of specific substance produced (e.g., grain, straw, total dry matter) per unit area (Soil Science Society of America, 2008). Grain yield refers to the weight of cleaned and dried grains harvested from a unit area. For rice, grain yield is usually expressed either in kilograms per hectare (kg ha^{-1}) or in metric tons per hectare (Mg ha^{-1}) at 13% or 14% moisture. Yield of a crop is determined by management practices, which will maintain the productive capacity of a crop ecosystem. These practices include the use of crop genotypes, water management, the use of fertilizers, and control of insects, diseases, and weeds (Fageria, 2014). Use of P in adequate amount is one of the most important factors that determine crop yields. Several examples of improvement in crop yield with the addition of P are provided in this section.

Fageria et al. (2012) studied the influence of P on the grain yield of 30 dry bean genotypes (Table 2.12). Grain yield varied from 0.66 g plant^{-1} produced by genotype BRS Embaixador to 2.11 g plant^{-1} produced by genotype BRS Pontal, with a mean value of 1.47 g plant^{-1} at the low P level (0 mg kg^{-1}). Similarly, grain yield at the high P level (200 mg kg^{-1}) varied from 5.33 g plant^{-1} produced by genotype BRS Embaixador to 10.78 g plant^{-1} produced by genotype BRS Pontal, with a mean value of 8.61 g plant^{-1}. Overall, the increase in grain yield was about sixfold at the high P level as compared to the low P level. Fageria (1989b), Fageria (2002), and Tang et al. (2004) reported grain yield differences among dry bean genotypes at different P levels.

Fageria et al. (2013a) also studied the influence of P on the grain yield of 20 upland rice genotypes (Table 2.13). P level × genotype interaction for grain yield was significant, indicating genotype differently responded to varying P levels. At the low P rate, grain yield varied from 7.58 to 17.6 g plant^{-1}, with a mean yield of 10.9 g plant^{-1}. The lowest grain yield–producing genotype was BRA032051, and the maximum grain-producing genotype was BRA052015 g plant^{-1} at the low P level. The difference in grain yield between the lowest and highest yield-producing genotypes was 132% at the low P level. At the high P level, grain yield varied from 10.49 g plant^{-1} produced by

TABLE 2.12
Grain Yield (g Plant^{-1}) of 30 Dry Bean Genotypes at the Two P Levels

Genotypes	0	200 mg P kg^{-1}
1. Aporé	1.07h–j	8.64d–h
2. Pérola	1.68a–f	9.36b–f
3. BRSMG Talisma	1.49b–i	8.83d–h
4. BRS Requinte	1.65a–f	8.48d–j
5. BRS Pontal	2.11a	10.78a
6. BRS 9435 Cometa	1.57b–h	9.72a–d
7. BRS Estilo	1.71a–e	9.42b–e
8. BRSMG Majestoso	1.46b–i	7.98g–j
9. CNFC 10429	1.45b–i	9.59a–e
10. CNFC 10408	1.97ab	9.02c–g
11. CNFC 10467	1.62a–g	8.15f–j
12. CNFC 10470	1.89a–c	10.33ab
13. Diamante Negro	1.55b–h	9.05c–g
14. Corrente	1.72a–d	10.17a–c
15. BRS Valente	1.17e–j	9.47b–e
16. BRS Grafite	1.01ij	8.35e–j
17. BRS Campeiro	1.10g–j	8.72d–h
18. BRS 7762 Supermo	1.54b–i	8.86d–h
19. BRS Esplendor	1.59a–h	8.49d–j
20. CNFP 10104	1.22d–i	8.63d–i
21. Bambuí	1.53b–i	7.66h–l
22. BRS Marfim	1.41c–i	9.07b–g
23. BRS Agreste	1.56b–h	8.90c–h
24. BRS Pitamda	1.14f–j	7.68h–l
25. BRS Verede	1.07h–j	7.36il
26. EMGOPA Ouro	1.41c–i	7.87g–j
27. BRS Radiante	1.48b–i	7.29g–j
28. Jalo Precoce	1.55b–h	6.53m
29. BRS Executivo	1.75a–d	8.39e–j
30. BRS Embaixador	0.66j	5.33m
Average	1.47b	8.61a
F-test		
P levels (P)	**	
Genotype (G)	**	
P × G	**	
CVP (%)	9.71	
CVG (%)	5.84	

Source: Fageria, N.K. et al., *Commun. Soil Sci. Plant Anal.*, 43, 2752, 2012.

**Significant at the 1% probability level. Means followed by the same letter within the same column or same line (P levels) are not significantly different at the 5% probability level by Tukey's test.

TABLE 2.13
Grain Yield (g Plant^{-1}) of Upland Rice Genotypes as Influenced by P Levels

Genotypes	25	200 mg P kg^{-1}
BRA01506	12.93abcde	17.63abcde
BRA01596	12.16bcdef	15.53cdefg
BRA01600	8.31ef	15.39cdefg
BRA02535	9.10def	15.35cdefg
BRA02601	11.96bcdef	12.20fg
BRA032033	11.88bcdef	16.24abcdef
BRA032039	9.42cdef	10.49g
BRA032048	8.50ef	15.20cdefg
BRA032051	7.58f	14.99cdefg
BRA042094	9.58cdef	16.81abcdef
BRA042156	10.37bcdef	16.75abcdef
BRA042160	8.86def	13.47efg
BRA052015	17.59a	16.70abcdef
BRA052023	14.95ab	19.49abcd
BRA052033	8.81def	14.81defg
BRA052034	9.55cdef	19.40abcd
BRA052045	10.14bcdef	21.12ab
BRA052053	14.67abc	20.31abc
BRS Primavera	14.04abcd	21.35a
BRS Sertaneja	7.74ef	15.90bcdef
Average	10.91	16.46
F-test		
P level (P)	**	
Genotype (G)	**	
P × G	**	

Source: Fageria, N.K. et al., *J. Plant Nutr.*, 36, 1868, 2013a.
**Significant at the 1% probability level. Means within the same column followed by the same letter are not significantly different at the 5% probability level by Tukey's test.

genotype BRA032039 to 21.35 g plant^{-1} produced by genotype BRS Primavera. The difference in grain yield between the lowest and highest grain yield-producing genotypes was 104%. Overall, grain yield increase at the high P level was 52% as compared to the low P level, indicating high importance of P fertilization for upland rice in the Brazilian Oxisols. The response of upland rice to P fertilization in Brazilian Oxisol has been reported by Fageria et al. (1982), Fageria and Baligar (1997b), and Fageria and Baligar (2001b). The response of annual crops to P fertilization is due to the low level of P in the Oxisols and high P immobilization capacity (Fageria, 1989a,b; Fageria and Baligar, 1989).

2.2.1.6.2 Panicle Density and Pod Number

Yield components that are directly related to yield are panicle density in cereals like rice and pod number in legumes. These yield components are influenced by soil, plant, and climatic factors. Fageria and Santos (2008b) studied the influence of P on lowland rice yield panicle density (Table 2.14). Panicle density increased quadratically with P fertilizer rate ($Y = 276.11 + 0.70X - 0.00077X^2$, $R^2 = 0.75^{**}$). Variation in panicle number was about 75% with the addition of P. Significant increases in lowland rice panicle density with the addition of P were reported when soil test for Mehlich 1 P

TABLE 2.14
Influence of Phosphorus Fertilization on Panicle Number of Lowland Rice

P Rate (kg ha^{-1})	Panicle Number (m^{-2})
0	264.3
131	365.0
262	432.0
393	412.2
524	417.3
655	419.2
F-test	
Year (Y)	**
P rate (P)	**
Y × P	NS
CV (%)	10

Source: Fageria, N.K. and Santos, A.B., *Commun. Soil Sci. Plant Anal.*, 39, 873, 2008b.
Values are averages of 2 years' field experimentation.
** and NS Significant at the 1% probability level and nonsignificant, respectively.

in Brazilian Inceptisols was in the range of low (2.6–8.8 mg P kg^{-1}) and medium (8.8–13 mg P kg^{-1}) range (Fageria et al., 1997). Panicle density had a significant quadratic relationship with grain yield of lowland rice ($Y = -15{,}604.45 + 87.16X - 0.09X^2$, $R^2 = 0.87^{**}$). The significant contribution of a number of panicles per unit area in increasing lowland rice grain yield is widely reported (Yoshida, 1981; Fageria and Baligar, 1999; Fageria and Baligar, 2001a).

Fageria et al. (2012) also studied the influence of P on dry bean pods per plant (Table 2.15). A number of pods per plant were significantly influenced by P level, genotype treatments, and P × genotype interaction, which was also significant for this trait (Table 2.15). P × genotype interaction indicates that genotypes differently responded with a number of pods at the two P levels and selection is possible at low as well as at high P levels for this trait. Number of pods per plant varied from 1.25 to 2.83 at the low P level (0 mg kg^{-1}), with a mean value of 1.97 pods plant^{-1}. Similarly, at the high P level (200 mg kg^{-1}), number of pods per plant varied from 4.08 to 9.99, with a mean value of 6.63. Overall, an increase in number of pods per plant was 237% at the high P level, as compared to lower P level. Fageria and Santos (2008a) reported significant and positive associations with pods per plant and grain yield of dry bean.

2.2.1.6.3 Panicle Length and Seed Per Pod

Panicle density and seed per pod are important traits in determining grain yield in cereals and legumes. Fageria and Santos (2008b) studied the influence of P on panicle length of lowland rice (Table 2.16). Panicle length was significantly influenced by P rates, and the increase in panicle length responded quadratically to increasing P rates ($Y = 19.32 + 0.0059X - 0.0000031X^2$, $R^2 = 0.56^{**}$). Panicle length responded quadratically to grain yield ($Y = -122{,}564.60 + 11{,}227.76X - 246.52X^2$, $R^2 = 0.78^{**}$).

Number of seeds per pod varied from 2.00 to 4.92 at the low P level, with a mean value of 3.62 seeds per pod (Table 2.17). At the high P level, number of seeds per pod varied from 2.55 to 6.52 seeds per pod, with a mean value of 4.59 seeds per pod. Overall, the increase in number of seeds per pod was 27% at the high P level as compared to the low P level. An increase in the number of pods and number of seeds per pod in the dry bean with the addition of P fertilizer was

TABLE 2.15
Number of Pods Per Plant in Dry Bean Genotypes at the Two P Levels

Genotypes	0	200 mg P kg^{-1}
1. Aporé	1.33de	7.08c–i
2. Pérola	2.08a–e	6.41d–j
3. BRSMG Talisma	1.41c–e	7.50b–g
4. BRS Requinte	2.33a–c	6.67c–j
5. BRS Pontal	1.91a–e	7.25c–i
6. BRS 9435 Cometa	2.00a–e	4.41n
7. BRS Estilo	2.00a–e	5.91g–l
8. BRSMG Majestoso	2.08a–e	6.33e–j
9. CNFC 10429	2.16a–e	6.75c–j
10. CNFC 10408	2.25a–d	7.08c–i
11. CNFC 10467	2.25a–d	7.17c–i
12. CNFC 10470	2.17a–e	6.83c–j
13. Diamante Negro	2.41ab	8.08a–d
14. Corrente	2.08a–e	9.99ab
15. BRS Valente	1.83b–e	8.17a–c
16. BRS Grafite	1.83b–e	7.41b–h
17. BRS Campeiro	2.83a	9.33a
18. BRS 7762 Supermo	1.91a–e	6.25f–j
19. BRS Esplendor	1.91a–e	6.58c–j
20. CNFP 10104	2.00a–e	5.67in
21. Bambuí	1.41c–e	5.75h0m
22. BRS Marfim	2.41ab	6.66c–j
23. BRS Agreste	1.75b–e	8.00a–e
24. BRS Pitamda	2.25a–d	7.33b–i
25. BRS Verede	1.58b–e	6.17f–j
26. EMGOPA Ouro	2.33a–c	7.67a–f
27. BRS Radiante	2.00a–e	4.00n
28. Jalo Precoce	1.70b–e	5.17n
29. BRS Executivo	1.67b–e	4.41n
30. BRS Embaixador	1.25e	4.08n
Average	1.97b	6.63a
F-test		
P levels (P)	**	
Genotype (G)	**	
P × G	**	
CVP (%)	9.51	
CVG (%)	10.01	

Source: Fageria, N.K. et al., *Commun. Soil Sci. Plant Anal.*, 43, 2752, 2012.

**Significant at the 1% probability level. Means followed by the same letter within the same column or same line (P levels) are not significantly different at the 5% probability level by Tukey's test.

TABLE 2.16
Influence of P on Panicle Length in Lowland Rice

P Rate (kg ha^{-1})	Panicle Length (cm)
0	19.0
131	20.1
262	21.3
393	20.9
524	20.8
655	22.3
F-test	
Year (Y)	**
P rate (P)	**
Y × P	NS
CV (%)	5

Source: Fageria, N.K. and Santos, A.B., *Commun. Soil Sci. Plant Anal.*, 39, 873, 2008b.
Values are averages of 2 years' field experimentation.
** and NS Significant at the 1% probability level and nonsignificant, respectively.

also reported by Fageria (1989a), Fageria and Baligar (1996), and Fageria and Santos (2008a). Number of pods per plant was having a significant positive quadratic correlation with grain yield ($Y = -1.65 + 2.83X - 0.29X^2$, $R^2 = 0.31$**). Similarly, the number of seeds per pod was also having a significant and positive association with grain yield ($Y = -1.02 + 2.71X - 0.29X^2$, $R^2 = 0.36$**). Fageria and Santos (2008a) reported significant and positive correlations with pods per plant and seeds per pod with grain yield of dry bean.

2.2.1.6.4 Thousand Grain or Hundred Grain Weight

Thousand or hundred grain weight is generally used as a measure of yield components in cereals and legumes. This yield-determining trait is genetically controlled as well as influenced by environmental factors. Fageria and Santos (2008b) studied the influence of P on 1000 grain weight of lowland rice (Table 2.18). P significantly increased thousand grain weights. Thousand grain weight responded quadratically ($Y = 23.43 + 0.01X - 0.000014X^2$, $R^2 = 0.59$**) to increasing P rates. The variability in the thousand grain weight due to use of P was 59%. In addition, thousand grain weight was significantly correlated to grain yield ($Y = -30{,}651.68 + 1{,}822.60X - 18.40X^2$, $R^2 = 0.81$**). The positive significant relationship between thousand grain weight and grain yield of lowland rice shows the importance of P fertilization in lowland rice production in the Brazilian Inceptisols.

Similarly, hundred seed weight of dry bean was significantly influenced by P and genotype treatments (Table 2.19). Hundred seed weight varied from 14.6 to 41.2 g at the low P level, with a mean value of 21.9 g. At the high P level, hundred seed weight varied from 21.6 to 61.3 g, with a mean value of 30.4 g. An increase in hundred seed weight was 39% at the high P level as compared to the low P level. Hundred grain weight is controlled genetically and also influenced by environmental factors (Fageria and Santos, 2008a).

2.2.1.6.5 Spikelet Sterility

Spikelet sterility is an important yield component in rice, and reducing spikelet is one way to improve yield. Overall, the filled spikelet percentage is about 85% in rice, even under favorable conditions (Yoshida, 1981). The possibility of increasing rice yield by 15% exists if breeding eliminates

TABLE 2.17
Number of Seeds Per Pod in Dry Bean Genotypes at the Two P Levels

Genotypes	0	200 mg P kg^{-1}
1. Aporé	3.90a–g	4.58b–h
2. Pérola	3.31b–h	4.61b–h
3. BRSMG Talisma	4.62a–c	3.83f–j
4. BRS Requinte	4.02a–f	5.64a–d
5. BRS Pontal	4.91a	5.34a–e
6. BRS 9435 Cometa	3.17c–h	6.52a
7. BRS Estilo	3.95a–f	5.20a–f
8. BRSMG Majestoso	2.77e–h	4.41c–i
9. CNFC 10429	3.25b–h	5.18a–f
10. CNFC 10408	3.74a–g	4.36c–i
11. CNFC 10467	3.89a–g	4.67b–h
12. CNFC 10470	3.85a–g	4.62b–h
13. Diamante Negro	4.22a–e	5.18a–f
14. Corrente	3.92a–g	4.13e–i
15. BRS Valente	2.72e–h	4.33d–i
16. BRS Grafite	3.28b–h	4.02e–i
17. BRS Campeiro	2.00h	3.14ij
18. BRS 7762 Supremo	4.92a	5.77a–c
19. BRS Esplendor	4.86a	5.66a–d
20. CNFP 10104	3.46a–h	5.94ab
21. Bambuí	4.73ab	5.15a–f
22. BRS Marfim	3.15c–h	4.79b–g
23. BRS Agreste	4.34a–d	4.50c–i
24. BRS Pitamda	2.98d–h	4.60b–h
25. BRS Verede	4.36a–d	4.91b–g
26. EMGOPA Ouro	4.16a–e	4.28d–i
27. BRS Radiante	2.42gh	3.33h–j
28. Jalo Precoce	2.85d–h	3.53g–j
29. BRS Executivo	2.55f–h	3.14ij
30. BRS Embaixador	2.17h	2.55j
Average	3.62b	4.59a
F-test		
P levels (P)	**	
Genotype (G)	**	
P × G	**	
CVP (%)	11.76	11.12
CVG (%)		

Source: Fageria, N.K. et al., *Commun. Soil Sci. Plant Anal.*, 43, 2752, 2012.
**Significant at the 1% probability level. Means followed by the same letter within the same column or same line (P levels) are not significantly different at the 5% probability level by Tukey's test.

TABLE 2.18
Influence of P on Thousand Grain Weight of Lowland Rice

P Rate (kg ha^{-1})	1000 Grain Weight (g)
0	23.2
131	25.1
262	26.3
393	26.5
524	26.0
655	26.6
F-test	
Year (Y)	*
P rate (P)	**
Y × P	NS
CV (%)	5

Source: Fageria, N.K. and Santos, A.B., *Commun. Soil Sci. Plant Anal.*, 39, 873, 2008b.

Values are averages of 2 years' field experimentation.

** and NS Significant at the 1% probability level and nonsignificant, respectively.

spikelet sterility. An increase in photoassimilates during spikelet-filling growth stage is a method to improve spikelet-filling rate. Of the 15% unfilled spikiest, however, about 5%–10% are unfertilized and difficult to eliminate (Yoshida, 1981). When filled spikelet number is more than 85%, yield capacity or sink is limiting yield, and when ripened spikelet number is less than 80%, assimilate supply or source is yield limiting (Murata and Matsushima, 1975).

Tanaka and Matsushima (1963) reported that the amount of carbohydrates stored in the shoot at the flowering stage improved spikelet filling by acting as a buffer substance in a case where a plant was grown under unfavorable conditions. Hayashi (1995) reported improved spikelet filling by a large amount of carbohydrate accumulated during flowering; however, cultivar differences exist in the amount of carbohydrate accumulation at the flowering stage. Furthermore, Hayashi (1995) reported that accumulation of a large amount of carbohydrates in the shoot before flowering reduces spikelet degeneration.

Percentage of ripened spikiest decreases when the number of spikiest per unit area increased (Yoshida, 1981). An appropriate number of panicles or spikiest per unit area are needed to achieve maximum yield. During the spikelet ripening, about 70% of the N absorbed by the shoot will be translocated to the spikelet to maintain N content of the spikelet at a certain level. Hence, less N absorption until flowering may reduce N level in the spikelet, which induces higher spikelet sterility (Yoshida, 1981).

Fageria et al. (2013a) studied the influence of P on spikelet sterility of upland rice genotypes (Table 2.20). Spikelet sterility was significantly influenced by genotype and P level × genotype interactions. At the low P level, spikelet sterility varied from 5.73% to 37.27%, with a mean value of 16.7%. Similarly, at the high P level, spikelet sterility varied from 5.19% to 37.4%, with a mean value of 14.27%. Overall, higher P level reduced spikelet sterility to about 1.2-fold, as compared to low P level treatment. Spikelet sterility was having a significant linear negative association with grain yield ($Y = 15.7411 - 0.1330X$, $R^2 = 0.1529$**). Rice yield is determined by yield components, which include number of panicles, spikelet per panicle, weight of 1000 spikelets, and spikelet sterility or filled spikelet (Fageria, 2007). Appropriate combination of these yield components is needed to achieve higher yields.

TABLE 2.19
Hundred Seed Weight of 30 Dry Bean Genotypes at the Two P Levels

	Hundred Seed Weight (g)	
Genotypes	**0**	**200 mg P kg⁻¹**
1. Aporé	20.79b–d	26.80bc
2. Pérola	25.34b–d	31.80bc
3. BRSMG Talisma	22.99b–d	31.54bc
4. BRS Requinte	17.57cd	22.72bc
5. BRS Pontal	22.40b–d	27.88bc
6. BRS 9435 Cometa	25.25b–d	34.78bc
7. BRS Estilo	22.23b–d	30.95bc
8. BRSMG Majestoso	25.35b–d	28.69bc
9. CNFC 10429	21.00b–d	27.47bc
10. CNFC 10408	24.31b–d	29.48bc
11. CNFC 10467	18.88cd	24.50bc
12. CNFC 10470	22.62b–d	32.83bc
13. Diamante Negro	15.27cd	21.60c
14. Corrente	21.38b–d	27.38bc
15. BRS Valente	24.42b–d	26.73bc
16. BRS Grafite	16.90cd	28.06bc
17. BRS Campeiro	20.41b–d	29.99bc
18. BRS 7762 Supermo	16.55cd	24.67bc
19. BRS Esplendor	17.20cd	22.78bc
20. CNFP 10104	18.13cd	25.97bc
21. Bambuí	23.20b–d	26.04bc
22. BRS Marfim	18.75cd	28.49bc
23. BRS Agreste	21.30b–d	24.80bc
24. BRS Pitamda	17.23cd	23.00bc
25. BRS Verede	16.27cd	24.41bc
26. EMGOPA Ouro	14.60d	23.99bc
27. BRS Radiante	30.63ab	57.63a
28. Jalo Precoce	31.06ab	36.08b
29. BRS Executivo	41.19a	61.31a
30. BRS Embaixador	26.14bc	51.27a
Average	21.98b	30.45a
F-test		
P levels (P)	**	
Genotype (G)	**	
P × G	**	
CVP (%)	21.34	
CVG (%)	14.41	

Source: Fageria, N.K. et al., *Commun. Soil Sci. Plant Anal.*, 43, 2752, 2012.
**Significant at the 1% probability level. Means followed by the same letter within the same column or same line (P levels) are not significantly different at the 5% probability level by Tukey's test.

TABLE 2.20
Spikelet Sterility (%) of 20 Upland Rice as Influenced by P Fertilization

Genotypes	25	200 mg P kg^{-1}
BRA01506	13.02cdef	8.06efg
BRA01596	16.81bcde	12.10bcdefg
BRA01600	10.79cdef	8.41efg
BRA02535	15.72cdef	20.43bcd
BRA02601	13.47cdef	11.74cdefg
BRA032033	20.42bcd	18.68bcde
BRA032039	20.29bcd	22.49b
BRA032048	16.13cdef	15.52bcdefg
BRA032051	8.24ef	9.54efg
BRA042094	10.53def	9.77efg
BRA042156	11.69cdef	8.82efg
BRA042160	37.27a	37.39a
BRA052015	5.73f	5.19g
BRA052023	12.60cdef	16.09bcdef
BRA052033	21.41bc	15.86bcdef
BRA052034	12.22cdf	7.74fg
BRA052045	27.34ab	12.19bcdefg
BRA052053	9.44ef	9.86defg
BRS Primavera	16.33cdef	14.52bcdefg
BRS Sertaneja	34.42a	21.02bc
Average	16.69	14.27
F-test		
P level (P)	NS	
Genotype (G)	**	
P × G	**	

Source: Fageria, N.K. et al., *J. Plant Nutr.*, 36, 1868, 2013a.
*,**, and NS Significant at the 5% and 1% probability level and nonsignificant, respectively. Means within the same column followed by the same letter are not significantly different at the 5% probability level by Tukey's test.

2.2.2 Physiological Functions

Physiological functions such as photosynthesis; improvement in GHI; improvement in the bioavailability of N, P, and K; improvement in grain quality; and acceleration in grain maturation are significantly affected by P fertilization in crop plants. These physiological functions are significantly and positively related to grain yield. Hence, they have special importance in crop production systems.

2.2.2.1 Essential for Photosynthesis

Photosynthesis is one of the most spectacular natural physiological phenomena in the green plants. Stoskopf (1981) stated that photosynthesis is the source of all primary energy for mankind and the very essence of agriculture, the basis of crop production. Stoskopf (1981) also reported that the air we breathe is comprised of 21% O, an element essential for life on earth. O is a direct product of photosynthesis. A hectare of land producing a crop can add 5–7 mg of O to the atmosphere, which is sufficient to supply 12 people for a year, and simultaneously removes

7–9 mg of carbon dioxide from the atmosphere (Stoskopf, 1981). Photosynthetic process in the green plants can be defined by the following equation:

$$6CO_2 + 12H_2O \text{ (sunlight energy)} \leftrightarrow C_6H_{12}O_6 + 6O_2 + 6H_2O$$

During the process of photosynthesis, carbohydrate or glucose sugar ($C_6H_{12}O_6$) is formed and O and water are released. Carbohydrate is used for the formation of dry matter or grain of food, feed, and fiber crops. Photosynthetic efficiency of plants is lower when there is deficiency of essential plant nutrients and water or plants are infested with diseases and insects. Adequate rate of P is very important for the process of photosynthesis since it determines LAI in plants (Fageria et al., 1982). Although LAI is adopted to measure the size of the photosynthetic system, it is recognized that other plant parts are capable of photosynthesis and may account for a sizable portion of total dry matter production. In addition to leaf blade, photosynthesis occurs in all green plant parts, including the stem or culm, leaf sheath, awans, glumes, and pods (Stoskopf, 1981). Growth and development of all these plant parts are also affected by mineral nutrition, including P.

2.2.2.2 Improves Grain Harvest Index

Grain harvest index (GHI) is the index that measures the portioning of carbohydrates in the shoot and grain. It can be calculated by the following equation:

$$\text{GHI} = \frac{\text{Grain yield}}{\text{Grain plus straw yield}}$$

GHI was introduced by Donald (1962), and since then it has been considered to be an important trait for yield improvement in field crops. Donald and Hamblin (1976) discussed the relationships between harvest index and yield and concluded that this was an important index for improving crop yields. Thomson et al. (1997) reported greater seed yields of faba bean with higher grain harvest indices. Morrison et al. (1999) examined the physiological differences associated with seed yield increases of soybean in Canada within groups of cultivars released from 1934 to 1992. They concluded that the increase in seed yield with year of release was significantly correlated with increases in harvest index (0.5% per year), photosynthesis, and stomatal conductance and decreases in LAI. They further concluded that present-day cultivars are more efficient at producing and allocating C resources to seeds than their predecessors.

GHI is calculated by the help of the following equation: GHI = (grain yield/(grain + straw yield)). Values for GHI in cereals and legumes are normally <1. Although GHI is a ratio, it is often expressed as a percentage. Generally, dry matter has positive associations with grain yield (Rao et al., 2002), and N is important for improving GHI. Snyder and Carlson (1984) reviewed GHI for selected annual crops and noted variations from 0.40 to 0.47 for wheat, from 0.23 to 0.50 for rice, from 0.20 to 0.47 for bunch-type peanut (*Arachis hypogaea* L.), and from 0.39 to 0.58 for dry bean. GHI values of modern crop cultivars are commonly higher than old traditional cultivars for major field crops (Ludlow and Muchlow, 1990). Cox and Cherney (2001) reported mean GHI values of 0.50 for 23 forage corn hybrids. Miller et al. (2003) reported GHI values of 0.39 for pea (*Pisum sativum* L.), 0.37 for lentil (*Lens culinaris* Medik.), 0.41 for chickpea (*Cicer arietinum* L.), 0.28 for mustard (*Brassica juncea* L.), and 0.38 for wheat grown on loamy soil. Winter and Unger (2001) reported that sorghum GHI values varied from 0.39 to 0.45 depending on the type of tillage system utilized. Rice GHI values varied among cultivars, locations, seasons, and ecosystems and ranged from 0.35 to 0.62, indicating the importance of GHI for yield simulation (Kiniry et al., 2001). Rao et al. (2002) reported GHI values of soybean (*Glycine max*) ranging from 0.37 to 0.45 with a genotypic mean of 0.43. Rao and Bhagsari (1998) reported similar ranges for GHI values for soybean grown in Georgia. Lopez-Bellido et al. (2000) reported that GHI values for wheat varied from 0.41 to 0.45

(mean value of 0.44) depending on tillage methods, crop rotation, and N rate. Limit to which harvest index can be increased is considered to be about 0.60 (Austin et al., 1980). Cultivars with low harvest indices would indicate that further improvement in partitioning of biomass would be possible. Cultivars with harvest indices between 0.50 and 0.60 would probably not benefit by increasing harvest index (Sharma and Smith, 1986).

Genetic improvement in annual crops such as wheat, barley, corn, oat, rice, and soybean has been reported due to the improvement in dry weight as well as GHI (Austin et al., 1980; Wych and Rasmusson, 1983; Wych and Stuthman, 1983; Feil, 1992; Peng et al., 2000). Peng et al. (2000) reported that genetic gain in rice cultivars released before 1980 was mainly due to the improvement in GHI, while increases in total biomass were associated with yield trends for cultivars developed after 1980. Cultivars developed after 1980 had relatively high GHI values, and further improvement in GHI was not achieved. They also reported that further increases in rice yield potential would likely occur through increasing biomass production rather than increasing GHI.

P fertilization significantly increased GHI of field crops. Fageria et al. (2012) studied the influence of P on GHI of 30 dry bean genotypes (Table 2.21). GHI varied from 0.37 to 0.54, with a mean value of 0.48 at the low P level. At the high P level, GHI varied from 0.49 to 0.81, with a mean value of 0.70. GHI had a significant linear association with grain yield ($Y = -0.70 + 9.70X$, $R^2 = 0.50^{**}$). GHI has been improved significantly in the modern crop cultivars and consequently grain yields. Sinclair (1998) reported that the GHI is an important trait associated with the dramatic increases in crop yields that have occurred in the twenty-first century.

Fageria et al. (2013a) studied the influence of P on GHI of upland rice (Table 2.22). A significant P level × genotype interaction occurred for GHI (Table 2.22), indicating differential responses of upland rice genotypes to P levels. At the low P level, GHI varied from 0.30 to 0.56, with a mean value of 0.50. At higher P level, GHI varied from 0.40 to 0.59, with a mean value of 0.52. Overall, GHI increased by 4% with the application of 200 mg P kg^{-1} of soil as compared to 25 mg P kg^{-1} of soil. Snyder and Carlson (1984) reviewed GHI for selected annual crops and noted variations from 0.23 to 0.50 for rice. GHI values of modern crop cultivars are commonly higher than old traditional cultivars for major field crops (Ludlow and Muchlow, 1990). Mae (1997) reported that the GHI of traditional rice cultivars is about 0.30 and 0.50 for improved, semidwarf cultivars.

Rice GHI values varied among cultivars, locations, seasons, and ecosystems and ranged from 0.35 to 0.62, indicating the importance of this variable for yield simulation (Kiniry et al., 2001). Amano et al. (1993) reported GHI of 0.67 with japonica F1 hybrid rice in Yunnan Province, South China. Osaki et al. (1991) reported GHI of 0.39 for standard old cultivar and 0.47 for modern high-yielding cultivar in Japan. Fageria and Baligar (2005) reported GHI across 20 upland rice genotypes, ranging from 0.43 to 0.50.

2.2.2.3 Positive Interactions with Nitrogen, Phosphorus, and Potassium Fertilization

P has positive interactions with N and K if applied in adequate rate and proper proportion (Fageria, 2014). Figure 2.18 shows the influence of N, P, and K fertilization on grain yield of upland rice. Grain yield varied from 0 to 23.01 g $plant^{-1}$ in the $N_0P_0K_0$ and $N_2P_2K_2$ treatments, respectively, with a mean value of 6.78 g $plant^{-1}$. Plants that did not receive P fertilization but received adequate rate of N and K did not produce panicle or grain. Therefore, P is a yield-limiting nutrient in highly weathered Brazilian Oxisol. Fageria and Baligar (1997a) and Fageria and Baligar (2001b) have reported similar results. Grain yield results also showed that there is a strong positive interaction among N, P, and K fertilization in upland rice production. This type of interaction is widely reported in the literature (Wilkinson et al., 2000). They reported that increasing N rates increased the demand for other nutrients, especially P and K, with higher yields obtained at the highest rates of N, P, and K. Wilson (1993) concluded that the plants' response to one nutrient depends on the sufficiency level of other nutrients. Yield reductions were reported when high levels of one nutrient were combined with low levels of the other nutrients (Wilkinson et al., 2000). Alleviating the yield-depressing effect of excessive macronutrient supply involved removing the limitation of a

TABLE 2.21
Grain Harvest Index of 30 Dry Bean Genotypes as Influenced by P Application Rates

Genotypes	0	200 mg P kg^{-1}
1. Aporé	0.49a–c	0.78a–c
2. Pérola	0.49a–c	0.68e–l
3. BRSMG Talisma	0.48a–d	0.73b–h
4. BRS Requinte	0.48a–d	0.73b–h
5. BRS Pontal	0.49a–c	0.75b–f
6. BRS 9435 Cometa	0.45a–d	0.74b–g
7. BRS Estilo	0.49a–c	0.73b–h
8. BRSMG Majestoso	0.44a–d	0.69d–l
9. CNFC 10429	0.47a–d	0.79a–c
10. CNFC 10408	0.53ab	0.81ab
11. CNFC 10467	0.45a–d	0.62m
12. CNFC 10470	0.47a–d	0.70d–j
13. Diamante Negro	0.47a–d	0.69e–l
14. Corrente	0.54a	0.83a
15. BRS Valente	0.53ab	0.62m
16. BRS Grafite	0.45a–d	0.67g–l
17. BRS Campeiro	0.40cd	0.77a–d
18. BRS 7762 Supermo	0.50a–c	0.72c–i
19. BRS Esplendor	0.49a–c	0.67f–l
20. CNFP 10104	0.43b–d	0.67f–l
21. Bambuí	0.50a–c	0.72c–i
22. BRS Marfim	0.49a–c	0.74b–g
23. BRS Agreste	0.45a–c	0.64l
24. BRS Pitamda	0.50a–c	0.65il
25. BRS Verede	0.46a–d	0.62m
26. EMGOPA Ouro	0.52ab	0.76a–e
27. BRS Radiante	0.51a–c	0.69d–l
28. Jalo Precoce	0.49a–c	0.56n
29. BRS Executivo	0.48a–d	0.66h–l
30. BRS Embaixador	0.37d	0.49n
Average	0.48b	0.70a
F-test		
P levels (P)	**	
Genotype (G)	**	
P × G	**	
CVP (%)	10.91	
CVG (%)	4.73	

Source: Fageria, N.K. et al., *Commun. Soil Sci. Plant Anal.*, 43, 2752, 2012.
**Significant at the 1% probability level. Means followed by the same letter within the same column or same line (P levels) are not significantly different at the 5% probability level by Tukey's test.

TABLE 2.22
Grain Harvest Index of 20 Upland Rice Genotypes as Influenced by P Levels

	Grain Harvest Index	
Genotypes	**Low P Level (25)**	**High P Level (200 mg kg^{-1})**
BRA01506	0.56a	0.58ab
BRA01596	0.55a	0.59a
BRA01600	0.51a	0.55abcd
BRA02535	0.41b	0.49bcde
BRA02601	0.49ab	0.49cde
BRA032033	0.50ab	0.51abcd
BRA032039	0.49ab	0.40e
BRA032048	0.57a	0.57abc
BRA032051	0.57a	0.55abcd
BRA042094	0.49ab	0.49cde
BRA042156	0.52a	0.53abcd
BRA042160	0.30c	0.40e
BRA052015	0.54a	0.53abcd
BRA052023	0.50ab	0.45de
BRA052033	0.53a	0.53abcd
BRA052034	0.52a	0.54abcd
BRA052045	0.52a	0.57abc
BRA052053	0.55a	0.55abcd
BRS Primavera	0.55a	0.53abcd
BRS Sertaneja	0.41b	0.48cde
Average	0.50	0.52
F-test		
P level (P)	NS	
Genotype (G)	**	
P × G	**	

Source: Fageria, N.K. et al., *J. Plant Nutr.*, 36, 1868, 2013a.

** and NS Significant at the 1% probability level and nonsignificant, respectively. Means within the same column followed by the same letter are not significantly different at the 5% probability level by Tukey's test.

low supply of other nutrients. Growth of upland rice at different N, P, and K levels is reported (Figures 2.19 through 2.21). Similarly, growth of upland rice roots has significant positive interaction with N, P, and K (Figure 2.22).

2.2.2.4 Improves Grain Quality

Application of P in adequate amount and proportion with other essential nutrients improves grain quality in crops. P-deficient crop plants produced shriveled poor-quality seeds (Clark, 1993). When P is present in optimum amount in the soils, plants produced bump and heavy-weight grains.

2.2.2.5 Accelerates Grain Maturation

P supply in adequate amount along with other essential nutrients such as N and K accelerates crop maturation of grains. They reported a 10- to 12-day delay in upland rice crop maturation in plots that did not receive P as compared to plots that received adequate amount of P grown on a Brazilian Oxisol.

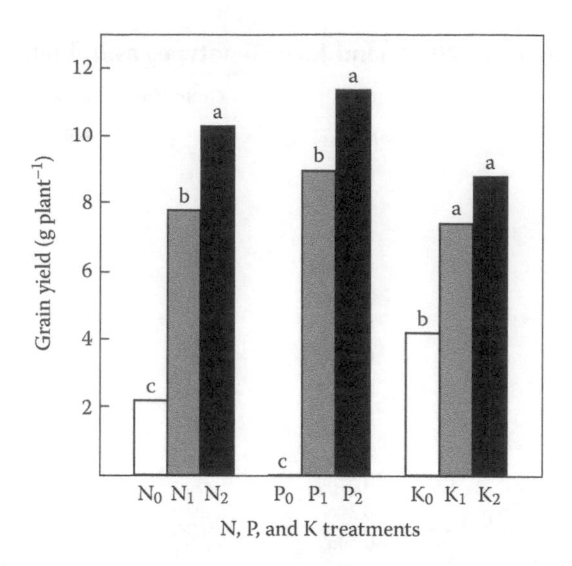

FIGURE 2.18 Grain yield of upland rice as influenced by nitrogen (N), phosphorus (P), and potassium (K) fertilization. Levels of N, P, and K were 0 (N_0), 150 (N_1), and 300 (N_2) mg kg^{-1}; 0 (P_0), 100 (P_1), and 200 (P_2) mg kg^{-1}; and 0 (K_0), 100 (K_1), and 200 (K_2) mg kg^{-1}. (From Fageria, N.K. and Oliveira, J.P., *J. Plant Nutr.*, 37, 1586, 2014.)

FIGURE 2.19 Upland rice plants without nitrogen (N) and with 200 mg phosphorus (P) and potassium (K) kg^{-1} of soil (left pot), with 150 mg N and 200 mg P and K kg^{-1} of soil (middle pot), and with 300 mg N and 200 mg P and K kg^{-1} of soil (right). (From Fageria, N.K. and Oliveira, J.P., *J. Plant Nutr.*, 37, 1586, 2014.)

FIGURE 2.20 Upland rice plants without phosphorus (P) and with 300 mg nitrogen (N) and 200 mg potassium (K) kg^{-1} of soil (left pot), with 100 mg P and with 300 mg N and 200 mg K kg^{-1} of soil (center pot), and with 200 mg P and with 300 mg N and 200 mg K kg^{-1} of soil (right pot). (From Fageria, N.K. and Oliveira, J.P., *J. Plant Nutr.*, 37, 1586, 2014.)

FIGURE 2.21 Upland rice plants without potassium (K) and with 300 mg nitrogen (N) and 200 mg phosphorus (P) kg^{-1} of soil (left pot), with 100 mg K and with 300 mg N and 200 mg K kg^{-1} of soil (center pot), and with 200 mg K and with 300 mg N and 200 mg P kg^{-1} of soil (right pot). (From Fageria, N.K. and Oliveira, J.P., *J. Plant Nutr.*, 37, 1586, 2014.)

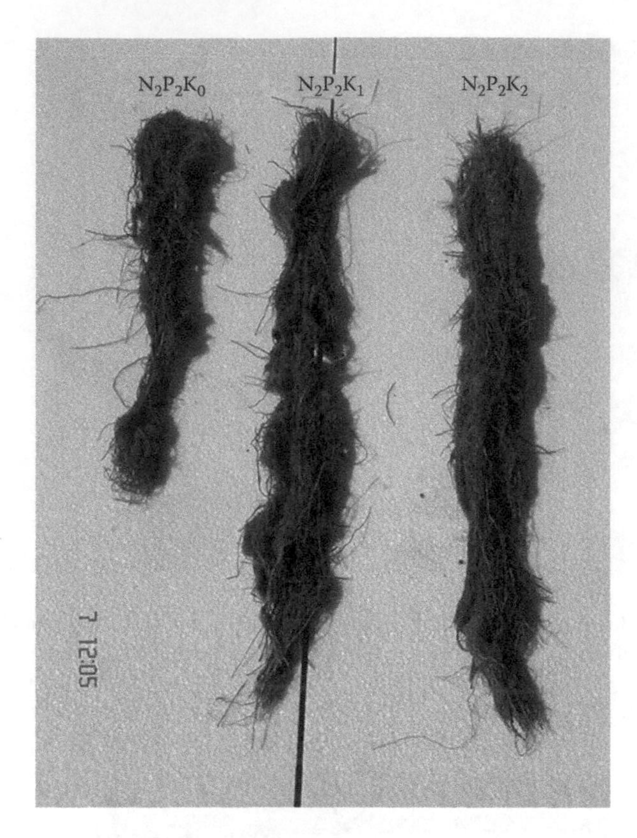

FIGURE 2.22 Root growth of upland rice at different levels of nitrogen (N), phosphorus (P), and potassium (K). Levels of N, P, and K were $N_2 = 300$ mg N kg^{-1}, $P_2 = 200$ mg P kg^{-1}, and $K_0 = 0$ mg K kg^{-1}, $K_1 = 100$ mg K kg^{-1} and $K_2 = 200$ mg K kg^{-1}. (From Fageria, N.K. and Oliveira, J.P., *J. Plant Nutr.*, 37, 1586, 2014.)

2.2.3 Biochemical Functions

Biochemical functions of P are important to the metabolic processes of the plants, which are related to growth and development. P is required for energy transfer processes in plants. In addition, P in adequate amount improves biological fixation in the legumes.

2.2.3.1 Role in Plant Metabolisms

P actively participated in the plant metabolic processes, which are responsible for growth and development. P deficiency is generally attributed to reduction of most metabolic processes, including cell division and expansion, respiration, and photosynthesis (Terry and Ulrich, 1973). Phosphate esters represent the metabolic machinery of the cells. Up to 50 individual esters formed from phosphate sugars and alcohols have been identified, about 10 of which, including glucose 6-phosphate and phosphoglyceraldehyde, are present in relatively high concentrations in cells. Most phosphate esters are part of metabolic pathways of biosynthesis and degradation (Marschner, 1995).

Regulatory function of P in photosynthesis and carbohydrate metabolism of leaves can be considered to be one of the major factors limiting growth, particularly during reproductive stage (Marschner, 1995). Level of P supplied during this period regulates the starch/sucrose ratio in the source leaves and the portioning of photosynthates between the source leaves and the reproductive organs (Giaquinta and Quebedeaux, 1980; Marschner, 1995). P is important in C metabolism in plants. A first step in C fixation is the capture of light energy by the leaf. Lauer et al. (1989) reported that leaves of soybean that were deficient or low in P had paraheliotropism or avoided sunlight by keeping the leaf edge toward the sun during daylight. Plants treated with normal P levels exhibited diaheliotropism or tracked the sun during the daylight hours (Blevins, 1994). Carboxylation efficiency decreased in low P plants as were many variables in photosynthesis (Lauer et al., 1989). In a soybean leaf, an adequate P concentration is required for the breakdown of starch and transport of triosephosphate from the chloroplast and ultimately sucrose from the cell (Giaquinta et al., 1985; Lauer et al., 1989).

2.2.3.2 Role in Energy Transfer Processes

P plays a major role in the energy transfer processes in plants. P is a component of ATP, phosphoenolpyruvate (PEP), nicotinamide adenine dinucleotide diphosphate (NADPH), and other biochemical properties that use the phosphate bond in energy utilization and storage. An important role for P in the control of ATPase activity may have an impact on many different aspects of metabolism in plants, since so many metabolites move into cells via cotransport with protons that were pumped into the apoplast by the H^+-ATPase (Blevins, 1994).

2.2.3.3 Role in Plant Structure

P plays an important role in the formation of plant structure. It is a component of some important structures like phospholipids in membranes and phosphorylated sugars and proteins and is an integral part of deoxyribonucleic acid (DNA) and ribonucleic acid (RNA). The DNA molecule is a carrier of genetic information, and RNA is the structures responsible for the translation of genetic information (Marschner, 1995). In both DNA and RNA, P forms a bridge between ribonucleoside units to form macromolecules. P is responsible for the strongly acidic nature of nucleic acids and thus for the exceptionally high cation concentration in DNA and RNA structures (Marschners, 1995). It is also a component of ATP, PEP, and NADPH, which are important enzymes in energy transfer process in plants (Blevins, 1994).

2.2.3.4 Improves Biological Nitrogen Fixation

Biological N fixation is the second most important natural phenomenon after photosynthesis in food and fiber production for mankind. It not only gives enormous economic benefits to growers but also is responsible for reducing environmental pollution. Dinitrogen (N_2) fixation is defined as the conversion of molecular N (N_2) to ammonia and subsequently to organic N utilization in biological processes (Soil Science Society of America, 2008). Total terrestrial biological N fixation quantity is reported to be 175 × 106 mg per year as compared with 77 × 106 mg per year produced by fertilizer industries (Brady and Weil, 2002). Quantity of N_2 fixed by legumes depends on species and environmental conditions. Grain legumes in symbiosis with rhizobia fix up to 450 kg N_2 ha^{-1} (Unkovich and Pate, 2000). Crop species fixing low amounts of N_2 are chickpea (0–141 kg ha^{-1}), dry bean (0–165 kg ha^{-1}), and lentil (5–191 kg ha^{-1}) (Lupwayi and Kennedy, 2007). A larger amount of N_2 is fixed by lupin (19–327 kg ha^{-1}), faba bean (12–330 kg ha^{-1}), and soybean (12–330 kg ha^{-1}) (Lupwayi and Kennedy, 2007). Soybean, a legume planted on nearly 30 million ha annually in the United States, can fulfill most of its N requirements via biological N fixation. Similarly, Brazil is the second largest soybean-producing country in the world after the United States. In Brazil farmers do not apply chemical N fertilizers in soybean. Most of the soybean N needs are made by inoculation with appropriate rhizobia.

N-fixing bacteria have relatively high requirement for P, S, Mo, and Fe, because these nutrients either are part of the nitrogenase molecule or are needed for its synthesis and use (Brady and Weil, 2002). Marschner (1995) reported that the effect of P on partitioning is also responsible, in part, for the insufficient photosynthate supply to nodulated roots of phosphorus-deficient legumes and the occurrence of N deficiency as a dominant symptom in N_2-fixing legumes receiving deficient levels of P.

Maximum benefits from N_2 fixation depend on soil P availability, with 33% of the world's arable land limited in P (Sanchez and Uehara, 1980). Acid-weathered soils of the tropics and subtropics are particularly prone to P deficiency. Even where P fertilizer is adequate, <15% of the P may be taken up by plants in the first year (Holford, 1998). Plants dependent on symbiotic N_2 fixation have ATP requirements for nodule development and function (Ribet and Drevon, 1996) and need additional P for signal transduction and membrane biosynthesis. P concentrations in the nodule are often significantly higher than those in shoot or root tissue (Israel, 1987). Al-Niemi et al. (1997) suggested that bacteroides can be P limited even when plants have received otherwise adequate P levels. Given this requirement for symbiosis, approaches leading to improved P acquisition and use in legumes need further study (Graham and Vance, 2003).

2.3 CONCLUSIONS

P is essential for growth and development of higher plants. An adequate supply of available P in soil is associated with increased root growth, which means root can explore more soil for nutrients and moisture. The roles or functions of P in crop production are divided into two principal groups: improvement in morphological traits of crop plants and participation in the physiological and biochemical processes in the plants. Important morphological traits that are affected by P nutrition are plant height, tillering, dry weight of shoot, leaf area index, root growth, source, and sink capacity. Physiological processes of plants that are affected by P nutrition are enzymes, photosynthesis, GHI, bioavailability of N and K, and grain quality. Biochemical processes that are affected by P are biological N fixation, photorespiration, and metabolic pathways. Most of these morphological, physiological, and biochemical processes are related positively to grain yield in cereals and legumes. Adequate mineral nutrition of P improves grain yield of crop plants. A deficiency of P will slow overall plant growth, delay maturity, and reduce yields.

REFERENCES

Al-Niemi, T. S., M. L. Kahn, and T. R. McDermott. 1997. Phosphorus metabolism in the bean *Rhizobium tropici* symbiosis. *Plant Physiol.* 113:1233–1242.

Amano, T., Q. Ahu, Y. Wang, N. Inoue, and H. Tanaka. 1993. Case studies on high yields of paddy rice in Jiangsu Province, china: I Characteristics of grain production. *Jpn. J. Crop Sci.* 62:267–274.

Arnon, D. I. and P. R. Stout. 1939. The essentiality of certain elements in minute quantity for plants with special reference to copper. *Plant Physiol.* 145:371–375.

Atkinson, D. 1991. Influence of root system morphology and development on the need for fertilizers and the efficiency of use. In: *Plant Roots: The Hidden Half*, eds. Y. Eshel and U. Kafkaki, pp. 411–451. New York: Marcel Dekker.

Austin, R. B., J. Bingham, R. D. Blackwell, L. T. Evans, M. A. Ford, C. L. Morgan, and M. Taylor. 1980. Genetic improvements in winter wheat yields since 1900 and associated physiological changes. *J. Agric. Sci.* 94:675–689.

Baligar, V. C. and N. K. Fageria. 2007. Agronomy and physiology of tropical cover crops. *J. Plant Nutr.* 30:1287–1339.

Baligar, V. C., N. K. Fageria, and Z. L. He. 2001. Nutrient use efficiency in plants. *Commun. Soil Sci. Plant Anal.* 32:921–950.

Baligar, V. C., N. K. Fageria, A. Q. Paiva, A. Silveira, A. W. V. Pomella, and R. C. R. Machado. 2006. Light intensity effects on growth and micronutrient uptake by tropical legume cover crops. *J. Plant Nutr.* 29:1959–1974.

Barber, S. A. 1995. *Soil Nutrient Bioavailability: A Mechanistic Approach*, 2nd edn. New York: Wiley.

Bennett, W. F. 1993. Plant nutrient utilization and diagnostic plant symptoms. In: *Nutrient Deficiencies and Toxicities in Crop Plants*, ed. W. F. Bennett, pp. 1–7. St. Paul, MN: The American Phytopathological Society, APS Press.

Black, A. L. and F. H. Siddoway. 1977. Hard red and durum spring wheat responses to seeding date and NP-fertilization on fallow. *Agro J.* 69:885–888.

Blevins, D. G. 1994. Uptake, translocation, and function of essential mineral elements in crop plants. In: *Physiology and Determination of Crop Yield*, eds. K. J. Boote, J. M. Bennett, T. R. Sinclair, and G. M. Paulsen, pp. 259–275. Madison, WI: SA, CSSA, and SSSA.

Brady, N. C. and R. R. Weil. 2002. *The Nature and Properties of Soils*, 13th edn. Upper Saddle River, NJ: Prentice Hall.

Brown, R. H. 1984. Growth of green plants. In: *Physiological Basis of Crop Growth and Development*, ed. M. B. Tesar, pp. 153–174. Madison, WI: ASA.

Bundy, L. G., H. Tunney, and A. D. Halvorson. 2005. Agronomic aspects of phosphorus management. In: *Phosphorus: Agriculture and the Environment*, ed. L. K. Al-Amoodi, pp. 685–727. Madison, WI: ASA, CSSA and SSSA.

Caradus, J. R. 1990. Mechanisms improving nutrient use by crop and herbage legumes. In: *Crops as Enhancers of Nutrient Use*, eds. V. C. Baligar and R. R. Duncan, pp. 253–311. San Diego, CA: Academic Press.

Christopher, S. F., R. Lal, and U. Mishra. 2009. Regional study of no-till effects on carbon sequestration in the Midwest United States. *Soil Sci. Soc. Am. J.* 73:207–216.

Clark, R. B. 1993. Sorghum. In: *Nutrient Deficiencies and Toxicities in Crop Plants*, ed. W. F. Bennett, pp. 21033. St. Paul, MN: APS Press, The American Phytopathology Society.

Cox, W. J. and D. J. R. Cherney. 2001. Influence of brown midrib, leafy, and transgenic hybrids on corn forage production. *Agron. J.* 93:790–796.

Dofing, S. M. and M. G. Karlsson. 1993. Growth and development of uniculm and conventional tillering barley lines. *Agron J.* 85:58–61.

Donald, C. M. 1962. In search of yield. *J. Aust. Inst. Agric. Sci.* 28:171–178.

Donald, C. M. and J. Hamblin. 1976. The biological yield and harvest index of cereals as agronomic and plant breeding criteria. *Adv. Agron.* 28:361–405.

Epstein, E. and A. J. Bloom. 2005. *Mineral Nutrition of Plants: Principles and Perspectives*. Sunderland, MA: Sinauer Associate, Inc. Publishers.

Evans, L. T. and I. F. Wardlaw. 1976. Aspects of the comparative physiology of grain yield in cereals. *Adv. Agron.* 28:301–359.

Fageria, N. K. 1989a. Effects of phosphorus on growth, yield and nutrient accumulation in the common bean. *Trop. Agric.* 66:249–255.

Fageria, N. K. 1989b. *Tropical Soils and Physiological Aspects of Crops*. Brasilia, Brazil: EMBRAPA.

Fageria, N. K. 1992. *Maximizing Crop Yields*. New York: Marcel Dekker.

Fageria, N. K. 1998. Phosphorus use efficiency by bean genotypes. *Rev. Bras. Eng. Agric. Amb.* 2:128–131.

Fageria, N. K. 2002. Nutrient management for sustainable dry bean production in the tropics. *Commun. Soil Sci. Plant Anal.* 33:1537–1575.

Fageria, N. K. 2005. Soil fertility and plant nutrition research under controlled conditions: Basic principles and methodology. *J. Plant Nutr.* 28:1975–1999.

Fageria, N. K. 2007. Yield physiology of rice. *J. Plant Nutr.* 30:843–879.

Fageria, N. K. 2009. *The Nutrient Use of Crop Plants*. Boca Raton, FL: CRC Press.

Fageria, N. K. 2013. *The Role of Plant Roots in Crop Production*. Boca Raton, FL: CRC Press.

Fageria, N. K. 2014. *Mineral Nutrition of Rice*. Boca Raton, FL: CRC Press.

Fageria, N. K. and V. C. Baligar. 1989. Response of legumes and cereals to phosphorus in solution culture. *J. Plant Nutr.* 12:1005–1019.

Fageria, N. K. and V. C. Baligar. 1996. Response of lowland rice and common bean grown in rotation to soil fertility levels on a varzea soil. *Fertilizer Res.* 45:13–20.

Fageria, N. K. and V. C. Baligar. 1997a. Response of common bean, upland rice, corn, wheat, and soybean to soil fertility of an Oxisol. *J. Plant Nutr.* 20:1279–1289.

Fageria, N. K. and V. C. Baligar. 1997b. Upland rice genotypes evaluation for phosphorus use efficiency. *J. Plant Nutr.* 20:499–509.

Fageria, N. K. and V. C. Baligar. 1999. Yield and yield components of lowland rice as influenced by timing of nitrogen fertilization. *J. Plant Nutr.* 22:23–32.

Fageria, N. K. and V. C. Baligar. 2001a. Lowland rice response to nitrogen fertilization. *Commun. Soil Sci. Plant Anal.* 32:1405–1429.

Fageria, N. K. and V. C. Baligar. 2001b. Improving nutrient use efficiency of annual crops in Brazilian acid soils for sustainable crop production. *Commun. Soil Sci. Plant Anal.* 32:1303–1319.

Fageria, N. K. and V. C. Baligar. 2005. Enhancing nitrogen use efficiency in crop plants. *Adv. Agron.* 88:97–185.

Fageria, N. K., V. C. Baligar, and B. A. Bailey. 2005. Role of cover crops in improving soil and row crop productivity. *Commun. Soil Sci. Plant Anal.* 36:2733–2757.

Fageria, N. K., V. C. Baligar, and R. B. Clark. 2002. Micronutrients in crop production. *Adv. Agron.* 77:185–268.

Fageria, N. K., V. C. Baligar, and R. B. Clark. 2006. *Physiology of Crop Production*. New York: The Haworth Press.

Fageria, N. K., V. C. Baligar, and C. A. Jones. 2011. *Growth and Mineral Nutrition of Field Crops*, 3rd edn. Boca Raton, FL: CRC Press.

Fageria, N. K., V. C. Baligar, and Y. C. Li. 2009. Differential soil acidity tolerance of tropical legume cover crops. *Commun. Soil Sci. Plant Anal.* 40:1148–1160.

Fageria, N. K., V. C. Baligar, A. Moreira, and L. A. C. Moraes. 2013c. Soil phosphorus influence on growth and nutrition of tropical legume cover crops in acid soil. *Commun. Soil Sci. Plant Anal.* 44:3340–3364.

Fageria, N. K., M. P. Barbosa Filho, and J. R. P. Carvalho. 1982. Response of upland rice to phosphorus fertilization on an Oxisol. *Agron. J.* 74:51–56.

Fageria, N. K., E. M. Castro, and V. C. Baligar. 2004. Response of upland rice genotypes to soil acidity. In: *The Red Soils of China: Their Nature, Management and Utilization*, eds. M. J. Wilson, Z. He, and X. Yand, pp. 219–237. Dordrecht, the Netherlands: Kluwer Academic Publishers.

Fageria, N. K., E. P. B. Ferreira, and A. M. Knupp. 2015. Micronutrients use efficiency in tropical cover crops as influenced by phosphorus fertilization. *Revista Caatinga* 28(1): 130–137.

Fageria, N. K., A. B. Heinemann, and R. A. Reis Jr. 2014. Comparative efficiency of phosphorus for upland rice production. *Commun Soil Sci. Plant Anal.* 45:1–22.

Fageria, N. K., L. C. Melo, J. P. Oliveira, and A. M. Coelho. 2012. Yield and yield components of dry bean genotypes as influenced by phosphorus fertilization. *Commun. Soil Sci. Plant Anal.* 43:2752–2766.

Fageria, N. K., O. P. Morais, and A. B. Santos. 2010. Nitrogen use efficiency in upland rice genotypes. *J. Plant Nutr.* 33:1696–1711.

Fageria, N. K., O. P. Morais, and M. J. Vasconcelos. 2013a. Upland rice genotypes evaluation for phosphorus use efficiency. *J. Plant Nutr.* 36:1868–1880.

Fageria, N. K. and A. Moreira. 2011. The role of mineral nutrition on root growth of crop plants. *Adv. Agron.* 110:251–331.

Fageria, N. K., A. Moreira, E. P. B. Ferreira, and A. M. Knupp. 2013b. Potassium use efficiency in upland rice genotype. *Commun. Soil Sci. Plant Anal.* 44:2656–2665.

Fageria, N. K. and J. P. Oliveira. 2014. Nitrogen, phosphorus and potassium interactions in Upland Rice. *J. Plant Nutr.* 37:1586–1600.

Fageria, N. K. and A. B. Santos. 1998. Phosphorus fertilization for bean crop in lowland soil. *Rev. Bras. Eng. Agric. Amb.* 2:124–127.

Fageria, N. K. and A. B. Santos. 2008a. Yield physiology of dry bean. *J. Plant Nutr.* 31:983–1004.

Fageria, N. K. and A. B. Santos. 2008b. Lowland rice response to thermophosphate fertilization. *Commun. Soil Sci. Plant Anal.* 39:873–889.

Fageria, N. K., A. B. Santos, and V. C. Baligar. 1997. Phosphorus soil test calibration for lowland rice on an Inceptisol. *Agron. J.* 89:737–742.

Fageria, N. K., N. A. Slaton, and V. C. Baligar. 2003. Nutrient management for improving lowland rice productivity and sustainability. *Adv. Agron.* 80:63–152.

Feil, B. 1992. Breeding progress in small grain cereals: A comparison of old and modern cultivars. *Plant Breed.* 108:1–11.

Gahoonia, T. S. and N. E. Nielsen. 2004. Root traits as tools for creating phosphorus efficient crop varieties. *Plant Soil* 260:47–57.

Giaquinta, R. T. and B. Quebedeaux. 1980. Phosphate-induced changes in assimilate partitioning in soybean leaves during pod filling. *Plant Physiol.* 65(Suppl):119.

Giaquinta, R. T., B. Quebedeaux, N. L. Sadler, and V. R. Francheschi. 1985. Assimilate partitioning in soybean leaves during seed filling. In: *World Soybean Research Conference*, 3rd edn., August 12–17, 1984, ed. R. Shibles, pp. 729–738. Ames, IA: Westview Press.

Grabau, L. J., D. G. Blevins, and H. C. Minor. 1986. Phosphorus nutrition during seed development: Leaf senescence, pod retention and seed weight of soybean. *Plant Physiol.* 82:1013–1018.

Graham, P. H. and C. P. Vance. 2003. Legumes: Importance and constraints to greater use. *Plant Physiol.* 131:872–877.

Grant, C. A., D. N. Flaten, D. J. Tomasiewicz, and S. C. Sheppard. 2001. The importance of early season phosphorus nutrition. *Can. J. Plant Sci.* 81:211–224.

Gravois, K. A. and R. S. Helms. 1992. Path analysis of rice yield and yield components as affected by seeding rate. *Agron. J.* 88:1–4.

Handa, K. 1995. Differentiation and development of tiller buds. In: *Science of the Rice Plant: Physiology*, Vol. 2, eds. T. Matsuo, K. Kumazawa, R. Ishii, K. Ishihara, and H. Hirata, pp. 61–65. Tokyo, Japan: Food and Agriculture Policy Research Center.

Hayashi, H. 1995. Translocation, storage and partitioning of photosynthetic products. In: *Science of the Rice Plant: Physiology*, Vol. 2, eds. T. Matsuo, K. Kumazawa, R. Ishii, K. Ishihara, and H. Hirata, pp. 546–5655. Tokyo, Japan: Food and Agriculture Policy Research Center.

Hinsinger, P., A. G. Bengough, D. Vetterlein, and I. M. Young. 2009. Rhizosphere: Biophysics, biogeochemistry and ecological relevance. *Plant Soil* 321:117–152.

Holford, I. C. R. 1998. Soil phosphorus: Its measurement and uptake by plants. *Aust. J. Soil Sci. Res.* 35:227–[illegible]

Israel, D. W. 1987. Investigation of the role of phosphorus in symbiotic dinitrogen fixation. *Plant Physiol.* 84:835–840.

Jaleel, C. A., P. Manivannan, A. Wahid, M. Farooq, H. J. Al-Juburi, R. Somasundarama, and R. P. Vam. 2009. Drought stress in plants: A review on morphological characteristics and pigments composition. *Int. J. Agric. Biol.* 11:100–105.

Jennings, P. R., W. R. Coffman, and H. E. Kauffman. 1979. *Rice Improvement.* Los Bãnos, Philippines: International Rice Research Institute.

Kiniry, J. R., G. McCauley, Y. Xie, and J. G. Arnold. 2001. Rice parameters describing crop performance of four U S. cultivars. *Agron. J.* 93:1354–1361.

Lauer, M. J., S. G. Pallardy, D. G. Blevins, and D. D. Randall. 1989. Whole leaf carbon exchange characteristics of phosphate deficient soybeans. *Plant Physiol.* 91:848–854.

Liao, M., I. R. P. Filley, and J. A. Palta. 2004. Early vigorous growth is a major factor influencing nitrogen uptake in wheat. *Funct. Plant Biol.* 31:121–129.

Lopez-Bellido, L., R. J. Lopez-Bellido, J. E. Castillo, and F. J. Lopez-Bellido. 2000. Effects of tillage, crop rotation, and nitrogen fertilization on wheat under rainfed mediterranean conditions. *Agron. J.* 92:1054–1063.

Ludlow, M. M. and R. C. Muchlow. 1990. A critical evaluation of traits for improving crop yields in water-limited environments. *Adv. Agron.* 43:107–153.

Lupwayi, N. Z. and A. C. Kennedy. 2007. Grain legumes in northern Great Plains: Impacts on selected biological processes. *Agron. J.* 99:1700–1709.

Mae, T. 1997. Physiological nitrogen efficiency in rice: Nitrogen utilization, photosynthesis, and yield potential. *Plant Soil* 196:201–210.

Marschner, H. 1995. *Mineral Nutrition of Higher Plants*, 2nd edn. New York: Academic Press.

Masle, J. 1985. Competition among tillers in winter wheat: Consequences for growth and development of the crop. In: *Wheat Growth and Modeling*, eds. W. Day and R. K. Atkin, pp. 33–54. New York: Plenum Press.

Mengel, K., E. A. Kirkby, H. Kosegarten, and T. Appel. 2001. *Principles of Plant Nutrition*, 5th edn. Dordrecht, the Netherlands: Kluwer Academic Publishers.

Miller, P. R., Y. Gan, B. G. McConkey, and C. L. McDonald. 2003. Pulse crops for the northern Great Plains; I. Grain productivity and residual effects on soil water and nitrogen. *Agron. J.* 95:972–979.

Morrison, M. J., H. D. Voldeng, and E. R. Cober. 1999. Physiological changes from 58 years of genetic improvement of short-season soybean cultivars in Canada. *Agron. J.* 91:685–689.

Muchow, R. C. 1985. Canopy development in grain legumes grown under different soil water regimes in a semi-arid environment. *Field Crops Res.* 11:99–109.

Murata, Y. and S. Matsushima. 1975. Rice. In: *Crop Physiology: Some Case Histories*, ed. L. T. Evans, pp. 73–99. London, U.K.: Cambridge University Press.

Murayama, N. 1995. Development and senescence of an individual plant. In: *Science of the Rice Plant: Physiology*, Vol. 2, eds. T. Matsuo, K. Kumazawa, R. Ishii, K. Ishihara, and H. Hirata, pp. 119–178. Tokyo, Japan: Food and Agriculture Policy Research Center.

Narayanan, S., R. M. Aiken, P. V. V. Prasda, Z. Xin, G. Paul, and J. Yu. 2014. A simple quantitative model to predict leaf area index. *Agron. J.* 106:219–226.

Nerson, H. 1980. Effects of population density and number of ears on wheat yield and its components. *Field Crops Res.* 3:225–234.

Noulas, C., M. Liedgens, P. Stamp, I. Alexiou, and J. M. Herrera. 2010. Subsoil root growth of field grown spring wheat genotypes (*Triticum aestivum* L.) differing in nitrogen use efficiency. *J. Plant Nutr.* 33:1887–1903.

Osaki, M., K. Morikawa, M. Yoshida, T. Shinano, and T. Tadano. 1991. Productivity of high-yielding crops. I. Comparison of growth and productivity among high-yielding crops. *Soil Sci. Plant Nutr.* 37:331–339.

O'Toole, J. C. and W. L. Bland. 1987. Genotypic variation in crop plant root systems. *Adv. Agron.* 41:91–145.

Ottis, B. V. and R. E. Talbert. 2005. Rice yield components as affected by cultivar and seedling rate. *Agron. J.* 97:1622–1625.

Peet, M. M., A. Bravo, D. H. Wallace, and J. L. Ozbun. 1977. Photosynthesis, stomatal resistance, and enzyme activities in relation to yield of field grown dry bean varieties. *Crop Sci.* 17:287–293.

Peng, S., R. C. Laza, R. M. Visperas, A. L. Sanico, K. G. Cassman, and G. S. Khush. 2000. Grain yield of rice cultivars and lines developed in the Philippines since 1966. *Crop Sci.* 40:307–314.

Power, J. F. and J. Alessi. 1978. Tiller development and yield of standard and semi-dwarf spring wheat varieties as affected by nitrogen fertilizer. *J. Agric. Sci.* 90:97–108.

Rajcan, I. and M. Tollenaar. 1999. Source: Sink ratio and leaf senescence in maize: I. Dry matter accumulation and partitioning during grain filling. *Field Crops Res.* 60:245–253.

Rao, M. S. S. and A. S. Bhagsari. 1998. Variation between and within maturity groups of soybean genotypes for biomass, seed yield, and harvest index. *Soybean Genet. Newsl.* 25:103–106.

Rao, M. S. S., B. G. Mullinix, M. Rangappa, E. Cebert, A. S. Bhagsari, V. T. Sapra, J. M. Joshi, and R. B. Dadson. 2002. Genotype X environment interactions and yield stability of food-grade soybean genotypes. *Agron. J.* 94:72–80.

Ribet, J. J. and J. Drevon. 1996. The phosphorus requirement of N_2 fixing and urea-fed *Acacia mangium*. *New Phytol.* 132:383–390.

Rowden, R., D. Gardiner, P. C. Whiteman, and E. S. Wallis. 1981. Effects of planting density on growth, light interception and yield of a photoperiod insensitive *Cajanus cajan*. *Field Crops Res.* 4:201–213.

Sanchez, P. A. and G. Uehara. 1980. Management considerations for acid soils with high phosphorus fixation capacity. In: *The Role of Phosphorus in Agriculture*, ed. R. C. Dinauer, pp. 471–515. Madison, WI: ASA, CSSA, and SSSA.

Sharma, R. C. and E. L. Smith. 1986. Selection for high and low harvest index in three winter wheat populations. *Crop Sci.* 26:1147–1150.

Sinclair, T. R. 1998. Historical changes in harvest index and crop nitrogen accumulation. *Crop Sci.* 38:638–643.

Snyder, F. W. and G. E. Carlson. 1984. Selecting for partitioning of photosynthetic products in crops. *Adv. Agron.* 37:47–72.

Soil Science Society of America. 2008. *Glossary of Soil Science Terms*. Madison, WI: Soil Science Society of America.

Stoskopf, N. C. 1981. *Understanding of Crop Production*. Reston, VA: Reston Publishing Company, Inc.

Tanaka, T. and S. Matsushima. 1963. Analysis of yield-determining process and the application to yield prediction and culture improvement of lowland rice. *Proc. Crop Sci. Soc. Jpn.* 32:35–38.

Tang, C., J. J. Drevon, B. Jaillard, G. Souche, and P. Hinsinger. 2004. Proton release of two genotypes of bean (*Phaseolus vulgaris* L.) as affected by N nutrition and P deficiency. *Plant Soil* 26:59–68.

Terry, N. and A. Ulrich. 1973. Effects of phosphorus deficiency on the photosynthesis and respiration of leaves in sugar beet. *Plant Physiol.* 51:43–47.

Thomson, B. D., K. H. M. Siddique, M. D. Barr, and J. M. Wilson. 1997. Grain legume species in low rainfall Mediterranean-type environments: I. Phenology and seed yield. *Field Crops Res.* 54:173–187.

Tsialtas, J. T. and N. Maslaris. 2008. Evaluation of leaf area prediction model proposed for sunflower. *Photosynthetica* 46:294–297.

Unkovich, M. J. and J. S. Pate. 2000. An appraisal of recent field measurements of symbiotic N_2 fixation by annual legumes. *Field Crops Res.* 65:211–228.

Wallace, D. H., J. L. Ozbum, and H. M. Munger. 1972. Physiological genetics of crop yield. *Adv. Agron.* 24:97–146.

Watson, D. J. 1958. The dependence of net assimilation rate on leaf area index. *Ann. Bot.* 22:37–54.

Wells, J. M. and J. M. Norman. 1991. Instrument for indirect measurement of canopy architecture. *Agron. J.* 83:818–825.

Wilkinson, S. R., D. L. Grunes, and M. E. Sumner. 2000. Nutrient interactions in soil and plant nutrition. In: *Handbook of Soil Science*, ed. M. E. Sumner, pp. 89–111. Boca Raton, FL: CRC Press.

Wilson, J. B. 1993. Macronutrient (NPK) toxicity and interactions in the grass *Festuca ovina*. *J. Plant Nutr.* 16:1151–1159.

Winter, S. R. and P. W. Unger. 2001. Irrigated wheat grazing and tillage effects on subsequent dryland grain sorghum production. *Agron. J.* 93:504–510.

Wych, R. D. and D. C. Rasmusson. 1983. Genetic improvement in malting barley cultivars since 1920. *Crop Sci.* 23:1037–1040.

Wych, R. D. and D. D. Stuthman. 1983. Genetic improvement in Minnesota adapted oat cultivars since 1923. *Crop Sci.* 23:879–881.

Yin, X., A. H. C. M. Schapendonk, M. J. Kropff, M. V. Oijen, and P. S. Bindrabn. 2000. A genetic equation for nitrogen-limited leaf area index and its application in crop growth models for predicting leaf senescence. *Ann. Bot.* 85:579–585.

Yoshida, S. 1972. Physiological aspects of grain yield. *Annu. Rev. Plant Physiol.* 23:437–464.

Yoshida, S. 1981. *Fundamentals of Rice Crop Science*. Los Bãnos, Philippines: IRRI.

3 Diagnostic Techniques for Phosphorus Requirements in Crop Plants

3.1 INTRODUCTION

Modern high-input and high-output agricultural production systems have significantly increased food supply worldwide, but are associated with high resource consumption, especially water and nutrients (FAO, 2013; Cui et al., 2014). China is a typical example of high fertilizer consuming countries. In the 2011 agricultural season, China consumed 50.5 million mt of N + P_2O_5 + K_2O, which accounted for 30% of the global consumption of essential plant nutrients (Cui et al., 2014). China's food demand is projected to increase by 30%–50% by 2030 (Zhang et al., 2011), due to population growth, rising income, and changing dietary habits. Similar trend in resin extraction used to determine soil test ds in food consumption were observed in other developing countries like India, South Africa, Brazil, and Indonesia. Future world food production should increase continuously because population growth will continue through 2050, with unprecedented rates in urbanization. Consequently, several problems will occur, including (1) limited cultivable land, (2) declining soil fertility, (3) depleted forests, and (4) fragile economies (Ringius, 2002; Snapp et al., 2002, FAO, 2011; Kadyampaken, 2014).

Nutrient deficiency or toxicity in crop plants is related to the morphology and chemical composition of the soil, which often represents a serious constraint for crop production and land development (Dudal, 1976). Nutrient deficiency or toxicity has attracted the attention of crop production scientists for many years. Approximately one-fourth of the world's soils are considered to have some kind of mineral stresses (Dudal, 1976). This estimation does not include the effects of pollution from human activities on the environment in which plants grow, nor does it include the depletion effects of intensive agriculture on the mineral element reserves in the soil (Hale and Orcutt, 1987). Deficiency of an element results from a number of conditions, including concentration, forms of the element, soil biogeochemical processes in which it becomes available to the plants, content of soil moisture, and soil pH (Hale and Orcutt, 1987). Soil conditions causing nutrient deficiencies in crop plants are presented in Table 3.1.

In modern agriculture, maximizing and sustaining crop yields are the main objectives. One of the major problems constraining the development of an economically successful agriculture is nutrient deficiency for crop production. As much as 50% of the increase in crop yields worldwide during the twentieth century was due to the adoption of chemical fertilizers (Fageria and Baligar, 2005). Soil infertility (natural element deficiencies or unavailability) is probably the single most important factor limiting crop yields worldwide. Nutrient diagnostic technique (sufficiency or deficiency) is important for crop production. In addition, nutritional disorders (deficiency/toxicity) limit crop yields in all types of soils around the world. For example, in Brazilian Oxisols and Inceptisols, deficiency of N, P, K, and Zn limited yield of almost all field crops (Fageria, 2009, 2013, 2014, 2015). The importance of P application in increasing yield of lowland rice (*Oryza glaberrima*) genotypes grown on a Brazilian Inceptisol is presented in Figure 3.1. Yield of lowland rice increased with increasing levels of P until an adequate level in the soil was achieved (Figure 3.1). Similarly, a significant increase in grain yield, shoot dry weight, number of pods, and hundred grain weight occurred in dry bean (*Phaseolus vulgaris* L.) grown on Brazilian Oxisol (Figures 3.2 and 3.3).

TABLE 3.1
Soil Conditions That Induce Nutrient Deficiencies in Crop Plants

Nutrient	Conditions Inducing Deficiency[a]
Nitrogen	Excess leaching with heavy rainfall, low organic matter content of soils, burning of crop residues.
Phosphorus	Acidic, organic, leached, and calcareous soils, high rate of liming.
Potassium	Sandy, organic, leached, and eroded soils; high liming application, intensive cropping system.
Calcium	Acidic, alkali, or sodic soils.
Magnesium	Similar to calcium.
Sulfur	Low organic matter content of soils; use of N and P fertilizers containing no sulfur, burning of crop residues.
Iron	Calcareous soils; soils high in P, Mn, Cu, or Zn; high rate of liming.
Zinc	Highly leached acidic soils, calcareous soils, high levels of Ca, Mg, and P in the soils.
Manganese	Calcareous silt and clay, high organic matter, calcareous soils.
Copper	Sandy soils, high liming rate in acid soils
Boron	Sandy soils, naturally acidic leached soils, alkaline soils with free lime.
Mo	Highly podzolized soils; well-drained calcareous soils.

[a] Material is collected from various references and also based on the authors' professional experiences.

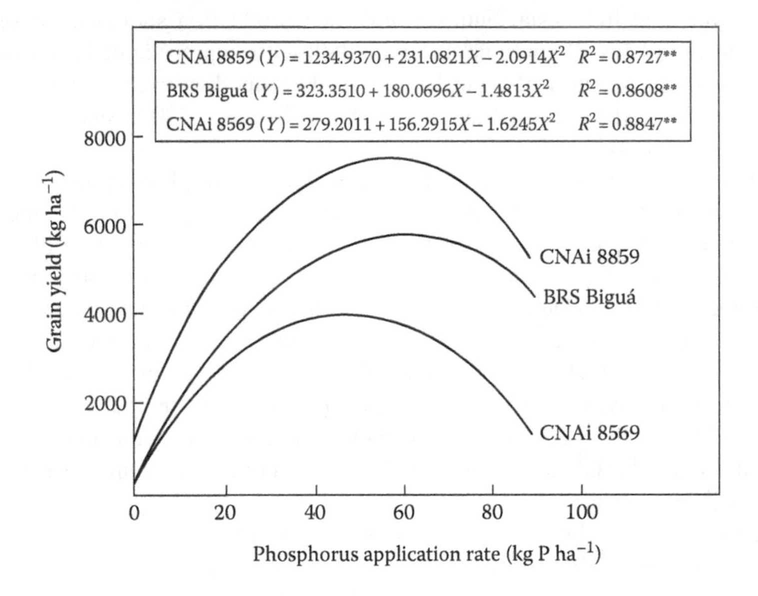

FIGURE 3.1 Response of lowland rice genotypes to P fertilization. (From Fageria, N.K. et al., *J. Plant Nutr.*, 31, 1121, 2008a.)

To obtain optimum yields of crops, nutrient disorders must be overcome. The first step in this direction is to identify the nutritional disorder.

Concepts of mineral nutrient requirements and the methods used to determine their deficiency or sufficiency were studied extensively in the twentieth century. However, there are many unanswered questions (Hale and Orcutt, 1987). Technologies that harness more food production per unit of natural resource would be worth evaluating. In this context, identifying nutrient deficiency and taking necessary measures to correct it are fundamental to achieve food security around the world. The objective of this chapter is to discuss P deficiency or toxicity diagnostic techniques in crop plants. This information is needed for taking corrective measures to overcome P deficiency and maximizing crop yields.

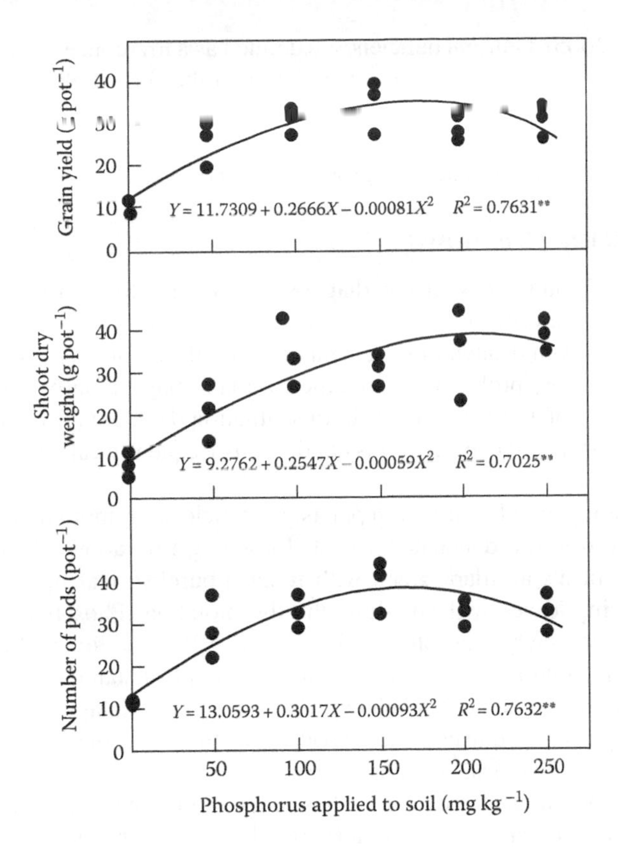

FIGURE 3.2 Dry bean grain yield, shoot dry weight and number of pods as affected by P fertilization. (From Fageria, N.K. and Baligar, V.C., *J. Plant Nutr.*, 2016, DOI: 10.1080/01904167.2016.1143489.)

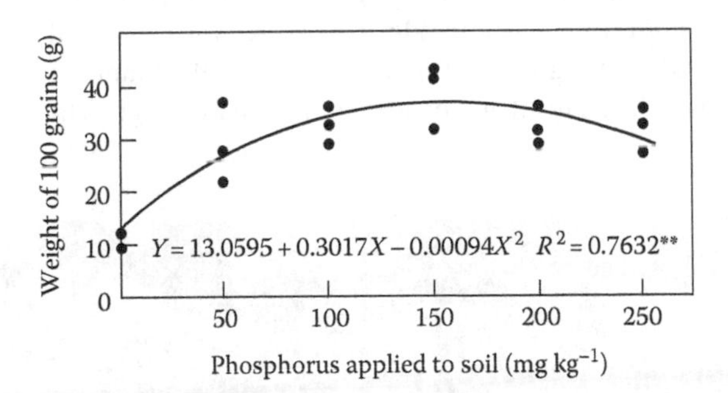

FIGURE 3.3 Influence of P rate on hundred grain weight of dry bean. (From Fageria, N.K. and Baligar, V.C., *J. Plant Nutr.*, 2016, DOI: 10.1080/01904167.2016.1143489.)

3.2 PHOSPHORUS DEFICIENCY, SUFFICIENCY, AND TOXICITY DIAGNOSTIC TECHNIQUES

Nutrient diagnostic technique in crop plants refers to identifying whether a given nutrient is deficient, sufficient, or in excess in the soil and plant tissues. Nutrient deficiency and/or sufficiency diagnostic technique is important to achieve optimum crop production. Since in modern agriculture sustaining desired crop yields is the main objective, supply of adequate amount of plant nutrients is essential to achieve these objectives. Furthermore, one of the major constraints in the development of an economically successful agriculture is nutrient deficiency for crop production

(Fageria and Baligar, 2005). Nutrient deficiency is defined as a low concentration of an essential element that reduces plant growth and prevents completion of the normal plant life cycle (Soil Science Society of America, 2008). Four diagnostic techniques are common in the field of mineral nutrition which can be used to identify nutritional disorders in crop plants: visual symptoms of deficiency or toxicity, soil test, plant tissue test, and crop growth response.

3.2.1 Visual Deficiency Symptoms

Visual symptoms are the most inexpensive diagnostic technique for nutritional disorders, as compared to other methods. However, it requires experience on the part of the observer, since deficiency symptoms are confused with drought, insects and disease infestation, herbicide damage, soil salinity, and inadequate drainage problems. Sometimes, a plant may be on borderline with respect to deficiency and adequacy of a given nutrient. In this situation there are no visual symptoms, but the plant is not producing at its full capacity. This condition is frequently called *hidden hunger* (Fageria and Baligar, 2005).

Phosphorus is a highly mobile nutrient in plants. Its deficiency symptoms start with orange color of the older leaves. However, if deficiency persists for a longer duration, whole plants are affected. Phosphorus-deficient plants are dark green with reddish-purplish leaf tips and margins on older leaves. Newly emerging leaves will not show the discoloration. Phosphorus-deficient plants are smaller and grow more slowly than plants with adequate P. In addition, P deficiency in cereals reduces plant height and tillering. This may appear as a thin plant stand. In legumes, the number of branches is reduced if there is acute P deficiency. Phosphorus also delays maturation and blooming. Phosphorus deficiency is favored in compacted soils, soils that are cool and wet, and plant roots that are injured with insects, herbicides, and other stresses.

Authors of this book conducted several studies related to P nutrition of crop plants. Growth of three cover crops was significantly reduced at 0 mg P kg^{-1} as compared to 100 and 200 mg P kg^{-1} soil (Figures 3.4 through 3.6). However, response of cover crops to soil P levels were different. Lowland rice to P fertilization at two growth stages, that is, at vegetative growth stage (left) and at physiological maturity (right), is presented in Figure 3.7. Rice plants did not produce tillering as well as panicles at the 0 mg P kg^{-1} level. In addition, there was growth of four dry bean genotypes at two P levels (i.e., 0 and 200 mg kg^{-1}) (Figures 3.8 and 3.9). Plants at 0 mg P kg^{-1} level had P deficiency symptoms as yellowing of lower leaves and reduced growth as compared to 200 mg P kg^{-1} soil level.

FIGURE 3.4 Growth of *Crotalaria spectabilis* cover crops at three levels of soil P.

FIGURE 3.5 Growth of *Crotalaria ochroleuca* cover crops at three levels of soil P.

FIGURE 3.6 Growth of *Cajanus cajan* cover crops at three levels of soil P.

Phosphorus deficiency reduces yield and yield components in crop plants. Fageria et al. (2010) studied response of dry bean genotypes to P fertilization (Table 3.2). Grain yield and straw yield of 20 dry bean genotypes were significantly influenced by P and genotype treatments (Table 3.2). Genotype × P interactions were also significant for grain and straw yield. Significant interaction was observed between soil P levels × genotype for grain yield and shoot dry weight, indicating differential response of genotypes to soil P levels (Table 3.2). Grain yield of 20 genotypes varied from 0.63 g plant^{-1} produced by genotype CNFP 10103 to 3.70 g plant^{-1} produced by genotype CNFP 8000, with a mean grain yield of 1.92 g plant^{-1} at the 25 mg P kg^{-1} of soil level. Similarly, grain yield at the higher P level (200 mg P kg^{-1}) varied from 3.67 g plant^{-1} produced by genotype CNFP 10120 to 10.18 g plant^{-1} produced by genotype CNFP 10104, with a mean grain yield of 7.81 g plant^{-1}. Overall, the increase in grain yield was 307% with an increase in soil applied P from 25 to 200 mg P kg^{-1} soil. Fageria et al. (2011a) and Fageria and Baligar (1997) reported an increase of grain yield in dry bean with the addition of P in Brazilian Oxisols. These authors reported that response of annual

FIGURE 3.7 Growth of lowland rice at two P levels and at two growth stages. At left, vegetative growth stage and at right, physiological maturity growth stage.

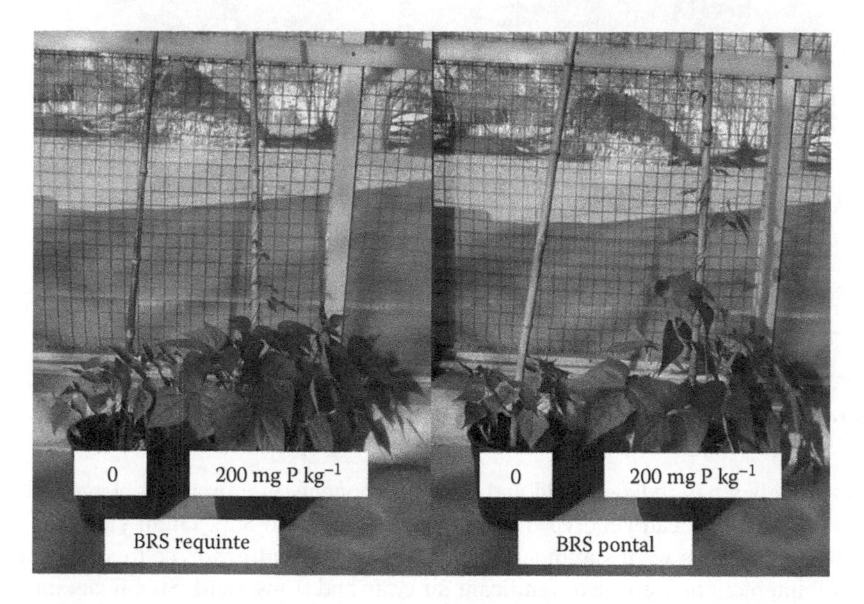

FIGURE 3.8 Growth response of dry bean genotypes BRS requinte and BRS pontal to low and high levels of soil P.

crops to addition of P in Oxisols was associated with low natural level of soil P and higher P immobilization capacity of these soils. High P immobilization capacity of such soils is related to clay, Fe, and Al contents, and applied P is precipitated as Ca, Fe, and Al phosphates, also chemically bonded to these cations at the surfaces of the soil minerals and adsorption on to clay (Mokwunye and Chien, 1980; Van Riemsdijk et al., 1984). Higgs et al. (2000) reported that a 30%–50% increase in world food grain production since the 1950s was attributable to fertilizer use, including P.

FIGURE 3.9 Growth response of dry bean genotypes BRS 9435 cometa and BRS estilo to low and high levels of soil P.

Fageria et al. (2010) studied influence of P on shoot dry weight of 20 dry bean genotypes (Table 3.2). At the low P level (25 mg P kg^{-1}), shoot dry weight of 20 genotypes varied from 0.33 g $plant^{-1}$ produced by genotype CNFC 10438 to 2.49 g $plant^{-1}$ produced by genotype CNFC 10444, with a mean yield of 1.62 g $plant^{-1}$. At the higher P level (200 mg P kg^{-1}), shoot dry weight varied from 2.96 g $plant^{-1}$ produced by genotype CNFC 10467 to 10.79 g $plant^{-1}$ produced by genotype CNFP 10120, with a mean shoot dry weight of 5.70 g $plant^{-1}$. Overall increase in shoot dry weight was 252% at the higher soil P level as compared to the low soil P level. Fageria et al. (2011a) and Fageria and Baligar (1997) also reported significant increases in dry bean shoot dry weight with increasing soil P levels. Similarly, Fageria (1989) reported significant quadratic increases in shoot dry weight of dry bean when soil applied P rates were increased from 0 to 200 mg P kg^{-1} in a Brazilian Oxisol. Significant differences in shoot dry weight of dry bean genotypes grown on Brazilian Oxisol have been reported (Fageria, 1998).

Yield components of annual crops were also reduced by P deficiency. Soil P level and genotype significantly affected pods per plant, seeds per pod, and 100 grain weight (Table 3.3). Interactions for soil P with genotypes had no significant effects on pods per plant and seeds per pod, indicating that the magnitude of response for these two yield components did not vary with the variation in soil P levels. However, significant soil P × genotype interaction occurred for 100 grain weight; therefore, values of this yield component are presented at the two soil P levels (Table 3.3). Pods per plant varied from 2.79 produced by genotype CNFC10438 to 5.75 produced by genotype CNFP 8000. Mean values of pods per plant were 4.05 for 20 genotypes. The increase in pods per plant was twofold between highest and lowest pod-producing genotypes. Seeds per pod varied from 3.24 (produced by genotype CNFP 10120) to 4.67 (produced by genotype CNFP 8000), with a mean value of 3.99 seeds per pod for 20 genotypes. Overall genotype CNFP 8000 produced highest pods per plant and highest seed per plant. Fageria et al. (2006) stated that pods per plant and seeds per pods in legumes are genetically controlled but differences existed among genotypes. Wallace et al. (1972) also reported that genotypes differed significantly in the physiological processes that determine yield. They reported that identification of these physiological components of yield and their genetic controls should make it possible to plan crosses to maximize segregation of genotypes possessing the physiological complementation and balance required for high yield, thereby leading to more rapid and predictable yield improvement.

At the low soil P level, the 100 grain weight varied from 19.6 g (produced by genotype CNFP 10093) to 34.9 g (produced by genotype CNFC 10410), with a mean value of 26.9 g. At the high soil P level, 100 grain weight varied from 20.7 g (produced by genotype CNFC 10438) to 33.5 g (produced by genotype CNFC 9461), with a mean value of 27.9 g. Such variations in 100 grain weight in dry bean genotypes have been reported (Fageria, 2009). Increase in 100 grain weight at higher soil P was about 4% compared to that at the low soil P level.

TABLE 3.2
Grain Yield and Shoot Dry Weight of 20 Dry Bean Genotypes at Two Levels of Soil P

Genotype	Grain Yield (g plant^{-1})		Shoot Dry Weight (g Plant^{-1})	
	25	200	25	200 mg P kg^{-1}
CNFC 10467	2.69abc	6.70abc	2.31ab	2.96c
CNFP 8000	3.70a	8.29ab	2.36ab	4.01bc
CNFC 10455	1.29abc	6.43abc	2.14ab	6.61abc
CNFP 10035	3.12ab	9.49a	1.62abc	5.75bc
CNFC 10410	3.11ab	9.11a	1.34abc	5.29abc
CNFP 10076	2.38abc	7.53abc	1.98ab	6.73abc
CNFC 10432	1.64abc	8.62ab	1.31abc	4.37bc
CNFP 10093	1.17bc	8.61ab	0.91bc	5.28bc
CNFC 10408	1.83abc	7.47abc	1.45abc	6.21abc
CNFP 10103	0.63c	7.89abc	1.83abc	5.96bc
CNFC 9461	2.05abc	9.23a	1.40abc	5.03bc
CNFP 10104	2.74abc	10.18a	1.89abc	6.78abc
CNFC 10429	1.37abc	9.92a	1.61abc	4.51bc
CNFP 10109	2.18abc	8.94a	1.60abc	4.97bc
CNFC 10431	1.79abc	7.74abc	1.98ab	7.10abc
CNFP 10120	1.34abc	3.67c	1.58abc	10.79a
CNFC 10438	0.89bc	4.40bc	0.33c	3.12c
CNFP 10206	1.32abc	9.03a	0.92bc	5.29bc
CNFC 10444	0.92bc	6.97abc	2.49a	5.01bc
CNFC 10470	2.22abc	5.89abc	1.36abc	8.24ab
Average	1.92	7.81	1.62	5.70
F-test				
P level (P)	**		**	
Genotype (G)	**		**	
P × G	**		**	

Source: Fageria, N.K. et al., *J. Plant Nutr.*, 33, 2167, 2010.
**Significant at the 1% probability level. Means within the same column followed by the same letter are not significantly different at the 5% probability level by Tukey's test.

Grain harvest index (GHI) was significantly influenced by soil P levels, genotype, and soil P × genotypes interaction (Table 3.4). At the low P level, GHI values varied from 0.26 for genotype CNFP 10103 to 0.73 for genotype CNFC 10438, with a mean value of 0.53. At the high soil P level, GHI values varied from 0.26 for genotype CNFP 10120 to 0.70 for genotype CNFC 10467, with a mean value of 0.58. Overall, the increase in GHI was about 9% at the high soil P level, as compared to that at the low soil P level. GHI varies with crop genotypes but was also influenced by environmental factors (Wallace et al., 1972).

Phosphorus deficiency also reduced root growth of crop plants. Influence of P on the root growth of corn (*Zea mays*), lowland rice, and upland rice is presented in Figures 3.10 through 3.13. Root growth of three crops reduced drastically at 0 mg P kg^{-1} as compared to the higher P level. Fageria et al. (2010) studied the influence of P on root growth of six dry bean genotypes (Table 3.5). A significant effect of P level and genotypes on root growth occurred, and soil P × genotype interaction was significant for root dry weight (Table 3.5). Significant P × G interaction clearly indicates that genotypes produced different root dry weight at the two soil P levels. Root dry weight varied from 0.49 to 0.79 g plant^{-1} at the low soil P level and 1.01 to 2.26 g plant^{-1} at the high soil P level.

TABLE 3.3
Number of Pods, Seeds per Pod, and 100 Grain Weight of 20 Dry Bean Genotypes at Two Soil P Levels

			100 Grain Weight (g)	
Genotype	Pods Plant^{-1}	Seeds Pod^{-1}	25	200 mg P kg^{-1}
CNFC 10467	3.79ab	4.12abc	27.94ab	26.60abcd
CNFP 8000	5.75a	4.67a	22.18b	21.41cd
CNFC 10455	3.54ab	3.31bc	29.95ab	32.68a
CNFP 10035	4.46ab	4.51ab	30.15ab	30.20ab
CNFC 10410	4.58ab	4.04abc	34.88a	30.81ab
CNFP 10076	4.50ab	3.88abc	25.13ab	27.27abcd
CNFC 10432	4.04ab	4.42abc	26.65ab	29.29ab
CNFP 10093	4.13ab	4.23abc	19.58b	27.53abcd
CNFC 10408	4.17ab	3.53abc	29.18ab	29.78ab
CNFP 10103	3.38b	3.36bc	20.12b	29.36ab
CNFC 9461	3.63ab	4.35abc	29.11ab	33.49a
CNFP 10104	4.58ab	4.28abc	27.09ab	31.26ab
CNFC 10429	4.50ab	3.98abc	28.73ab	28.79abc
CNFP 10109	4.29ab	4.36abc	27.63ab	26.92abcd
CNFC 10431	4.00ab	3.87abc	27.32ab	29.29abc
CNFP 10120	3.38b	3.24c	25.96ab	21.32cd
CNFC 10438	2.79b	4.06abc	27.40ab	20.74d
CNFP 10206	4.58ab	4.05abc	25.82ab	24.85bcd
CNFC 10444	3.58ab	3.70abc	24.29ab	26.93abcd
CNFC 10470	3.25b	3.91abc	28.31ab	28.81abc
Average	4.05	3.99	26.87	27.86
F-test				
P level (P)	**	**	**	
Genotype (G)	**	**	**	
P × G	NS	NS	**	

Source: Fageria, N.K. et al., *J. Plant Nutr.*, 33, 2167, 2010.
** and NS Significant at the 1% probability level and nonsignificant, respectively. Means within the same column followed by the same letter are not significantly different at the 5% probability level by Tukey's test.

Increasing soil P from low to high increased root weight by 166%. At the low soil P level, genotypes CNFC 10431 produced highest root dry weight, whereas at the high soil P level genotype CNFC 10120 produced maximal root dry weight. Inter- and intraspecies differences in root dry matter accumulations have been well documented and are influenced by environmental factors (Baligar et al., 1998; Fageria et al., 2006). Baligar et al. (1998) also reported that dry root weight of dry bean increased significantly and quadratically with increasing P rate in a Brazilian Oxisol. A summary of P deficiency symptoms for food crops is presented in Table 3.6.

3.2.2 Visual Toxicity Symptoms

Nutrient toxicity is defined as the quality, state, or degree of harmful effect from an essential nutrient in excess concentration in the plant. Nutrient toxicity symptoms in crop plants always start in the lower leaves and spread to whole plants when toxicity persists for a longer duration. When essential plant nutrients are absorbed by plants in excess amount they cause nutritional imbalances with other nutrients and will result in poor growth and reduced yield. Clark (1993) reported that in

TABLE 3.4
Grain Harvest Index (GHI) of 20 Dry Bean Genotypes at Two Soil P Levels

	GHI	
Genotype	**25**	**200 mg P kg^{-1}**
CNFC 10467	0.55abc	0.70a
CNFP 8000	0.62a	0.69a
CNFC 10455	0.48abc	0.49bc
CNFP 10035	0.65a	0.63ab
CNFC 10410	0.70a	0.63ab
CNFP 10076	0.53abc	0.53abc
CNFC 10432	0.51abc	0.66ab
CNFP 10093	0.56ab	0.62ab
CNFC 10408	0.54abc	0.55abc
CNFP 10103	0.26c	0.57abc
CNFC 9461	0.55abc	0.65ab
CNFP 10104	0.55abc	0.60abc
CNFC 10429	0.47abc	0.69ab
CNFP 10109	0.58a	0.64ab
CNFC 10431	0.46abc	0.52abc
CNFP 10120	0.46abc	0.26d
CNFC 10438	0.73a	0.59abc
CNFP 10206	0.59a	0.64ab
CNFC 10444	0.27bc	0.59abc
CNFC 10470	0.62a	0.42cd
Average	0.53	0.58
F-test		
P level (P)	**	
Genotype (G)	**	
P × G	**	

Source: Fageria, N.K. et al., *J. Plant Nutr.*, 33, 2167, 2010.
**Significant at the 1% probability level. Means within the same column followed by the same letter are not significantly different at the 5% probability level by Tukey's test.

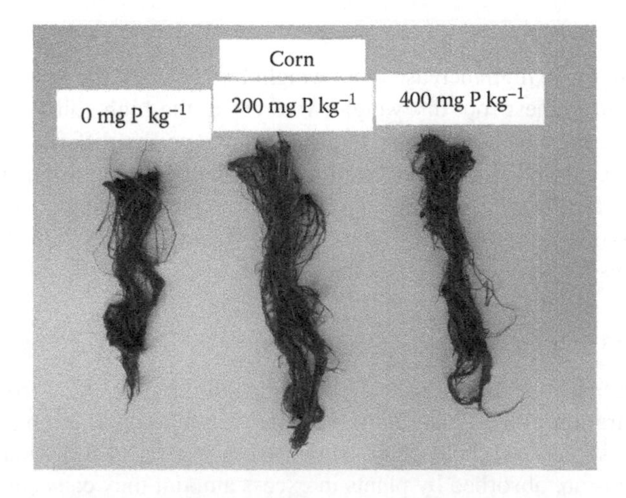

FIGURE 3.10 Root growth of corn at three P levels.

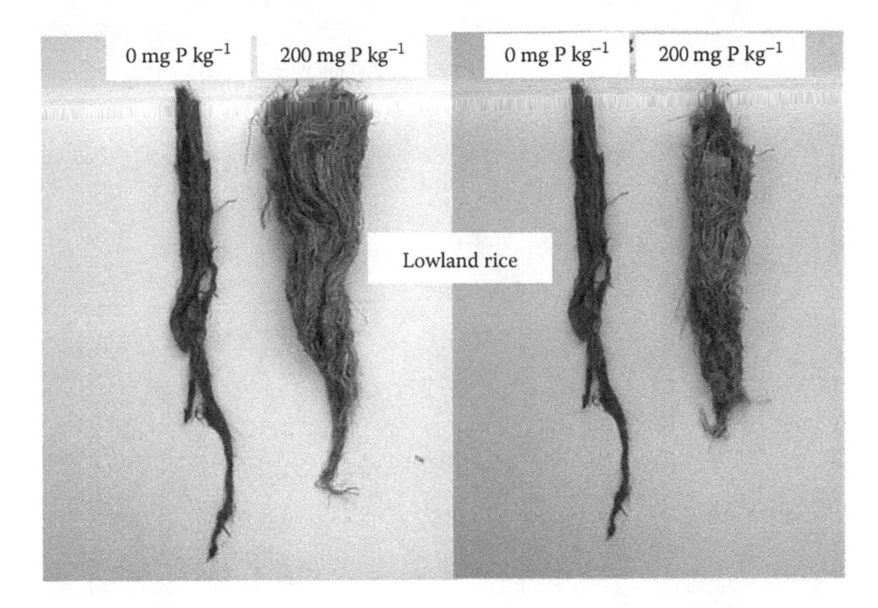

FIGURE 3.11 Root growth of lowland at two P levels and two P sources. Left simple superphosphate and right polymer-coated simple superphosphate. (From Fageria, N.K. et al., *Commun. Soil Sci. Plant Anal.*, 43, 2752, 2012.)

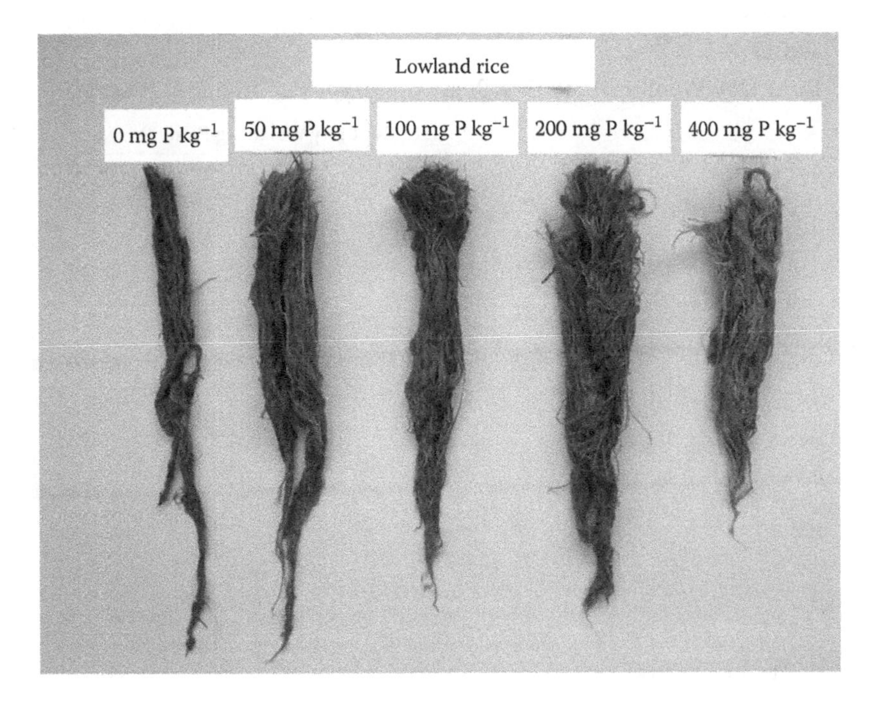

FIGURE 3.12 Root growth of lowland rice at different levels of monoammonium phosphate fertilization. (From Fageria, N.K. et al., *Commun. Soil Sci. Plant Anal.*, 43, 2752, 2012.)

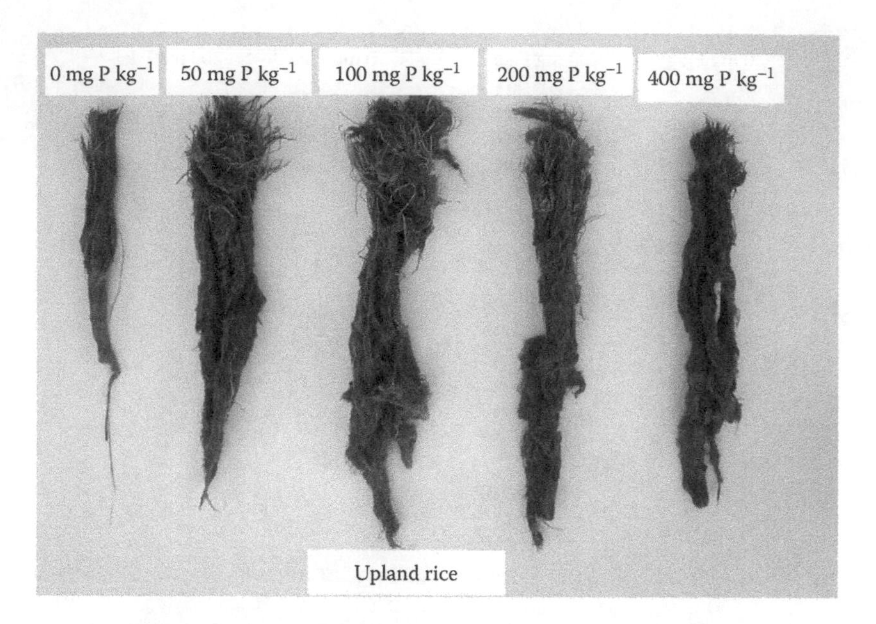

FIGURE 3.13 Root growth of upland rice at different levels of simple superphosphate fertilizer. (From Fageria, N.K. et al., *Commun. Soil Sci. Plant Anal.*, 45, 1, 2014b.)

TABLE 3.5
Root Dry Weight of Six Dry Bean Genotypes at Two Soil P Levels

	Root Dry Weight (g Plant^{-1})	
Genotype	**25**	**200 mg P kg^{-1}**
CNFP 10103	0.54a	1.63ab
CNFP 10104	0.68a	1.87ab
CNFC 10429	0.61a	1.01b
CNFC 10431	0.79a	1.79ab
CNFP 10120	0.59a	2.26a
CNFC 10470	0.49a	1.34ab
Average	0.62	1.65
F-test		
P level (P)	**	
Genotype (G)	**	
P × G	*	

Source: Fageria, N.K. et al., *J. Plant Nutr.*, 33, 2167, 2010.

*,**Significant at the 5% and 1% probability levels, respectively. Means within the same column followed by the same letter are not significantly different at the 5% probability level by Tukey's test.

TABLE 3.6
Summary of Phosphorus Deficiency Symptoms in Principal Field Crops

Crop Species	Description of Deficiency Symptoms	References
Rice (*Oryza sativa*)	Reduced tillering, leaf area, root growth, plants are dark green with erect leaves and reduced grain yield. Leaves of P-deficient plants turned reddish-yellow color.	Fageria et al. (2011a)
Corn (*Zea mays*)	Plants are dark green with reddish-purplish tips and leaf margins. Plant growth is reduced. Grain yield of corn is significantly reduced in P-deficient soils. Root growth of P-deficient corn plants significantly reduced.	Voss (1993)
Sorghum (*Sorghum bicolor*)	Phosphorus deficiency is characterized by stunted, spindly plants with low vigor and dark green leaves, which have overtones of dark red coloration. Because P is readily remobilized in plants, older sheaths and leaves first show red pigmentation, which progresses to younger leaves if deficiency persists for a longer duration. Patterns and sharpness of red pigmentation may be used to distinguish P deficiency from the natural red pigmentation that often appears on leaves of many sorghum genotypes. Phosphorus-deficient leaf tips and interveinal tissue show redness that progresses toward the base, veinal tissue, and midrib; eventually the whole leaf is covered with a uniform red color. Boundaries between affected and unaffected tissue are usually distinct. If the deficiency continues, leaves turn pale brown and die. Leaves of P-deficient young plants often appear more erect and sometimes leathery.	Clark (1993)
Wheat (*Triticum* spp.)	Unlike N-deficient plants, P-deficient plants maintain their green color and may be darker green than plants with sufficient P. However, plant growth of P-deficient plants is slow and maturity is delayed. Leaf tips die back when P deficiency is very severe and foliage of some cultivars may display shades of purple or red. Older tissues, such as in older leaves, are first to display P deficiency symptoms since the nutrient is translocated to metabolically active sites.	Wiese (1993)
Sugarcane (*Saccharum* spp.)	Phosphorus-deficient older leaves die back. Leaf blades become dark green to blue-green. Red or purple colors often appear, particularly at tips and margins exposed to direct sunlight. The leaves are slender. Older leaves turn yellow and eventually die back from tips and along the margins. Stalks are shorter and slender. Internode length is significantly reduced toward the top of the plant. Plant vigor and tillering are reduced. Sugarcane yield is significantly reduced in P-deficient soils.	Gascho et al. (1993)
Sugar beet (*Beta vulgaris* L.)	An overall stunting of the plant and a gradual deepening of the green color of foliage are the only visible symptoms of P deficiency in sugar beet. As the deficiency becomes more severe, the deep green color often assumes a metallic luster ranging from dull grayish-green to almost bluish-green. The purpling that is often associated with P deficiency in other crops is seldom.	Ulrich et al. (1993)
Soybeans (*Glycine max*)	Soybeans require a large amount of P for pod setting and N fixation. The chief P deficiency symptoms are retarded growth and affected plants are spindly and have small leaflets. Phosphorus-deficient plants turn dark green or bluish green. The leaf blade may curl up and appear pointed. Blooming and maturity are delayed. The dark green color of the leaves gives the impression that the plants are quite healthy.	Sinclair (1993)

(*Continued*)

TABLE 3.6 (*Continued*)
Summary of Phosphorus Deficiency Symptoms in Principal Field Crop

Crop Species	Description of Deficiency Symptoms	References
Peanuts (*Arachis hypogaea* L.)	Deficiency of P results in poor growth, because the availability of carbohydrates to the cells is limited. Leaf size is markedly reduced, and plant maturity is often delayed. Affected leaves may first become bluish-green; however, as leaves mature, they become thickened and are characterized by a dull, dark green color. In time, the older leaves turn orange-yellow, and veins become reddish-brown from the accumulation of anthocyanin pigments. Accumulation may also occur in the stems. Eventually, the entire leaf becomes brown and finally drops. Deficiency of P also reduces flower production and fertilization. Pod yield reduces accordingly. Symptoms usually appear within 4 weeks after sowing.	Smith et al. (1993)
Cotton (*Gossypium*)	Phosphorus deficiency symptoms are not distinct in cotton. Plants are stunted, leaves become darker green than normal, flowering is delayed, and boll retention is poor.	Cassman (1993)
Onions (*Allium cepa* L.)	Phosphorus deficiency in onions is related to reduced growth, delayed maturity, and a high percentage of thick necks at harvest. Deficient plants are initially recognized by the dual green color of the leaves. Leaf tips then wilt and eventually die without the yellowing associated with N and K deficiency. The necrosis advances toward the base of the leaves, sometimes leaving islands of green within the yellow or brown necrotic tissue. In the final stage, the dead tissues turn black.	Bender (1993)
Tomato (*Solanum lycopersicum*)	Plants deficient in P grow slowly, and maturity is delayed. Seedlings growth is stunted, especially during cool weather. Leaves become dark green with purple interveinal tissue on the underside of the leaf. Stems become slender, fibrous, and hard.	Wilcox (1993)
Dry bean (Fabaceae)	Reduced in growth of tops as well as roots. Older leaves become pale yellow and with severe deficiency dry and fall down. Stems are thin and internodes are shortened. Vegetative growth stage may increase and flowering phase delayed and shortened, with consequently reduced yield. The number of aborted flowers is often high. Few pods form and yield may be reduced.	Hall and Schwartz (1993); Fageria et al. (2011a)
Potato (*Solanum tuberosum*)	Phosphorus-deficient plants appear somewhat stunted and have a darker green color than normal plants. As P deficiency increases in severity, leaf roll, the upward cupping of the leaf blades occurs and reveals the gray-green color of the lower surface.	Ulrich (1993)

sorghum (*S. bicolor* L. *Moench*) genotypes red-speckling intensity increased when higher levels of P were applied in nutrient solution. Clark (1993) also reported that at higher P levels, Fe deficiency occurred in the younger leaves of sorghum genotypes in addition to re-speckling on lower leaves. Smith et al. (1993) reported that excess application of P induced Zn deficiency in cotton (*Gossypium hirsutum*) and yield is reduced.

3.2.3 Soil Test

In a broad sense, soil testing is any chemical or physical measurement that is made on a soil. The main objective of soil testing is to measure soil nutrient status and lime requirements in order to make fertilizer and lime recommendations for profitable farming. Soil testing is an important tool in high-yield farming but it produces optimum results only when it is used in conjunction with other good farming practices. *There is good evidence that the competent use of soil tests can make a valuable contribution to the more intelligent management of the soil.* This statement was made by the USA National Soil Test Workgroup in its 1951 report and is still applicable today (Fageria and Baligar, 2005).

Soil testing involves collecting soil samples, preparation for analysis, chemical or physical analysis, interpretation of analysis results, and finally making fertilizers and lime recommendations for the crops to be grown. A detailed description of these soil testing components can be found in several articles.

Use of soil analysis as a fertilizer recommendation method is based on the relationship between the amount of nutrient extracted from the soil by chemical methods and crop yield. When a soil analysis test indicates a low level of a particular nutrient in a given soil, application of that nutrient is expected to increase crop yield. Generally, nutrient analysis is arbitrarily classified as very low, low, adequate, high, and excess. Under very low nutrient level, relative crop yield is expected to be less than 70% and larger application of fertilizer for soil-building purposes is required. After the application of the nutrient, growth response is expected to be dramatic and profitable. Under the low fertility level, relative yield is expected to be 70%–90%. Under this situation annual application of fertilizer is necessary to produce maximum response and increase soil fertility. Increased yield justifies the cost of fertilization. When soil analysis test indicates an adequate level, relative crop yield under this situation is expected to be 90%–100%. Normal annual applications to produce maximum yields are recommended. In this case more fertilizer may increase yields slightly but the added yield would not justify the expense of the additional fertilizers. Under a high level of nutrient, there is no increase in yield. Under this situation a small application is used to maintain soil nutrient level. The amount suggested may be doubled and applied in alternative years. When soil test indicates very high or excess of a nutrient, yield may be reduced due to toxicity or imbalances of nutrients. Under this situation there is no need to apply nutrient until the level reduces to a low range. To obtain such nutrient level and yield relationship, it is necessary to conduct fertilizer yield trials in several locations in a given agroecological region for a specific crop. Some specific recommendations for soil analysis are summarized here.

1. Soil samples must be representative of the land area in question. It is recommended taking a minimum of one composite sample per 12–15 ha for lime and fertilizer recommendations. A representative soil sample is composed of 15–20 subsamples from a uniform field with no major variation in slope, drainage, or past fertilizer history. Any of these listed factors, if changed, will have an effect on the number of samples and unit area from which the sample is obtained.
2. Depth of sampling for mobile nutrients like N should be of 60 cm and for immobile nutrients like P, K, Ca, and Mg, 15–20 cm sampling depth can provide satisfactory results. For pasture crops, a sampling depth of 10 cm is normally sufficient to evaluate nutrient status and making lime and fertilizer recommendations.
3. Selecting appropriate extractants. Three extracting solutions, 0.05 N $HC1$ + 0.025 N H_2SO_4 (Mehlich 1), 0.03 N NH_4F + 0.025 N $HC1$ (Bray-P1), and 0.5 N $NaHCO_3$ at pH 8.5 (Olsen) are the most commonly used extractants for P and are generally adequate to cover the broad

range of soils. Commonly used extractants for K, Ca, and Mg are double acid (Mehlich 1), 1M NH4Ac at pH 7, and NaOAc at pH 4.8. Multi-element extracting reagents are replacing the more familiar single-element extractants. After mixing with an appropriate aliquot of soil, the obtained extract is assayed by an inductively coupled plasma emission (ICP-AES) spectrometer. A flow injection analyzer (FIA) is another multi-element analyzer capable of assaying these soil extracts.

4. Optimum soil test values for macro- and micronutrients vary from soil to soil, crop to crop, and extractant to extractant. But normally >10 mg P kg^{-1}, >50 mg K kg^{-1}, >600 mg Ca kg^{-1}, >120 mg kg^{-1}, and >12 mg S kg^{-1} can produce satisfactory results for most soils and crops. For micronutrients, the critical values reported are Fe 2.5–5 mg kg^{-1}, Mn 4–8 mg kg^{-1}, Zn 0.8–3 mg kg^{-1}, B 0.1–2 mg kg^{-1}, Cu 0.5–2 mg kg^{-1}, and Mo 0.2–0.5 mg kg^{-1}, respectively.
5. The pH of agricultural soils is in the range of 4–9. It is difficult to define optimum pH values of different plant species. Most food crops grow well in acid soils if pH is around 6.0. Lime is considered the foundation of crop production or workhouse in acid soils.
6. Fertilizer field trials are the important part of the overall diagnostic process of nutrient disorder, since the only method to establish the fertilizer requirements accurately is by field experiments, under which the entire interaction of soil, plant, climate, and management factors can occur.

3.2.4 Plant Tissue Test

Plant analysis is based on the concept that the concentration of an essential element in a plant or part of the plant indicates the soil's ability to supply that nutrient (Fageria and Baligar, 2005). This means it is directly related to the quantity in the soil that is available to the plant. For annual crops, the primary objective of plant analysis is to identify nutritional issues or to determine or monitor the nutrient status during the growing season. If deficiency is identified early in the growth stage of a crop, a correction may occur during the current season. Otherwise, steps should be taken to correct nutrient deficiencies in the next cropping cycle. Plant analysis can be useful for the prediction of nutrient needs in perennial crops, usually for the year following the time of sampling and analysis. Like soil analysis, plant analysis also involves plant sampling, plant tissue preparation, analysis, and interpretation of analytical results. All these steps are important for a meaningful plant analysis program.

Many factors such as soil, climate, plant, and their interaction affect the absorption of nutrients by growing plants. However, the concentrations of the essential nutrients are maintained within rather narrow limits in plant tissues. Such consistency is thought to arise from the operation of delicate feedback systems, which enable plants to respond in a homeostatic fashion to environmental fluctuations.

For the interpretation of plant analysis results, a critical nutrient concentration concept was developed. This concept is widely used now in interpretation of plant analysis results for nutritional disorders diagnostic purposes. Critical nutrient concentration is usually designated as a single point within the bend of a quadratic curve when crop yield is plotted against nutrient concentration where the plant nutrient status shifts from deficient to adequate. The critical nutrient concentration has been defined in several ways: (1) the concentration that is just deficient for maximum growth; (2) the point where growth is 10% less than the maximum; (3) the concentration where plant growth begins to decrease; and (4) the lowest amount of element in the plant accompanying the highest yield. These definitions are similar, but not identical. A good relationship exists between concentration of a nutrient and yield of a given crop. For example, in central part of Brazil, significant correlations ($R^2 = 0.46^{**}$ and $R^2 = 0.53^{**}$) were reported between P concentration in straw and grain of lowland rice (*Oryza sativa* L.) from field experiments (Figures 3.14 and 3.15). Similarly, P uptake

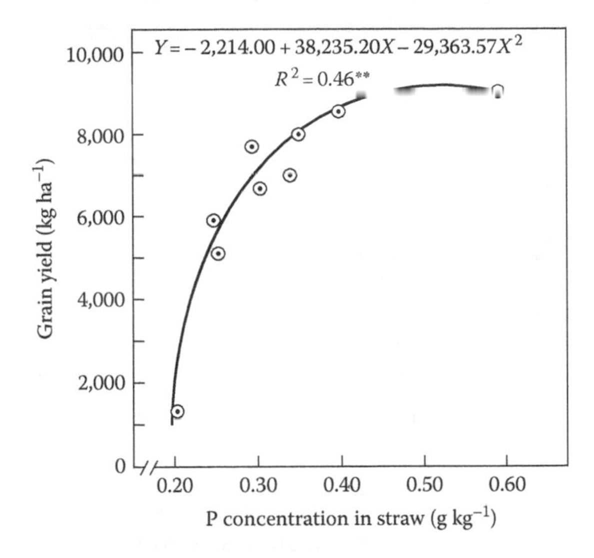

FIGURE 3.14 Relationship between P concentration in straw and grain yield of lowland rice.

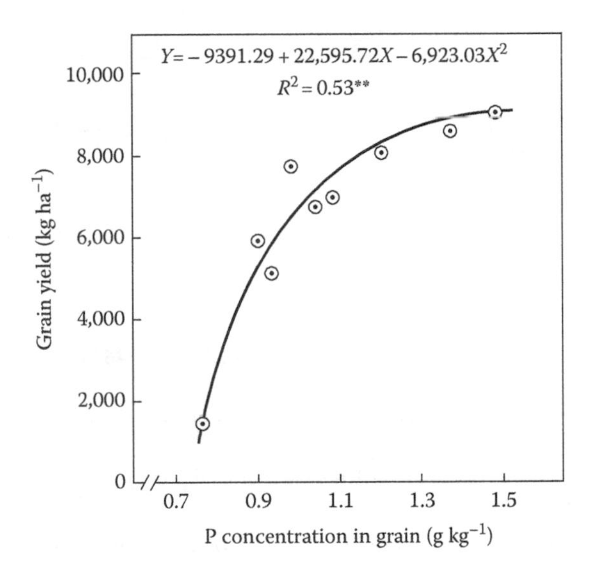

FIGURE 3.15 Relationship between P concentration in grain and grain yield of lowland rice.

(N concentration × dry weight) in straw as well as in grain had a significant quadratic relationship with grain yield in lowland rice (Figures 3.16 and 3.17). Variability in grain yield due to P uptake in straw and grain was 61% and 94%, respectively.

3.3 PHOSPHORUS CONCENTRATION IN PLANTS

Phosphorus concentration in plant is defined as the content per unit of dry weight. It is generally expressed in g kg^{-1}. Plants require a determined minimum concentration of P for optimum growth and development. Critical nutrient concentration is defined as the concentration in the plant, or specific plant part, above which additional plant growth response slows. Crop yield, quality, or performance is less than optimum when the concentration is below the critical concentration

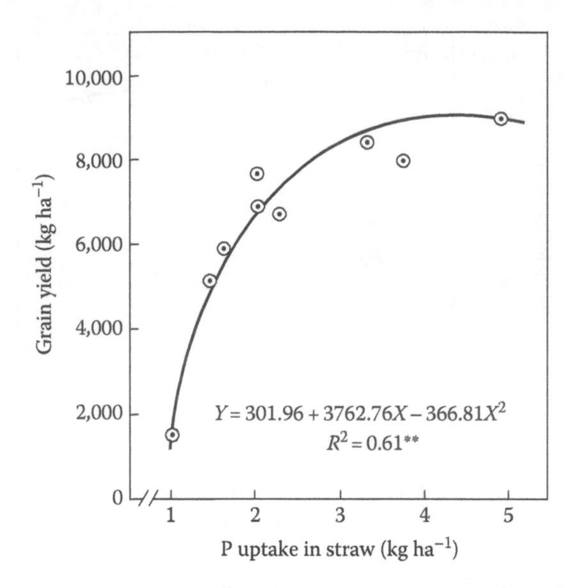

FIGURE 3.16 Relationship between P uptake in straw and gran yield of lowland rice.

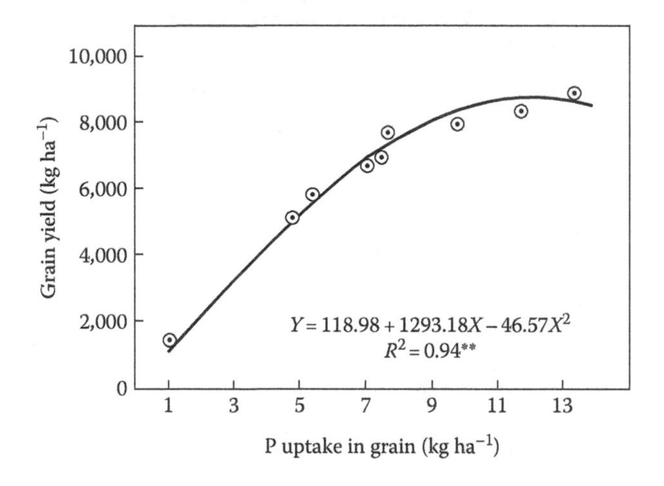

FIGURE 3.17 Relationship between P uptake in grain and grain yield of lowland rice.

(Soil Science Society of America, 2008). Fageria (2014) defined a critical nutrient concentration as the concentration at which 90% or 95% of maximum relative yield is obtained. A hypothetical relationship between nutrient concentration in plant tissue and relative yield is presented in Figure 3.18. According to Ulrich and Hills (1973), the most useful calibration curve is one in which the transition zone is sharp; that is, there is a narrow range in nutrient concentration between plants that are deficient and those that receive an optimum supply of nutrients.

Reference concentrations of mineral nutrients in specific plant parts are determined and used as a guide to indicate how well plants are supplied with essential plant nutrients at a certain time. Such reference concentrations provide a tool to assist the agronomist in evaluating nutrient disorders and in improving fertilization in the present or succeeding crops (Ulrich and Hills, 1973). Ulrich and Hills (1973) reported that the concentration of a nutrient within the plant at any particular moment is an integrated value of all the factors that have influenced the nutrient concentration up to the time of sampling. Rauschkolb et al. (1984) stated that an attractive feature

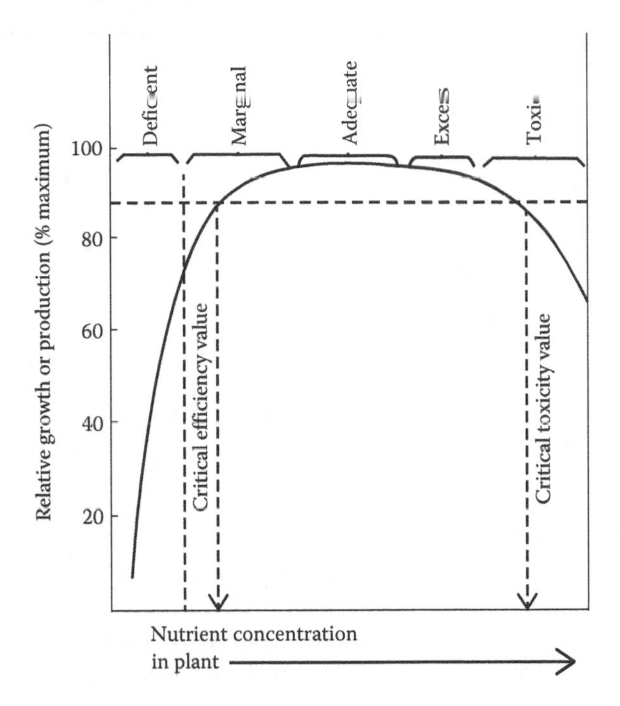

FIGURE 3.18 Relationship between nutrient concentration in plant tissue and relative growth or yield. (From Fageria, N.K. et al., *Growth and Mineral Nutrition of Field Crops*, 3rd edn., CRC Press, Boca Raton, FL, 2011a.)

of tissue test is that the plant root system tends to integrate spatial variability of soil N supplying power over a relatively large field volume.

Several factors (climate, soils, and plants) affect nutrient concentration in plant tissues. Among these factors, plant part sampled and growth stage should be considered. Generally, nutrient concentration in the plant tissues decreases with the advancement of the plant age. In the young actively growing plants, P is most abundant in the actively growing tissue. By the time plants have attained about 25% of their total dry weight, they may have accumulated as much as 75% of their total nutrient requirements. Therefore, most crops require significant quantities of P during the early stages of growth. For example cereal crops will often take up to 75% of their P requirement within 40 days after crop emergence.

Phosphorus concentration in the shoot or straw of upland rice, dry bean, corn, and soybean (*Glycine max*) during crop growth cycles is presented in Figure 3.19. Phosphorus concentration in the shoot of upland rice, dry bean, corn, and soybean decreased significantly in a quadratic fashion with the advancement of plant age (Figure 3.19). Phosphorus concentration in the shoot of upland rice was 3.5 g kg^{-1} at 19 days after sowing and decreased to 0.7 g kg^{-1} at harvest. Phosphorus concentration in the shoot of upland rice decreased almost linearly up to 100 days after sowing and was more or less constant during grain filling growth stage. In dry bean, shoot P concentration was 2.8 g kg^{-1} at 15 days after sowing and decreased to 1 g kg^{-1} at harvest. In corn, shoot P concentration at 18 days after sowing was 3.3 g kg^{-1} and dropped to 0.8 g kg^{-1} at harvest. Similarly, in soybean shoot P concentration at 27 days of growth was 2.9 g kg^{-1} and dropped to 1.7 at harvest. Variation in P concentration with the advancement of plant age in rice was 92%. In dry bean, variation in P concentration was 76%, in corn 89%, and in soybean 65% due to advancement of plant age. Higher variability in P concentration in cereals as compared to legumes indicates higher growth rate of upland rice and corn as compared to dry bean and soybean.

At the early stage of growth P concentration in the shoot of upland rice and corn was higher, as compared to dry bean and soybean. However, at harvest it was inverse. Decrease in plant nutrient

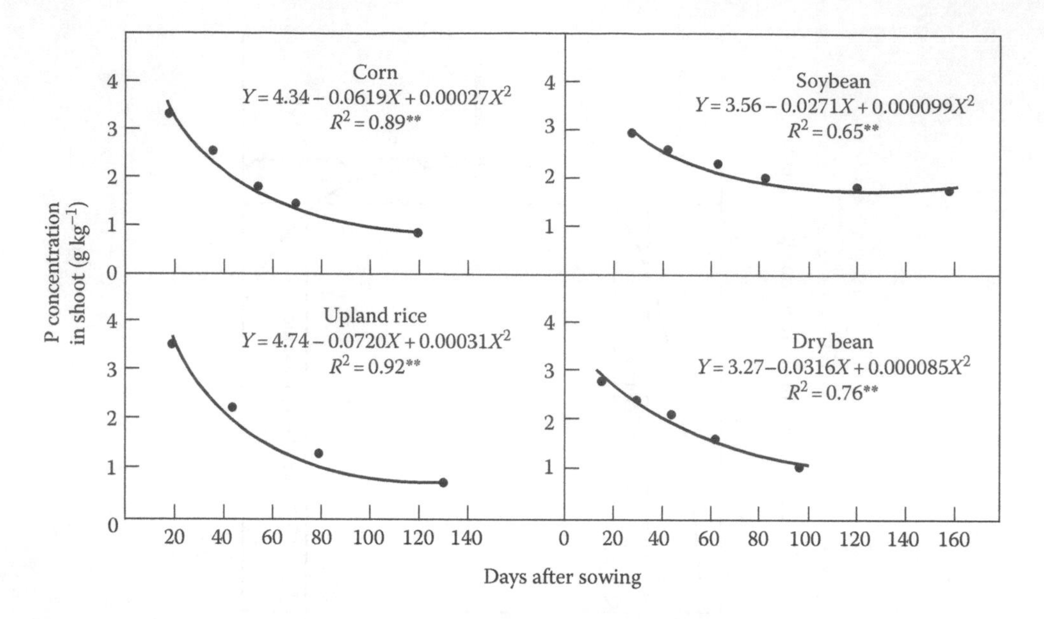

FIGURE 3.19 Phosphorus concentration in the shoot (straw) of corn, soybean, upland rice, and dry bean during growth cycle. Values are averages of 2-year field experimentation. (From Fageria, N.K. et al., *J. Plant Nutr.*, 36, 2013, 2013b.)

concentration including P with the advancement of plant age is widely reported in the literature (Fageria, 1992; Osaki, 1995; Fageria et al., 2006). Decrease in P concentration with the advancement of plant age was associated with an increase in shoot dry weight. Decreases in nutrient concentration with the advancement of plant age are known as the dilution effect in the mineral nutrition (Fageria et al., 2011a). Significant variation in P concentration with the advancement of plant age in four crop species indicates necessity of at least three tissue sampling for P deficiency or sufficiency diagnostic purposes. These sampling can be in the beginning of vegetative growth stage, during middle of crop growth cycle, and last sampling may be conducted at harvest.

Phosphorus concentration in the grain of upland rice and corn was about 2 g kg^{-1} and in the grain of dry bean and soybean was 3.2 and 5.9 g kg^{-1}, respectively (Figure 3.20). Osaki et al. (1991) reported higher P harvest index (P accumulation in grain/P accumulation in grain plus shoot) and P concentration in the seeds of soybean as compared to rice and corn. The higher P concentration in the grain of legumes compared to cereals indicates higher demand of legume for P compared

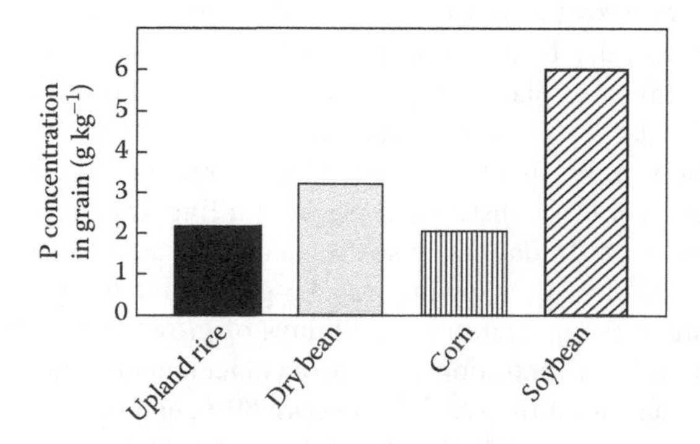

FIGURE 3.20 Phosphorus concentration in the grain of upland rice, dry bean, corn, and soybean. Values are averages of 2-year field experimentation for each crop. (From Fageria, N.K. et al., *J. Plant Nutr.*, 36, 2013, 2013b.)

to cereals (Fageria et al., 2006). Fageria (2009) suggested that legume seeds have higher nutritional value (P) compared to cereals.

Fageria et al. (2014a) studied concentrations of macro- and micronutrients, including P in the shoots of sixteen 46-day-old tropical cover crops. Macro- and micronutrient concentrations averaged across three soil pH are presented in Table 3.7. Carbon (C) concentration in shoots of legume crop species was significantly affected by soil pH and cover crop species, but their interactions were nonsignificant, indicating that the cover crops used had similar responses to soil pH levels. Overall, across the three soil pH levels (5.1–7.0), C concentration among 16 crop species varied from 390 g kg^{-1} for Showy crotalaria to 429 g kg^{-1} for black pigeon pea, with a mean value of 415 g kg^{-1}. Variability in C concentration among the cover crop species is associated with different growth habits, especially root growth and nutrient uptake mechanisms under similar growing conditions (Fageria, 2007; Fageria and Moreira, 2011).

Nitrogen (N) concentration in the shoots of 16 cover crop species was significantly influenced by cover crop species and the interaction between soil pH and cover crop species. Reuter and Robinson (1997) reported that N concentration in legume cover crops varied from 25 to 50 g kg^{-1}, depending on crop species. Results of this current study were within this range. Overall, N concentration at high soil pH (7.0) was 11% higher than at low soil pH (5.1). Liming increased N concentration in the cover crop species studied. Positive effect of liming on N uptake by dry bean in a Brazilian Oxisol has been reported by Fageria et al. (2007).

Phosphorous concentrations in shoots of 16 cover crop species varied from 1.74 to 4.14 g kg^{-1}, with a mean value of 2.71 g kg^{-1}. Reuter and Robinson (1997) reported that P concentration in tropical legume crops varied from 1.5 to 6.0 g kg^{-1}. In our study soil pH had no significant influence on P concentrations in 16 cover crops.

Calcium (Ca) concentrations varied from 10.3 to 20.2 g kg^{-1}. Reuter and Robinson (1997) reported that Ca concentration in tropical legume crops varied from 7 to 30 g kg^{-1}. Overall, Ca concentration increased with increasing pH as expected. Calcium concentrations in shoots of cover crops were significantly influenced by crop species and their interactions with soil pH. Magnesium (Mg) concentration averaged over pH ranges of 5.1–7.0 varied from 3.17 to 5.66 g kg^{-1}, with a mean value of 3.98 g kg^{-1}. Reuter and Robinson (1997) reported that Mg concentration in tropical legume crops varied from 3 to 5 g kg^{-1}. The results of the present study are within this range. Only cover crop species had significant effects on Mg concentrations of the shoot. In the current study, overall concentrations of N, P, Ca, and Mg were within sufficient to adequate range (Jones et al., 1991; Reuter and Robinson, 1997).

Zinc (Zn) concentrations in the shoots of 16 cover crop species were significantly influenced by soil pH, cover crop species, and their interactions. However, iron (Fe) concentrations in the shoots of cover crops were influenced only by cover crop species. Reuter and Robinson (1997) reported that Zn concentration in the shoot of tropical legume cover crops varied from 17 to 50 mg kg^{-1}. In the current study, Fe concentration in shoots of 16 cover crop species varied from 18.3 to 174.0 mg kg^{-1}, with a mean value of 68.2 mg kg^{-1}. Reuter and Robinson (1997) reported that Fe concentration in the shoots of tropical legume cover crops varied from 50 to 300 mg kg^{-1}. Our results related to Fe concentration fall within this range.

Manganese (Mn) concentration in the shoots of 16 cover crop species was significantly influenced by soil pH, cover crop species, and their interactions; however, the concentration of boron (B) was significantly influenced only by cover crop species. Manganese concentration in shoots of 16 cover crop species averaged over all soil pH, varied from 48.1 to 160.9 mg kg^{-1} with a mean value of 83.4 mg kg^{-1}, and B concentration varied from 42.0 to 69.1 mg kg^{-1}, with a mean value of 57.0 mg kg^{-1}. Reuter and Robinson (1997) reported that Mn concentration in the shoots of cover crop species varied from 50 to 300 mg kg^{-1} and B concentration varied from 25 to 112 mg kg^{-1}. Results of Mn and B concentrations in the shoot of cover crop species of the present study fall within this range. In the current study, the overall concentrations of Mn and B were in the sufficient to adequate range, whereas concentration of Zn and Fe were within the low to deficiency range (Reuter and Robinson, 1997; Jones et al., 1991). Manganese and Zn concentrations were reduced with increasing pH.

TABLE 3.7
Macro- and Micronutrient Concentration in the Tops of 16 Tropical Cover Crops

	C	N	P	Ca	Mg	Zn	Fe	Mn	B
Cover Crop Species			(g kg^{-1})[a]				(mg kg^{-1})		
Short-flowered crotalaria	402.99	29.11	3.52	13.85	4.36	9.63	18.28	59.12	61.16
Sunn hemp	410.81	28.47	2.49	11.90	4.70	10.72	31.20	75.98	47.69
Smooth crotalaria	417.54	39.13	3.21	10.79	4.20	19.12	43.40	160.94	69.09
Showy crotalaria	390.23	35.00	2.85	20.17	4.15	13.00	34.43	85.42	65.99
Crotalaria Ochroleuca	415.52	37.04	2.69	10.32	5.66	17.14	45.13	93.53	63.76
Calopo	419.28	28.61	3.02	12.81	3.31	5.18	153.08	48.08	60.97
Black jack bean	427.13	34.74	3.00	15.01	3.65	13.48	54.57	92.23	49.62
Bicolor pigeon pea	426.98	36.12	3.19	14.73	3.86	16.77	63.78	92.80	59.12
Black pigeon pea	428.84	30.99	2.35	13.21	3.68	14.93	62.24	86.14	51.55
Mulato pigeon pea	413.84	28.40	2.24	15.44	3.98	15.49	55.41	58.87	54.57
Lablab	416.17	30.22	1.74	15.36	4.22	14.80	59.31	92.63	49.26
Mucuna bean ana	417.31	22.34	1.96	15.91	3.88	13.48	75.07	84.97	51.23
Black mucuna bean	424.32	24.88	1.77	10.65	3.17	11.54	45.65	66.33	42.00
Gray mucuna bean	415.43	24.90	2.15	17.31	3.54	16.14	59.42	50.35	51.69
White Jack bean	412.80	33.40	2.97	12.39	3.87	20.41	115.59	100.06	64.92
Brazilian lucerne	406.83	33.00	4.14	16.78	3.48	21.86	174.01	86.22	68.67
Average	415.38	31.02	2.71	14.16	3.98	14.61	68.16	83.36	56.96
F-test									
Soil pH (S)	*	NS	NS	NS	NS	*	NS	**	NS
Cover crop species (C)	**	**	**	**	**	**	**	**	**
S × C	NS	**	NS	**	NS	**	NS	**	NS

Source: Fageria, N.K. et al., *J. Plant Nutr.*, 37, 294, 2014a.

*,**, and NS Significant at the 5% and 1% probability level and no significant, respectively.

[a] Values averaged over across three soil pH (pH 5.1, 6.5, and 7.0).

Increasing soil pH reduces solubility and ionic concentrations of Mn and Zn in soil solution (Fageria et al., 2002). Therefore, increasing soil pH reduces the plant availability of the micronutrients, and this is reflected in lower concentrations of Zn and Fe in shoots of cover crops (Table 3.7).

3.4 PHOSPHORUS UPTAKE IN PLANTS

Nutrient uptake is calculated by multiplication of concentration with dry weight and is generally expressed in kg ha^{-1} for macronutrients and g ha^{-1} for micronutrients. Values of nutrient uptake are used to know the soil fertility depletion with the cultivation or growing of a particular crop species. Such information is useful to replenish the soil fertility for the succeeding crops. Green revolution started during 1960s, which increased yields of most grain crops. However, modern crop cultivars remove two to three times as much P, as compared to the cultivars prior to the green revolution period. For example, wheat (*Triticum* spp.) in the United Kingdom removed about 7 kg P ha^{-1} in 1950, 13 kg P ha^{-1} in 1975, and 20 kg P ha^{-1} in 1995 (Edwards et al., 1997; Smil, 2000). Typical harvests now take up (in grain and straw) between 15 and 35 kg P ha^{-1} in cereals, 15–25 kg P ha^{-1} in leguminous and root crops, and 5–15 kg P ha^{-1} in vegetables and fruits (Pierzynski and Loan, 1993; Smil, 2000). Highest rates can be over 45 kg P ha^{-1} for corn, sugar beets (*Beta vulgaris* L.), and sugarcane (*Saccharum officinarum*) (Smil, 2000). The total, based on separate calculations for all major field crops, shows that the global crop harvest (including

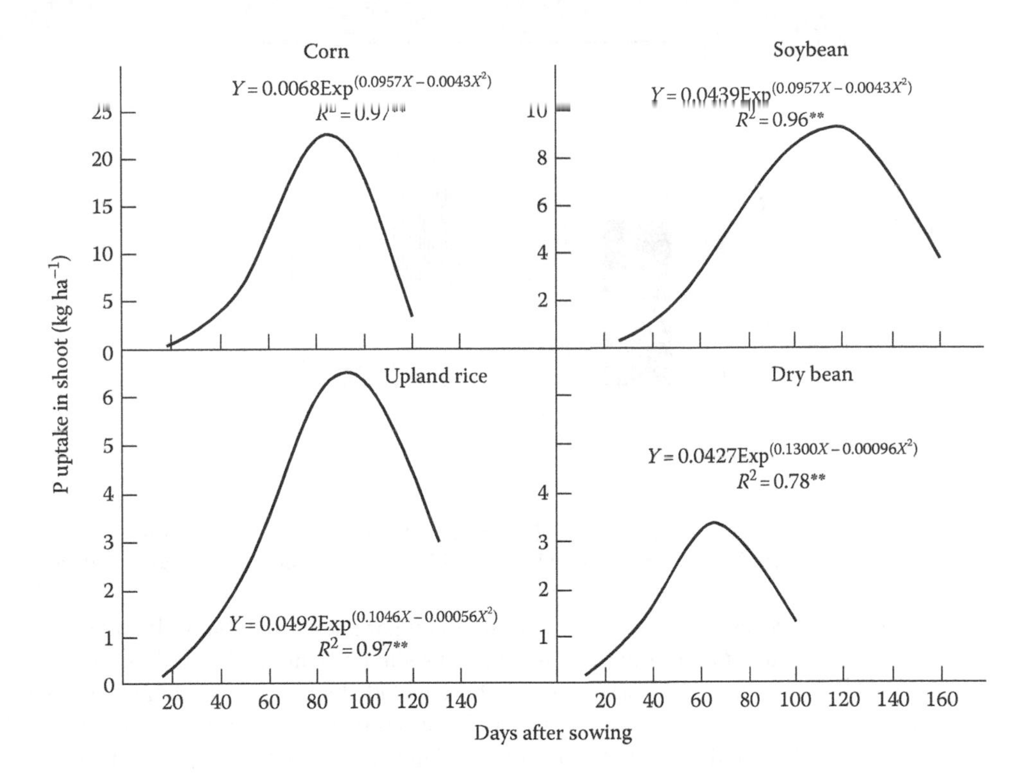

FIGURE 3.21 Phosphorus uptake in the shoot of corn, soybean, upland rice, and dry bean as influenced by plant age. Values are averages of 2-year field experimentation for each crop. (From Fageria, N.K. et al., *J. Plant Nutr.*, 36, 2013, 2013b.)

forages grown on arable land but not the phytomass produced on permanent pastures) assimilates annually about 12 million mt of P in crops and their residues (Smil, 2000). Cereals and legumes account for most of the flux, containing 0.25%–0.45% P in their grains (only soybean have 0.6% P), and mostly only 0.05%–0.1% P in their straws (Smil, 1999).

Fageria et al. (2013b) studied P uptake in the shoot of four crop species during their growth cycles. Phosphorus uptake was significantly and quadratically increased with the advancement of plant age (Figure 3.21). In upland rice, P uptake in shoot increased almost linearly with the advancement of plant age up to 90 days after sowing and then decreased. Maximum P uptake in rice shoots was 6.5 kg ha^{-1} at 90 days after sowing. At harvest rice shoots contain about 3 kg P ha^{-1}. In dry bean shoot P uptake was also linear up to 66 days after sowing and then decreased. Maximum P uptake in dry bean at 66 days after sowing attained a level of 3.3 kg ha^{-1} and at harvests the P level in the shoot was 1.5 kg ha^{-1}. In corn shoot P uptake was maximum at 86 days after sowing and dropped to 4 kg ha^{-1} at harvest. In soybean shoot P uptake increased up to 120 days after sowing and having a value of 9.2 kg ha^{-1} at this growth stage. At harvest P uptake value in soybean dropped to 3.8 kg ha^{-1}. Uptake of P in the shoot of four crop species followed shoot dry weight accumulation pattern with the advancement of plant age (Fageria and Barbosa Filho, 2008). A decrease in P uptake of shoot of four crop species after certain age was associated with translocation of P to the grain. Fageria et al. (2006) reported similar decreases in P uptake of shoot of cereals and legume crops.

Phosphorus uptake in the shoot of four crop species was in the order of corn > soybean > upland rice > dry bean. Higher P uptake in the corn shoot as compared to other three crop species was due to higher dry weight of shoot of corn (Fageria and Barbosa Filho, 2008). Variability of P uptake in the shoot of upland rice due to advancement of plant age was 97%. For dry bean the variability in P uptake in shoot was 78% with the advancement of plant age. Variability of P for corn was 97% and

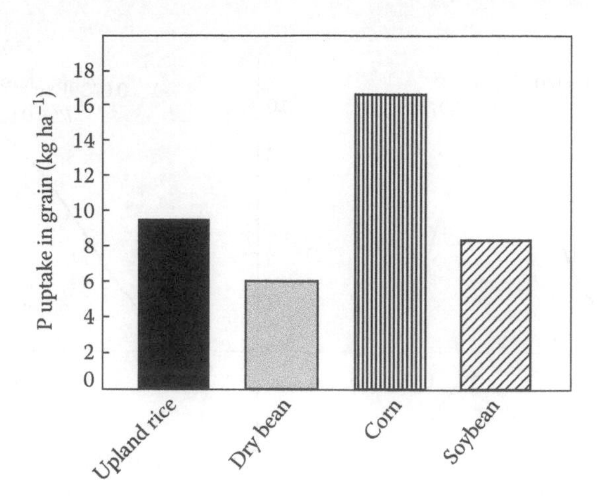

FIGURE 3.22 Phosphorus uptake in the grain of upland rice, dry bean, corn, and soybean. Values are averages of 2-year field experimentation for each crop. (From Fageria, N.K. et al., *J. Plant Nutr.*, 36, 2013, 2013b.)

for soybean 96%. This means that plant age is an important parameter in defining P uptake in crop species and should be taken into account for tissue analysis in nutrient uptake studies. Osaki et al. (1991), Fageria et al. (2011), and Fageria et al. (2006) concluded similarly in relation to P uptake in annual crop species. Phosphorus uptake in grain of four crop species also varied (Figure 3.22). The P uptake in grain followed the pattern of corn > upland rice > soybean > dry bean. This may be associated with grain yield (Fageria et al., 2013b), except soybean. In soybean, grain yield was lower than dry bean but P uptake was higher than dry bean. This may be related to higher concentration of P in the grain of soybean compared to dry bean (Fageria et al., 2013b).

Fageria (2001) studied nutrient uptake in four crop species (upland rice, dry bean, soybean, and corn) grown on a Brazilian Oxisol (Table 3.8). Among macronutrients, requirement of N and K was higher for all the crop species compared to P. These results suggest that internal P use efficiency (nutrient uptake per unit grain produced) was higher compared to N and K in the four crop species.

Fageria et al. (2014a) studied the uptake of nutrients in sixteen 46-day-old tropical legume cover crops. Macro- and micronutrient uptake and C/N ratio averaged across three soil pH are presented in Table 3.9. Total uptake of C per plant was significantly influenced by soil pH levels, cover crop species, and their interactions. Overall, C uptake decreased with increasing soil pH. This is associated with the decrease in shoot dry weight with increasing soil pH (Fageria et al., 2014a). Nitrogen uptake was significantly influenced by soil pH and cover crop species, and it varied from 54.0 to 603.3 mg plant^{-1}, with a mean value of 206 mg plant^{-1}. Carbon/nitrogen (C/N) ratio was significantly influenced by soil pH, crop species, and their interactions, indicating a significant variability in C/N ratio of cover crop species under different soil pH levels (Table 3.9). Fageria (2007) reported that C/N ratio of legume cover crops varied from 11 to 28, with a mean value of 16. Results of our study fall within this range. Overall, C/N ratio decreased with decreasing soil pH. Carbon/nitrogen ratio (C/N ratio) of crop residues plays an important role in the release or immobilization of soil N because plant tissue is a primary source and sinks for C and N (Fageria, 2007). When plant residues having C/N ratio greater than 20 are incorporated into the soil, available soil N is immobilized during the first few weeks of residue decomposition (Dinnes et al., 2002). This occurs since the microbial populations that decompose plant residues increase their biomass in response to added C. In aerobic soils, C/N ratios of <20 for organic residues are required for net mineralization to occur (Fageria, 2007). Zebarth et al. (2009) and Kumar and Goh (2000) reported that incorporation of low C/N ratio residues generally results in net mineralization, whereas high C/N ratio residues result in net immobilization. Under field conditions, the break point between net mineralization and immobilization is commonly a C/N ratio of 20–30 (Kumar and Goh, 2000).

TABLE 3.8
Uptake of Macro- and Micronutrients by Upland Rice, Common Bean, Corn, and Soybean Grown on Brazilian Oxisols

Nutrients	Straw	Grain	Total	Required to Produce 1 t of Grain
Upland rice				
Nitrogen (kg ha^{-1})	56	70	126	28
Phosphorus (kg ha^{-1})	3	9	12	3
Potassium (kg ha^{-1})	150	56	206	45
Calcium (kg ha^{-1})	23	4	27	6
Magnesium (kg ha^{-1})	14	5	19	4
Zinc (g ha^{-1})	161	138	299	65
Copper (g ha^{-1})	35	57	92	20
Iron (g ha^{-1})	654	117	771	169
Manganese (g ha^{-1})	1319	284	1603	351
Boron (g ha^{-1})	53	30	83	18
Common bean				
Nitrogen (kg ha^{-1})	19	68	87	45
Phosphorus (kg ha^{-1})	1	6	7	4
Potassium (kg ha^{-1})	25	36	61	32
Calcium (kg ha^{-1})	16	6	22	11
Magnesium (kg ha^{-1})	7	4	11	6
Zinc (g ha^{-1})	29	74	103	54
Copper (g ha^{-1})	8	22	30	16
Iron (g ha^{-1})	268	144	412	215
Manganese (g ha^{-1})	73	27	100	52
Boron (g ha^{-1})	20	14	34	18
Corn				
Nitrogen (kg ha^{-1})	72	127	199	24
Phosphorus (kg ha^{-1})	4	17	21	3
Potassium (kg ha^{-1})	153	34	187	23
Calcium (kg ha^{-1})	33	8	41	5
Magnesium (kg ha^{-1})	20	9	29	4
Zinc (g ha^{-1})	184	192	376	46
Copper (g ha^{-1})	53	14	67	8
Iron (g ha^{-1})	2048	206	2254	274
Manganese (g ha^{-1})	452	82	534	65
Boron (g ha^{-1})	103	43	146	18
Soybean				
Nitrogen (kg ha^{-1})	33	91	124	86
Phosphorus (kg ha^{-1})	4	8	12	9
Potassium (kg ha^{-1})	30	28	58	40
Calcium (kg ha^{-1})	33	6	39	27
Magnesium (kg ha^{-1})	14	10	24	16
Zinc (g ha^{-1})	43	78	121	84
Copper (g ha^{-1})	53	31	84	58
Iron (g ha^{-1})	778	190	968	671
Manganese (g ha^{-1})	193	32	225	156
Boron (g ha^{-1})	22	21	43	30

Source: Fageria, N.K., *Rev. Bras. Eng. Agric. Amb.*, 5, 416, 2001.

TABLE 3.9
Macro- and Micronutrient Uptake in the Tops of 16 Tropical Cover Crops

Cover Crop Species	C/N Ratio	C	N	P	Ca	Mg	Zn	Fe	Mn	B
		(mg plant^{-1})[a]					(µg plant^{-1})			
Short-flowered crotalaria	14.17	929.51	62.10	7.49	29.10	8.66	28.75	39.48	187.98	139.75
Sunn hemp	14.55	3333.79	230.33	20.96	91.58	35.29	97.18	283.35	782.29	378.86
Smooth crotalaria	10.88	1185.77	109.42	9.35	30.64	11.62	63.47	120.91	583.60	194.25
Showy crotalaria	11.27	1809.59	161.01	13.26	91.79	18.91	64.56	163.43	444.86	297.71
Crotalaria Ochroleuca	11.40	3138.66	281.23	19.51	78.02	41.77	148.54	371.16	958.57	500.64
Calopo	15.18	1374.25	92.70	9.38	39.58	10.31	23.97	482.20	190.07	202.50
Black jack bean	12.35	2350.24	193.03	16.48	78.49	19.33	88.09	306.62	575.43	269.09
Bicolor pigeon pea	12.07	2234.22	189.97	16.80	72.35	18.53	108.43	358.03	586.41	298.15
Black pigeon pea	14.20	2232.71	162.75	12.64	65.03	17.71	80.48	318.30	494.52	265.39
Mulato pigeon pea	15.26	2290.70	151.35	11.25	73.84	19.85	68.51	258.34	342.48	303.65
Lablab	14.87	4245.76	299.84	17.17	140.81	38.45	145.22	616.23	1123.69	514.25
Mucuna bean ana	20.00	3581.79	175.59	16.29	129.15	30.57	135.31	531.54	907.24	448.09
Black mucuna bean	19.05	5155.66	271.73	20.64	122.19	36.14	141.14	456.83	943.71	511.15
Gray mucuna bean	17.19	4657.12	258.49	26.42	163.47	36.14	99.24	516.72	677.99	594.34
White jack bean	12.99	7557.34	603.32	54.72	215.65	70.02	364.03	2157.40	1934.30	1223.39
Brazilian lucerne	12.54	600.98	54.00	5.77	25.66	4.63	47.03	236.57	287.20	118.68
Average	14.25	2917.38	206.05	17.38	90.46	26.12	106.51	451.07	688.77	391.24
F-test										
Soil pH (S)	*	**	**	*	**	*	**	NS	**	**
Cover crop species (C)	**	**	**	**	**	**	**	**	**	**
S × C	**	**	NS	NS	NS	*	**	NS	**	NS

Source: Fageria, N.K. et al., *J. Plant Nutr.*, 37, 294, 2014a.

*,**, and NS Significant at the 5% and 1% probability level and no significant, respectively.

[a] Values averaged over across three soil pH (pH 5.1, 6.5, 7.0).

However, the amount of N mineralized from a crop residue varies not only with the C/N ratio but also with the composition of the residue (Thorup-Kristensen et al., 2003).

Uptake of P, Ca, and Mg was influenced significantly by soil pH and cover crop species. Magnesium uptake was also significantly influenced by soil pH × cover crop interaction. Overall, Mg uptake decreased slightly with increasing soil pH. Uptake of P varied from 5.77 to 54.7 mg plant^{-1}, with a mean value of 17.4 mg plant^{-1}. Interspecies variations in macronutrient uptake in tropical legumes have been reported (Baligar et al., 2008). Overall (species and pH) uptake of macronutrients was in the order of C > N > Ca > Mg > P. Reuter and Robinson (1997) reported a similar uptake pattern in tropical legume cover crops. In our study, macronutrient accumulation in all the cover crop species was parallel to dry matter accumulation. Brazilian lucerne, with low dry matter, accumulated the lowest amount of macronutrients per plant, and White jack bean with high dry matter, accumulated highest levels of macronutrients (Fageria et al., 2014a). Baligar et al. (2008) have reported significant variability in shoot dry weight and nutrient uptake among cover crop species and the uptake of macronutrient was N > Ca > P > Mg, and such variability is associated with different growth habits and amount of dry matter accumulated in the shoot of the cover crop species.

Total uptake of Mn and Zn per plant was influenced significantly by soil pH cover crop species and their interactions. Uptake of Fe, however, was only significantly influenced by cover crop species, and B uptake was influenced significantly by soil pH and cover crop species. Overall, uptake of Zn and Mn decreased with increasing soil pH (Fageria et al., 2014a). This may be associated with a decrease in shoot dry weight with increasing soil pH. Increasing soil pH reduces the micronutrient solubility, ionic concentrations in soil solution, and mobility of B, Fe, Mn, and Zn, and consequently reduces the acquisition of these elements by plants (Fageria et al., 2002). In the current study, uptake of micronutrients in the shoots of cover crop species was in the order of Mn > Fe > B > Zn. Baligar et al. (2006) have reported similar micronutrient uptake patterns in several tropical legume cover crops. Fageria et al. (2002) concluded that accumulation of essential micronutrients in plant follows the order of Mn > Fe > Zn > B > Cu and Mo. The higher uptake of Mn and Fe may be associated with higher levels of Mn and Fe in the Oxisol. Baligar et al. (2006) reported higher uptake of Mn and Fe as compared to other micronutrients in tropical legume cover crops. Interspecies variations for micronutrient uptake in legumes have been reported (Baligar et al., 2001, 2006; Fageria et al., 2002).

Average across three soil pH, white jack bean accumulated highest amounts of macro- and micronutrients, as compared to other cover crops. This may be related to higher dry matter accumulation by white jack bean. Similarly, Brazilian lucerne accumulated minimum amounts of macro- and micronutrients, as compared to other cover crops. This may be related to lower dry matter accumulation by these cover crops.

3.5 PHOSPHORUS HARVEST INDEX

Phosphorus harvest index (PHI) is defined as partitioning of total plant P into grain. It is calculated by using the following equation:

$$\text{PHI} = \frac{\text{PU in grain}}{\text{PU in grain plus straw}}$$

where

PHI is the P harvest index

PU is the P uptake in the grains

The amount of P remobilized from storage tissues is important in grain P use efficiency, varies among genotypes, and appears to be under genetic control (Clark, 1990). In calculating P harvest index, the

TABLE 3.10
Phosphorus Harvest Index of Four Crop Species

Crop Species	Phosphorus Harvest Index
Upland rice	0.77
Corn	0.79
Dry bean	0.86
Soybean	0.89

Source: Fageria, N.K., *The Use of Nutrients in Crop Plants*, CRC Press, Boca Raton, FL, 2009.

P uptake of aboveground plant parts (grain + straw) is considered, whereas roots are not included. This index is very useful in measuring P partitioning in crop plants, which provides an indication of how efficiently the plant utilized the acquired P for grain production. PHI is positively related to grain yield in lowland rice (PHI versus grain yield (Y) = $12.72 + 0.89X - 0.0029X^2$, $R^2 = 0.76^{**}$) (Fageria et al., 2013a). Fageria (2009) determined the PHI of upland rice, corn, dry bean, and soybean (Table 3.10). PHI of legumes was higher as compared to that of cereals (Table 3.10). This means that P requirements are higher for legumes. Phosphorus concentrations are typically much higher in grain than in straw at maturity of most food grain crops. Breeding for high yield in the past has increased the GHI in all crops (Sinclair, 1998), but PHI has not increased to the same extent, which means that a decrease in P concentrations in the grain occurred. Batten (1986) reported a 27% decrease in grain P from diploid to hexaploid wheat grown in pots with adequate P fertilization, while GHI more than doubled and PHI increased by only 15%. This trend was also presented in a field study by Calderini et al. (1995) for wheat cultivars released between 1920 and 1990, but a container study of cultivars released between 1840 and 1983 had increased HI but not decreased grain P (Jones et al., 1989). Veneklaas et al. (2012) reported that modern crop cultivars use P more efficiently than older cultivars, as a result mainly of improvements in GHI which are related to plant structure and C allocation traits.

3.6 PHOSPHORUS REQUIREMENTS OF CEREALS

Rice, corn, wheat, sorghum, and barley are important cereal crops worldwide. These cereals are staple food for a large portion of world population, feed for animals, and ration for poultry industries. Cereal production is much higher worldwide, as compared to legumes. In the last few decades, cereal production has increased significantly, especially rice, corn, and wheat. This increase is related to genetic improvement and better management practices, especially the use of adequate amount of fertilizers. Knowledge of P requirements of these cereals is important to improve their yield and quality.

After N, P is the second most important nutrient requiring large applications as a fertilizer to maintain high productivity of cereals (Clark, 1990). Clark (1990) reported that 60% of the P fertilizer used in the United States is used for cereal production. Many kinds of P compounds are used as a P sources but inorganic fertilizers are major sources of P for crop production. Recovery efficiency of P is less than 20% in most agroecosystems due to higher P immobilization capacity of soils (Fageria, 2009, 2014). Phosphorus immobilization (fixation and precipitation) is higher in acidic soils due to the presence of Al and Fe oxides.

Many soil factors affect the amount of P that will be available to plant roots for uptake. Such factors as solution P, buffering capacity, distribution along the soil profile, moisture, and temperature have been discussed by Barber (1980) and Ozanne (1980). Many plant factors are also important for P uptake, such as root growth, which varies with crop species and genotypes with intraspecific differences, which would be expected since plant species vary so extensively in growth habit and dimensions (Barber, 1980). Both inter- and intraspecies variations in P nutrition have been recognized among cereal species and genotypes (Clark, 1990). Interspecies differences would be expected since plant species vary so extensively in growth habit and dimensions (Barber, 1980). Phosphorus requirements for optimum yields vary with different crops. For example, wheat requires less P than canola due to the lower protein content of the seed.

3.6.1 Rice

Rice (*Oryza sativa* L.) is an important food crop for a large proportion of the world's population. Rice is the staple food in the diet of the population of Asia, Latin America, and Africa. Rice is cultivated on all the continents except Antarctica, over an area of more than 161 million ha (production of about 680 million mt), but most rice production takes place in Asia (Jena and Mackill, 2008; Kumar and Ladha, 2011). Rice occupies about 23% of the total area under cereal production in the world (Wassmann et al., 2009; Jagadish et al., 2010). Historical importance of rice in Asia is so significant that it has supported many civilizations in the river deltas of India, China, and Southeast Asia and has become deeply intertwined with the cultures in these regions (Krishnan et al., 2011). More than 90% of rice is produced and consumed in Asia (Grewal et al., 2011). For thousands of years since its domestication, Asia rice has been cultivated in diverse agro-ecosystems to meet different human demands (Xiong et al., 2011). This has resulted in tremendous genetic diversity in rice around the world, as shown by different molecular tools such as the analysis of restriction fragment length polymorphism (Zhang et al., 1992), and simple sequence repeats. As a consequence, many rice varieties with different characteristics have arisen under natural and human selection (Vaugham et al., 2007). Yan et al. (2010) studied genetic diversity in the U.S. Department of Agriculture (USDA) rice world collection and concluded that germplasm accessions obtained from the southern Asia, Southeast Asia, and Africa were highly diversified, while those from North America and western and eastern Europe had the lowest diversity.

3.6.1.1 Adequate Phosphorus Level in Rice Plant Tissues

Plant age is a very important factor in determining the adequate P level in rice plant tissues (Table 3.11). As the age advances, P concentration decreases. This is related to increase in dry weight of the plants. In mineral nutrition this is known as "dilution effects" (Fageria, 2014). Adequate level of plant tissue can serve as approximate values or guidelines because the values vary with crop genotypes and management practices adopted (Table 3.11).

3.6.1.2 Adequate Phosphorus Fertilizer Rate

Adequate P rate for rice crop varies with initial soil P level, soil types, rice cultivar planted, yield level, and management practices adopted. Adequate P rate for upland grown on a Brazilian Oxisol is shown in Figures 3.23 through 3.25. Response to applied P was quadratic when P was applied in

TABLE 3.11
Adequate Plant Tissue Level of P for Rice

Plant Age	Plant Part Analyzed	Adequate Level (g kg^{-1}) or (%)
35 days after sowing	Whole shoot	10.0–20.0 (1.0–1.5)
55 days after sowing	Whole shoot	5.9–6.6 (0.59–0.66)
73 days after sowing	Whole shoot	2.7–3.0 (0.27–0.30)
112 days after sowing	Whole shoot	1.7–2.5 (0.17–0.25)
Maturity	Straw	1.5–1.8 (0.15–0.18)

Source: Reuter, D.J. and Robinson, J.B., *Plant Analysis: An Interpretation Manual*, CSIRO, Collingwood, Victoria, Australia, 1997.

Values in the parenthesis are in percent.

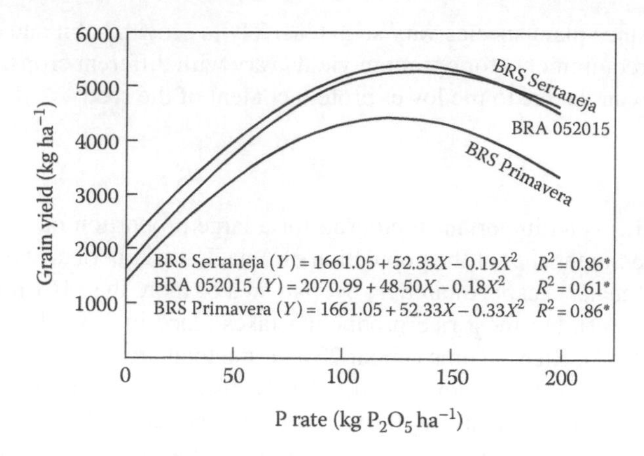

FIGURE 3.23 Response of three upland rice genotypes to P fertilization. (From Fageria, N.K. (Ed)., *The Role of Plant Roots in Crop Production*, CRC Press, 2012.)

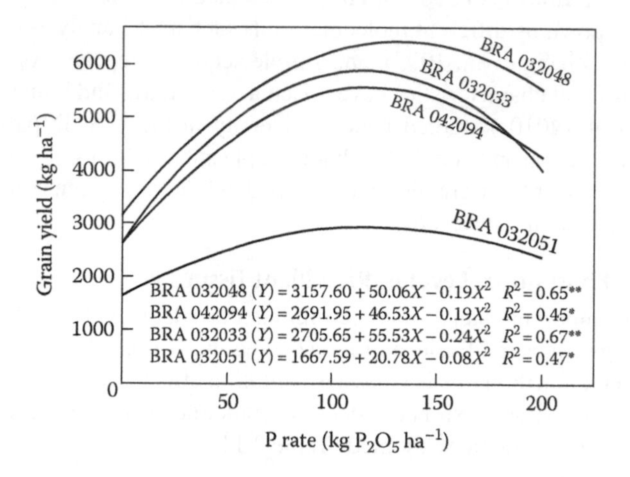

FIGURE 3.24 Response of four upland rice genotypes to P fertilization. (From Fageria, N.K. (Ed)., *The Role of Plant Roots in Crop Production*, CRC Press, 2012.)

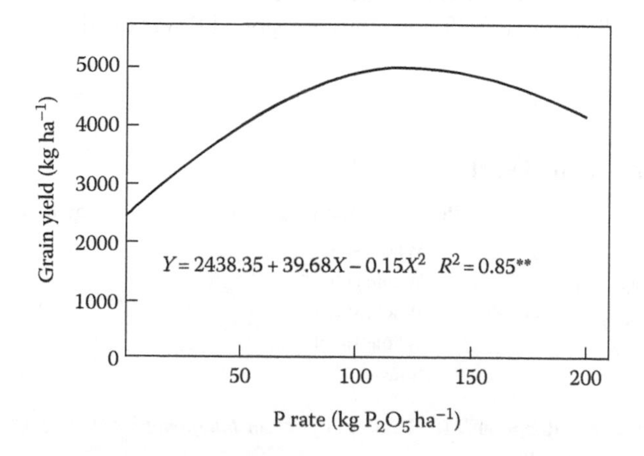

FIGURE 3.25 Response of upland rice to P fertilization. Values are averages of 12 upland rice genotypes. (From Fageria, N.K. (Ed)., *The Role of Plant Roots in Crop Production*, CRC Press, 2012.)

TABLE 3.12
Grain Yield of 12 Genotypes as Influenced by P Fertilization

	P Rate (kg P ha^{-1})				
Genotype	**0**	**22**	**44**	**66**	**88**
BRS Jaçnã	932.9	5903.2	7373.3	6750.0	5681.2
CNAi 8860	1153.4	588.8	6361.2	5470.7	4874.7
CNAi 8879	1174.9	5517.8	6500.3	4736.2	5038.8
BRS Fronteira	1259.0	5185.2	6190.5	4979.3	4182.0
CNAi 8880	1738.0	5346.7	4794.7	5477.8	4994.7
CNAi 8886	1698.3	5377.8	6017.0	5377.1	4318.0
CNAi 8885	1289.0	5139.8	5853.0	5046.3	4854.2
CNAi 8569	58.0	3482.8	3684.0	3402.7	1565.0
BRS Guará	248.9	5100.2	6163.3	5303.3	4826.2
BRS Alvorada	64.2	4435.7	4802.8	4255.2	3764.7
BRS Jaburu	41.6	3779.8	5384.7	4704.3	4441.5
BRS Biguá	128.6	3844.3	5717.5	5026.7	5005.5
F-Test					
Year (Y)			**		
P rate (P)			**		
Y × P			NS		
Genotype (G)			**		
Y × G			**		
CV (%)			24.3		

Source: Fageria, N.K. et al., *J. Plant Nutr.*, 34, 1087, 2011b.
** and NS Significant at the 1% probability level and nonsignificant, respectively.

the range of 0 to 200 kg P_2O_5 ha^{-1}. Average values of the 12 genotypes (Figure 3.25) had maximum grain yield at the addition of 132 kg P_2O_5 ha^{-1} or kg P ha^{-1} ($P_2O_5/2.29$ = 58 kg P ha^{-1}). Original P level of the experimental area was 2.0 mg P kg^{-1} by the Mehlich 1 extracting solution.

Similarly, Fageria et al. (2011b) also studied P requirement of 12 lowland rice genotypes grown on a Brazilian Inceptisol. Year and P rate significantly influenced grain yield. However, year × P rate interaction for grain yield was not significant; therefore, mean values of 2 years were pooled (Table 3.12). Grain yields of 12 genotypes increased quadratically with increasing P rate in the range 0–88 kg P ha^{-1} (Table 3.13). An increase in grain yield of genotypes was associated with an increase in panicle number and shoot dry weight with increasing P rate (Fageria et al., 2011b). Mean grain yield of 12 lowland rice genotypes was obtained with the application of 54 kg P ha^{-1} or 124 kg P_2O_5 ha^{-1} ($P \times 2.29 = P_2O_5$) applied in the furrow (Figure 3.26). Initial P level of the experimental site was 2.3 mg kg^{-1} as estimated by Mehlich 1 method.

3.6.2 Wheat and Barley

Wheat (*Triticum aestivum* L.) and barley (*Hordeum vulgare* L.) together constitute the world's most important cereal crops. Wheat is the leading cereal in terms of total world production. Barley is the world's fourth most important cereal crop after wheat, rice, and corn (Fageria et al., 2011a). Wheat and barley cereals contribute about 41% of the world production of important cereal crops. Leading wheat- and barley-producing countries in the world are China, India, the United States, Canada, Russia, Australia, France, the United Kingdom, Germany, Argentina, Turkey, and Pakistan. These cereals are important sources of food for human consumption and feed for livestock (Fageria et al., 2011a).

TABLE 3.13
Regression Equations Showing the Relationships between P Rate (*X*) and Grain Yield (*Y*) of 12 Lowland Rice Genotypes

Genotype	Regression Equation	R^2	P Rate for Maximum Grain Yield (kg ha^{-1})
BRS Jaçnã	$Y = 1234.94 + 231.06X - 2.09X^2$	0.87**	55
CNAi 8860	$Y = 1655.52 + 182.48X - 1.68X^2$	0.82**	54
CNAi 8879	$Y = 1657.66 + 172.19X - 1.60X^2$	0.73**	54
BRS Fronteira	$Y = 1564.98 +177.11X - 1.72X^2$	0.80**	51
CNAi 8880	$Y = 2148.82 +120.44X - 1.03X^2$	0.67**	58
CNAi 8886	$Y = 1973.31 + 163.50X - 1.16X^2$	0.74**	70
CNAi 8885	$Y = 1656.87 + 156.74X - 1.42X^2$	0.73**	55
CNAi 8569	$Y = 279.20 + 156.29X - 1.62X^2$	0.88**	48
BRS Guará	$Y = 659.75 + 205.91X -1.86X^2$	0.89**	55
BRS Alvorada	$Y= 500.62 + 170.98X - 1.57X^2$	0.82**	54
BRS Jaburu	$Y = 255.94 + 177.80X - 1.48X^2$	0.88**	60
BRS Biguá	$Y = 323.35 + 180.07X - 1.48X^2$	0.86**	61
Average of 12 genotypes	$Y = 1156.88 + 175.02X - 1.61X^2$	0.90**	54

Source: Fageria, N.K. et al., *J. Plant Nutr.*, 34, 1087, 2011b.
**Significant at the 1% probability level.
Values are averages of 2-year field trial.

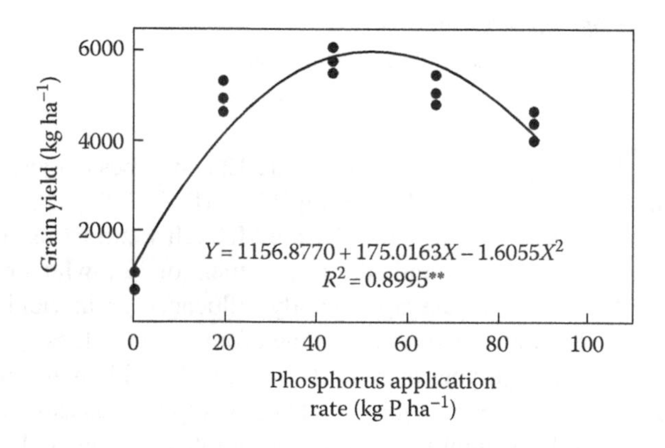

FIGURE 3.26 Response of lowland rice to P fertilization. Values are averages of 2-year field experimentation and 12 genotypes. (From Fageria, N.K. et al., *J. Plant Nutr.*, 34, 1087, 2011b.)

3.6.2.1 Adequate Phosphorus Level in Wheat and Barley Plant Tissues and Soil Test to Apply Phosphorus Fertilizer

Plant tissue analysis is one of the important nutrient deficiency/sufficiency techniques. Adequate P levels both in wheat and barley are presented in Table 3.14. In addition to plant tissue test, soil test is another important criterion to apply adequate rate of P in crops. Soil test results need to be interpreted properly to make fertilizer recommendations. To determine P application rates, a soil test is required to establish soil P index. Soil test index for P are grouped as very low, low, medium, and sufficient. Approximate soil test index and P application rate for winter cereals (wheat and barley) are given in Table 3.15.

TABLE 3.14
Adequate Level of P in the Plant Tissues of Wheat and Barley

Plant Age	Plant Part Analyzed	Adequate Level (g kg^{-1}) or (%)
Wheat		
Tillering	Leaf blade	3.5–4.9 (0.35–0.49)
Shooting	Leaf blade	3.2–4.0 (0.32–0.40)
Heading	Whole tops	2.1–5.0 (0.21–0.50)
Flowering	Leaf blade	2.5–3.4 (0.25–0.34)
Barley		
Tillering	Leaves	5.0–6.8 (0.5–0.68)
Shooting	Leaves	4.2–4.8 (0.42–0.48)
Heading	Whole tops	2.0–5.0 (0.20–0.50)
Flowering	Leaves	3.1–4.2 (0.31–0.50)

Source: Fageria, N.K. et al., *Growth and Mineral Nutrition of Field Crops*, 3rd edn., CRC Press, Boca Raton, FL, 2011a.
Values inside the parentheses are in percentage.

TABLE 3.15
Soil P Test, Their Interpretation, and P Requirements for Winter Cereals

Soil P Test (mg kg^{-1})	Interpretation	P Rate (kg ha^{-1})
0–3.0	Very low	50 (115)
3.1–6.0	Low	40 (90)
6.1–10.0	Medium	30 (70)
>10.1	Sufficient	20 (45)

Values inside the parentheses are approximate in P_2O_5.

3.6.3 Corn

Corn (*Zea mays* L.), known in much of the world as maize, is the world' third most important cereal after wheat and rice. Corn is grown primarily for grain and secondarily for fodder and raw material for industrial processes. Grain is used for both human and animal consumption. Vegetative parts of the plant are cut green and either dried or utilized as silage for animal feed. Domestication and selection of corn probably began in central or southwestern Mexico about 7000 years ago (Goodman, 1988). Date of origin for cultivated maize in the highlands of central Mexico is unknown, but Palomero Toluqueño was considered an ancient indigenous race by Wellhausen et al. (1952). Kato (1984) proposed a multicenter domestication of maize, with two of the four centers in central Mexico. Highland maize probably came from the higher-altitude centers. Excavations at Teotihuacán, in the Valley of Mexico, have uncovered ears with the characteristics of Cónico (Wellhausen et al., 1952), indicating that maize similar to current types was important for the ancient civilizations of highland Mexico and has been cultivated at altitudes above 2000 m for a millennium (Eagles and Lothrop, 1994). Distinguishable groups of cultivars arose in Mexico and Central America, in the northeastern United States, on the northern coast of South America, in the Andes, and in central Brazil. Spanish and Portuguese quickly distributed corn throughout the world in the sixteenth century (Jones, 1985). Corn is currently grown in more countries than any other cereal and has produced the largest grain yield of any cereal.

In the early 1990s more than 50% of the total world area planted with corn was in Latin America, Africa, and Asia, but probably less than 35% of the total world grain corn production was in these areas (Russell, 1991). Except for a few countries in these three continents, mean yields per hectare are very low. Chief corn-producing countries are the United States, Russia, Romania, the former Yugoslavia, Hungary, Italy, China, Brazil, Mexico, South Africa, Argentina, India, and Indonesia (Fageria et al., 2011a). Main reason for its wide distribution is that corn has many advantages, which include its high yield per unit of labor and per unit of area. It is a compact, easily transportable source of nutrition. Its husks give protection from birds and rain. It can be harvested over a long period, can be stored well, and can even be left dried in the field until harvesting is convenient. Corn grain has been traditionally used for direct human consumption, but at present the major use of corn is as animal feed. In the United States, approximately 75% of the grain is used as an animal feed and 20% as a source of industrial products (Tollenaar and Dwyer, 1998). It provides numerous useful food products, and it is frequently preferred to sorghum and the millets (*pancilin menhouse*) (Jones, 1985). Thomison et al. (2003) and Tanaka and Maddonni (2008) reported that in feed rations of livestock and poultry, grain with high oil concentration is preferred because of its energy value and as a substitute for animal fats. Chemical composition of corn grain is about 77% starch, 2% sugar, 9% protein, 5% fat, 5% pentosan, and 2% ash (Purseglove, 1985; Maddonni and Otegui, 2006).

3.6.3.1 Adequate Phosphorus Level in Corn Plant Tissues and Soil Test to Apply Phosphorus Fertilizer

Plant tissue test provides useful information on nutritional status of a crop plant at a defined growth stage. However, sufficient experimental results are required for adequate interpretation of the tissue analysis data. Sufficiency level of P in the corn tissues is presented in Table 3.16. In addition to plant tissue test, soil test is another important criterion to recommend P fertilizer for corn. Soil test interpretation for P and relative P application rate are presented in Table 3.17. These soil tests are designated as very low, low, medium, and sufficient or high. When soil tests show very low or low level of P in the soil, crop response to applied P is expected, provided some other factors are not yield limiting.

TABLE 3.16
Adequate Plant Tissue Level of P for Corn

Plant Age	Plant Part Analyzed	Adequate Level (g kg^{-1}) or (%)
30–45 days after sowing	Whole shoot	4.0–8.0 (0.4–0.8)
Tasseling to initial silk	Ear leaf	2.5–4.5 (0.25–0.45)
Silking	Ear leaf	2.3–2.5 (0.23–0.25)

Source: Reuter, D.J. and Robinson, J.B., *Plant Analysis: An Interpretation Manual*, CSIRO, Collingwood, Victoria, Australia, 1997.

Values inside the parentheses are in percentage.

TABLE 3.17
Soil P Test, Their Interpretation, and P Requirements for Corn

Soil P Test (mg kg^{-1})	Interpretation	P_2O_5 Rate (kg ha^{-1})
0–6.0	Very low	140
7.0–15.0	Low	100
16.0–40	Medium	60
>40	Sufficient	40

Source: Raij, B.V. et al., Liming and fertilizer recommendations for the state of São Paulo, Technical Bulletin No. 100, Campinas Institute of Agronomy, São Paulo, Brazil, 1985.

Values inside the parentheses are approximate in P_2O_5. Yield expected more than 8 mg ha^{-1} and resin used as an extracting agent.

3.6.4 Sorghum

Sorghum is ranked fifth among cereals behind corn, rice, wheat, and barley. Sorghum is the major cereal of rainfed agriculture in the semiarid tropics (SAT). Grain sorghum is a major crop grown under semiarid conditions in the United States and other parts of the world (Bandaru et al., 2006), and it is a dietary staple of more than 500 million people in more than 30 countries (National Research Council, 1996). Over 55% of the global sorghum production is in the SAT, and of the total SAT production, Asia and Africa contribute about 65%, of which 34% is harvested in India (Sahrawat et al., 1996). Cultivated sorghum originated in northeast Africa, where the greatest diversity of types exists. Areas of origin are most likely now occupied by Ethiopia and part of Sudan, from which sorghum probably spread to West Africa (Doggett, 1970). There is evidence of sorghum in Assyria by 700 BC and in India and Europe by AD 1 (Eastin, 1983). Cultivated sorghums were first introduced to America and Australia about 100 years ago. Domestication and cultivation of sorghum has been practiced worldwide, especially in the top 10 sorghum producing countries. Production (in millions of mt) in 2007 was mainly in the United States of America (12.6), Nigeria (9.1), India (7.2), Mexico (6.2), Sudan (5.8), Argentina (2.8), China (2.4), Ethiopia (2.2), Burkina Faso (1.6), and Brazil (1.4) (FAO, 2009, http://faostat.fao.org/site/339/default.aspx). Sorghum is the basic cereal food in parts of Asia and Africa, while in the United States and Europe it serves mainly as feed for poultry and livestock. Sorghum stems and foliage are often used as animal fodder, and in some areas, the stems are used as building material and fuel. Some sorghums have sweet, juicy stems that contain up to 10% sucrose and are chewed or used to produce syrup. Sorghum is also widely used for brewing beer, particularly in Africa, and it is among the most widely adapted of the warm-season cereals with potential for biomass and fuel production. High-energy sorghum consists of hybrids of grain and sweet sorghum types that are currently being developed for both grain and biomass production (Hons et al., 1986). They provide slightly lower grain yields than conventional grain sorghums but produce large amounts of stover with high carbohydrate concentrations.

3.6.4.1 Adequate Phosphorus Level in Sorghum Plant Tissues and Soil Test to Apply Phosphorus Fertilizer

Plant tissue analysis is an important tool in the diagnosis of nutrient deficiency or sufficiency in crop plants. Phosphorus sufficient concentration data in the tissue of sorghum are presented in Table 3.18. However, plant analysis criteria for annual crops varies widely with plant age, plant part, and environmental conditions. As a result, it is not possible to state for a specific crop and nutrient that any one set of criteria or criterion is necessarily optimum over another. Readers should examine the data assembled in this chapter and select tests which best suit their own situation (Reuter and Robinson, 1997). Soil test analysis data and P recommendations for sorghum are presented in Table 3.19. These results are applicable to Brazilian conditions for sorghum production. However, values can also be used for other climatic conditions with slight modifications.

TABLE 3.18
Adequate Plant Tissue Level of P for Sorghum

Plant Age	Plant Part Analyzed	Adequate Level (g kg^{-1}) or (%)
30 days after sowing	Whole shoot	2.2–2.9 (0.22–0.29)
48 days after sowing	Whole shoot	0.31–0.36 (0.31–0.36)
Vegetative and early flowering	Third leaf blade below head	2.0–5.0 (0.20–0.50)
Full heading	Younger mature leaf	2.0–5.0 (0.20–0.50

Source: Reuter, D.J. and Robinson, J.B., *Plant Analysis: An Interpretation Manual*, CSIRO, Collingwood, Victoria, Australia, 1997.

Values inside the parentheses are in percentage.

TABLE 3.19
Soil P Test, Their Interpretation, and P Requirements for Sorghum

Soil P Test (mg kg^{-1})	Interpretation	P_2O_5 Rate (kg ha^{-1})
0–6	Very low	70
7–15	Low	50
16–40	Medium	30
>40	Sufficient	20

Source: Raij, B.V. et al., Liming and fertilizer recommendations for the state of São Paulo, Technical Bulletin No. 100, Campinas Institute of Agronomy, São Paulo, Brazil, 1985.

Values inside the parentheses are approximate in P_2O_5. Yield expected more than >3 mg ha^{-1} and resin used as an extracting agent.

3.7 PHOSPHORUS REQUIREMENT OF LEGUMES

Legumes, broadly defined by their unusual flower structure, podded fruit, and the ability of 88% of the species examined to date to form nodules with rhizobia (Faria et al., 1989; Graham and Vance, 2003), are second only to the Gramineae in their importance to humans. Area planted, grain, and oil production by important crops, including legumes, are presented in Table 3.20. The 670–750 genera and 18,000–19,000 species of legumes include important grain, pasture, and agroforestry species (Polhill et al., 1981). Grain and forage legumes are grown on some 180 million ha, or 12%–15% of the Earth's arable land area (Table 3.20). They account for 27% of the world's primary crop production, with grain legumes alone contributing to 33% of the dietary protein of humans (Graham and Vance, 2003). In rank order, dry bean, pea (*Pisum sativum*), chickpea (*Cicer arietinum*), broad bean, pigeon pea (*Cajanus cajan*), cowpea (*Vigna unguiculata*), and lentil (*Lens culinaris*) constitute the primary dietary legumes (National Academy of Science, 1994).

Legumes such as soybean and peanuts (*Arachis hypogaea*) provide more than 30% of the world's processed vegetable oil (Table 3.20). These two legumes are also rich sources of dietary protein for chicken and pork industries (Graham and Vance, 2003). In addition, biological N fixation is another spectacular natural phenomenon associated with legumes. Biological N fixation reduces the cost of

TABLE 3.20
Area Planted, Production of Grain, and Oil of Important Food Crops

Crop Species	Grain Yield (mg × 10^6)	Area Planted (ha × 10^6)	Crop Species	Oil Production (mg × 10^6)
Corn (*Zea mays*)	609	138	Soybean	26.8
Rice (*Oryza sativa*)	590	152	Peanut	5.3
Wheat (*Triticum* spp.)	583	214	Canola	12.6
Barley (*Hordeum vulgare*)	141	54	Palm	23.9
Grain legumes	275	160	Sunflower	9.1
Forage legumes	605	20	Cotton	4.1
Potatoes (*Solanum tuberosum*)	308	19	Olive	2.7
Cassava (*Manihot esculenta*)	179	20	Coconut	3.6
	—	—	Corn	2.0
Total	3320	777	Total	90.1

Source: Food and Agricultural Organization of the United Nations (FAO), Database, 2013, http://www.apps.fao.org/collections.

crop production. The ability of legumes to sequester C has been used as a means to offset increase in atmospheric CO_2 levels while enhancing soil quality and tilth (Graham and Vance, 2003).

3.7.1 Dry Bean

Dry bean, also known as common bean, is an important seed legume crop and supplies a large part of the daily protein requirement of the people of South America, the Caribbean, Africa, and Asia (Fageria et al., 2011a). Dry bean is a principal source of protein for more than 500 million people in Latin America and Africa. When consumed as snap beans, it is an important dietary source of vitamins and minerals in Asia (Yan et al., 1995; Fageria, 2002, 2006; Fageria et al., 2008). Dry bean seeds are rich in protein (~20%). Though dry bean protein is deficient in sulfur-containing amino acids, it complements cereals and other carbohydrate-rich foods in providing near-perfect nutrition to people of all ages. Moreover, a regular intake of beans helps lower cholesterol and cancer risks (Singh, 1999). Beans are also one of the best non-meat sources of Fe, providing 23%–30% of daily recommended levels from a single serving (Perla et al., 2003; Shimelis and Rakshit, 2004). Santalla et al. (2001) reported that common bean is potentially the most valuable source of plant protein for human consumption in many parts of South Europe and contributes significantly to the sustainability of traditional cropping systems.

Among major food legumes, dry bean is the third most important worldwide, superseded only by soybean and peanuts. Land area devoted to bean production in developing countries has increased steadily in the last several decades (CIAT, 1992). However, production has not kept pace with population growth, and significant yield increases are required in Latin America and Africa to satisfy expected demand (Fageria and Barbosa Filho, 2008). Bean production in developing countries is often on marginal land, and few developing countries have significant reserves of arable land that can be opened to bean cultivation, so increased bean production will largely have to come about through increased yield per hectare rather than expansion of area under cultivation.

Mean bean yields in most developing countries are less than 20% of yield potential, indicating that substantial improvement in bean production could be realized by increasing yields per unit land are a (Yan et al., 1995). For example, in Brazil the mean common bean yield is less than 1000 kg ha^{-1}. In contrast, experimental yields of more than 3000 kg ha^{-1} are frequently reported (Fageria, 2006, 2008; Fageria and Barbosa Filho, 2008; Fageria et al., 2008b). Mean yields of common bean are less than 1 mg ha^{-1} in most developing countries and less than 1.5 mg ha^{-1} in most developed countries (Laing et al., 1984; Fageria, 2006; Fageria and Barbosa Filho, 2008). Low yields are associated with water deficit, high incidence of diseases and insects, and limited use of inorganic fertilizers. Minimal research has been devoted to improving productivity of these crops. Given the importance of bean as human food sources, information related to plant tissue test for P, and P fertilizer recommendations is reviewed in this section.

3.7.1.1 Adequate Phosphorus Level in Dry Bean Plant Tissues and Soil Test to Apply Phosphorus Fertilizer

Adequate plant tissue test for P during growth cycle of dry bean is presented in Table 3.21. These values can serve as guideline for the sufficiency level of P in the plant part analyzed. Similarly, P fertilizer recommendations for dry bean based on soil test results and calibration with grain yield are presented in Figure 3.27 and Table 3.22. When extractable soil P increased, grain yield of dry bean grown on an Inceptisol of central Brazil increased quadratically (Figure 3.27).

3.7.2 Soybean

Soybean is one of the most important legume crops in the world. Approximately one-third of the world's edible oils and two-thirds of the world's protein meal are derived from soybean (Golbitz, 2004). Brazil and the United States are the largest soybean-producing countries in the world followed

TABLE 3.21
Adequate Plant Tissue Level of P for Dry Bean

Growth Stage	Plant Part Analyzed	Adequate Level (g kg^{-1}) or (%)
Vegetative	YML	2.5–6.0 (0.25–0.6)
Onset of flowering	YML	2.5–5.0 (0.25–0.50)
10% flowering	YMB	4.0–4.5 (0.4–0.45)
10% flowering	YMB + P	2.5–3.5 (0.25–0.35)

Source: Reuter, D.J. and Robinson, J.B., *Plant Analysis: An Interpretation Manual*, CSIRO, Collingwood, Victoria, Australia, 1997.

Values in the parenthesis are in percentage.

YML, young mature leaf; YMB, youngest upper most mature leaf blade; YMB + P, youngest upper most mature leaf blade + petiole.

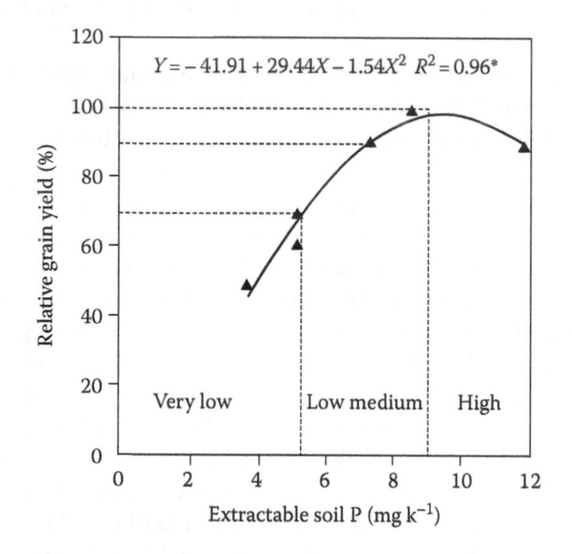

FIGURE 3.27 Relative grain yield response of dry bean to Mehlich 1 extractable soil P in an acid Inceptisol of central Brazil. (From Fageria, N.K. and Santos, A.B., *Braz. J. Agric. Eng. Ambien.*, 2, 119, 2008; Fageria, N.K. and Baligar, V.C., *Handbook Soil Acidity*, 359, 2003.)

TABLE 3.22
Soil P Test, Their Interpretation, and P Requirements for Dry Bean

Soil P Test (mg kg^{-1})	Interpretation	P_2O_5 Rate (kg ha^{-1})
0–5.3	Very low	150
5.3–7.1	Low	100
7.1–9.0	Medium	75
>9.0	Sufficient	50

Source: Fageria, N.K. and Santos, A.B., *Braz. J. Agric. Eng. Ambien.*, 2, 119, 1998.

Mehlich 1 extracting solution was used to determine soil test P.

by Argentina. In Brazil, soybean is one of the most important crops for export as well as national consumption. It is grown throughout the country from South to North and West to East. Central part of Brazil, locally known as "Cerrado" region, is a dominant area in soybean production. Most of the soils in this region are Oxisols (46%), Ultisols (15%), and Entisols (15%), with low natural soil fertility, high aluminum saturation, and high P fixation capacity (Fageria and Stone, 1999). Although low fertility is characteristic of acid soils, these vast areas have a large proportion of favorable topography for agriculture, adequate temperatures for plant growth throughout the year, sufficient moisture availability year-round in 70% of the region, and for 6–9 months in the remaining 30% of the region (Narro et al., 2001). When the chemical constraints are eliminated by liming and using adequate amounts of fertilizers, the productivity of Oxisols and Ultisols is among the highest in the world (Sanchez and Salinas, 1981).

3.7.2.1 Adequate Phosphorus Level in Soybean Plant Tissues and Soil Test to Apply Phosphorus Fertilizer

Adequate P level in the plant tissue test of soybean is presented in Table 3.23, and adequate soil P test level, their interpretation, and P fertilizer recommendations are presented in Table 3.24.

3.7.3 Peanuts

The peanut, commonly known as groundnut, earthnut, monkey nut, pindar, or goober, is both an oilseed crop and a food grain legume (Krapovickas, 1969). One of the world's most important oilseed crop, along with soybean, cottonseed (*Gossypium herbaceum*), rapeseed (*Brassica napus*), and sunflower (*Helianthus annuus*), it is also a rich source of vegetable protein and is grown in 107 countries around the world (Wynne and Gregory, 1981; Upadhyaya et al., 2005). Approximately 53% of the total global production of peanut is crushed for high-quality edible oil, 32% for confectionary consumption, and the remaining 15% is used for feed and seed production (Dwivedi et al., 2003). Peanut seeds contain 25%–30% protein, about 50% oil, 20% carbohydrate, and 5% fiber and ash. Properties of peanut oil are determined by the fatty acid composition. Approximately 90% of peanut oil is composed of palmitic acid (16 carbons and no double bonds: 16:0), oleic acid (18:1), and linoleic acid (18:2). Although many studies have identified genetic differences in fatty acid composition in peanuts, most have examined a limited number of genotypes (Knauft and Wynne, 1995).

Cultivated peanut occurs throughout the tropical and temperate regions of the world; however, wild species of *Arachis* are found only in South America, specifically the countries of Brazil, Argentina, Bolivia, Paraguay, and Uruguay (Singh and Simpson, 1994). The genus *Arachis* contains a rich diversity of plant types. Both annuals and perennials are known, and although most species reproduce by seed, some are rhizomatous and reproduce largely through vegetative means.

TABLE 3.23
Adequate Plant Tissue Level of P for Soybean

Growth Stage	Plant Part Analyzed	Adequate Level (g kg^{-1}) or (%)
20 days after sowing	Whole shoot	3.3–4.0 (0.33–0.40
25–33 days after sowing	Whole shoot	2.5–3.0 (0.25–0.30)
Early flowering	LB	>2.0 (0.2)
Early pods	LB	>1.5 (0.15)

Source: Reuter, D.J. and Robinson, J.B., *Plant Analysis: An Interpretation Manual*, CSIRO, Collingwood, Victoria, Australia, 1997.

Values in the parenthesis are in percentage. LB, leaf blade (excluding sheath or petiole).

TABLE 3.24
Soil P Test, Their Interpretation, and P Requirements for Soybean

Soil P Test (mg kg^{-1})	Interpretation	P_2O_5 Rate (kg ha^{-1})
0–6.0	Very low	100
7.0–15.0	Low	80
16.0–40	Medium	60
>40	Sufficient	40

Source: Raij, B.V. et al., Liming and fertilizer recommendations for the state of São Paulo, Technical Bulletin No. 100, Institute of Agronomy, Campinas, São Paulo, Brazil, 1985.

Values in parenthesis are approximate in P_2O_5. Yield expected more than 3 mg ha^{-1} and resin extraction used to determine soil test P.

Species occur in ecozones as different as poorly drained, swampy areas near sea level, dry areas, and mountainous regions at elevations up to 1600 m.

Arachis hypogaea L. originated in southern Bolivia or northern Argentina (Gregory et al., 1980). It is generally cultivated for human food and oil around the world. It is also used as a fodder for cattle in many Asian countries (Ramakrishna et al., 2006). Peanut was probably brought to Africa from Brazil by the Portuguese early in the sixteenth century and somewhat later was transported by the Spanish from the west coast of South America to Asia. Peanut may have reached the United States by way of slave ships from West Africa, although precisely when and where it was introduced is not known (Gibbons, 1980).

India, China, Indonesia, Myanmar, and Vietnam have the largest peanut-growing areas in Asia, while in Africa the major producers are Nigeria, Senegal, Sudan, Democratic Republic of Congo, Chad, Mozambique, Zimbabwe, Burkina Faso, Uganda, and Mali. In the Western Hemisphere, the United States, Brazil, Argentina, and Mexico are the leading peanut producers. Peanut production in the United States occurs from humid areas of Georgia and Florida to arid areas of the southern High Plains of Texas (Kiniry et al., 2005). In the United States, Valencia peanuts for the in-shell market are predominantly grown in eastern New Mexico and West Texas (Dwivedi et al., 2008).

Seventy percent of the world's peanut production occurs in the semiarid tropics, where drought, diseases, and insects are the main yield-limiting factors. Smart (1994) reviewed global production practices and noted that they vary considerably. In the United States, Australia, and portions of South America, the crop is grown with intensive management, generally with high levels of mechanical and chemical inputs. In parts of Africa and Southeast Asia the crop is grown in mixtures with other species, mainly to provide food and cooking oil for the farmer. In many countries the crop is grown in monoculture as a cash crop, primarily for export. The intensity of management varies considerably around the world, depending on the economic return for the crop or the role of peanuts in farm subsistence. Altering plant population and row pattern can affect crop yield, quality factors, and pest development (Lanier et al., 2004). In the United States, peanut is generally grown in single rows spaced 91–102 cm apart; however, research suggests that pod yield can be increased by growing peanut in twin rows (18–23 cm spacing) on beds spaced 91–102 cm apart (Jordan et al., 2001; Lanier et al., 2004).

3.7.3.1 Adequate Phosphorus Level in Soybean Plant Tissues and Soil Test to Apply Phosphorus Fertilizer

Adequate levels of P in the plant tissues of peanuts at different growth stages are presented in Table 3.25. These values can be used in the interpretation of plant tissue results for P in the peanuts. Similarly, P soil test result interpretation and P recommendations for peanut are presented in Table 3.26.

TABLE 3.25
Adequate Plant Tissue Level of P for Peanuts

Growth Stage	Plant Part Analyzed	Adequate Level (g kg^{-1}) or (%)
25–33 days after sowing	Whole shoot	2.8–4.2 (0.28–0.42)
50–57 days after sowing	Whole shoot	1.9–2.4 (0.19–0.24)
70 days after sowing	YMB	3.2 (0.32)
Pre-flowering or flowering	YMB	2.5–5.0 (0.25–0.50)
R2 (pegging)	YMB	2.4–4.0 (0.24–0.40)

Source: Reuter, D.J. and Robinson, J.B., *Plant Analysis: An Interpretation Manual*, CSIRO, Collingwood, Victoria, Australia, 1997.
Values in the parenthesis are in percentage.
YMB, young mature blade.

TABLE 3.26
Soil P Test, Their Interpretation, and P Requirements for Peanuts

Soil P Test (mg kg^{-1})	Interpretation	P_2O_5 Rate (kg ha^{-1})
0–6.0	Very low	80
7.0–15.0	Low	60
16.0–40	Medium	40
>40	Sufficient	20

Source: Raij, B.V. et al., Liming and fertilizer recommendations for the state of São Paulo, Technical Bulletin No. 100, Campinas Institute of Agronomy, São Paulo, Brazil, 1985.
Yield expected more than 3 mg ha^{-1} and resin extraction used to determine soil test P.

3.8 PHOSPHORUS NUTRITION OF COTTON, SUGARCANE, AND SUGAR BEET

Cotton is an important fiber crop worldwide. Similarly, sugarcane and sugar beet are important sugar crops in the tropical and temperate climates. Knowledge of P nutrition of these crops is fundamental to improve yields and quality of lint and sugar.

3.8.1 Adequate Phosphorus Level in Cotton, Sugarcane, and Sugar Beet Plant Tissues and Soil Test to Apply Phosphorus Fertilizer

According to the United Nations Food and Agriculture Organization, the ten countries that produced the most cotton lint in 2007 (in million mt) were China (7.6), India (4.4), the United States (4.2), Pakistan (2.0), Brazil (1.4), Uzbekistan (1.1), Turkey (1.0), Syria (0.4), Turkmenistan (0.3), and Greece (0.3) (FAO, 2009). Historically extra-long staple production has been dominated by Egypt, Sudan, and the former USSR, while India, Pakistan, and China produced virtually all the short-staple cotton (Phillips, 1976). Although cotton is mostly grown for fiber, the seeds are also important. Cottonseed oil is used for culinary purposes, and the oil cake residue is a protein-rich feed for ruminant livestock. Adequate plant tissue level of P for cotton is presented in Table 3.27, and soil test P results interpretation and P recommendations are presented in Table 3.28.

Sugarcane is an important economic crop in the tropics and subtropics due to its high sucrose content and increasing interest in its bioenergy potential (Gilbert et al., 2007). Sugarcane (*Saccharum* spp. hybrid) is the world's most important sugar crop (Bakker, 1999; Cheeroo-Nayamuth et al., 2000). Sugarcane is an erect, very robust, tillering, perennial C_4 grass and is grown primarily for sugar

TABLE 3.27
Adequate Plant Tissue Level of P for Cotton

Growth Stage	Plant Part Analyzed	Adequate Level (g kg^{-1}) or (%)
42 days after sowing	Whole shoot	3.0 (0.30)
Vegetative to flowering	YMB	2.5–5.0 (0.25–0.50)
Flowering-boll development	YML	3.0–5.0 (0.30–0.50)

Source: Reuter, D.J. and Robinson, J.B., *Plant Analysis: An Interpretation Manual*, CSIRO, Collingwood, Victoria, Australia, 1997.
Values in the parenthesis are in percentage.
YMB, young mature blade; YML, youngest mature leaf.

TABLE 3.28
Soil P Test, Their Interpretation, and P Requirements for Cotton

Soil P Test (mg kg^{-1})	Interpretation	P_2O_5 Rate (kg ha^{-1})
0–6	Very low	100
7–15	Low	80
16–40	Medium	60
41–80	Sufficient	40

Source: Raij, B.V. et al., Liming and fertilizer recommendations for the state of São Paulo, Technical Bulletin No. 100, Campinas Institute of Agronomy, São Paulo, Brazil, 1985.
Yield expected more than 3 mg ha^{-1} and resin extraction used to determine soil test P.

(sucrose), but molasses, ethyl alcohol, and fiber (bagasse) are important by-products. Commercial sugarcane cultivars are complex interspecies hybrids of up to five species of *Saccharum robustum*; *Saccharum sinense, Saccharum barberi, Saccharum officinarum*, the "noble" canes; and *Saccharum spontaneum*, a freely tillering wild species used as a source of vigor and disease resistance (Lingle and Tew, 2008; Wang et al., 2008). The "noble" canes may have been selected from *S. robustum* by Stone Age cultures in New Guinea. They were spread throughout the Pacific and Southeast Asia prior to the arrival of Europeans. Cane was taken by the Spanish and Portuguese to the New World to form the basis of sugarcane culture in the sixteenth century. In the late eighteenth century, more desirable cultivars of *S. officinarum* were introduced. Modern sugarcane breeding began at the end of the nineteenth century, when viable true seeds were discovered (Jones, 1985).

Brazil, India, and China are the three largest sugarcane producing countries, with an annual production of 739, 341, and 126 million mt, respectively, in 2013. Other major producers (and their production in millions of mt in 2005) include Thailand (50), Pakistan (47), Mexico (45), Colombia (40), Australia (38), Philippines (31) and the United States (26) (FAO, 2009).

Brazil has a long tradition of growing sugarcane. In sixteenth century, it was the world's major source of sugar (Courtenay, 1980; Hartemink, 2008). Currently, sugarcane is a major source of ethanol, with one hectare of sugarcane yielding at 82 mg ha^{-1} producing about 7000 L of ethanol (Hartemink, 2008). Brazil currently produces about 31% of global production, and it is the largest producer, consumer, and exporter of ethanol for fuel (Andrietta et al., 2007; Hartemink, 2008). Sugar and ethanol industry were valued to reach about 17% of Brazil's agricultural output (Valdes, 2007; Hartemink, 2008). Cultivation of sugarcane for bioethanol is increasing, and the area under sugarcane is expanding in Brazil. Adequate levels of P in the plant tissue of sugarcane are presented in Table 3.29. Phosphorus fertilizer recommendations based on soil test are presented in Table 3.30.

TABLE 3.29
Adequate Plant Tissue Level of P for Sugarcane

Growth Stage	Plant Part Analyzed	Adequate Level (g kg^{-1}) or (%)
Rapid growth	TVD	2.2–3.0 (0.22–0.30)
3- to 6-month plant	TVD	2.1–3.5 (0.21–0.35)
7-month ratoon	TVD	2.1–3.0 (0.21–0.30)

Source: Reuter, D.J. and Robinson, J.B., *Plant Analysis: An Interpretation Manual*, CSIRO, Collingwood, Victoria, Australia, 1997.
Values in the parenthesis are in percentage.
TVD, top visible dewlap (sugarcane).

TABLE 3.30
Soil P Test, Their Interpretation, and P Requirements for Sugarcane

Soil P Test (mg kg^{-1})	Interpretation	P_2O_5 Rate (kg ha^{-1})
0–6	Very low	120
7–15	Low	100
16–40	Medium	80
41–80	Sufficient	60

Source: Raij, B.V. et al., Liming and fertilizer recommendations for the state of São Paulo, Technical Bulletin No. 100, Campinas Institute of Agronomy, São Paulo, Brazil, 1985.
Yield expected more than 3 mg ha^{-1} and resin extraction used to determine soil test P.

Sugar beet is a member of the *Chenopodiaceae* family. It has 11–13 species in Europe and Asia (Letschert, 1993; Mabberley, 1997). Agronomically, sugar beet is one of the four cultural types of *Beta vulgaris* L.: sugar beet, red beet or garden beet, Swiss chard, and fodder beet (Ulrich et al., 1993). At an early growth stage, shoots of all cultural types, especially Swiss chard, may serve as a leafy vegetable, and at the later growth stages serve as livestock forage. Fleshy root and tops of the red beet are excellent vegetables (Ulrich et al., 1993). Sugar beet is a biennial plant that is agriculturally important because of its ability to store sucrose to high concentrations in its storage root. Ulrich et al. (1993) reported that the commercial production of beet sugar has been an outstanding achievement scientifically and economically as an alternative source of sugar when other supplies are insecure. Crystalline sugar was a scarce luxury in the Western world before the seventeenth century (Campbell, 1984). Originally, all sugar came from sugarcane grown in the tropics but at the present time, beet sugar accounts for nearly half the total world production of the refined product (Campbell, 1984). Sugar is extracted from the beet in factories, using a process similar to that for sugarcane.

Sugar yield in the storage roots depends on the pathway photosynthate is partitioned within in the crop and is the product of the total amount of dry matter produced during growth, the proportion allocated to the storage root, and the proportion of the storage root dry matter accumulated as sucrose (Bell et al., 1996). In addition, the efficiency of the sugar extraction process is dependent on the concentration of solutes other than sucrose (K, Na, amino acids, and glycine betaine), and the interrelationships among accumulation of sucrose. These so-called impurities are important determinants of root quality. Propagation of sugar beet is always from seed. Regarding origin and domestication, there is no archeological records exists for preclassical times; linguistic records place leafy forms of the cultivated beet to the eighteenth century BC in Babylonia (Siemonsma and Piluek, 1993; Zohary and Hopf, 1993). Sugar beet is used for ethanol production in Europe. The largest ethanol-producing factory from sugar beet is located in France. White varieties of sugar beet are used for ethanol production rather than red varieties. White varieties roots are larger as compared to red sugar beet varieties.

TABLE 3.31
Adequate Plant Tissue Level of P for Sugar Beet

Growth Stage	Plant Part Analyzed	Adequate Level (g kg^{-1}) or (%)
50–60 days after sowing	ML	3.5–6.0 (0.35–0.60)
50–80 days after sowing	L	4.5–11.0 (0.45–1.10)
Maturity	Whole shoot	>2.1 (0.21)

Source: Reuter, D.J. and Robinson, J.B., *Plant Analysis: An Interpretation Manual*, CSIRO, Collingwood, Victoria, Australia, 1997.
Values in the parenthesis are in percentage.
TVD, top visible dewlap (sugarcane).

TABLE 3.32
Soil P Test, Their Interpretation, and P Requirements for Sugar Beet

Soil P Test (mg kg^{-1})	Interpretation	P_2O_5 Rate (kg ha^{-1})
0–40	Very low to medium	400
>40	Sufficient	200

Source: Raij, B.V. et al., Liming and fertilizer recommendations for the state of São Paulo, Technical Bulletin No. 100, Campinas Institute of Agronomy, São Paulo, Brazil, 1985.
Yield expected more than 3 mg ha^{-1} and resin extraction used to determine soil test P.

Europe is the largest producer of sugar beet followed by North and Central America (Simpson and Conner-Ogrorzaly, 1995). Sugar beet roots contain 7%–10% carbohydrates (sucrose), 1.5%–2% protein, and small quantities of fat, ash, and fiber. Roots contain a lower mineral and vitamin content than most other vegetables. Red color is produced by betanins (red betacyanins). Geosmin causes the earthy smell (Austin et al., 1991). To produce good yields of sugar beet, growers need to plant early to a uniform stand, meet the water and fertilizer requirements of the crop, prevent weed competition, and control pests. Adequate plant tissue test of P for sugar beet is presented in Table 3.31. Phosphorus fertilizer recommendations based on soil test are presented in Table 3.32.

3.9 PHOSPHORUS NUTRITION OF TUBER AND ROOT CROPS

Tuber crops like potato (*Solanum tuberosum*) and root crops like cassava (*Manihot esculenta*) and sweet potato (*Ipomoea batatas*) are important food crops worldwide. Potato is eaten as a food crop in all continents. The importance of these tuber and roots crops as a food is enormous due to their higher starch content. Starch content in the roots of cassava and sweet potato ranges between 65% and 90% of the total dry matter, as a result of a long period of starch deposition. Patterns of starch accumulation are specific to the species and are related to the particular pattern of differentiation of the organ (Preiss and Sivak, 1996). Root crops can be a source of calorie supplement along with cereals. A large population of tropics depends on root crops for calorie supplements. For example, roasted cassava flour is daily eaten with rice and bean by people of northeastern region of Brazil. Use of root crop as a food is not only restricted to tropics. China and Japan make extensive use of the sweet potato, even though these countries lie mostly within temperate zones. Similarly, sugar beet roots are eaten by Europeans as a cooked salad or vegetables. There is a large number of root crops, and it is not possible to discuss all of them in one chapter. Hence, the ecophysiology of major root crops, that is, sugar beet, cassava, sweet potato, and carrot, will be discussed in this chapter because they are largely used as food crops in developed as well as developing countries.

3.9.1 Adequate Phosphorus Level in Potato, Cassava, and Sweet Potato Plant Tissues and Soil Test to Apply Phosphorus Fertilizer

Plant tissue test is an important criterion to know whether a given nutrient is deficient or sufficient in the plants at a given growth stage. Sufficient levels of P in the plant tissues of potato are presented in Table 3.33. Similarly, soil test is commonly used to make fertilizer recommendations for annual crops. Soil test of P for potato and P fertilizer recommendations are presented in Table 3.34. Similarly, adequate P level in the plant tissues of cassava is presented in Table 3.35, and soil P test and P fertilizer recommendations for cassava are presented in Table 3.36. Adequate P level in the plant tissues of sweet potato is presented in Table 3.37, and P soil test and P fertilizer recommendations for this crop are presented in Table 3.38.

TABLE 3.33
Adequate Plant Tissue Level of P for Potato

Growth Stage	Plant Part Analyzed	Adequate Level (g kg^{-1}) or (%)
42 days after emergence	UMB + P	2–4
Early flowering	UMB + P	3.5–5.5
Tubers half grown	UMB + P	2–4

Source: Fageria, N.K. et al., *Growth and Mineral Nutrition of Field Crops*, 3rd edn., CRC Press, Boca Raton, FL, 2011a.
UMB + P, youngest mature leaf blade + petiole.

TABLE 3.34
Soil P Test, Their Interpretation, and P Requirements for Potato

Soil P Test (mg kg^{-1})	Interpretation	P_2O_5 Rate (kg ha^{-1})
0–15	Low	300
16–40	Medium	200
>40	Sufficient	100

Source: Raij, B.V. et al., Liming and fertilizer recommendations for the state of São Paulo, Technical Bulletin No. 100, Campinas Institute of Agronomy, São Paulo, Brazil, 1985.
Yield expected more than 15 mg ha^{-1} and resin extraction used to determine soil test P.

TABLE 3.35
Adequate Plant Tissue Level of P for Cassava

Growth Stage	Plant Part Analyzed	Adequate Level (g kg^{-1}) or (%)
28 days after sowing	Whole shoot	4.7–6.6 (0.47–0.66)
98 days after sowing	YMB	3.3–4.1 (0.33–0.41)
3–4 months	YMB	3.6–5.0 (0.36–0.50)
3–5 months	YMB	4.2–4.7 (0.42–0.47)

Source: Reuter, D.J. and Robinson, J.B., *Plant Analysis: An Interpretation Manual*, CSIRO, Collingwood, Victoria, Australia, 1997.
Values in the parenthesis are in percentage.
YMB, young mature blade.

TABLE 3.36
Soil P Test, Their Interpretation, and P Requirements for Cassava

Soil P Test (mg kg^{-1})	Interpretation	P_2O_5 Rate (kg ha^{-1})
0–6	Very low	80
7–15	Low	60
16–40	Medium	40
>40	Sufficient	20

Source: Raij, B.V. et al., Liming and fertilizer recommendations for the state of São Paulo, Technical Bulletin No. 100, Campinas Institute of Agronomy, São Paulo, Brazil, 1985.
Yield expected about 25–30 mg ha^{-1} and resin extraction used to determine soil test P.

TABLE 3.37
Adequate Plant Tissue Level of P for Sweet Potato

Growth Stage	Plant Part Analyzed	Adequate Level (g kg^{-1}) or (%)
4th leaf	Leaves	2.3 (0.23)
Mid growth	Young mature leaf	2.0–3.0 (0.20–0.30)
At harvest	Tubers	1.2 (0.12)

Source: Sanchez, C.A., *Handbook of Plant Nutrition*, eds. A.V. Barker and D.J. Pilbeam, CRC Press, Boca Raton, FL, 2007, pp. 51–90.
Values in the parenthesis are in percentage.
YMB, young mature blade.

TABLE 3.38
Soil P Test, Their Interpretation, and P Requirements for Sweet Potato

Soil P Test (mg kg^{-1})	Interpretation	P_2O_5 Rate (kg ha^{-1})
0–6	Low	80
7–15	Medium	60
>15	Sufficient	40

Source: Raij, B.V. et al., Liming and fertilizer recommendations for the state of São Paulo, Technical Bulletin No. 100, Campinas Institute of Agronomy, São Paulo, Brazil, 1985.
Resin extraction used to determine soil test P.

3.10 CONCLUSIONS

Phosphorus is the second most important nutrient in crop production after nitrogen. Deficiency of P is widely reported in cereals, legumes, and other crops of economic importance. Under these situations, nutrient diagnostic techniques have special importance in rational use of P fertilizers and maximizing crop yields. Important nutrient deficiency or sufficiency diagnostic techniques are visual deficiency symptoms, soil test, plant tissue test, and crop response to applied nutrients. Phosphorus is a mobile nutrient in plants; therefore, its deficiency symptoms first appear in the older leaves. Plant leaves deficient in P become orange-reddish colored due to an above-normal level of anthocyanin, a red plant pigment, which accumulates when metabolic processes are disrupted. The effect is first evident on leaf tips, and then progresses toward the base. Eventually, the leaf tip dies. Soil test is widely used as a P deficiency/sufficiency diagnostic technique in making fertilizer recommendations. The success of soil test as a tool in determining the P status of the soils for a crop requires a soil test and plant yield calibration data to interpret soil analysis results. Plant tissue test and crop responses to applied nutrients are the most expensive methods of identifying P deficiency or sufficiency in crop plants. However, they are essential and should be used along with other techniques to improve the overall P availability and use efficiency by crops. Successful application of all the diagnostic techniques in correcting P deficiency in crop plants can increase the profitability and minimize environmental impact of fertilization.

REFERENCES

Andrietta, M. G. S., S. R. Andrietta, C. Steckelberg, and E. N. A. Stupiello. 2007. Bioethanol: Brazil, 30 years of Proalcool. *Int. Sugar J.* 109:195–200.

Austin, D. F., R. K. Jansson, and G. W. Wolfe. 1991. Convolvulaceae and *Cyclas*: A proposed hypothesis on the origins of the plant/insect relationship. *Trop. Agric.* 68:162–170.

Bakker, H. 1999. *Sugarcane Cultivation and Management*. Dordrecht, the Netherlands: Kluwer Academic Publishers.

Baligar, V. C., N. K. Fageria, and M. Elrashidi. 1998. Toxicity and nutrient constraints on root growth. *HortScience* 33:960–965.

Baligar, V. C., N. K. Fageria, and Z. L. He. 2001. Nutrient use efficiency in plants. *Commun. Soil Sci. Plant Anal.* 32:921–950.

Baligar, V. C., N. K. Fageria, A. Q. Paiva, A. Silveira, A. W. J. O. de Souza Jr, E. Lucena, J. C. Faria, R. Cabral, A. W. V. Pomella, and J. Jorda Jr. 2008. Light intensity effects on growth and nutrient-use efficiency of tropical legume cover crops. In: *Towards Agroforestry Design: An Ecological Approach*, eds. S. Jose and A. M. Gordon, pp. 67–79. New York: Springer Science Publisher.

Baligar, V. C., N. K. Fageria, A. Q. Paiva, A. Silveira, A. W. V. Pomella, and R. C. R. Machado. 2006. Light intensity effects on growth and micronutrient uptake by tropical legume cover crops. *J. Plant Nutr.* 29:1959–1974.

Bandaru, V., B. A. Stewart, R. L. Baumhardt, S. Ambati, C. A. Robinson, and A. Schlegel. 2006. Growing dryland grain sorghum in clumps to reduce vegetative growth and increase yield. *Agron. J.* 98:1109–1120.

Barber, S. A. 1980. Soil–plant interactions in the phosphorus nutrition of plants. In: *The Role of Phosphorus in Agriculture*, ed. R. C. Dinauer, pp. 591–615. Madison, WI: ASA, CSSA, and SSSA.

Batten, G. D. 1986. The uptake and utilization of phosphorus and nitrogen by diploid tetraploid and hexaploid wheats. *Ann. Bot.* 58:49–59.

Bell, C. I., G. F. J. Milford, and R. A. Leigh. 1996. Sugar beet. In: *Photoassimilate Distribution in Plants and Crops*, eds. E. Zamski and A. A. Schaffer, pp. 691–707. New York: Marcel Dekker.

Bender, D. A. 1993. Onions. In: *Nutrient Deficiencies and Toxicities in Crop Plants*, ed. W. F. Bennett, pp. 131–135. St. Paul, MN: The American Phytopathological Society.

Calderini, D. F., S. Torres-Leon, and G. A. Slafer. 1995. Consequences of wheat breeding on nitrogen and phosphorus yield, grain nitrogen and phosphorus concentration and associated traits. *Ann. Bot.* 76:315–322.

Campbell, K. G. K. 1984. Sugarbeet. In: *Evolution of Crop Plants*, ed. N. W. Simmonds, pp. 25–28. London, U.K.: Longman.

Cassman, K. G. 1993. Cotton. In: *Nutrient Deficiencies and Toxicities in Crop Plants*, ed. W. F. Bennett, pp. 111–119. St. Paul, MN: The American Phytopathological Society.

Cheeroo-Nayamuth, F. C., M. J. Robertson, M. K. Wegener, and A. R. H. Nayamuth. 2000. Using a simulation model to assess potential and attainable sugarcane yield in Mauritius. *Field Crops Res.* 66:225–243.

CIAT. 1992. Trends in CIAT commodities, 1992. Working Document 111. Cali, Colombia: CIAT.

Clark, R. B. 1990. Physiology of cereals for mineral nutrient uptake, use, and efficiency. In: *Crops as Enhancers of Nutrient Use*, eds. V. C. Baligar and R. R. Duncan, pp. 131–209. New York: Academic Press.

Clark, R. B. 1993. Sorghum. In: *Nutrient Deficiencies and Toxicities in Crop Plants*, ed. W. F. Bennett, pp. 21–26. St. Paul, MN: The American Phytopathological Society.

Courtenay, P. 1980. *Plantation Agriculture*, 2nd edn. London, U.K.: Bell & Hyman.

Cui, Z., Z. Dou, X. Chen, X. Ju, and F. Zhang. 2014. Managing agricultural nutrients for food security in China: Past, present, and future. *Agron. J.* 106:191–198.

Dinnes, D. L., D. L. Karlen, D. B. Jaynes, T. C. Kaspar, J. L. Hatfield, T. S. Colvin, and C. A. Cambardella. 2002. Nitrogen management strategies to reduce nitrate leaching in tile-drained Midwestern soils. *Agron. J.* 94:153–171.

Doggett, H. 1970. *Sorghum*. London, U.K.: Longman.
Dudal, R. 1976. Inventory of the major soils of the world with special reference to mineral stress hazards. In: *Plant Adaption to Mineral Stress in Problem Soils*, ed. M. J. Wright, pp. 3–13. Ithaca, NY: Cornel University.
Dwivedi, S. L., J. H. Crouch, S. N. Nigam, M. F. Ferguson, and A. H. Paterson. 2003. Molecular breeding of groundnut for enhanced productivity and food security in the semi-arid tropics: Opportunities and challenges. *Adv. Agron.* 153–221.
Dwivedi, S. L., N. Puppla, H. D. Upadhyaya, N. Manivannan, and S. Singh. 2008. Developing a core collection of peanut specific to Valencia market type. *Crop Sci.* 48:625–632.
Eagles, H. A. and J. E. Lothrop. 1994. Highland maize from central Mexico—Its origin, characteristics, and use in breeding programs. *Crop Sci.* 34:11–19.
Eastin, J. D. 1983. Sorghum. In: *Potential Productivity of field Crops under Different Environments*, ed. IRRI, pp. 181–204. Los Banos, Philippines: IRRI.
Edwards, A. C., P. J. A. Withers, and T. J. Sims. 1997. *Are Current Fertilizer Recommendation Systems for Phosphorus Adequate*?. New York: Fertilizer Society.
Fageria, N. K. 1989. Effects of phosphorus on growth, yield and nutrient accumulation in the common bean. *Trop. Agric.* 66:249–255.
Fageria, N. K. 1992. *Maximizing Crop Yields*. New York: Marcel Dekker.
Fageria, N. K. 1998. Phosphorus use efficiency by bean genotypes. *Rev. Bras. Eng. Agric. Amb.* 2:128–131.
Fageria, N. K. 2001. Reponse of upland rice, dry bean, corn and soybean to base saturation in cerrado soil. *Rev. Bras. Eng. Agric. Amb.* 5:416–424.
Fageria, N. K. 2002. Nutriet management for sustainable dry bean production in the tropics. *Commun. Soil Sci. Plant Anal.* 29:1219–1228.
Fageria, N. K. 2006. Liming and copper fertilization in dry bean production on an Oxisol in no-tillage system. *J. Plant Nutr.* 29:1219–1228.
Fageria, N. K. 2007. Green manuring in crop production. *J. Plant Nutr.* 30:691–719.
Fageria, N. K. 2008. Optimum soil acidity indices for dry bean production on an Oxisol in no-tollage system. *Commun. Soil Sci. Plant Anal.* 39:845–857.
Fageria, N. K. 2009. *The Use of Nutrients in Crop Plants*. Boca Raton, FL: CRC Press.
Fageria, N. K. 2013. *The Role of Plant Roots in Crop Production*. Boca Raton, FL: CRC Press.
Fageria, N. K. 2014. *Mineral Nutrition of Rice*. Boca Raton, FL: CRC Press.
Fageria, N. K. 2015. *Nitrogen Management in Crop Production*. Boca Raton, FL: CRC Press.
Fageria, N. K. and V. C. Baligar. 1997. Response of common bean, upland rice, corn, wheat and soybean to soil fertility of an Oxisol. *J. Plant Nutr.* 20:1279–1289.
Fageria, N. K. and V. C. Baligar. 2003. Fertility management of tropical acid soils for sustainable crop production. In: *Handbook Soil Acidity*, ed. Z. Rengel, pp. 359–385. New York: Marcel Dekker.
Fageria, N. K. and V. C. Baligar. 2005. Nutrient availability. In: *Encyclopedia of Soils in the Environment*, ed. D. Hillel, pp. 63–71. San Diago, CA: Elsevier.
Fageria, N. K., V. C. Baligar, and R. B. Clark. 2002. Micronutrients in crop production. *Adv. Agron.*77:185–268.
Fageria, N. K., V. C. Baligar, and R. B. Clark. 2006. *Physiology of Crop Production*. New York: The Haworth Press.
Fageria, N. K., V. C. Baligar, and C. A. Jones. 2011a. *Growth and Mineral Nutrition of Field Crops*, 3rd edn. Boca Raton, FL: CRC Press.
Fageria, N. K., V. C. Baligar, and Y. C. Li. 2008a. The role of nutrient efficient plants in improving crop yields in the twenty first century. *J. Plant Nutr.* 31:1121–1157.
Fageria, N. K., V. C. Baligar, and Y. C. Li. 2014a. Nutrient uptake and use efficiency by tropical legume cover crops at varying pH of na Oxisol. *J. Plant Nutr.* 37:294–311.
Fageria, N. K., V. C. Baligar, A. Moreira, and T. A. Portes. 2010. Dry bean genotypes evaluation for growth, yield components and phosphorus use efficiency. *J. Plant Nutr.* 33:2167–2181.
Fageria, N. K., V. C. Baligar, and R. W. Zobel. 2007. Yield, nutrient uptake, and soil chemical properties as influenced by liming and boron application in common bean in a no-tillage system. *Commun. Soil Sci. Plant Anal.* 38:1637–1653.
Fageria, N. K. and M. P. Barbosa Filho. 2008. Growth and zinc uptake and use efficiency in food crops. *J. Plant Nutr.* 31:983–1004.
Fageria, N. K., A. B. Heinemann, and R. A. Reis Jr. 2014b. Comparative efficiency of phosphorus for upland rice production. *Commun. Soil Sci. Plant Anal.* 45:1–22.
Fageria, N. K., A. M. Knupp, and M. F. Moraes. 2013a. Phosphorus nutrition of lowland rice in tropical lowland soil. *Commun. Soil Sci. Plant Anal.* 44:2932–2940.

Fageria, N. K., L. C. Melo, J. P. Oliveira, and A. M. Coelho. 2012. Yield and yield components of dry bean genotypes as influenced by phosphorus fertilization. *Commun. Soil Sci. Plant Anal.* 43:2752–2766.

Fageria, N. K. and A. Moreira. 2011. The role of mineral nutrition on root growth of crop plants. *Adv. Agron.* 110:251–331.

Fageria, N. K., A. Moreira, and A. B. Santos. 2013b. Phosphorus uptake and use efficiency in field crops. *J. Plant Nutr.* 36:2013–2022.

Fageria, N. K. and A. B. Santos. 1998. Phosphorus fertilization for bean crop in lowland soil. *Braz. J. Agric. Eng. Ambien.* 2:119–246.

Fageria, N. K., A. B. Santos, and A. B. Heinemann. 2011b. Lowland rice genotypes evaluation for phosphorus use efficiency in tropical lowland. *J. Plant Nutr.* 34:1087–1095.

Fageria, N. K. and L. F. Stone. 1999. *Acidity Management of Cerrado and Varzea Soils of Brazil.* Santo Antonio de Goias, Brazil: National Rice and Bean Research Center of EMBRAPA.

Fageria, N. K., L. F. Stone, and A. Moreira. 2008b. Liming and manganese influence on common bean yield, nutrient uptake, and changes in soil chemical properties of an Oxisol under no-tillage system. *J. Plant Nutr.* 31:1723–1735.

FAO. 2009. Cotton. Rome, Italy: FAO, http://faostat.fao.org/site/339/default.aspx (verified October 17, 2009).

FAO. 2011. Conservation agriculture. Rome, Italy: FAO, www.fao.org/ag/ca/index.html (accessed July 15, 2013).

FAO. 2013. FAOSTAT data base-Agricultural production. Rome, Italy: FAO, http:faostat3.fao.org/faostat-gateway/go/to/home/E (accessed November 4, 2013).

Faria, S. M., G. P. Lewis, J. I. Sprent, and J. M. Sutherland. 1989. Occurrence of nodulation in the leguminosae. *New Phytol.* 111:607–619.

Gascho, G. J., D. L. Anderson, and J. E. Bowen. 1993. Sugarcane. In: *Nutrient Deficiencies and Toxicities in Crop Plants*, ed. W. F. Bennett, pp. 37–42. St. Paul, MN: The American Phytopathological Society.

Gibbons, R. W. 1980. Adaption and utilization of groundnuts in different environments and farming systems. In: *Advances in Legume Science*, eds. R. J. Summerfield and A. H. Bunting, pp. 483–493. Kew, U.K.: Royal Botanical Gardens.

Gilbert, R. A., C. R. Rainbolt, D. R. Morris, and A. C. Bennett. 2007. Morphological responses of sugarcane to long term flooding. *Agron. J.* 99:1622–1628.

Golbitz, P. 2004. *Soya and Oilseed Bluebook.* Bar Harbor, ME: Soyatech. Inc.

Goodman, M. M. 1988. The history and evaluation of maize. *CRC Crit. Rev. Plant Sci.* 7:197–220.

Graham, P. H. and C. P. Vance. 2003. Legumes: Importance and constraints to greater use. *Plant Physiol.* 131:872–877.

Gregory, W. C., A. Krapovickas, and M. P. Gregory. 1980. Structures, variation, evolution and classification in *Arachis*. In: *Advances in Legume Science*, eds. R. J. Summerfield and A. H. Bunting, pp. 469–481. Kew, U.K.: Royal Botanic Gardens.

Grewal, D., C. Manito, and V. Bartolome. 2011. Doubled haploids generated through another culture from crosses of elite *indica* and *japonica* cultivars and/or lines of rice: Large scale production, agronomic performance, and molecular characterization. *Crop Sci.* 51:2544–2553.

Hale, M. G. and D. M. Orcutt. 1987. *The Physiology of Plants under Stress*. New York: John Wiley & Sons.

Hall, R. and H. F. Schwartz. 1993. Common bean. In: *Nutrient Deficiencies and Toxicities in Crop Plants*, ed. W. F. Bennett, pp. 143–147. St. Paul, MN: The American Phytopathological Society.

Hartemink, A. E. 2008. Sugarcane for bioethanol: Soil and environmental issues. *Adv. Agron.* 99:125–182.

Higgs, B., A. E. Johnston, J. L. Salter, and C. J. Dawson. 2000. Some aspects of achieving sustainable phosphorus use in agriculture. *J. Environ. Qual.* 29:80–87.

Hons, F. M., R. F. Moresco, R. P. Wiedenfeld, and J. H. Cothren. 1986. Applied nitrogen and phosphorus effects on yield and nutrient uptake by high-energy sorghum produced for grain and biomass. *Agron. J.* 78:1063–1078.

Jagadish, S. V. K., J. Cairns, R. Lafitte, T. R. Wheeler, A. H. Price, and P. Q. Craufurd. 2010. Genetic analysis of heat tolerance at anthesis in rice. *Crop Sci.* 50:1633–1641.

Jena, K. K. and D. J. Mackill. 2008. Molecular markers and their use in marker-assisted selection in rice. *Crop Sci.* 48:1266–1276.

Jones, C. A. 1985. *C_4 Grasses and Cereals: Growth, Development, and Stress Response*. New York: Wiley.

Jones, G. P. D., G. J. Blair, and R. S. Jessop. 1989. Phosphorus efficiency in wheat—A useful selection criterion. *Field Crops Res.* 21:257–264.

Jones, J. B. Jr, B. Wolf, and H. A. Mills. 1991. *Plant Analysis Handbook.* Athens, GA: Micro-Macro Publisher, Inc.

Jordan, D. L., J. B. Beam, P. D. Johnson, and J. F. Spears. 2001. Peanut response to prohexadione calcium in three seeding rate row pattern planting systems. *Agron. J.* 93:232–236.

Kadyampaken, D. M. 2014. Soil, water, and nutrient management options for climate change adaption in southern Africa. *Agron. J.* 106:100–110.

Kato, Y. A. T. 1984. Chromosome morphology and the origin of maize and its races. *Evol. Biol.* 17:219–255.

Kiniry, J. R., C. E. Simpson, A. M. Schubert, and J. D. Reed. 2005. Peanut leaf area index, light interception, radiation use efficiency, and harvest index at three sites in Texas. *Field Crops Res.* 91:297–306.

Knauft, D. A. and J. C. Wynne. 1995. Peanut breeding and genetics. *Adv. Agron.* 55:393–445.

Krapovickas, A. 1969. The origin, variability and spread of the groundnut (*Arachis hypogaea*). In: *The Domestication and Exploitation of Plants and Animals*, eds. P. J. Ucko and G. W. Dimbley, pp. 427–444. London, U.K.: Duckworth.

Krishnan, P., B. Ramakrishnan, K. R. Reddy, and V. R. Reddy. 2011. High-temperature effects on rice growth, yield, and grain quality. *Adv. Agron.* 111:87–206.

Kumar, K. and K. M. Goh. 2000. Crop residues and management practices: Effects on soil quality, soil nitrogen dynamics, crop yield, and nitrogen recovery. *Adv. Agron.* 68:197–319.

Kumar, V. and J. K. Ladha. 2011. Direct seeding of rice: Recent development and future research needs. *Adv. Agron.* 111:297–413.

Laing, D. R., P. J. Jones, and J. H. C. Davis. 1984. Common bean (*Phaseolus vulgaris* L.). In: *The Physiology of Tropical Field Crops*, eds. P. R. Goldsworthy and N. M. Fisher, pp. 305–351. New York: Wiley.

Lanier, J. E., D. L. Jordan, J. F. Spears, R. Wells, P. D. Johnson, J. S. Barnes, C. A. Hurt, R. L. Brandenburg, and J. E. Bailey. 2004. Peanut response to planting pattern, row spacing and irrigation. *Agron. J.* 96:1066–1072.

Letschert, J. P. W. 1993. *Beta* section *Beta*: Biogeographical patterns of variation, and taxonomy, PhD thesis. Wageningen, Germany: Wageningen Agriculture University, Papers 91, pp. 1–55.

Lingle, S. E. and T. L. Tew. 2008. A comparison of growth and sucrose metabolism in sugarcane germplasm from Louisiana and Hawaii. *Crop Sci.* 48:1155–1163.

Mabberley, D. J. 1997. *The Plant Book*, 2nd edn. Cambridge, U.K.: Cambridge University Press.

Maddonni, G. A. and M. E. Otegui. 2006. Intra-specific competition in maize: Contribution of extreme plant hierarchies to grain yield, grain yield components and kernel composition. *Field Crops Res.* 97:155–166.

Mokwunye, A. U. and S. H. Chien. 1980. Reaction of partially acidulated phosphate rock with soils from the tropics. *Soil Sci. Soc. Am. J.* 44:477–482.

Narro, L., S. Pandey, C. D. Leon, F. Salazar, and M. P. Arias. 2001. Implication of soil acidity tolerant maize cultivars to increase production in developing countries. In: *Plant Nutrient Acquisition: New Perspectives*, eds. A. Ae, J. Arihara, K. Okada, and A. Srinivasan, pp. 447–463. Tokyo, Japan: Springer.

National Academy of Science. 1994. *Biological Nitrogen Fixation*. Washington, DC: National Academy Press.

National Research Council. 1996. *Low Crops of Africa*, Volume I. Grains. Washington, DC: National Academy Press.

Osaki, M. 1995. Ontogenetic changes of N, P, and K contents in individual leaves of field crops. *Soil Sci. Plant Nutr.* 41:429–438.

Osaki, M., K. Morikawa, T. Shinano, M. Urayama, and T. Tadano. 1991. Productivity of high-yielding crops. II. Comparison of N, P, K, ca, and Mg accumulation and distribution among high-yielding crops. *Soil Sci. Plant Nutr.* 37:445–454.

Ozanne, P. G. 1980. Phosphorus nutrition of plants: A general treatise. In: *The Role of Phosphorus in Agriculture*, ed. R. C. Dinauer, pp. 559–589. Madison, WI: ASA, CSSA, and SSSA.

Perla, O., A. B. Luis, G. S. Sonia, P. B. Maria, T. Juscelino, and P. L. Octavio. 2003. Effect of processing and storage time on in vitro digestibility and resistant starch content of two bean varieties. *J. Sci. Agric.* 83:1283–1288.

Phillips, L. L. 1976. Cotton. In: *Evolution of Crop Plants*, ed. N. W. Simmonds, pp. 196–200. London, U.K.: Longman.

Pierzynski, G. M. and T. J. Loan. 1993. Crop, soil, and management effects of phosphorus test levels. *J. Prod. Agric.* 6:513–520.

Polhill, R. M., P. H. Raven, and C. H. Stirton. 1981. Evaluation and systematics of the leguminosae. In: *Advances in Legume Systematics, Part 1*, eds. R. M. Polhill and P. H. Raven, pp. 1–26. Kew, U.K.: Royal Botanical Gardens.

Preiss, J. and M. N. Sivak. 1996. Starch synthesis in sinks and sources. In: *Photoassimilate Distribution in Plants and Crops: Source–Sink Relationships*, eds. E. Zamski and A. A. Schafeer, pp. 63–96. New York: Marcel Dekker.

Purseglove, J. W. 1985. *Tropical Crops: Monocotyledons*. New York: Longman.

Raij, B. V., N. M. Silva, O. C. Bataglia, J. A. Quaggio, R. Hiroce, H. Cantarella, R. B. Junior, A. R. Dechen, and P. E. Trani. 1985. Liming and fertilizer recommendations for state of São Paulo, Technical Bulletin No. 100. São Paulo, Brazil: Campinas Institute of Agronomy.

Ramakrishna, A., H. M. Tam, S. P. Wani, and T. D. Long. 2006. Effect of mulch on soil temperature, moisture, weed infestation and yield of groundnut in northern Vietnam. *Field Crops Res.* 95:115–125.

Rauschkolb, R. S., T. L. Jackson, and A. I. Dow. 1984. Management of nitrogen in the Pacific States. In: *Nitrogen in Crop Production*, ed. R. D. Hauck, pp. 765–777. Madison, WI: ASA.

Reuter, D. J. and J. B. Robinson. 1997. *Plant Analysis: An Interpretation Manual*, 2nd edn. Collingwood, Victoria, Australia: CSIRO.

Ringius, L. 2002. Soil carbon sequestration and the CDM: Opportunities and challenges for Africa. *Clim. Change* 54:471–495.

Russell, W. A. 1991. Genetic improvement of maize yields. *Adv. Agron.* 46:245–298.

Sahrawat, K. L., G. Pardhasaradhi, T. J. Rego, and M. H. Rahman. 1996. Relationship between extracted phosphorus and sorghum yield in a Vertisol and an Alfisol under rainfed cropping. *Fertiliz. Res.* 44:23–26.

Sanchez, C. A. 2007. Phosphorus. In: *Handbook of Plant Nutrition*, eds. A. V. Barker and D. J. Pilbeam, pp. 51–90. Boca Raton, FL: CRC Press.

Sanchez, P. A. and J. G. Salinas. 1981. Low-input technology for managing Oxisols and Ultisols in tropical America. *Adv. Agron.* 34:280–406.

Santalla, M., A. P. Rodino, P. A. Casquero, and A. M. Ron. 2001. Interaction of bush bean intercropped with field and sweet maize. *Eur. J. Agron.* 15:185–196.

Shimelis, E. A. and S. K. Rakshit. 2004. Proximate composition and physico-chemical properties of improved dry bean varieties grown in Ethopia. *Food Sci. Technol.* 38:331–338.

Siemonsma, J. J. and K. Piluek. 1993. *Plant Resources of South-Easr Asia No. 8. Vegetables*. Wageningen, the Netherlands: Pudoc Scientific.

Simpson, B. B. and M. Conner-Ogorzaly. 1995. *Economic Botany: Plant in Our World*, 2nd edn. New York: McGraw Hill.

Sinclair, J. B. 1993. Soybeans. In: *Nutrient Deficiencies and Toxicities in Crop Plants*, ed. W. F. Bennett, pp. 99–103. St. Paul, MN: The American Phytopathological Society.

Sinclair, T. 1998. Histrorical changes in harvest index and crop nitrogen accumulation. *Crop Sci.* 38:638–643.

Singh, A. K. and C. E. Simpson. 1994. *The Groundnut Crop: A Scientific Basis for Improvement*. London, U.K.: Chapman & Hall.

Singh, S. P. 1999. *Common Bean Improvement in the Twenty First Century*. Dordrecht, the Netherlands: Kluwer Academic Publishers.

Smart, J. 1994. The groundnut in farming systems and the rural economy: A global view. In: *The Groundnut Crop: A Scientific Basis for Improvement*, ed. J. Smart, pp. 664–699. London, U.K.: Chapman & Hall.

Smil, V. 1999. Crop residues: Agricultural largest harvest. *BioScience* 49:299–308.

Smil, V. 2000. Phosphorus in the environment: Natural flows and human interferences. *Annu. Rev. Energy Environ.* 25:53–88.

Smith, D. H., M. A. Wells, D. M. Porter, and F. R. Cox. 1993. Peanuts. In: *Nutrient Deficiencies and Toxicities in Crop Plants*, ed. W. F. Bennett, pp. 105–110. St. Paul, MN: The American Phytopathological Society.

Snapp, S. S., D. D. Rohrbach, F. Simtpwe, and H. A. Freeman. 2002. Sustainable soil management options for Malawi: Can smallholder farmers grow more legumes? *Agric. Ecosyst. Environ.* 91:159–174.

Soil Science Society of America. 2008. *Glossary of Soil Science Terms*. Madison, WI: Soil Science Society of America.

Tanaka, W. and G. A. Maddonni. 2008. Pollen source and post-flowering source/sink ratio on maize kernel weight and oil concentration. *Crop Sci.* 48:666–677.

Thomison, P. R., A. B. Geyer, L. D. Lotz, H. J. Siegrist, and T. L. Dobbels. 2003. Top cross high oil corn production: Select grain quality attributes. *Agron. J.* 95:147–154.

Thorup-Kristensen, K., J. Magid, and J. S. Jensen. 2003. Catch crops as green manures as biological tools in nitrogen management in temperate zones. *Adv. Agron.* 79:227–302.

Tollenaar, M. and L. M. Dwyer. 1998. Physiology of maize. In: *Crop Yield, Physiology and Process*, eds. D. L. Smith and C. Hamel, pp. 169–204. New York: Springer Verlag.

Ulrich, A. 1993. Potato. In: *Nutrient Deficiencies and Toxicities in Crop Plants*, ed. W. F. Bennett, pp. 149–156. St. Paul, MN: The American Phytopathological Society.

Ulrich, A. and F. J. Hills. 1973. Plant analysis as an aid in fertilizing sugar crops. Part 1. Sugarbeet. In: *Soil Testing and Plant Analysis*, eds. L. M. Walsh and J. D. Beaton, pp. 271–288. Madison, WI: Soil Science Society of America.

Ulrich, A., J. T. Moraghan, and E. D. Whitney. 1993. Sugar beet. In: *Nutrient Deficiencies and Toxicities in Crop Plants*, ed. W. F. Bennett, pp. 91–98. St. Paul, MN: The American Phytopathological Society.

Upadhyaya, H. D., B. P. M. Swamy, P. V. K. Goudar, B. Y. Kullaiswamy, and S. Singh. 2005. Identification of diverse groundnut germplasm through multienvironment evaluation of a core collection for Asia. *Field Crops Res.* 93:293–299.

Valdes, C. 2007. Ethanol demand driving the expansion of Brazils sugar industry. *Sugar Sweeteners Outlook* 249:31–38.

Van Riemsdijk, W. H., L. J. M. Baumans, and F. A. M. De Hann. 1984. Phosphate sorption by soils I. A model for phosphate reaction with metal-oxide in soil. *Soil Sci. Soc. Am. J.* 48:537–541.

Vaugham, D. A., E. Balazs, and J. S. Heslop-Harrison. 2007. From crop domestication to super-domestication. *Ann. Bot.* 100:893–901.

Veneklaas, E. J., H. Lambers, J. Bragg, P. M. Finnegan, C. E. Lovelock, W. C. Plaxton, C. A. Price, W. R. Scheible, M. W. Shane, P. J. White, and J. A. Raven. 2012. Opportunities for improving phosphorus use efficiency in crop plants. *New Physiol.* 195:306–320.

Voss, R. D. 1993. Corn. In: *Nutrient Deficiencies and Toxicities in Crop Plants*, ed. W. F. Bennett, pp. 11–19. St. Paul, MN: The American Phytopathological Society.

Wallace, D. H., J. L. Ozbun, and H. M. Munger. 1972. Physiological genetics of crop yield. *Adv. Agron.* 24:97–146.

Wang, L. P., P. A. Jackson, X. Lu, Y. H. Fan, J. W. Foreman, X. K. Chen, H. H. Deng, C. Fu, L. Ma, and K. S. Aitken. 2008. Evaluation of sugarcane X saccharum spontaneum progeny for biomass composition and yield components. *Crop Sci.* 48:951–961.

Wassmann, R., S. V. K. Jagadish, S. Heuer, A. Ismail, E. Redona, R. Serraj, R. K. Singh, S. Heuer, A. Ismail, and K. Sumfleth. 2009. Climate change affecting rice production: The physiological and agronomic basis for possible adaptation strategies. *Adv. Agron.* 101:59–122.

Wellhausen, E. J., L. M. Roberts, and E. Harnandez. 1952. *Races of Maize in Mexico*. Cambridge, MA: The Bussey Institue of Harvard University.

Wiese, M. V. 1993. Wheat. In: *Nutrient Deficiencies and Toxicities in Crop Plants*, ed. W. F. Bennett, pp. 27–33. St. Paul, MN: The American Phytopathological Society.

Wilcox, G. E. 1993. Tomato. In: *Nutrient Deficiencies and Toxicities in Crop Plants*, ed. W. F. Bennett, pp. 137–141. St. Paul, MN: The American Phytopathological Society.

Wynne, J. C. and W. C. Gregory. 1981. Peanut breeding. *Adv. Agron.* 34:39–72.

Xiong, Z. Y., S. J. Zhang, B. V. Ford-Lloyd, X. Jin, Y. Wu, H. X. Yan, P. Liu, X. Yang, and B. R. Lu. 2011. Latitudinal distribution and differentiation of rice germplasm: Its implication in breeding. *Crop Sci.* 51:1050–1058.

Yan, W. G., H. Agrama, M. Jia, R. Fjellstrom, and A. McClung. 2010. Geographic description of genetic diversity and relationships in the USDA rice world collection. *Crop Sci.* 50:2406–2417.

Yan, X., J. P. Lynch, and S. E. Beebe. 1995. Genetic variation for phosphorus efficiency of common bean in contrasting soil types: I Vegetative response. *Crop Sci.* 35:1086–1093.

Zebarth, B. J., W. J. Arsenault, S. Moorehead, H. T. Kunelius, and M. Sharifi. 2009. Italian ryegrass management effects on nitrogen supply to a subsequent potato crop. *Agron. J.* 101:1573–1580.

Zhang, F. S., Z. L. Cui, M. S. Fan, W. Zhang, X. Chen, and R. Jiang. 2011. Integrated soil-crop system management: Reducing environmental risk while increasing crop productivity and improving nutrient use efficiency in China. *J. Environ. Qual.* 40:1051–1057.

Zhang, Q. F., M. A. Saghai-Maroof, T. Y. Lu, and B. Z. Shen. 1992. Genetic diversity and differentiation of *indica* and *japonica* rice detected by RFLP analysis. *Theor. Appl. Genet.* 83:495–499.

Zohary, D. and M. Hopf. 1993. *Domestication of Plants in the Old World*. Oxford, U.K.: Clarendon Press.

4 Phosphorus Use Efficiency in Crop Plants

4.1 INTRODUCTION

Phosphorus is considered as the second most important nutrient in crop production after nitrogen (N). P deficiency is considered a major factor in crop production, especially in the tropics and subtropics (Ramaekers et al., 2010). In addition, as compared with other nutrients, P is the least mobile and least available to plants in most soil conditions (Schachtman et al., 1998; Hinsinger, 2001). P is required by plants for photosynthesis, respiration, seed production, root growth, and many other physiological and biochemical functions (Bundy et al., 2005). An adequate, steady supply of P is needed for normal plant growth and development. P has more influence on agricultural ecosystems, except for N (Brady and Weil, 2002). In addition, a large part of P is removed from the harvested grain of crops and very small part is retained in the straw to recycle. Hence, application of P fertilizers is essential to obtain higher crop yields and also to sustain a cropping system.

Adequate P nutrition is the basis of soil fertility and crop management. Maximum economic yield of crops is possible with the adequate level of P in the soil along with other essential nutrients (Fageria et al., 2011a; Fageria, 2013). P deficiency is more severe in highly weathered acid soils as compared to other soils. Most P-deficient soils are classified as Oxisols and Ultisols in the U.S. Soil Taxonomy. These soils are naturally low in P as well as having high immobilization capacity of P. P immobilization capacity of these soils is associated with the presence of aluminum (Al) and iron (Fe) oxides (Fageria and Baligar, 2003). Due to these reasons, recovery efficiency of fertilizer applied P in these acid soils is less than 20% (Fageria et al., 2013).

Growth and productivity of field crops is affected by their genetic potential and the environment in which they are grown. Soil P is one of the most important environmental factors affecting crop productivity. P must be available in proper amount and balance to achieve the maximum economic yield of crops. In addition, use of high-yield potential cultivars and control of diseases, insects, and weeds are also important components of crop production to obtain higher yields.

Because of the fundamental importance of P as an essential nutrient in plants, crop production can be severely limited when P is deficient due to inadequate supply in the soil. In their natural conditions, many of the major world soil groups do not have adequate P to support intensive crop production. This can lead to reduced crop yields, and P-deficient plants may suffer from biotic and abiotic stresses (Bundy et al., 2005). Modern production agriculture requires efficient, sustainable, and environmentally sound management practices (Fageria and Baligar, 2005). P is one of the most limiting nutrients for crop production in many of the world agricultural areas, and its efficient use is important for the economic sustainability of cropping systems. In addition, efficient use of P reduces the costs of crop production as well as environmental impacts. Furthermore, annual crops such as cereals, legumes, and oil seeds provide about 60% of the dietary energy for the world's growing population (FAO, 2010). Improving the efficiency of P fertilizer use for crop growth requires enhanced P use efficiency (PUE) (acquisition as well as uptake). The objective of this chapter is to discuss PUE in crop plants. Management practices adopted to improve PUE are discussed in Chapter 7.

4.2 DEFINITION OF MINERAL NUTRIENT EFFICIENCY AND NUTRIENT-EFFICIENT PLANTS

Providing the definition of mineral nutrient efficiency and nutrient-efficient plants is important to understand PUE. Mineral nutrient efficiency has been defined differently in the literature. In broad terms, efficiency is the ratio of outputs to inputs or outputs divided by inputs. Higher ratio systems are more efficient than lower ratio systems. Clark (1990) reported that several terms used for nutrient efficiency are absorption efficiency, acquisition efficiency, agronomic efficiency (AE), apparent nutrient recovery efficiency, assimilation efficiency, distribution efficiency, economic yield efficiency, efficient quotient, efficiency ratio, metabolic efficiency, mobilization efficiency, nutrient harvest index, photosynthetic efficiency, utilization efficiency (UE), and many other efficiency terms are also used. Regardless of the many concepts and definitions, efficiency needs to be defined. In the opinion of the authors of this book, nutrient efficiency is defined as the higher economic part of the plant produced with the unit amount of nutrient applied or uptook in the plant tissues. Similarly, there are several definitions of efficient plants (Table 4.1). But the best definition of nutrient-efficient plant is "that produced higher economic yield with a determined quantity of applied or absorbed nutrient as compared to other or standard plant under similar growing conditions."

4.3 TYPE OF PHOSPHORUS USE EFFICIENCY AND MATHEMATICAL EQUATIONS FOR THEIR CALCULATION

Evaluation of PUE is useful to differentiate plant species, genotypes, and cultivars for their ability to absorb and utilize nutrients for maximum yields. PUE is based on (1) uptake efficiency (acquisition from soil, influx rate into roots, influx kinetics, radial transport in roots based on root parameters

TABLE 4.1
Definitions of Nutrient-Efficient Plants

Definition	Reference
Nutrient-efficient plant is defined: a plant that absorbs, translocates, or utilizes more of a specific nutrient than another plant under conditions of relatively low nutrient availability in the soil or growth media.	Soil Science Society of America (2008)
The nutrient efficiency of a genotype (for each element separately) is defined as the ability to produce a high yield in a soil that is limiting in that element for a standard genotype.	Graham (1984)
Nutrient efficiency of a genotype/cultivar is defined as the ability to acquire nutrients from a growth medium and/or to incorporate or utilize them in the production of shoot and root biomass or utilizable plant material (grain).	Blair (1993)
An efficient genotype is one that absorbs relatively high amounts of nutrients from soil and fertilizer, produces a high grain yield per unit of absorbed nutrient, and stores relatively little nutrients in the straw.	Isfan (1993)
Efficient plants are defined as those that produce more dry matter or have a greater increase in harvested portion per unit time, area, or applied nutrient, have fewer deficiency symptoms, or have greater incremental increases and higher concentrations of mineral nutrients than other plants grown under similar conditions or compared to a standard genotype.	Clark (1990)
Efficient germplasm requires less nutrients than an inefficient one for normal metabolic processes.	Gourley et al. (1994)
Efficient plant is defined as one that produces higher economic yield with a determined quantity of applied or absorbed nutrient compared to other or a standard plant under similar growing conditions.	Fageria et al. (2008)

per length and uptake is also related to the amounts of particular nutrient applied or present in soil), (2) incorporation efficiency (transports to shoot and leaves), and (3) UE (based on remobilization, whole plant, i.e., root and shoot parameters) (Baligar et al., 2001).

According to the opinion of the authors of this book, PUE in crop plants can be defined or grouped into five efficiencies. These are AE, which is defined as the grain yield produced per unit of P applied; physiological efficiency (PE), which is defined as the biological yield (grain plus straw) produced per unit of P uptake in grain plus straw; agrophysiological efficiency (APE), which is defined as the grain yield produced per unit of P uptake in the grain plus straw; apparent recovery efficiency (ARE), which is defined as the P uptake in grain plus straw per unit of P added; and UE, which is defined as the PE × ARE. For the extensive coverage of these efficiencies, readers are referred to Fageria (1992), Baligar et al. (2001), Fageria and Baligar (2005), Fageria et al. (2013), and Fageria (2013, 2014). The efficiency can be calculated by the following equations:

$$\text{AE}\left(\text{kg kg}^{-1}\right) = \frac{\text{GY}_{\text{pf}} \text{ in kg} - \text{GY}_{\text{pu}} \text{ in kg}}{\text{P rate applied in kg}}$$

where

AE is the agronomic efficiency
GY_{pf} is the grain yield of P fertilized plot in kg
GY_{pu} is the grain yield of P unfertilized plot in kg

$$\text{PE}\left(\text{kg kg}^{-1}\right) = \frac{\text{BY}_{\text{pf}} \text{ in kg} - \text{BY}_{\text{puf}} \text{ in kg}}{\text{BPU}_{\text{pf}} \text{ in kg} - \text{BPU}_{\text{puf}} \text{ in kg}}$$

where

PE is the physiological efficiency
BY_{pf} is the biological yield (grain plus straw) of P fertilized plot in kg
BY_{puf} is the biological yield (grain plus straw) of P unfertilized plot in kg
BUP_{pf} is the biological P uptake (grain plus straw) of P fertilized plot in kg
BPU_{puf} is the biological P uptake (grain plus straw) of P unfertilized plot in kg

$$\text{ARE}\left(\%\right) = \frac{\text{BPU}_{\text{pf}} \text{ in kg} - \text{BPU}_{\text{puf}} \text{ in kg}}{\text{P rate applied in kg}} \times 100$$

where

ARE is the apparent recovery efficiency
BPU_{pf} is the biological P uptake (grain plus straw) of P fertilized plot in kg
BPU_{puf} is the biological P uptake (grain plus straw) of P unfertilized plot in kg

$$\text{UE}\left(\text{kg kg}^{-1}\right) = \text{PE} \times \text{ARE}$$

where

UE is the utilization efficiency
PE is the physiological efficiency
ARE is the apparent recovery efficiency

When PUE is determined under controlled or greenhouse conditions, values of P efficiency are expressed in mg mg^{-1}. Definitions and equations to calculate PUE under controlled conditions are presented in Table 4.2.

TABLE 4.2
Definitions and Methods of Calculating PUE under Controlled or Greenhouse Conditions

PUE	Definitions and Formulas for Calculation
AE	The AE is defined as the economic production obtained per unit of nutrient applied. It can be calculated by AE (mg mg^{-1}) = $G_f - G_u/N_a$, where G_f is the grain yield of the fertilized pot (mg), G_u is the grain yield of the unfertilized pot (mg), and N_a is the quantity of nutrient applied (mg).
PE	PE is defined as the biological yield obtained per unit of nutrient uptake. It can be calculated by PE (mg mg^{-1}) = $BY_f - BY_u/N_f - N_u$, where BY_f is the biological yield (grain plus straw) of the fertilized pot (mg), BY_u is the biological yield of the unfertilized pot (mg), N_f is the nutrient uptake (grain plus straw) of the fertilized pot, and N_u is the nutrient uptake (grain plus straw) of the unfertilized pot (mg).
APE	APE is defined as the economic production (grain yield in case of annual crops) obtained per unit of nutrient uptake. It can be calculated by APE (mg mg^{-1}) = $G_f - G_u/N_{uf} - N_{uu}$, where G_f is the grain yield of fertilized plot (mg), G_u is the grain yield of the unfertilized pot (mg), N_{uf} is the nutrient uptake (grain plus straw) of the fertilized pot (mg), N_{uf} is the nutrient uptake (grain plus straw) of unfertilized pot (mg).
ARE	ARE is defined as the quantity of nutrient uptake per unit of nutrient applied. It can be calculated by ARE (%) = $(N_f - N_u/N_a) \times 100$, where N_f is the nutrient uptake (grain plus straw) of the fertilized pot (mg), N_u is the nutrient uptake (grain plus straw) of the unfertilized plot (mg), and N_a is the quantity of nutrient applied (mg).
UE	Nutrient UE is the product of PE and ARE. It can be calculated by UE (mg mg^{-1}) = PE × ARE

Source: Fageria, N.K. et al., *Commun. Soil Sci. Plant Anal.*, 44, 2932, 2013.

In addition to earlier-mentioned five PUEs, nutrient efficiency ratio (NER) was suggested by Gerloff and Gableman (1983) to differentiate genotypes into efficient and inefficient nutrient utilizers. NER can be calculated by the following equation:

$$\text{NER}\left(\text{kg kg}^{-1}\right) = \frac{\text{Economic yield in kg}}{\text{Nutrient uptake in plant tissue in kg}}$$

The kg kg^{-1} unit is used for expressing field experimental results of nutrient use efficiency (NUE), and mg mg^{-1} unit is used to express controlled conditions or greenhouse results. Detailed discussion of this efficiency is reported by Gerloff and Gableman (1983), Baligar et al. (1990), Clark and Duncan (1991), Blair (1993), and Baligar et al. (2001).

4.3.1 Experimental Results

Fageria and Barbosa Filho (2007) determined AE, PE, APE, ARE, and UE efficiencies in lowland rice (*Oryza glaberrima*) grown on a Brazilian Inceptisol (Table 4.3). Across P rates, 10.3 kg rice grain was produced with the application of 1 kg P. Similarly, 509 kg dry matter (straw plus grain) was produced with the accumulation of 1 kg P in the grain plus straw. In the case of APE, across P rates, 324 kg grain yield was produced with the accumulation of 1 kg P in the grain plus straw. Mean recovery efficiency was 4.3%, and UE was 22.4 kg grain yield with the utilization of 1 kg P. The highest efficiency is usually obtained with the first increment of nutrient, with additional increments providing smaller increases (Fageria et al., 1997). Singh et al. (2000) reported that APEs in lowland rice varied from 235 to 316 kg grain per kg P. Similarly, Witt et al. (1999) reported an APE value of 385 kg grain per kg P when all production factors were at normal levels. Sahrawat and Sika (2002) reported apparent recovery of applied P in the range of 4.8%–11% by rice in an Ultisol. Low recovery efficiency may be associated with the high rate of P fixation in this soil by Fe and Al oxides (Abekoe and Saharawat, 2001).

TABLE 4.3
PUE in Lowland Rice under Different P Rates

P Rate (kg ha^{-1})	AE (kg kg^{-1})	PE (kg kg^{-1})	APE (kg kg^{-1})	RE (%)	UE (kg kg^{-1})
131	15.5	604.4	300.9	6.3	39.6
262	12.7	536.8	269.5	5.1	27.1
393	10.5	521.8	477.4	3.8	19.7
524	6.8	443.6	277.8	3.4	14.8
655	6.2	439.3	296.6	2.7	10.9
Average	10.3	509.2	324.4	4.3	22.4

Regression analysis

P rate (X) vs. AE (Y) = 17.66 – 0.0178X, R^2 = 0.52**

P rate (X) vs. PE (Y) = 636.22 – 0.3232X, R^2 = 0.23^{NS}

P rate (X) vs. APE (Y) = 324.70 – 0.00025X, R^2 = 0.02^{NS}

P rate (X) vs. recovery efficiency (Y) = 6.93 – 0.0067X, R^2 = 0.43**

P rate (X) vs. UE (Y) = 43.27 – 0.0529X, R^2 = 0.50**

Source: Fageria, N.K. and Barbosa Filho, M.P., *Commun. Soil Sci. Plant Anal.*, 38, 1289, 2007.

Values are averaged across 2 years.

AE, agronomic efficiency; PE, physiological efficiency; APE, agrophysiological efficiency; RE, recovery efficiency; UE, utilization efficiency.

**,^{NS}Significant at the 5% and 1% probability levels and nonsignificant, respectively.

Fageria et al. (2011b) studied the AE of 12 lowland rice genotypes grown on a Brazilian Inceptisol (Figure 4.1). AE of 12 genotypes varied significantly. Genotype BRS Jaçana produced maximum PUE, and genotypes CNAi 8569 produced minimum PUE. PUE expressed as AE decreased quadratically with increasing P rate (22–88 kg ha^{-1}) (Figure 4.2). At 22 kg P ha^{-1}, AE was 186 kg grain produced per kg P applied and was reduced to 40 kg grain produced per kg P applied at 88 kg P ha^{-1} (Figure 4.2). With the increasing P rates, grain yield was increased but PUE decreased due to low plant capacity in the absorption and utilization of P (Fageria et al., 2003). Decrease in PUE is also associated with relative decrease in grain yield with successive increment in P rates (Fageria, 1992).

Fageria (2014) studied the AE of 10 upland rice genotypes grown on a Brazilian Oxisol (Figure 4.3). Genotypes differ significantly in AE. Genotype BRA 032046 produced maximum PUE (about 73 kg grain per kg P applied), and genotype BRA 0322051 produced minimum PUE (about 22 kg grain per kg P applied). Variation in PUE has been reported by Fageria and Santos (2002) and Fageria et al. (2011a,b) in lowland as well as upland rice. In another field experiment, Fageria et al. (2014c) studied the AE of 5 upland rice (Table 4.4). PUE, defined as kg grain produced per kg P applied, varied from 7.31 produced by genotype BRA032051 to 26.3 produced by genotype BRA02601, with a mean value of 16.9 (Table 4.4). Higher value of PUE in genotype BRA02601 was associated with high yield of this genotype to P fertilization. For example, at low P level, this genotype produced 3033 kg ha^{-1} grain, and at higher P level, the grain yield was 5340 kg ha^{-1}. The lowest PUE was associated with low response of genotypes to P fertilization. The lowest PUE-producing genotype BRA032051 produced 3506 kg ha^{-1} grain at low P level and 4145 kg ha^{-1} grain at high P level. NUE generally decreased with decreasing response to applied nutrients (Fageria, 1992).

Fageria (2014) studied the PUEs of 18 upland rice genotypes under controlled conditions (Table 4.5). PUE calculated as AE, PE, APE, ARE, and UE varied significantly among upland rice genotypes, except physiological and utilization efficiencies (Table 4.5). Across the 18 genotypes, AE was 63 mg grain produced per mg P applied and PE was 927 dry matter produced (grain + straw) per mg P accumulated in grain plus straw. Similarly, APE was 358 mg grain produced per mg P accumulated in grain plus straw, and ARE was 18%. UE was 155 mg grain plus straw produced

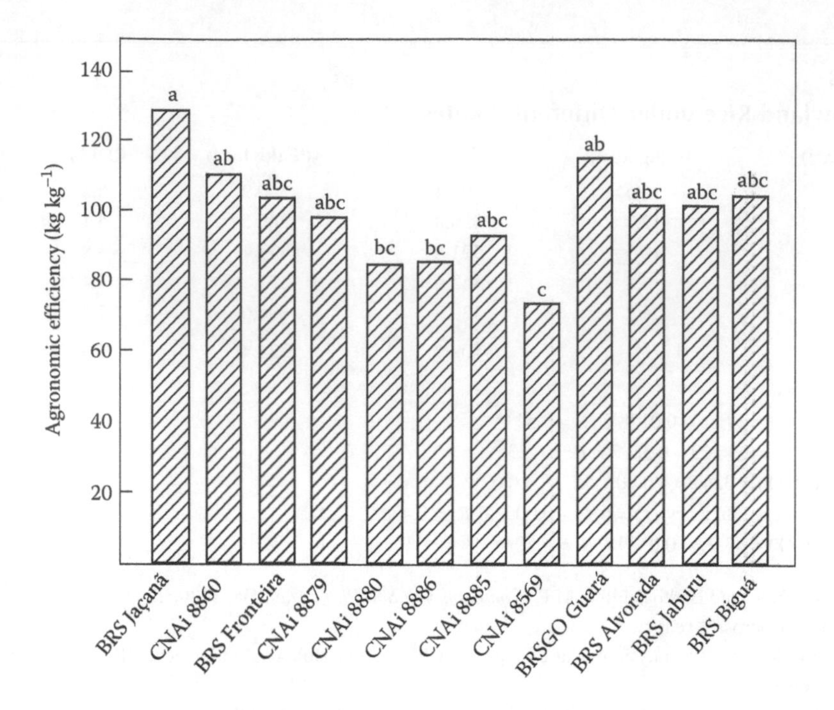

FIGURE 4.1 AE of 12 lowland rice genotypes. (From Fageria, N.K. et al., *J. Plant Nutr.*, 34, 1087, 2011b.)

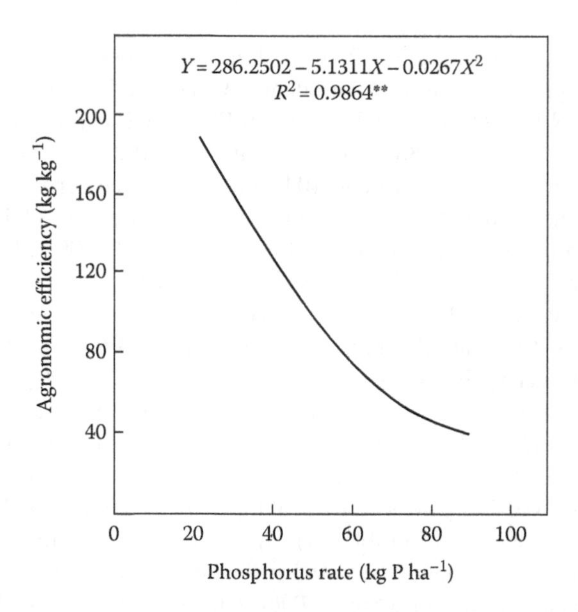

FIGURE 4.2 Relationship between phosphorus rate and AE of lowland rice. Values are averages of 12 genotypes and 2 years' field trials. (From Fageria, N.K. et al., *J. Plant Nutr.*, 34, 1087, 2011b.)

per mg P applied. Fageria et al. (2004b) reported that the efficiency of P use varies among rice genotypes. PUEs are generally higher than use efficiencies for N and potassium (K) (Fageria et al., 2004b), except recovery efficiency. Low recovery efficiency of P is associated with high P fixation capacity of Brazilian Oxisols (Goedert, 1989).

Fageria et al. (2013) calculated five PUEs of lowland rice under different P levels (Table 4.6). All the PUEs were significantly decreased with increasing P rates except PE. Across P rates,

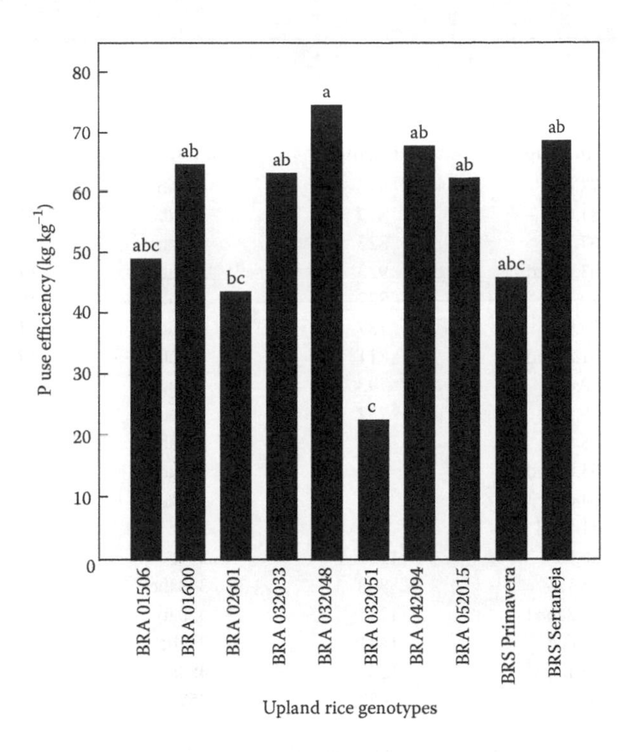

FIGURE 4.3 PUE of 10 upland rice genotypes. (From Fageria, N.K., *Mineral Nutrition of Rice*, CRC Press, Boca Raton, FL, 2014.)

TABLE 4.4
PUE of Five Upland Rice Genotypes

Genotype	PUE (kg Grain/kg P Applied)
BRA01596	19.30ab
BRA01600	15.88b
BRA02535	15.34b
BRA02601	26.51a
BRA032051	7.31c
Average	16.87

Source: Fageria, N.K. et al., *J. Plant Nutr.*, 37, 633, 2014c.
Means within the same column, followed by the same letter, do not differ significantly at the 5% probability level by Tukey's test.

AE was 100 mg grain produced per mg P applied, and PE was 1038 mg biological yield (straw plus grain) per unit of P accumulated. APE was 405 mg grain produced per mg of P accumulated in the grain and straw across P rates. ARE was 24.2%, and UE was 256 mg grain plus straw produced per mg of P utilized across P rates. PUE data are limited; therefore, we cannot compare our results with the published ones. However, Baligar and Bennett (1986) reported that the recovery of fertilizer P by crops that are planted immediately after the soluble fertilizer applications amounts to only 10%–30% of the quantity applied to the soil. Remaining 70%–90% P have been accounted for by microbial assimilation, precipitation by cations in the soil solution, or adsorption on the clay matrix. Our results of P recovery efficiency are within this range, especially at adequate P levels.

TABLE 4.5
PUE by Upland Rice Genotypes

Genotype	AE ($mg\ mg^{-1}$)	PE ($mg\ mg^{-1}$)	APE ($mg\ mg^{-1}$)	ARE (%)	UE ($mg\ mg^{-1}$)
CRO 97505	73.4ab	945	424ab	17.4abc	159.9
BRS Liderança	81.6a	817	382abc	21.3ab	173.7
BRS Curinga	67.1abc	823	347abc	19.3abc	158.8
CNAs 8938	63.0abcd	923	365abc	17.3abc	159.4
CNAs 8960	72.5ab	827	381abc	19.1abc	156.8
BRS Colosso	79.9a	737	378abc	21.2ab	156.1
CNAs 8824	61.8abcd	811	350abc	17.7abc	143.8
CNAs 8957	72.6ab	793	352abc	20.7abc	162.3
CRO 97442	68.8abc	732	375abc	18.5abc	137.4
CNAs 8817	68.6abc	847	370abc	18.5abc	155.7
BRS Aroma	53.8abcd	917	314abc	17.3abc	154.0
CNAs 8950	64.0abc	834	350abc	18.2abc	151.2
BRS Talento	41.8bcd	801	240c	16.8abc	131.8
BRS Caripuna	26.7d	1187	237c	11.0c	128.7
Primavera	76.7ab	808	348abc	22.3a	178.6
Canastra	59.3abcd	1211	450ab	13.2abc	156.9
Maravilha	34.6cd	1441	294bc	11.6bc	164.4
Carisma	62.1abcd	1237	483a	13.8abc	155.7
Average	62.7	927.3	357.8	17.5	154.7

Source: Fageria, N.K., *Mineral Nutrition of Rice*, CRC Press, Boca Raton, FL, 2014.
The P source was triple superphosphate.
AE, agronomic efficiency; PE, physiological efficiency; APE, agrophysiological efficiency; APE, apparent recovery efficiency; UE, utilization efficiency. Definitions and methods of calculating these efficiencies are given in Table 4.1. Means followed by the same letter within the same column are not significantly different at the 5% probability level by Tukey's test.

TABLE 4.6
Influence of P Fertilization on PUE in Lowland Rice

P Rate ($mg\ kg^{-1}$)	AE ($mg\ mg^{-1}$)	PE ($mg\ mg^{-1}$)	APE ($mg\ mg^{-1}$)	ARE (%)	UE ($mg\ mg^{-1}$)
50	184.3	1105.5	449.4	41.2	454.3
100	109.5	1154.5	433.2	25.2	289.7
150	88.9	1069.7	426.7	20.8	222.1
200	66.0	948.7	357.8	18.5	175.4
250	54.7	915.9	359.3	15.2	139.1
Average	100.7	1038.9	405.3	24.2	256.1
F-test	**	**	*	**	**
CV(%)	12	7	11	8	6

Regression analysis
P rate (X) vs. AE (Y) = $253.8250 - 1.6739X + 0.00356X^2$, $R^2 = 0.9275$**
P rate (X) vs. PE (Y) = $1114.4740 + 0.5420X - 0.00571X^2$, $R^2 = 0.5791$**
P rate (X) vs. APE (Y) = $481.9905 - 0.51126X$, $R^2 = 0.4382$**
P rate (X) vs. RE (Y) = $55.5134 - 0.3525X + 0.00078X^2$, $R^2 = 0.9293$**
P rate (X) vs. UE (Y) = $602.9245 - 3.5355 + 0.00649X^2$, $R^2 = 0.9475$**

Source: Fageria, N.K. et al., *Commun. Soil Sci. Plant Anal.*, 44, 2932, 2013.
AE, agronomic efficiency; PE, physiological efficiency; APE, agrophysiological efficiency; ARE, apparent recovery efficiency; UE, utilization efficiency.
*,**Significant at the 5% and 1% probability levels, respectively.

Fageria et al. (2014a) studied the NUE of tropical legume cover crops including PUE. Cover crop species differed significantly in PUE (Table 4.7). Overall, PUE varied from 248 for Brazilian lucerne (*Stylosanthes guianensis* (Aubl.) Sw.) to 597.57 for black mucuna bean (*Mucuna pruriens*) mg mg^{-1}, with a mean value of 427 mg mg^{-1}. PUE was maximal among macronutrients. Overall, it was 12.5-fold higher for N use efficiency, 177-fold higher for carbon (C) use efficiency, 5.5-fold higher for calcium (Ca) use efficiency, and 1.6-fold higher for magnesium (Mg) use efficiency. Higher PUE in crop plants, including legumes, has been reported by Fageria et al. (2006). Since P acquisition by plants rarely exceeds 20% of the total fertilizer P applied, higher internal plant use efficiency is important for practical purposes or crop production (Friesen et al., 1997).

Fageria et al. (2014b) studied the PUE in five tropical legume cover crops at three P levels (Table 4.8). PUE was significantly influenced by cover crops, and P × cover crop interaction was significant for PUE (Table 4.8). PUE decreased with the increase in P rate, suggesting increased root dry weight with increasing P rates as compared to the control treatment (Table 4.9). Root dry weight at 100 and 200 mg P kg^{-1} levels increased as compared to control treatments of two cover crops (Figures 4.4 and 4.5). Fageria (1992) reported that NUE decreased with the increase in nutrient rates in crop plants.

Fageria and Baligar (2014) studied the AE, PE, and ARE of 14 tropical cover crops. AE is defined as the amount of dry matter produced by cover crops per unit P applied and was significantly

TABLE 4.7
Nutrient Use Efficiency in the Tops of 16 Tropical Cover Crops

	C	N	P	Ca	Mg	Zn	Fe	Mn	B
Cover Crop Species			(mg mg^{-1})^a				(mg μg^{-1})		
Short-flowered crotalaria	2.48	35.15	299.10	75.18	248.73	189.55	68.99	26.32	17.19
Sunn hemp	2.43	35.42	430.54	88.33	225.87	188.48	56.09	32.57	22.79
Smooth crotalaria	2.40	26.06	322.44	93.06	242.92	63.39	26.44	21.39	15.98
Showy crotalaria	2.56	28.88	363.66	50.17	246.32	180.29	36.28	21.51	16.44
Ochroleuca crotalaria	2.41	27.46	425.39	98.00	180.57	155.10	25.51	18.70	16.89
Calopo	2.39	36.15	339.63	80.29	308.88	734.71	7.90	29.13	18.19
Black Jack bean	2.34	28.93	348.64	70.19	289.45	178.48	20.32	14.23	21.82
Bicolor pigeon pea	2.34	28.29	326.49	70.73	273.85	287.76	20.25	15.23	18.98
Black pigeon pea	2.33	33.13	434.69	81.46	312.03	111.21	17.31	16.20	19.89
Mulato pigeon pea	2.42	36.88	514.19	78.81	284.19	226.95	25.44	23.10	19.91
Lablab	2.40	35.69	596.75	74.99	289.16	71.96	19.81	21.27	22.65
Mucuna bean ana	2.40	47.88	556.43	68.31	288.55	114.57	17.50	19.67	21.01
Black mucuna bean	2.36	44.77	597.57	100.71	331.44	401.25	31.35	22.98	25.74
Gray mucuna bean	2.41	41.28	591.96	69.47	301.62	378.45	26.13	37.44	20.09
White Jack bean	2.42	31.39	446.06	86.50	281.36	89.96	17.19	13.39	18.64
Brazilian lucerne	2.46	30.83	248.15	60.73	290.76	58.45	7.41	43.04	15.40
Average	2.41	34.26	427.61	77.93	274.73	251.60	26.50	23.51	19.48
F-test									
Soil pH (S)	*	*	NS	**	**	NS	NS	**	NS
Cover crop species (C)	**	**	**	**	**	**	**	NS	**
S × C	NS	**	NS	NS	NS	NS	NS	NS	NS

Source: Fageria, N.K. et al., *J. Plant Nutr.*, 37, 294, 2014a.

[a] Values averaged over across three soil pH (pH 5.1–7.0). NUE = dry wt of shoot produced per unit wt of nutrient absorbed.

[*,**, and NS] Significant at the 5% and 1% probability level and nonsignificant, respectively.

TABLE 4.8
PUE (mg mg^{-1}) in Five Tropical Legume Cover Crops

	P Fertilizer Rate (mg P kg^{-1})[a]		
Cover Crops	**0**	**100**	**200**
Sunn hemp	1119.75ab	992.93a	910.03a
Pigeon pea	918.33b	1271.85a	833.12a
Lablab	1177.13ab	837.26a	456.32a
Gray mucuna bean	1215.08ab	910.07a	434.77a
White jack bean	1960.67a	702.86a	456.87a
Average	1278.19a	942.99ab	618.22b

Source: Fageria, N.K. et al., *Commun. Soil Sci. Plant Anal.*, 45, 555, 2014b.

[a] Means followed by the same letter within the same column are statistically not significant at the 5% probability level by Tukey's test. For comparison of means of P levels, same letter in the same line under different P levels is statistically not significant at the 5% probability level by Tukey's test.

$$\text{PUE}\left(\text{mg mg}^{-1}\right) = \frac{\text{Root dry wt. in mg}}{\text{P uptak in mg}}$$

TABLE 4.9
Root Dry Weight (g Plant^{-1}) of Five Tropical Legume Cover Crops as Influenced by P Fertilization

	P Level (mg kg^{-1})		
Cover Crops	**Low P (0)**	**Medium P (100)**	**High P (200)**
Sunn hemp	0.17c	0.83ab	0.64c
Pigeon pea	0.16c	0.51b	0.13d
Lablab	0.13c	1.14a	0.96b
Gray mucuna bean	0.53b	0.83ab	1.42a
White jack bean	0.77a	1.12a	0.93b
Average	0.35b	0.89a	0.82a
F-test			
P level (P)	**		
Cover crops (CC)	**		
P × CC	**		

Source: Fageria, N.K. et al., *Commun. Soil Sci. Plant Anal.*, 45, 555, 2014b.

**Significant at the 1% probability level. Means followed by the same letter within the same column are statistically not significant at the 5% probability level by Tukey's test. For comparison average values of P levels, same letter in the same line under different P levels, are statistically not significant at the 5% probability level by Tukey's test.

influenced by P, cover crops, and P × cover crop interactions (Table 4.10). Significant P × cover crop interactions indicate that cover crops were having different responses with increasing P levels. AE values also varied with the change in P levels. At 100 mg P kg^{-1} level, AE varied from 1.82 mg mg^{-1} produced by C1 (*Crotalaria breviflora*) to 18.2 mg mg^{-1} produced by C12 (*Mucuna aterrima*), with a mean value of 7.02 mg mg^{-1}. AE values varied from 0.14 mg mg^{-1} produced by C6¨ (*Calopogonium mucunoides*) to 11.81 mg mg^{-1} produced by C12 (*M. aterrima*), with a mean value of 4.85 mg mg^{-1}

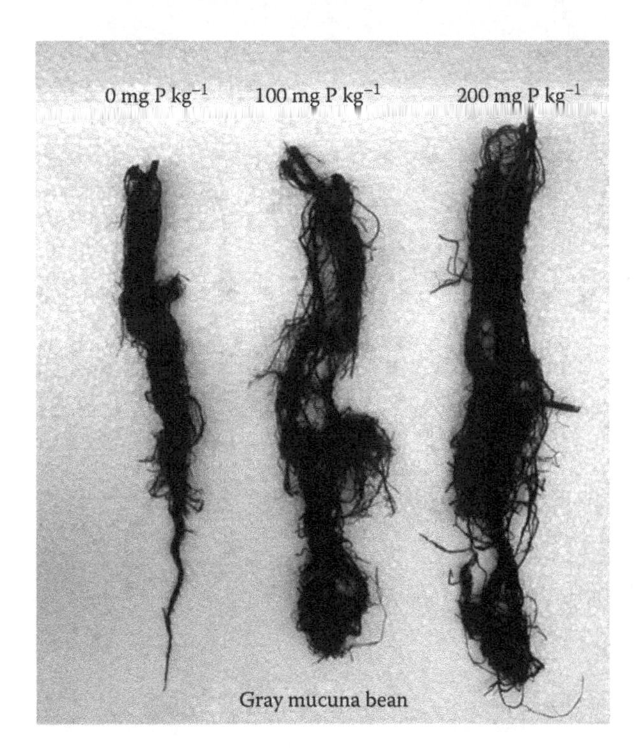

FIGURE 4.4 Root growth of gray mucuna bean at three P levels grown on Brazilian Oxisol. (From Fageria, N.K. et al., *Commun. Soil Sci. Plant Anal.*, 45, 555, 2014b.)

FIGURE 4.5 Root growth of white jack bean at three P levels grown on Brazilian Oxisol. (From Fageria, N.K. et al., *Commun. Soil Sci. Plant Anal.*, 45, 555, 2014b.)

TABLE 4.10
AE of Tropical Legume Cover Crops as Influenced by P Fertilization

Cover Crop Species	AE at 100 mg P kg^{-1} (mg mg^{-1})	AE at 200 mg P kg^{-1} (mg mg^{-1})
Crotalaria breviflora	1.82e	1.25ef
Crotalaria juncea L.	9.76bcd	7.34bc
Crotalaria mucronata	2.51e	1.19ef
Crotalaria spectabilis Roth	3.11cde	2.35ef
Crotalaria ochroleuca G. Don	4.78cde	4.59cde
Calopogonium mucunoides	2.97de	0.14f
Pueraria phaseoloides Roxb.	2.51e	1.90ef
Cajanus cajan L. Millspaugh	5.66cde	2.99def
Cajanus cajan L. Millspaugh	5.23cde	3.03def
Dolichos lablab L.	13.83ab	10.40ab
Mucuna deeringiana (Bort) Merr.	10.12bc	6.87bcd
Mucuna aterrima (Piper & Tracy) Holland	18.24a	11.81bcd
Mucuna cinereum L.	4.69cde	6.51bcd
Canavalia ensiformis L. DC.	13.09ab	7.70bc
Average	7.02a	4.85b
F-test		
P level (P)	**	
Cover crops (CC)	**	
P × CC	*	

Source: Fageria, N.K. and Baligar, V.C., *Commun. Soil Sci. Plant Anal.*, 45, 1, 2014.
*,**Significant at the 5% and 1% probability levels, respectively. Means followed by the same letter within the same column or in the same line under two P levels are statistically not significant at the 5% probability level by Tukey's test.

at 200 mg P kg^{-1} level. Variation in AE among cover crop species may be related to their different growth responses to applied P or shoot growth (Figure 4.6). Fageria (2009) reported different NUE in crop plants due to different growth responses.

AE was significantly lower at 200 mg P kg^{-1} as compared with 100 mg P kg^{-1} of soil. Decrease in AE was 45% at the high P level as compared with the low P level. NUE decreased in crop plants with the increase in nutrient levels reported by Fageria (1992, 2009). This decrease in AE at higher P level is related to relative decrease in crop response with increasing nutrient levels (Fageria, 1992).

PE of P (dry matter produced per unit of P uptake) was significantly affected by cover crop treatment (Table 4.11). Physiological PUE varied from 200 mg mg^{-1} produced by cover crop pigeon pea (*Cajanus cajan*) (mixed color) to 540 mg mg^{-1} produced by cover crop black mucuna bean, with a mean value of 338 mg mg^{-1}. The difference in PUE in crop plants is reported by Fageria (2009). Fageria (2009) also reported that among physiological efficiencies, PUE was maximum among macronutrients in legume crop plants and N use efficiency was minimum. These authors also reported that variation in PE in crop plants may be associated with different root growth, shoot growth, or utilization in shoot. Root dry weight variation of 14 cover crop species is presented (Figure 4.7).

ARE (nutrient uptake in shoot per unit of nutrient applied) of P was significantly influenced by cover crop (Table 4.12). Apparent PUE varied from 0.43% to 4.19%, with a mean value of 2%. In this study, P recovery efficiency was lowest as compared to other macronutrients, like N, K, Ca, and Mg (Fageria and Baligar, 2014). Lower P recovery efficiency in *C. breviflora* and higher P recovery efficiency in *Canavalia ensiformis* were associated with lower and higher dry weight of shoot of these

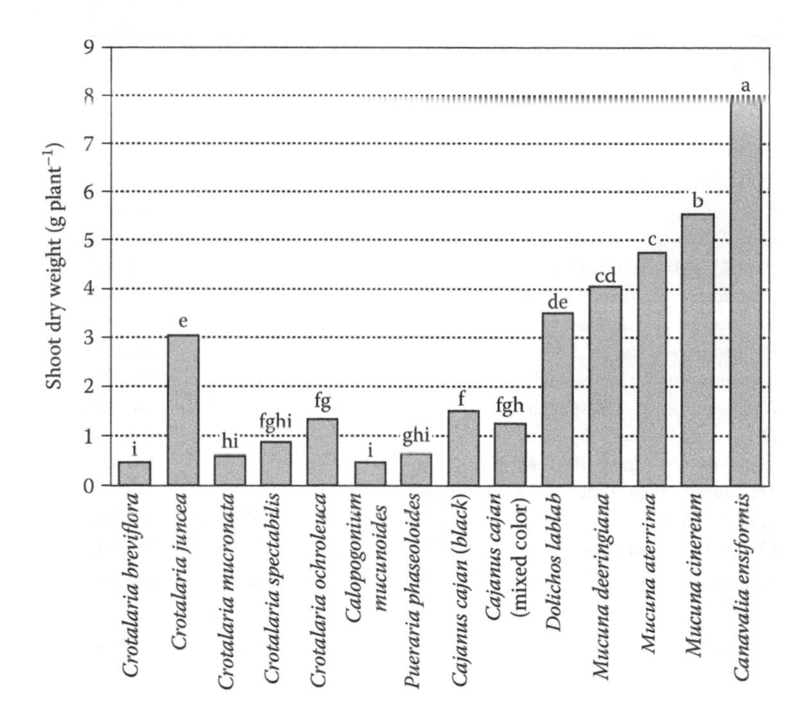

FIGURE 4.6 Shoot dry weight of 14 tropical legume cover crops. (From Fageria, N.K. and Baligar, V.C., *Commun. Soil Sci. Plant Anal.*, 45, 1, 2014.)

cover crop species (Figure 4.6). The lowest recovery efficiency of P among macronutrients may have been related to higher P immobilization capacity of Brazilian Oxisols (Fageria and Barbosa Filho, 1987; Fageria and Baligar, 2003, 2008).

4.4 PHOSPHORUS USE EFFICIENCY VERSUS CROP YIELD

Crop yield is the most important parameter to evaluate treatment effects, including nutrient rate or NUE. Fageria et al. (2013) studied the association among APE, UE, and grain yield of lowland rice. The positive relationship was significant. Equations were as follows: APE (X) vs. grain yield (Y) = $-96.1865 + 0.9938X - 0.00155X^2$, $R^2 = 0.32^*$ and UE (X) vs. grain yield (Y) = $10.8386 + 0.4622X - 0.00107X^2$, $R^2 = 0.75^{**}$. This indicates that agrophysiological and utilization efficiencies are responsible for 32% and 75% variation in grain yield, respectively. Fageria (2014) also studied the association between grain yield of upland rice genotypes and PUEs (Table 4.13). All the five PUEs have significant positive association with grain yield. Therefore, it can be concluded that improvement in PUE can improve yield of rice.

4.5 MECHANISMS RESPONSIBLE FOR VARIATION IN PHOSPHORUS USE EFFICIENCY IN CROP PLANTS

Nutrient uptake and use efficiency varied from nutrient to nutrient, crop species to crop species, and genotypes within crop species. However, certain broad processes are common for all nutrients and crop species. These processes are (1) nutrient acquisition from the growth medium, (2) absorption by roots, (3) translocation and distribution in the plant parts, and (4) utilization in growth and development (Gerloff and Gableman, 1983; Clark, 1990). Many soil factors affect P availability to plant roots and its uptake. Such factors are concentration of P in the soil solution, soil buffering capacity, P distribution in the soil profile, soil moisture content, soil temperature, and soil texture

TABLE 4.11
PE of P in Tropical Cover Crops as Influenced by P Fertilization

Cover Crops	PE of P (mg mg^{-1})
Crotalaria breviflora	345.99abc
Crotalaria juncea L.	388.66abc
Crotalaria mucronata	444.17ab
Crotalaria spectabilis Roth	355.90abc
Crotalaria ochroleuca G. Don	313.07abc
Calopogonium mucunoides	343.60abc
Pueraria phaseoloides Roxb.	440.65ab
Cajanus cajan L. Millspaugh	200.25c
Cajanus cajan L. Millspaugh	246.09bc
Dolichos lablab L.	403.68abc
Mucuna deeringiana (Bort) Merr.	260.45bc
Mucuna aterrima (Piper & Tracy) Holland	540.23a
Mucuna cinereum L.	205.70c
Canavalia ensiformis L. DC.	257.40bc
Average	338.98
F-test	
P level (P)	NS
Cover crops (CC)	**
P × CC	NS

Source: Fageria, N.K. and Baligar, V.C., *Commun. Soil Sci. Plant Anal.*, 45, 1, 2014. Values are across two P levels (i.e., 100 and 200 mg P kg^{-1} soil).

**, and NS Significant at the 1% probability level and nonsignificant, respectively. Means followed by the same letter within the same column are statistically not significant at the 5% probability level by Tukey's test.

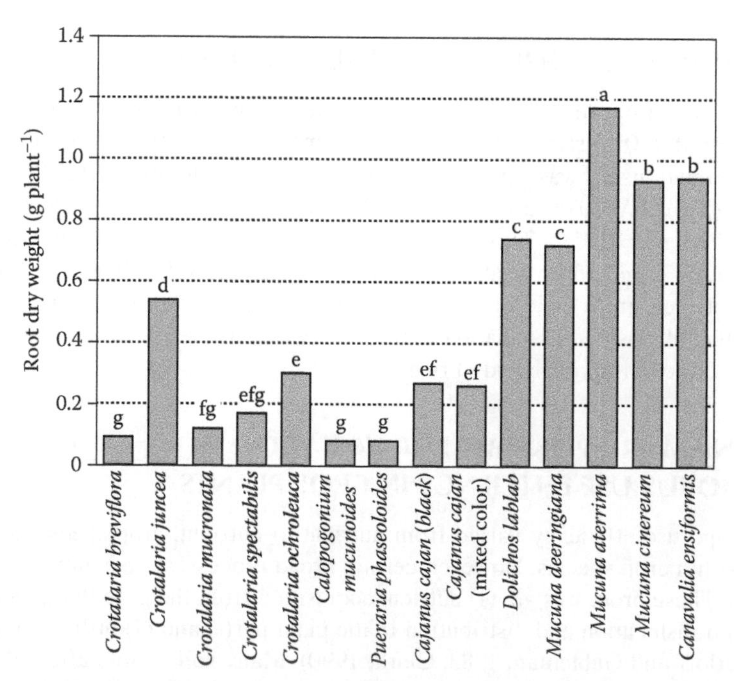

FIGURE 4.7 Root dry weight of 14 tropical cover crops. (From Fageria, N.K. and Baligar, V.C., *Commun. Soil Sci. Plant Anal.*, 45, 1, 2014.)

TABLE 4.12
ARE of P in Tropical Cover Crops as Influenced by P Fertilization

Cover Crops	ARE of P (%)
Crotalaria breviflora	0.43f
Crotalaria juncea L.	2.51bcd
Crotalaria mucronata	0.47f
Crotalaria spectabilis Roth	0.78ef
Crotalaria ochroleuca G. Don	1.48def
Calopogonium mucunoides	0.48f
Pueraria phaseoloides Roxb.	0.58f
Cajanus cajan L. Millspaugh	2.28bcde
Cajanus cajan L. Millspaugh	1.80cdef
Dolichos lablab L.	3.63ab
Mucuna deeringiana (Bort) Merr.	3.26abc
Mucuna aterrima (Piper & Tracy) Holland	3.05abc
Mucuna cinereum L.	3.20abc
Canavalia ensiformis L. DC.	4.19a
Average	2.01
F-test	
P level (P)	NS
Cover crops (CC)	**
P × CC	NS

Source: Fageria, N.K. and Baligar, V.C., *Commun. Soil Sci. Plant Anal.*, 45, 1, 2014.
Values are across two P levels (i.e., 100 and 200 mg P kg^{-1} soil).
**, and NS Significant at the 1% probability level and nonsignificant, respectively. Means followed by the same letter within the same column are statistically not significant at the 5% probability level by Tukey's test.

TABLE 4.13
Relationship between PUEs (*X*) and Grain Yield (*Y*) across 18 Upland Rice Genotypes

Plant Variable	Regression Equation	R^2
AE vs. grain yield	$Y = -0.015 + 1.20X$	0.99**
PE vs grain yield	$Y = 181.82 - 0.17X + 0.000051X^2$	0.34**
APE vs grain yield	$Y = -120.00 + 0.93X - 0.00103X^2$	0.60**
ARE vs grain yield	$Y = 1.43 + 4.21X$	0.67**
UE vs grain yield	$Y = -28.47 + 0.67X$	0.41**

Source: Fageria, N.K., *Mineral Nutrition of Rice*, CRC Press, Boca Raton, FL, 2014.
AE, agronomic efficiency; PE, physiological efficiency; APE, agrophysiological efficiency; APE, apparent recovery efficiency; UE, utilization efficiency.
**Significant at the 1% probability level.

(Barber, 1980; Ozanne, 1980; Clark, 1990; Fageria, 2013). Plant factor such as root morphology is also important in nutrient uptake by plants from the soil solution. Major part of the P move to plant roots by diffusion. Hence, root growth and concentration and movement of P in the soil–plant system will determine its uptake.

Significant variation exists among crop species and genotypes of the same species in nutrient uptake and utilization (Gerloff and Gableman, 1983; Baligar et al., 1990, 2001; Epstein and Bloom, 2005; Fageria and Baligar, 2005; Fageria, 2013, 2014). Nutrient uptake and utilization differences may be associated with optimum root geometry; ability of plants to take up sufficient nutrients from lower or subsoil concentrations; plants' ability to solubilize nutrients in the rhizosphere; better transport, distribution, and utilization within plants; and balanced source and sink relationships (Graham, 1984; Baligar et al., 2001; Fageria and Baligar, 2003). Antagonistic (uptake of one nutrient is restricted by another nutrient) and synergistic (uptake of one nutrient is enhanced by other nutrient) effects of nutrients on NUE among various plant species and cultivars within species have not been well explored (Fageria et al., 2008).

4.5.1 Root Geometry

Plants having vigorous and extensive root systems that can explore large soil volumes absorb more water and nutrients under stress conditions and increase crop yield and improve NUE (Merrill et al., 2002). Quantity of nutrient taken up by plants is largely influenced by root radius, mean root hair density, and length of root (Barber, 1995). Shape and extent of root systems influences the rate and pattern of nutrient uptake from soil. Vose (1984) states that rooting depth, lateral spreading, branching, and number of root hairs have a major impact on plant nutrition. Configuration of root system is influenced markedly by nutrient supply. Mineral excess and deficiency affect growth (dry mass, root to shoot ratio) and morphology (length, thickness, surface area, density) of roots and root hairs. Nutrient deficiency leads to finer roots. When plants are N deficient, their roots branch more in regions where the soil is locally enriched with N (Scott-Russell, 1977). Configuration (root and root hair abundance and density, distribution, effective radius, and elongation) of root systems, in relation to nutrient uptake, is extensively covered by Barley (*Hordeum vulgare*) (1970). C and N supplied by roots can be significant for maintaining or improving soil organic matter and influencing NUE (Sainju et al., 2005). A well-developed root system may play a dominant role in soil C and N cycles (Gale et al., 2000; Puget and Drinkwater, 2001) and may have relatively greater influence on soil organic C and N levels than the aboveground plant biomass (Boone, 1994; Norby and Cotrufo, 1998). Roots can contribute from 400 to 1460 kg C ha^{-1} during a growing season (Qian and Doran, 1996; Kuo et al., 1997). Liang et al. (2002) reported that maize roots contributed as much as 12% of soil organic C, 31% of water-soluble C, and 52% of microbial biomass C within a growing season. All of these chemical and biological changes in soils affected by root systems improve NUE in plants.

Cultivar differences in root size are quite common and have been related to differences in nutrient uptake (Caradus, 1990; Baligar et al., 1998; Fageria et al., 2006). Differences between white clover (*Trifolium repens* L.) populations and cultivars in P uptake per plant at low levels of P have been related to differences in root size and absolute growth rate (Caradus and Snaydon, 1986). There is widespread evidence for genotype diversity in root characteristics of many crops in response to environment and increasing interest in using this diversity to improve agricultural production and consequently NUE (Barber, 1994; Gregory, 1994). Mineral deficiency and toxicity, mechanical impedance, moisture stress, oxygen stress, and temperature have tremendous effects on root growth, development, and function and subsequently the ability of roots to absorb and translocate nutrients (Barber, 1995; Marschner, 1995; Baligar et al., 1998; Mengel et al., 2001). Mineral deficiency induces considerable variations in growth and morphology of roots, and such variations are strongly influenced by plant species and genotypes. Overall, the growth of the main axis is minimally affected by nutrient deficiency, but lateral branches and their elongation rates appear to be substantially reduced. Baligar et al. (1998) summarized the effects of various essential elements as follows: N deficiency increases root hair length, increases or has no effect on root hair density, and reduces branching, P deficiency increases overall growth of roots and root hair length, increases number of second-order laterals, and either increases or does not affect root hair density; and K and Ca deficiencies reduce root growth; however, higher Mg levels reduce the dry mass of roots. These nutrient stress factors on nutrient efficiency in plant have not been well explored. Baligar et al. (1998) state that low pH reduces root mass and length and root hair formation; in alkaline soils, ammonium toxicity causes severe root inhibition; and in general salinity leads to reduction in mass and length of roots and dieback of laterals.

4.5.2 Higher Rate of Nutrient Absorption at Low Rhizosphere Concentrations

Capacity of some plant species or genotypes within species to absorb nutrients at higher rate at low nutrient concentration of the growth medium is one of the mechanisms responsible for efficient nutrient use by plants. V_{max} and K_m values according to Michaelis–Menten kinetics or enzyme kinetics are generally used to explain the rate of ion influx in plant roots (Barber, 1995). According to this hypothesis, when nutrient uptake rate is plotted against nutrient concentration, a quadratic increase is obtained and maximum rate of uptake is designated as V_{max} (*Y*-axis). Half of the maximum velocity line touching the uptake rate curve and corresponding concentration on the *X*-axis is designated by K_m. Lower K_m values (higher affinity) indicate a higher uptake rate of plants for a determined nutrient at a low concentration. Uptake rates and K_m values of two genotypes are presented (Figure 4.8). Although the two genotypes have similar V_{max} values, genotype A has a lower K_m value than genotype B, and therefore genotype A will have higher uptake rates at low rhizosphere nutrient concentrations. In this case, genotype A is more efficient in nutrient uptake at lower rhizosphere nutrient concentration. K_m values for P uptake by various plant species and P uptake rate were in the order of peanut (*Arachis hypogaea* L.) > rice > alfalfa (*Medicago sativa* L.) > corn > Barley > wheat (Table 4.14).

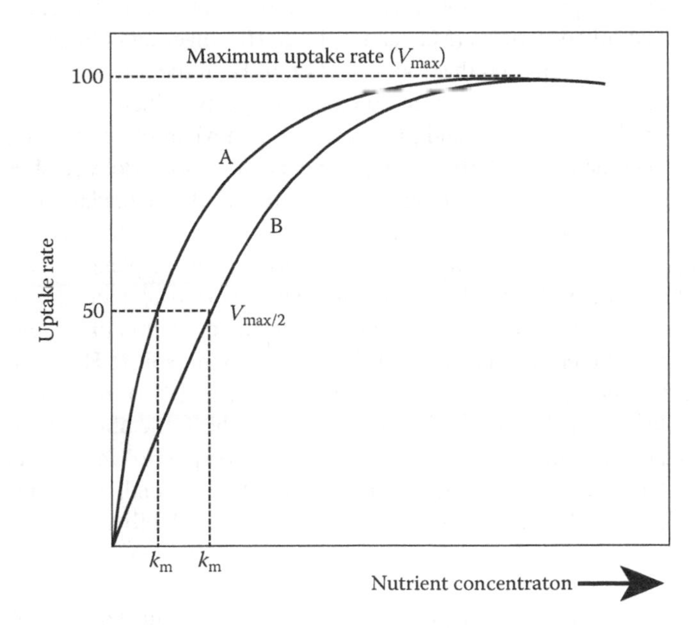

FIGURE 4.8 Hypothetical relationship between nutrient concentration and uptake rate in two genotypes and their K_m and V_{max} values. (From Fageria, N.K. et al., *J. Plant Nutr.*, 31, 1121, 2008.)

TABLE 4.14
Michaelis–Menten Constants for the Absorption of Phosphorus by Principal Crop Species in Nutrient Solution

Crop Species	Range of P Concentration (μM)	K_m in Mole (M) at Low Concentration	Reference
Barley (*Hordeum vulgare*)	1–1000	5.4×10^{-6}	Andrew (1966)
Peanut (*Arachis hypogaea L*)	0.03–400	0.6×10^{-6}	Alagarswamy (1971)
Alfalfa (*M. Sativa*)	100–1000	2.0×10^{-6}	Baligar (1987)
Alfalfa	1–500	4.3×10^{-6}	Andrew (1966)
Wheat (*Triticum* spp.)	0.1–1000	7.4×10^{-6}	Edwards (1970)
Rice (*Oryza sativa*)	0.1–161	2.5×10^{-6}	Fageria (1973)
Rice	0.6–161	1.4×10^{-6}	Fageria (1974)
Corn (*Zea mays*)	100–1000	2.2×10^{-6}	Baligar (1987)

4.5.3 Ability of Plant to Solubilize Nutrients in the Rhizosphere

Several chemical changes occur in the rhizosphere, due to plant roots and soil environmental interactions. Among these changes, pH, reduction–oxidation potential, rhizodeposition, nutrient concentrations, and root exudates are prominent. These chemical changes in the rhizosphere significantly influence nutrient solubility and uptake by plants. Soil pH is one of the most important chemical properties, influencing nutrient solubility. At lower pH (<5.5), availability of most micronutrients is higher except Mo and decreases with increasing soil pH. This decrease is mostly associated with adsorption and precipitation processes. Availability of N as well as P is lower at lower pH and is improved in a quadratic response with increasing pH to about 7.0. N availability increase is associated with improved activity of N turnover by bacteria. P availability is associated with neutralization of Al, manganese (Mn), and Fe compounds that immobilize this element at lower soil pH.

Acidification of the rhizosphere can solubilize several low soluble macronutrients (Riley and Barber, 1971; Barber, 1995) and micronutrients (Marschner, 1995; Hinsinger and Gilkes, 1996; Fageria et al., 2002). Bar-Yosef et al. (1980) reported that root excretion of H^+ at the root surface is a mechanism for enhancing zinc (Zn) uptake than excretion of complexing agents. When more cations are absorbed, H^+ ions are released in the rhizosphere and pH decreases, and when more anions are absorbed, OH^- ions are released and pH increases (Barber, 1995; Mengel et al., 2001). Release of H^+ and OH^- ions in the rhizosphere is associated with maintaining cation and anion balance in plants during the ion uptake process. Enhanced reducing activity at root surfaces has been noted as root-induced responses to Fe deficiency in dicotyledonous and nongraminaceous monocotyledonous plants (Marschner, 1995).

Root-induced rhizosphere chemical change has been reported to increase availability of P to pigeon pea (*Cajanus cajan* L. Millsp.) (Ae et al., 1990). Roots of this plant release piscidic acid, which complexes Fe, and thereby free some of the tightly bound soil P. Hence, pigeon pea is successfully grown in P-deficient tropical soils (Radin and Lynch, 1994). Keerthisinghe et al. (2001) reported that white lupin (*Lupinus albus* L.) and pigeon pea have the ability to access fixed P, and this is attributed to the exudation of organic acids into the rhizosphere. Under P-limiting conditions, white lupin exudes large quantities of citrate, and pigeon pea responds by increased exudation of malonic and piscidic acids. These organic acids increase the availability of P in acid soils, mainly by chelation of Al and Fe bound to P and by suppressing readsorption and precipitation of organic P. Major physical, chemical, and biological changes occurring in the rhizosphere are summarized in Figure 4.9. Discussions of chemical changes in the rhizosphere and nutrient availability are reported by Baligar et al. (1990), Darrah (1993), Barber (1995), Marschner (1995), Hinsinger (1998), Fageria and Stone (2006), and Fageria et al. (2002).

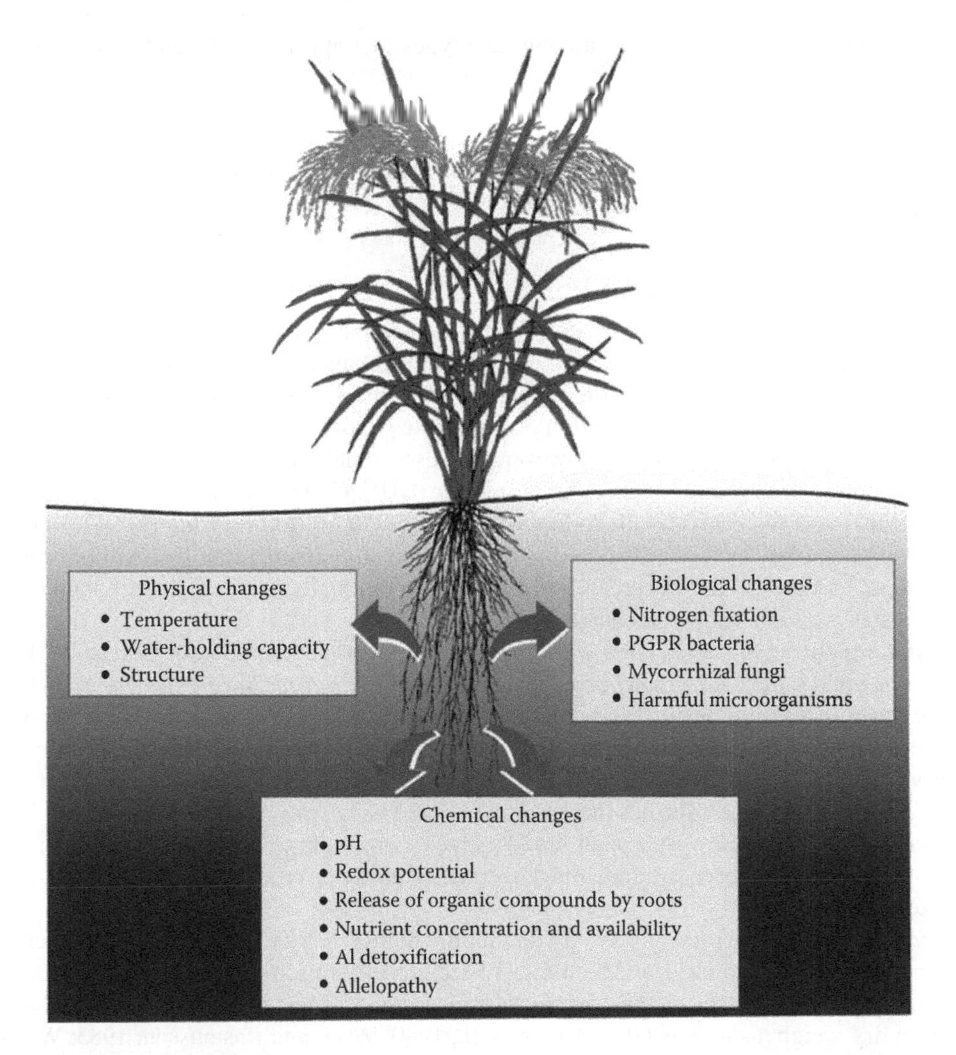

FIGURE 4.9 Major physical, chemical, and biological changes in the rhizosphere. (From Fageria, N.K. and Stone, L.F., *J. Plant Nutr.*, 29, 1327, 2006.)

4.5.4 Better Distribution and Utilization of Nutrients within Plants

Improved distribution of nutrients in parts of plant (root, shoot, and grain) reflects their use efficiency. In recent years, there have been major increases in the mean yields of most crops. Most of these increases in yields have been accompanied by an increase in plant tissue with a high nutrient content such as grain as compared to the lower nutrient content straw (Atkinson, 1990). Higher accumulation of N and P in grain improves yield and consequently leads to higher use efficiency of these nutrients (Fageria et al., 2006). Proportion of total plant N or P partitioned to grain is called N or P harvest index. Nutrient harvest index is defined as nutrient uptake in grain divided by nutrient uptake in grain plus straw. This index is very useful in measuring nutrient partitioning in crop plants, which provides an indication of how efficiently the plant utilizes acquired nutrients for grain production (Fageria and Baligar, 2005). High P harvest index is associated with efficient utilization of P (Fageria et al., 2008).

Schmidt (1984) reported that new cultivar development may need to be directed toward the production of genotypes that exploit inputs most efficiently, not on genotypes that have superior yield only when high production inputs are used. Amounts of P remobilization from storage tissues influence grain PUE, and this varies among genotypes and appears to be under genetic control (Fageria et al., 2008). Variation in nutrient harvest indices among crop species, or genotypes of the same species, is a useful trait in selecting crop genotypes for higher grain yield (Fageria and Baligar, 2005). Inter- and intraspecies differences in NUE of macro- and micronutrients for sorghum (*Sorghum bicolor* L.), maize, alfalfa (*Medicago sativa* L.), and red clover (*Trifolium pretense* L.) have been reported (Baligar and Fageria, 1997).

4.5.5 Allocation of Dry Matter within Plants

Improved distribution of dry matter in crop plants (shoot and grain) is generally associated with higher yields and consequently higher NUE. While the production and utilization of dry matter within a plant depend on each other, the regulation of the partitioning of dry matter into different plant parts is independent of the production of assimilates (Ho, 1988). Partitioning of assimilates is genetically determined in crop plants. However, it is also influenced by environmental factors. Dry matter distribution is measured by grain harvest index (GHI). GHI is the ratio of grain yield to total biological yield and calculated by the following equation: GHI = (grain yield/grain + straw yield). GHI was introduced by Donald (1962) and since has been considered a trait for yield improvement in field crops. Values for GHI in cereals and legumes are normally less than 1. Although GHI is a ratio, it is often expressed as a percentage.

Dry matter is positively associated with grain yield (Fageria et al., 2004a). Evans (1993, 1994) reported that yield increases in many cereals, legumes, and root crops during the twentieth century were due to an increase in harvest indices of these crops. Austin (1994) reported that in modern rice, wheat, and barley, cultivars are short in stature and can have a GHI near 0.50. In contrast, old cultivars are taller and have harvest indices of 0.30 or lower. Hay (1995) reported that GHI of grain crops, particularly cereals, has increased with increasing crop yields during the last 50 years of the twentieth century. However, plant breeders have not sought to raise GHI, and probably any increase in this trait has been an unplanned secondary effort of breeding for grain yield (Araujo and Teixeira, 2003).

GHI values of modern crop cultivars are commonly higher than those of old traditional cultivars for major field crops (Ludlow and Muchlow, 1990). Genetic improvement in annual crops such as wheat, barley, corn, oat, rice, and soybean (*Glycine max* L. Merr.) has been reported due to an increase in dry weight as well as GHI (Austin et al., 1980; Wych and Rasmusson, 1983; Wych and Stuthman, 1983; Cregan and Yaklich, 1986; Payne et al., 1986; Tollenaar, 1989; Feil, 1992; Peng et al., 2000). In potato (*Solanum tuberosum* L.), modern cultivars have plant dry weights 10 times that of the wild species (*Solanum demissum* L.). Harvest index (tuber dry weight as a proportion of plant weight) increased from 7% in wild species to 81% in modern cultivars (Inoue and Tanaka, 1978). Peng et al. (2000) reported that genetic gain in yield of rice cultivars released before 1980 was mainly due to improvement in GHI, while increases in total biomass were associated with yield trends for cultivars developed after 1980. Cultivars developed after 1980 had relatively high GHI values, but further improvement in GHI has not been achieved. These authors also reported that further increases in rice yield potential would likely occur through increasing biomass production rather than increasing GHI.

4.5.6 Balanced Source and Sink Relationship

Genetic and production physiological studies indicate that crop yield potential is high and is not fully exploited (Fageria et al., 2006). Balanced source and sink relationships are vital for higher yields and consequently higher NUE in crop plants. However, neither source nor sink manipulation alone can improve crop yield indefinitely (Ho, 1988). Most plants have the ability to buffer any imbalance between source and sink activity by storing carbohydrates during the periods of excess production and mobilization of these reserves when the demands of growth exceed the supply of carbohydrates available through current photosynthesis (Evans and Wardlaw, 1976). Both source and sink activities vary with plant development and are modified by environmental factors.

Biomass production in plants depends on photosynthesis. In the beginning of plant growth, leaves function as sinks, but with advancement of age, leaves serve as sources. Hence, leaves are the main site of photosynthesis and source of carbohydrates in plants; however, with advances of plant age, stems and inflorescence of some cereals contribute substantially to photosynthetic activity (Evans and Wardlaw, 1976). Evans and Wardlaw (1976) reported that photosynthesis by glumes and young grains of wheat constitutes an important source of assimilates and a means of recapturing respired CO_2. Ear photosynthesis throughout grain growth contributed 33% to grain growth requirements in one awned wheat cultivar and 20% in an unawned one (Evans and Rawson, 1970).

Panicles or heads in cereals, pods in legumes, and tubers in root crops are main sinks of photoassimilates. A small portion of photosynthetic product is also translocated to the roots. Growing organs of plant are active sinks, and these prevent the accumulation of photoassimilates in the sources, if source capacity is limited. Assimilated carbohydrates in the source and sink are lost through respiration, and this loss is reportedly half of the total C assimilated in photosynthesis (Evans and Wardlaw, 1976). In modern cultivars, source capacity has been more limiting to yield as compared to the older ones (Evans and Wardlaw, 1976). During the twentieth century, both source and sink have been improved in important annual crops, and this facilitated an improvement in yields (Ho, 1988). Capacity of dry matter production in leaves may either be higher or lower than the capacity of dry matter accumulation in other parts of the plant. At different times, either source- or sink-limiting situations may exist in crop production (Ho, 1988).

4.6 BREEDING CROP SPECIES/GENOTYPES FOR IMPROVED PHOSPHORUS USE EFFICIENCY

Breeding crop species or genotypes for improved PUE is an important strategy in modern agriculture due to its impact on cost of production and environmental pollution. Through plant breeding, the genetic yield potential of wheat, soybean, corn, and peanuts has been improved by 40%–100% within the twentieth century (Gifford et al., 1984; Ho, 1988). Genetic variability among crop species and genotypes for macro- and micronutrient use or requirement is documented (Clark and Duncan, 1991; Baligar et al., 2001; Fageria and Baligar, 2005; Hillel and Rosenzweig, 2005).

Considerable progress has been made in identifying crop species and genotypes within species for NUE, tolerance to elemental toxicity, and understanding possible mechanisms involved (Foy, 1984, 1992; Graham, 1984; Clark and Duncan, 1991; Marschner, 1995; Baligar et al., 2001; Blamey, 2001; Okada and Fischer, 2001; Fageria et al., 2003, 2006; Yang et al., 2004; Epstein and Bloom, 2005; Fageria and Baligar, 2005). Plant traits and characteristics expressing tolerance to essential nutrient deficiencies are numerous and have been reviewed (Baligar et al., 1990). Clark and Duncan (1991) suggested that juvenile stage of plant growth is more desirable to evaluate plants for mineral stress tolerance. They stated that yield (vegetative or grain/seed/fruit) is probably the most common trait used to evaluate plants for tolerance to soil mineral stresses. Progress has been limited in releasing crop cultivars expressing these traits.

Breeding of more efficient plants for major nutrients such as N, P, and K, which are required in large amounts by crop plants for maximum economic yield, requires special attention. Several field and greenhouse experiments using genotypes of rice, wheat, and common bean in Brazilian Inceptisols and Oxisols using different N and P rates have been conducted (Fageria, 1998; Fageria, 2000). In these studies, inter- and intraspecies differences were measured for growth and NUE and PUE. When P level in the soil extracted by Mehlich 1 solution was approximately 2 mg kg^{-1} of soil, most of the genotypes either did not produce or produced insignificant grain yield. Similarly, without addition of N fertilizers, rice genotypes produced very low grain yield. Therefore, the strategy should be using efficient crop genotypes with judicious use of N, P, and K fertilizers.

Although numerous studies have shown a wide range of genotypic differences among and within species for N, P, and K efficiency traits, the genetics of these plant responses are not well understood and appear to be complicated (Clark and Duncan, 1991). Most studies indicated a genetic control. Clark and Duncan (1991) reported that P efficiency traits are heritable and could be used to improve germplasm for P nutrition. A prime example of success has been with white clover in New Zealand (Caradus, 1990). Root growth, morphology, ion uptake, and use efficiency should be considered when plants are to be improved for mineral nutrition traits involving K in breeding programs (Pettersson and Jensen, 1983; Clark and Duncan, 1991). Yield has been classified as a character controlled by quantitative genetics, that is, one is influenced by multiple genes with the effects of individual genes normally unidentified (Wallace et al., 1972). Yield improvement by the use of nutrient-efficient genotypes deserves special attention in relation to identifying physiological components attributing cultivar differences in economic yield and to acquiring an understanding of their genetic control. High yields achieved in rice by incorporating short, erect, thick, dark-green leaves and short stiff stems clearly demonstrate the merit of including physiological component traits in plant breeding programs (Wallace et al., 1972; Fageria et al., 2006).

Richardson (2001) reported that soil P uptake can be increased by plant modification. Selection of plants for increased efficiency of P has been demonstrated with root morphology being particularly important (Lynch, 1995). Similarly, gene technologies offer opportunities for manipulating the structure and function of plant roots for improved acquisition of soil P (Richardson, 2001). Plant genes that regulate root branching have been isolated (Zhang and Forde, 1998), and the expression in plants of specific bacterial genes (i.e., encoding phytohormone activities) may offer new insights into the role of such genes in plant growth and development (Richardson, 2001). Cloning and characterization of plant and fungal phosphate transporter genes may provide new possibilities for increasing plant P uptake (Smith et al., 2000; Richardson, 2001).

Molecular biology technology can be used as an approach in the isolation, identification, localization, and laboratory reproduction of gene(s) carrying desirable nutrient efficiency traits (Clark and Duncan, 1991). In the twentieth century, genetic engineering techniques did not have a significant role in improving nutrient-efficient crop genotypes. However, its wide applicability or potential in the twenty-first century for improving nutrient efficiency in crop plants is highly predicted. In addition, recently, new possibilities have arisen to transfer desired traits (genes) not just between strains of the same species but even from one species to another, thus greatly enlarging the range of potential genetic resources available to agricultural scientists (Hillel and Rosenzweig, 2005).

Blair (2013) reported that the roots are the primary organs for the uptake of P from the soil and when breeding for low soil P conditions, root morphology and other root characteristics should be considered. Specifically, roots and root hairs, together with mycorrhizal fungi, are responsible for foraging through different layers of soil for P (Marschner, 1995). Since P is more concentrated in the upper soil layer, breeding for low P must consider root angle or gravitropism and the number of lateral roots versus basal roots in a root system for efficient P uptake (Liao et al., 2001). Total root length of each type and their diameter are important factors (Beebe et al., 2006; Cichy et al., 2009). In addition to root types, total P uptake depends on the total surface area of all types of roots, and therefore a high density of root hairs especially on lateral roots can add substantially to the absorptive surface of roots and their P uptake efficiency (Yang et al., 2004). This mechanism is common in cultivars from low P regions but requires a high investment in a large root system and many root hairs (Blair, 2013). Cultivar differences are notable, and measurement of this trait should become a standard in common bean (Ramaekers et al., 2010).

4.7 ENHANCING PHOSPHORUS USE EFFICIENCY IN CROP SPECIES/GENOTYPES

Crop production is widely limited by P deficiency and will probably increase in the future. In addition, there is an increasing awareness that there are limits to global rock phosphate reserves and that increasing the efficiency with which these reserves are used to produce crops is vital to maintain or increase current agricultural productivity (Cordell et al., 2009; Veneklaas et al., 2012). Enhancing PUE is important in modern crop production system to increase productivity, to reduce the cost of production, and to reduce soil degradation and environmental pollution. Management practices that can be adopted to improve PUE are divided into three groups: (1) improvement in climatic factors, (2) improvement in soil factors, and (3) improvement in plant factors. These practices should be adopted together or in integration rather than in isolation to achieve favorable PUEs in crop plants.

4.7.1 IMPROVING CLIMATIC FACTORS

Among climatic factors that influence P uptake by plant are solar radiation, soil temperature, and water availability to crops (Baligar et al., 2001). Solar radiation is directly related to photosynthesis in crop plants, which in turn affects plants' demands for nutrients. Quality of radiation and crop shading reduces crop growth rate and ion uptake (Fageria, 1992; Baligar et al., 2001). Soil temperature influences the rate of nutrient release from organic and inorganic reserves and the uptake by roots and subsequent translocation and utilization in plants (Baligar et al., 2001). Similarly, soil water content is very important for solubilization and transport of P to rhizosphere vicinity for uptake by plants. Importance of water in upland rice production in the central part of Brazil locally known as "Cerrado" region is presented (Figure 4.10). When rainfall was normal, upland rice produced about 5000 kg rice yield per ha. However, yield was significantly reduced when drought occurred around flowering stage (30 days before and 10 days after) as compared to normal precipitation. Grain yield differences of upland rice cultivars grown in the Cerrado region of Brazil at two P levels are presented (Table 4.15). In general, early-maturing cultivars have higher grain yields than late-maturing cultivars. Late-maturing cultivars suffered from drought, while early-maturing cultivars escaped drought effects.

Climatic variable cannot be changed, but cultivar selection and crop management must be tailored to prevailing climatic conditions. Soil temperature can be minimally modified by adopting

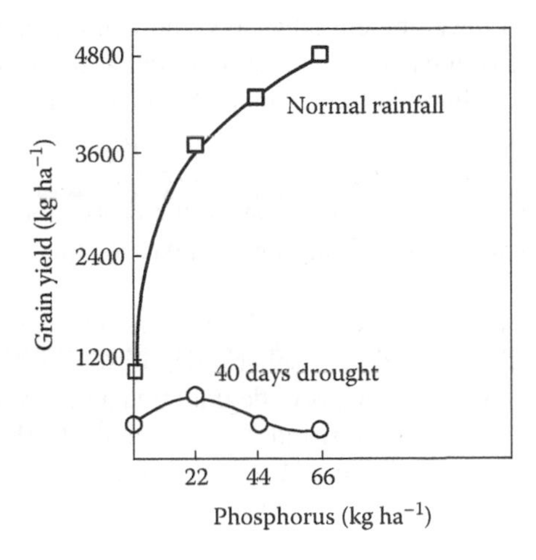

FIGURE 4.10 Response of upland rice to phosphorus fertilization and water availability. (From Fageria, N.K., *Pesq. Agropec. Bras.*, 15, 259, 1980.)

TABLE 4.15
Grain Yield of Upland Rice Cultivars Having Different Growth Cycles

Flowering in Days	Growth Cycle (Days)	GY at P 2.3 mg kg^{-1}	GY at 4.9 mg P kg^{-1}
85 (25)[a]	110	1591	2093
95 (15)	120	1216	1320
106 (930)	130	1071	1164
111 (2)	135	679	757

Source: Fageria, N.K., *Maximizing Crop Yields*, Marcel Dekker, New York, 1992.
GY, grain yield.
[a] Values inside the parentheses represent the number of cultivars tested.

conservation tillage and mulching practice. Soil moisture can be improved by artificial irrigation where rainfall is not adequate or not well distributed during the crop growth cycle. Soil management practices can improve water use efficiency (WUE) and consequently PUE.

Soil management practices that can improve WUE are conservation tillage, increased soil organic matter content, reduced length of fallow periods, contour farming, furrow dikes, control of plow pan, crop selection, and use of appropriate crop rotation (Nielsen et al., 2002; Stone and Schlegel, 2006; Fageria and Stone, 2013). It is possible to increase WUE by 25%–40% through soil management practices that involve tillage and by 15%–25% by modifying nutrient management practices (Hatfield et al., 2001). Precipitation use efficiency can be enhanced through adoption of intensive cropping systems in semiarid environments and increased plant populations in more temperate and humid environments (Hatfield et al., 2001; Fageria and Stone, 2013). Detailed discussion of these practices is described in Chapter 7.

4.7.2 Improving Soil Factors

Soil conditions that affect P availability include the physical, chemical, and biological properties of soil. Low level of soil-available nutrients is common in highly weathered Oxisols and Ultisols (Fageria and Baligar, 2008). These soils have a low natural level of P as well as high P immobilization capacity due to the presence of Fe and Al oxides and hydro-oxides. Soil erosion is responsible for the loss of top soil layers, which are rich in organic matter and P. Soil compaction and high bulk density affect root growth and development adversely and consequently P uptake. High mechanical impedance leads to loss of root caps and reduction in radial thickening primarily due to shorter and wider cells with the same volume in the cortex and a thicker cortex (Baligar et al., 1975). This may cause changes in cell structure of the endodermis and pericycle (Baligar et al., 1975; Bennie, 1996; Fageria, 2013). Such changes in the size and internal and external morphology of roots due to the adverse soil's physical conditions will influence the roots' ability to explore larger soil volume and reduce nutrient and water availability and uptake, leading to low NUE and lower yields (Baligar et al., 2001).

Leaching and crop removal of basic cations, N_2 fixation by legumes, use of heavy levels of organic and inorganic N fertilizers, and atmospheric deposition of N and sulfur oxide are major factors influencing soil acidification that leads to degradation and lower productivity and soil quality in temperate and tropical regions of the world (Sumner et al., 1991; Baligar et al., 2001). Acidic soils have phytotoxic levels of Al^{3+}, Mn^{2+}, and H^+ and deficiency of N, P, K, Ca, Mg, Mo, and Zn to support optimum plant growth (Sumner et al., 1991; Fageria and Baligar, 2008). These factors are responsible for reduced plant growth and lower PUE (Marschner, 1995; Fageria, 2013).

Use of lime and gypsum in acid soils, deep plowing, use of organic manures, and recycling crop residues can improve soil's physical properties and improve PUE in crop plants. Detailed discussion of these practices is described in Chapter 4.

4.7.3 Improving Plant Factors

Selection of improved genotypes of wheat, rice, corn, soybean, and peanuts and use of nutrients, especially N, have contributed significantly to yield increase in the second half of the twentieth century (Baligar et al., 2001). Variability in P uptake and use efficiency in crop species and genotypes is widely reported (Clark, 1990; Clark and Duncan, 1991; Baligar et al., 2001; Fageria, 2013, 2014). Variability may be related to root morphology and other plant factors such as higher photosynthetic rate per unit of P uptake. Therefore, selecting plants or genotypes with higher PUE may be an important strategy in improving PUE in crop plants.

4.7.4 Improving Biological Factors

Improving biological factors such as association of mycorrhizal fungi and inoculation with plant growth-promoting rhizobacteria (PGPR) are important strategies in increasing PUE in crop plants. Raaijmakers et al. (2009) and Hartmann et al. (2009) reported that the rhizosphere is a hot spot of microbial interactions as exudates released by plant roots are a main food source for microorganisms and a driving force of their population density and activity.

Plant rhizosphere is a dynamic environment in which many variables may influence the population structure, diversity, and activity of the microbial community. Plant species and soil type are important factors in determining the structure of microbial community present in the vicinity of plant roots (Garbeva et al., 2008). Soil microorganisms have important roles in soil quality and plant productivity (Hill et al., 2000). Microbial biomass in soil constitutes a pool of nutrients that has a rapid turnover as compared with organic matter (Bååth and Anderson, 2003). Therefore, quantitative and qualitative changes in the composition of soil microbial communities may serve as an important and sensitive indicator of both short- and long-term changes in soil health (Hill et al., 2000). Soil microbial communities may be strongly influenced by agricultural practices that change the soil environment (Lundquist et al., 1999; Wang et al., 2008).

4.7.4.1 Mycorrhizae

Association between certain fungi and roots of higher plants called mycorrhizae is known. Mycorrhizae are presumably important contributor to plant growth in most ecosystems (Troeh and Loynachan, 2009; Ortas and Akpinar, 2011; Shen et al., 2011). Brady and Weil (2002) reported that mycorrhizal structures have been reported in fossils of plants that lived some 400 million years ago, indicating that mycorrhizal infection may have played a role in the evolutionary adaptation of plants to the land environment. However, the topic of microbial colonization of plant roots has received special attention by the agricultural scientists in the recent years. This has happened due to the increased cost of agricultural inputs for crop production and concern for environmental pollution. In nature, most plant roots are colonized by bacteria, fungi, and other organisms for mutual or individual benefits. Root microbial association varies with plant species and is influenced by environmental conditions. Plant roots that are invaded by fungus may transform into mycorrhizal or fungal roots. Mycorrhiza was first used by the German scientist A. B. Frank in 1885 to describe the fungal hyphae closely associated with plant roots (Mengel et al., 2001). Mycorrhiza is a Greek word that means "fungus" and "root." When roots are colonized by mycorrhiza, the root morphology is modified; however, as long as a balance relationship is maintained, there are no pathological symptoms (Gerdemann, 1974). Marschner (1995) reported that as a rule, the fungus is partially or wholly dependent on the higher plant, whereas the plant may or may not benefit. In some instances, mycorrhizae are essential.

Improving mycorrhizal association with roots of crop plants can improve PUE of crop plants through the extension of the plant root system with mycorrhizal hyphae (Bucher, 2007; Ramaekers et al., 2010). Fungi arbuscular mycorrhiza (AM) is probably the most widespread terrestrial symbiosis and is formed by 70%–90% of land plant species (Parniske, 2008). In exchange for P and other nutrients supplied to the plant, the fungal symbiont obtains reduced C (Jacobsen et al., 2005; Ramaekers et al., 2010). Beneficial effects of AM fungi depend on P level in the soil. If the P level is low, the effects of fungi on P availability to plants are significant and vice versa (Kaeppler et al., 2000; Ramaekers et al., 2010).

Studies on P uptake from isotopically labeled media have shown that the external fungal hyphae are able to absorb P directly from the soluble P pools in the soil and translocate it to the host root (Wang et al., 2002). Hyphal inflow of P can differ with fungal species, and this may in part explain the differences in the efficiency of various fungi in promoting plant growth, but AM fungi may be more than just an extension of the plants' root system (Javaid, 2009). Besides hyphae that extend beyond the root depletion zone, various other mechanisms have been proposed to explain P uptake by mycorrhizal fungi, such as the kinetics of P uptake into hyphae that differ from those of roots either through a higher affinity (lower K_m) or a lower threshold concentration in which influx equals efflux (C_{min}) (Joner and Jakobsen, 1995). Plant roots' infection with mycorrhizal colonization depends on the original soil P concentration. Soils with lower P concentrations are subjected to increased mycorrhizal colonization in roots (Covacevich et al., 2007). In addition, roots infected with mycorrhiza may take P from nonlabile sources such as Fe and Al phosphates (insoluble inorganic P sources) in soil through the interaction between the roots and hyphae (Shibata and Yano, 2003).

4.7.4.2 Plant Growth-Promoting Rhizobacteria (PGPR)

Colonization of rhizosphere by microorganisms results in modifications in plant growth and development. Kloepper and Schroth (1978) introduced the term "PGPR" to designate these bacteria. PGPR have been divided into two classes according to whether they can affect plant growth either directly or indirectly (Bashan and Holguin, 1998). Direct influence is related to increased solubilization and uptake of nutrients and production of phytohormones. Soil microorganisms have important role in the cycling of many soil nutrients including bacteria, yeasts, actinomycetes, and mycorrhizal fungi that are reported to cause increases in plant-available P in the soil (Whitelaw, 2000). Raghu and MacRae (1966) and Whipps and Lynch (1986) reported that a high proportion of P-solubilizing microorganisms are concentrated in the rhizosphere of plants. Microorganisms in the rhizosphere obtain their nutrition from root exudates, plant mucigel, and root lysates. Rhizosphere microorganisms are normal heterotrophs but revive in an environment with high levels of nutrients, such as C and N, and tend to adapt rapidly to improvements of nutrient supply (Tinker, 1984; Whitelaw, 2000).

4.8 MOLECULAR AND GENETIC APPROACH TO IMPROVE PHOSPHORUS USE EFFICIENCY

Molecular and genetic approach is an important tool in increasing PUE of crop plants. However, its use in developing P-efficient crop plants is limited. PUE trait is very complex and governed by polygenic genes. Information on genetic variability in crop species and within species are enormous (Zhang, 2007; Fageria et al., 2008, 2011a; Hammond et al., 2008; Ramaekers et al., 2010; Fageria, 2013). Veneklaas et al. (2012) also reported that there is substantial genetic variation in traits associated with PUE in crop plant species and within species. Analysis of this variation has led to the identification of numerous genetic loci that influence PUE. Ramaekers et al. (2010) reported that tolerance to low P is a quantitative trait. An appropriate method to dissect its complex polygenic inheritance is through quantitative trait loci (QTL) analysis. A QTL is a region in the genome that is responsible for variation in the quantitative trait on the statistically significant association of phenotypic differences for the trait of interest with molecular markers that constitute the genetic map (Doerge, 2001; Ramaekers et al., 2010). Molecular markers found to be linked to the target trait can be used for selection in the breeding process (Ramaekers et al., 2010).

Nearly all QTL mapping has focused on traits associated with efficient P acquisition rather than efficient internal use of P (Veneklaas et al., 2012). These traits include high total plant P content, large root systems, improved root architecture, and exudation of phosphatases and organic acids into the rhizosphere (Veneklaas et al., 2012). Nevertheless, QTL that have the potential to influence internal PUE have been reported in several crop species (Vance, 2010). However, internal PUE is generally lower in plants with high P acquisition efficiency as a result of higher tissue P concentration, making it difficult to disentangle QTL that affect agronomic PUE generally from QTL that may specifically influence internal PUE (Veneklaas et al., 2012). Veneklaas et al. (2012) reported that QTL for internal PUE requires studies where P acquisition is equal and metabolically nonsaturating among cultivars.

A range of QTL have been identified for tolerance mechanisms to low P in various food crops (Kaeppler et al., 2000; Wissuwa et al., 2005; Beebe et al., 2006). In common bean, some of these have been notably correlated with low P adaption in the field despite being evaluated in hydroponic or greenhouse conditions (Yang et al., 2004; Ramaekers et al., 2010). An association between low P tolerance and Al toxicity QTL has been reported in common bean and could suggest the added importance of organic acid exudation (Lopez-Marin et al., 2009). Heuer et al. (2009) provided few examples in which attempts were made to further dissect a major QTL acquisition (*Pup 1*) in rice.

Several attempts have been made to improve P acquisition processes in food crops through genetic engineering or developing transgenic plants with specific bacterial, fungal, or plant genes (Ramaekers et al., 2010). Focus has been mainly on genes that enhance solubilization of P in the soil. Ramaekers et al. (2010) presented a summary of reported transgenic crop plants to improve PUE in barley, rice, clover, potato, soybean, alfalfa, tomato (*Solanum lycopersicum*), and tobacco (Nicotiana). Lopez-Buico et al. (2000) introduced a bacterial citrate synthase gene into tobacco and reported a two- to fourfold increase of citrate efflux by roots and superior growth and yield in low P alkaline soils of transgenic lines.

4.9 CONCLUSIONS

NUE, including P, is an important topic in the field of mineral nutrition of crop plants. PUE can be defined as AE, PE, APE, ARE, UE, or nutrient efficiency ratio, based on different methods of calculation. There are large variations in PUE among crop species and within species. This variation may be associated with differential soils and plant mechanisms. Variation in root growth is one of the most important plant factors responsible for different P uptake among crop species or genotypes within species. Improvement in NUE results in the reduction of the costs of crop production and minimization of environmental pollution. Nutrient efficiency can be improved with the improvement in the plant genetics and with the adoption of improved crop management practices. P recovery efficiency is less than 20% by most crop plants under most agroecological conditions. This low recovery efficiency is related to higher immobilization capacity of P by soils, especially when soils are acidic or alkaline in reactions. However, plant internal efficiency of P (grain yield per unit of P uptake in plant tissues) is much higher as compared to other essential plant nutrients. Improved knowledge of PUE genetics is urgently needed.

REFERENCES

Abekoe, M. K. and K. L. Saharawat. 2001. Phosphate retention and extractability in soils of the humid zone in West Africa. *Geoderma* 102:175–187.

Ae, N., J. Arihara, K. Okada, T. Yoshihara, and C. Johansen. 1990. Phosphorus uptake by pigeon pea and its role in cropping systems of the Indian subcontinent. *Science* 248:477–480.

Alagarswamy, G. 1971. Modelling of phosphate uptake by the groundnut plant (*Arachis hypogaea* L.). Doctoral thesis. Louvain, Belgium: Catholic University of Louvain.

Andrew, C. S. 1966. A kinetic study of phosphate absorption by excised roots of *Stylosanthes humilis, Phaseolus lathyroides, Desmodium uncinatum, Medicago sativa* and *Hordeum vulgare*. *Aust. J. Agric. Res.* 17:611–624.

Araujo, A. P. and M. G. Teixeira. 2003. Nitrogen and phosphorus harvest indices of common bean cultivars: Implications for yield quantity and quality. *Plant Soil* 257:425–433.

Atkinson, D. 1990. Influence of root system morphology and development on the need for fertilizers and the efficiency of use. In: *Crops as Enhancers of Nutrient Use*, eds. V. C. Baligar and R. R. Duncan, pp. 411–451. New York: Academic Press.

Austin, R. B. 1994. Plant breeding opportunities. In: *Physiology and Determination of Crop Yield*, eds. K. J. Boote, J. M. Bennett, T. R. Sinclair, and G. M. Paulsen, pp. 567–589. Madison, WI: ASA, CSSA, SSSA.

Austin, R. B., J. Bingham, R. D. Blackwell, L. T. Evans, M. A. Ford, C. L. Morgan, and M. Taylor. 1980. Genetic improvements in winter wheat yields since 1900 and associated physiological changes. *J. Agric. Sci.* 94:675–689.

Bååth, E. and T.-H. Anderson. 2003. Comparison of soil fungal/bacterial ratios in a pH gradient using physiological and PLFA-based techniques. *Soil Biol. Biochem.* 35:955–963.

Baligar, V. C. 1987. Phosphorus uptake parameters of alfalfa and corn as influenced by P and pH. *J. Plant Nutr.* 10:33–46.

Baligar, V. C. and O. L. Bennett. 1986. NPK-fertilizer efficiency—A situation analysis for the tropics. *Fertiliz. Res.* 10:47–164.

Baligar, V. C., R. R. Duncan, and N. K. Fageria. 1990. Soil–plant interactions on nutrient use efficiency in plants. An overview. In: *Crops as Enhancers of Nutrient Use*, eds. V. C. Baligar and R. R. Duncan, pp. 351–373. San Diego, CA: Academic Press.

Baligar, V. C. and N. K. Fageria. 1997. Nutrient use efficiency in acid soils: Nutrient management and plant use efficiency. In: *Plant Soil Interaction at Low pH*, eds. A. C. Monez, A. M. C. Furlani, R. E. Schaffert, N. K. Fageria, C. A. Roselem, and H. Cantarella, pp. 75–95. Campinas, Brazil: Brazilian Soil Science Society.

Baligar, V. C., N. K. Fageria, and M. A. Elrashidi. 1998. Toxicity and nutrient constraints on root growth. *HortScience* 3:960–965.

Baligar, V. C., N. K. Fageria, and Z. L. He. 2001. Nutrient use efficiency in plants. *Commun. Soil Sci. Plant Anal.* 32:921–950.

Baligar, V. C., V. E. Nash, M. L. Hare, and J. A. Price. 1975. Soybean root anatomy as influenced by soil bulk density. *Agron. J.* 67:842–844.

Barber, S. A. 1980. Soil–plant interactions in the phosphorus nutrition of plants. In: *The Role of Phosphorus in Agriculture*, ed. R. C. Dinauer, pp. 591–615. Madison, WI: ASA, CSSA, and SSSA.

Barber, S. A. 1994. Root growth and nutrient uptake. In: *Physiology and Determination of Crop Yield*, eds. K. J. Boote, J. M. Bennett, T. R. Sinclair, and G. M. Paulsen, pp. 95–99. Madison, WI: ASA, CSSA, SSSA.

Barber, S. A. 1995. *Soil Nutrient Bioavailability: A Mechanistic Approach*. New York: John Wiley & Sons Inc.

Barley, K. P. 1970. The configuration of the root system in relation to nutrient uptake. *Adv. Agron.* 22:159–201.

Bar-Yosef, B., S. Fishman, and H. Talpaz. 1980. A model of zinc movement to single roots in soils. *Soil Sci. Soc. Am. J.* 44:1272–1279.

Bashan, Y. and G. Holguin. 1998. Proposal for the division of plant growth-promoting rhizobacteria into two classifications: biocontrol–PGPB (plant growth–promoting bacteria) and PGPB. *Soil Biol. Biochem.* 30:1225–1228.

Beebe, S. E., M. Rojas-Pierce, X. Yan, M. W. Blair, F. Pedraza, F. Munoz, J. Tohme, and J. P. Lynch. 2006. Quantitative trait loci for root architecture traits correlated with phosphorus acquisition in common bean. *Crop Sci.* 46:413–423.

Bennie, A. T. 1996. Growth and mechanical impedance. In: *Plant Roots: The Hidden Half*, eds. Y. Waisel, A. Eshel, and U. Kafkafi, pp. 453–470. New York: Marcel Dekker.

Blair, G. 1993. Nutrient efficiency: What do we really mean. In: *Genetic Aspects of Plant Mineral Nutrition*, eds. P. J. Randall, E. Delhaize, R. A. Richards, and R. Munns, pp. 205–213. Dordrecht, the Netherlands: Kluwer Academic Publishers.

Blair, M. W. 2013. Breeding approaches to increasing nutrient use efficiency: Examples from common beans. In: *Improving Water and Nutrient Use Efficiency in Food Production Systems*, ed. Z. Rengel, pp. 161–175. Ames, IA: John Wiley & Sons.

Blamey, F. P. C. 2001. The role of root cell wall in aluminum toxicity. In: *Plant Nutrient Acquisition: New Perspectives,* eds. N. Ae, J. Arihara, K. Okada, and A. Srinivasan, pp. 201–226. Tokyo, Japan: Springer.

Boone, R. D. 1994. Light-fraction soil organic matter: Origin and contribution to net nitrogen mineralization. *Soil Biol. Biochem.* 26:1459–1468.

Brady, N. C. and R. R. Weil. 2002. *The Nature and Properties of Soils,* 13th edn. Upper Saddle River, NJ: Prentice Hall.

Bucher, M. 2007. Functional biology of plant phosphate uptake at root and mycorrhiza interfaces. *New Phytol.* 173:11–26.

Bundy, L. G., H. Tunney, and A. D. Halvorson. 2005. Agronomic aspects of phosphorus management. In: *Phosphorus: Agriculture and the Environment*, ed. L. K. Al-Almoodi, pp. 685–727. Madison, WI: ASA, CSSA and SSSA.

Caradus, J. R. 1990. Mechanisms improving nutrient use by crop and herbage legumes. In: *Crops as Enhancers of Nutrient Use*, eds. V. C. Baligar and R. R. Duncan, pp. 253–311. New York: Academic Press.

Caradus, J. R. and R. W. Snaydon. 1986. Plant factors influencing phosphorus uptake by white clover from solution culture. I. Populations differences. *Plant Soil* 93:153–163.

Cichy, K. A., M. W. Blair, and C. H. Galeano. 2009. QTL analysis of root architecture traits and low phosphorus tolerance in an Andean bean population. *Crop Sci.* 49:59–68.

Clark, R. B. 1990. Physiology of cereals for mineral nutrient uptake, use, and efficiency. In: *Crops as Enhancers of Nutrient Use*, eds. V. C. Baligar and R. R. Duncan, pp. 131–209. New York: Academic Press.

Clark, R. B. and R. R. Duncan. 1991. Improvement of plant mineral nutrition through breeding. *Field Crops Res.* 27:219–240.

Cordell, D., J. O. Drangert, and S. White. 2009. The story of phosphorus: Global food security and food for thought. *Glob. Environ. Change* 19:292–305.

Covacevich, F., H. E. Echeverria, and L. A. N. Aguirrezabal. 2007. Soil available phosphorus status determines indigenous mycorrhizal colonization of field and glasshouse grown spring wheat from Argentina. *Appl. Soil Ecol.* 35:109.

Cregan, P. B. and R. W. Yaklich. 1986. Dry matter and nitrogen accumulation and partitioning in selected soybean genotypes of different derivation. *Teo. Appl. Gentic.* 72:782–786.

Darrah, P. R. 1993. The rhizosphere and plant nutrition: A quantitative approach. *Plant Soil* 156:1–20.

Doerge, R. 2001. Mapping and analysis of quantitative trait loci in experimental populations. *Nat. Rev. Gen.* 3:43–52.

Donald, C. M. 1962. In search of yield. *J. Aust. Inst. Agric. Sci.* 28:171–178.

Edwards, D. G. 1970. Phosphate absorption and long-distance transport in wheat seedlings. *Aust. J. Biol. Sci.* 23:255–264.

Epstein, E. and A. J. Bloom. 2005. *Mineral Nutrition of Plants: Principles and Perspectives*, 2nd edn., Sunderland, MA: Sinauer Associations, Inc.

Evans, L. T. 1993. *Crop Evolution, Adaptation and Yield.* Cambridge, U.K.: Cambridge University Press.

Evans, L. T. 1994. Crop physiology: Prospects for the retrospective science. In: *Physiology and Determination of Crop Yield*, eds. K. J. Boote, J. M. Bennett, T. R. Sinclair, and G. M. Paulsen, pp. 19–35. Madison, WI: ASA, CSSA, SSSA.

Evans, L. T. and H. M. Rawson. 1970. Aspects of the comparative physiology of grain yield in cereals. *Adv. Agron.* 28:301–359.

Evans, L. T. and I. F. Wardlaw. 1976. Aspects of the comparative physiology of grain yield in cereals. *Adv. Agron.* 28:301–359.

Fageria, N. K. 1973. Uptake of nutrients by the rice plant from dilute solutions, Doctoral thesis. Louvain, Belgium: Catholic University of Louvain.

Fageria, N. K. 1974. Kinetics of phosphate absorption in intact rice plants. *Aust. J. Agric. Res.* 25:395–400.

Fageria, N. K. 1980. Upland rice response to phosphate fertilization as affected by water deficiency in Cerrado soils. *Pesq. Agropec. Bras.* 15:259–265.

Fageria, N. K. 1992. *Maximizing Crop Yields.* New York: Marcel Dekker.

Fageria, N. K. 1998. Phosphorus use efficiency by bean genotypes. *Revista Brasileira de Engenharia Agricola e Ambiental* 2:128–131.

Fageria, N. K. 2000. Potassium use efficiency of upland rice genotypes. *Pesquisa Agropecuaria Brasileira* 35:2115–2120.

Fageria, N. K. 2009. *The Use of Nutrients in Crop Plants.* Boca Raton, FL: CRC Press.

Fageria, N. K. 2013. *The Role of Plant Roots in Crop Production.* Boca Raton, FL: CRC Press.

Fageria, N. K. 2014. *Mineral Nutrition of Rice.* Boca Raton, FL: CRC Press.

Fageria, N. K. and V. C. Baligar. 2003. Fertility management of tropical acid soil for sustainable crop production. In: *Handbook of Soil Acidity*, ed. Z. Rengel. New York: Marcel Dekker.

Fageria, N. K. and V. C. Baligar. 2005. Enhancing nitrogen use efficiency in crop plants. *Adv. Agron.* 88:97–185.

Fageria, N. K. and V. C. Baligar. 2008. Ameliorating soil acidity of tropical Oxisols by liming for sustainable crop production. *Adv. Agron.* 99:345–399.

Fageria, N. K. and V. C. Baligar. 2014. Macronutrient use efficiency and changes in chemical properties of an Oxisol as influenced by phosphorus fertilization and tropical cover crops. *Commun. Soil Sci. Plant Anal.* 45:1–20.

Fageria, N. K., V. C. Baligar, and R. B. Clark. 2002. Micronutrients in crop production. *Adv. Agron.* 77:185–268.

Fageria, N. K., V. C. Baligar, and R. B. Clark. 2006. *Physiology of Crop Production*. New York: The Haworth Press.

Fageria, N. K., V. C. Baligar, and C. A. Jones. 2011a. *Growth and Mineral Nutrition of Field Crops*, 3rd edn. Boca Raton, FL: CRC Press.

Fageria, N. K., V. C. Baligar, and Y. C. Li. 2008. The role of nutrient efficient plants in improving crop yields in the twenty first century. *J. Plant Nutr.* 31:1121–1157.

Fageria, N. K., V. C. Baligar, and Y. C. Li. 2014a. Nutrient uptake and use efficiency by tropical legume cover crops at varying pH of an Oxisol. *J. Plant Nutr.* 37:294–311.

Fageria, N. K. and M. P. Barbosa Filho. 1987. Phosphorus fixation in Oxisol of central Brazil. *Fertiliz. Agric.* 94:33–37.

Fageria, N. K. and M. P. Barbosa Filho. 2007. Dry matter and grain yield, nutrient uptake, and phosphorus use efficiency of lowland rice as influenced by phosphorus fertilization. *Commun. Soil Sci. Plant Anal.* 38:1289–1297.

Fageria, N. K., M. P. Barbosa Filho, and L. F. Stone. 2004a. Phosphorus nutrition for bean production. In: *Phosphorus in Brazilian Agriculture*, eds. T. Yamada and S. R. S. Abdalla, pp. 435–455. São Paulo, Brazil: Brazilian Potassium and Phosphate Research Association.

Fageria, N. K., M. P. Barbosa Filho, L. F. Stone, and C. M. Guimarães. 2004b. Phosphorus nutrition in upland rice production. In: *Phosphorus in Brazilian Agriculture*, eds. T. Yamada and S. R.S. Abdalla, pp. 401–418. São Paulo, Brazil: Brazilian Phosphorus and Potassium Institute.

Fageria, N. K., A. M. Knupp, and M. F. Moraes. 2013. Phosphorus nutrition of lowland rice in tropical lowland soil. *Commun. Soil Sci. Plant Anal.* 44:2932–2940.

Fageria, N. K., O. P. Morais, A. B. Santos, and M. J. Vasconcelos. 2014c. Phosphorus use efficiency in upland rice genotypes under field conditions. *J. Plant Nutr.* 37:633–642.

Fageria, N. K., A. Moreira, L. C. A. Moraes, and M. F. Moraes. 2014b. Root growth, nutrient uptake, and nutrient use efficiency by roots of tropical legume cover crops as influenced by phosphorus fertilization. *Commun. Soil Sci. Plant Anal.* 45:555–569.

Fageria, N. K. and A. B. Santos. 2002. Lowland rice genotypes evaluation for phosphorus use efficiency. *J. Plant Nutr.* 25:2793–2802.

Fageria, N. K., A. B. Santos, and V. C. Baligar. 1997. Phosphorus Soil test calibration for lowland rice on an Inceptisol. *Agron. J.* 89:737–742.

Fageria, N. K., A. B. Santos, and A. B. Heinemann. 2011b. Lowland rice genotypes evaluation for phosphorus use efficiency in tropical lowland. *J. Plant Nutr.* 34:1087–1095.

Fageria, N. K., N. A. Slaton, and V. C. Baligar. 2003. Nutrient management for improving lowland rice productivity and sustainability. *Adv. Agron.* 80:63–152.

Fageria, N. K. and L. F. Stone. 2006. Physical, chemical, and biological changes in rhizosphere and nutrient availability. *J. Plant Nutr.* 29:1327–1356.

Fageria, N. K. and L. F. Stone. 2013. Water and nutrient use efficiency in food production in South America. In: *Improving Water and Nutrient Use Efficiency in Food Production System*, 1st edn., ed. Z. Rengel, pp. 275–296. Ames, IA: John Wiley & Sons.

FAO. 2010. FAO statistical yearbook 2010. WWW document URL http://www.Fao.Org/economic/ess/ess-publications/ess-yearbook/ess-yearbook.2010/yearbook2010-consumption/em. In(accessed December 12, 2011).

Feil, B. 1992. Breeding progress in small grain cereals: A comparison of old and modern cultivars. *Plant Breed.* 108:1–11.

Foy, C. D. 1984. Physiological effects of hydrogen, aluminum and manganese toxicities in acid soils. In: *Soil Acidity and Liming,* 2nd edn., ed. F. Adams, pp. 57–97. Madison, WI: SSSA, ASA and CSSA.

Foy, C. D. 1992. Soil chemical factors limiting plant root growth. *Adv. Soil Sci.* 19:97–149.

Friesen, D. K., I. M. Rao, R. J. Thomas, A. Obserson, and J. I. Sanz. 1997. Phosphorus acquisition and cycling in crop and pasture systems in low fertility tropical soils. In: *Plant Nutrition for Sustainable Crop Production and Environment*, eds. T. Ando, K. Fujita, T. Mae, H. Tsumoto, S. Mori, and J. Sekiya, pp. 493–498. Dordrecht, the Netherlands: The Kluwer Academic Publishers.

Gale, W. J., C. A. Cambardell, and T. B. Bailey. 2000. Root derived carbon and the formation and stabilization of aggregates. *Soil Sci. Soc. Am. J.* 64:201–207.

Garbeva, P., J. D. V. Elsas, and J. A. V. Veen. 2008. Rhizosphere microbial community and its response to plant species and soil history. *Plant Soil* 302:19–32.

Gerdemann, J. W. 1974. Mycorrhizae. In: *The Plant Root and Its Environment*, ed. E. W. Carson, pp. 205–217. Charlottesville, VA: University Press of Virginia.

Gerloff, G. C. and W. H. Gableman. 1983. Genetic basis of inorganic plant nutrition. In: *Inorganic Plant Nutrition: Encyclopedia and Plant Physiology New Series*, Vol. 15B, eds. A. Lauchi and R. L. Bieleski, pp. 453–480. New York: Springer Verlag.

Gifford, R. M., J. H. Thorne, W. Hitz, and R. T. Giaquinta. 1984. Crop productivity and photoassimilate partitioning. *Science* 225:801–808.

Goedert, W. J. 1989. Cerrado region: Agricultural potential and politics for its development. *Pesq. Agropec. Bras.* 24:1–17.

Gourley, C. J. P., D. L. Allan, and M. P. Russelle. 1994. Plant nutrient efficiency: A comparison of definitions and suggested improvement. *Plant Soil* 158:29–37.

Graham, R. D. 1984. Breeding for nutritional characteristics in cereals. In: *Advances in Plant Nutrition*, Vol. 1, eds. P. B. Tinker and A. Lauchi, pp. 57–102. New York: Praeger Publisher.

Gregory, P. J. 1994. Root growth and activity. In: *Physiology and Determination of Crop Yield*, eds. K. J. Boote, J. M. Bennett, T. R. Sinclair, and G. M. Paulsen, pp. 65–93. Madison, WI: ASA, CSSA, SSSA.

Hammond, J. P., M. R. Broadley, P. J. White, G. J. King, H. C. Bowen, R. Hayden, M. C. Meacham et al. 2008. Shoot yield drives phosphorus use efficiency in *Brassica oleracea* and correlates with root architecture traits. *J. Exp. Bot.* 60:1953–1968.

Hartmann, A., M. Schmid, D. V. Tuinen, and G. Berg. 2009. Plant driven selection of microbes. *Plant Soil.* 321:235–257.

Hatfield, J. L., T. J. Sauer, and J. H. Prueger. 2001. Managing soils to achieve greater water use efficiency: A review. *Agron. J.* 93:271–280.

Hay, R. K. M. 1995. Harvest index: A review of its use in plant breeding and crop physiology. *Ann. Appl. Biol.* 126:197–216.

Heuer, S., X. Lu, J. H. Chin, J. P. Tanaka, H. Kanamori, T. Matsumoto, T. D. Leon et al. 2009. Comparative sequence analysis of the major quantitative trait locus phosphate uptake (*Pup 1*) reveal a complex genetic structure. *Plant Biotechnol. J.* 7:456–471.

Hill, G. T., N. A. Mitkowski, L. Aldrich-Wolfe, L. R. Emele, D. D. Jurkonie, A. Ficke, S. Maldonado-Ramirez, S. T. Lynch, and E. B. Nelsona. 2000. Methods of assessing the composition and diversity of soil microbial communities. *Appl. Soil Ecol.* 15:25–36.

Hillel, D. and C. Rosenzweig. 2005. The role of biodiversity in agronomy. *Adv. Agron.* 88:1–34.

Hinsinger, P. 1998. How do plant roots acquire mineral nutrients? Chemical processes involved in the rhizosphere. *Adv. Agron.* 64:225–265.

Hinsinger, P. 2001. Bioavailability of soil inorganic P in the rhizosphere as affected by root-induced chemical changes: A review. *Plant Soil* 237:173–195.

Hinsinger, P. and R. J. Gilkes. 1996. Mobilization of phosphate from phosphate rock and alumina-sorbed phosphate by the roots of ryegrass and clover as related to rhizosphere pH. *Eur. J. Soil Sci.* 47:533–544.

Ho, L. C. 1988. Metabolism and compartmentation of imported sugars in sink organs in relation to sink strength. *Annu. Rev. Plant Physiol. Plant Mol. Biol.* 39:355–378.

Inoue, H. and A. Tanaka. 1978. Comparison of source and sink potentials between wild and cultivated potatoes. *J. Soil Sci. Manure Jpn.* 49:321–327.

Isfan, D. 1993. Genotypic variability for physiological efficiency index of nitrogen in oats. *Plant Soil* 154:53–59.

Jacobsen, I., M. E. Leggett, and A. E. Richardson. 2005. Rhizosphere microorganisms and plant phosphorus uptake. In: *Phosphorus in Agriculture and the Environment*, eds. L. K. Al-Amoodi, pp. 437–494. Madison, WI: ASA, CSSA, and SSSA.

Javaid, A. 2009. Arbuscular mycorrhizal mediated nutrition in plants. *J. Plant Nutr.* 32:1595–1618.

Joner, E. J. and I. Jakobsen. 1995. Growth and extracellular phosphatase activity of arbuscular mycorrhizal hyphae as influenced by soil organic matter. *Soil Biol. Biochem.* 27:1153–1159.

Kaeppler, S. M., J. L. Parke, S. M. Mueelle, L. Senior, C. Stuber, and W. F. Tracy. 2000. Variation among maize inbred lines and determination of QTL for growth at low phosphorus and responsiveness to arbuscular mycorrhizal fungi. *Crop Sci.* 40:358–364.

Keerthisinghe, G., F. Zapta, P. Chalk, and P. Hocking. 2001. Integrated approach for improved P nutrition of plants in tropical acid soils. In: *Plant Nutrition: Food Security and Sustainability of Agroecosystems*, eds. W. J. Horst, M. K. Schenk, A. Burkert, N. Claassen, H. Flessa, W. B. Frommer, H. Goldbach, H. W. Olfs, V. Romheld, B. Sattelmacher, U. Schmidhalter, S. Schubert, N. V. Wiren, and L. Wittenmayer, pp. 974–975. Dordrecht, the Netherlands: Kluwer Academic Publishers.

Kloepper, J. W. and M. N. Schroth. 1978. Plant growth-promoting rhizobacteria on radishes. In: *Proceedings of the Fourth International Conference on Plant Pathogenic Bacteria*, Vol. 2, pp. 879–882. Tours, France: Station de Pathologie Vegetale et de Phytobacteriologie.

Kuo, S., U. M. Sainju, and E. J. Jellum. 1997. Winter cover crop effects on soil organic carbon and carbohydrate. *Soil Sci. Soc. Am. J.* 61:145–152.

Liang, B. C., X. L. Wang, and B. L. Ma. 2002. Maize root-induced change in soil organic carbon pools. *Soil Sci. Soc. Am. J.* 66:845–847.

Liao, H., G. Rubio, and X. Yan. 2001. Effect of phosphorus availability on basal root shallowness in common bean. *Plant Soil* 232:69–79.

Lopez-Buico, J., O. M. Vaga, A. Guevara-Garcia, and L. Herrera-Estrella. 2000. Enhanced phosphorus uptake in transgenic tobacco plants that overproduce citrate. *Nature* 18:450–453.

Lopez-Marin, H. D., I. M. Rao, and M. W. Blair. 2009. Quantitative trait loci for root morphology trait under aluminum stress in common bean. *Theor. Appl. Gen.* 119:449–458.

Ludlow, M. M. and R. C. Muchlow. 1990. A critical evaluation of traits for improving crop yields in water-limited environments. *Adv. Agron.* 43:107–153.

Lundquist, E. J., K. M. Scow, L. E. Jackson, S. L. Uesugi, and C. R. Johnson. 1999. Rapid response of soil microbial communities from conventional, low input, and organic farming systems to a wet/dry cycle. *Soil Biol. Biochem.* 31:1661–1675.

Lynch, J. 1995. Root architecture and plant productivity. *Plant Physiol.* 109:7–13.

Marschner, H. 1995. *Mineral Nutrition of Higher Plants*, 2nd edn. New York: Academic Press.

Mengel, K., E. A. Kirkbay, H. Kosegarten, and T. Appel. 2001. *Principles of Plant Nutrition*. 5th edn. Dordrecht, the Netherlands: Kluwer Academic Publishers.

Merrill, S. D., D. L. Tanaka, and J. D. Hanson. 2002. Root length growth of eight crop species in Haplustoll soils. *Soil Sci. Soc. Am. J.* 66:913–923.

Nielsen, D. C., M. F. Vigil, and R. L. Anderson. 2002. Cropping system influence on planting water content and yield of winter wheat. *Agron. J.* 94:962–967.

Norby, R. J. and M. F. Cotrufo. 1998. A question of litter quality. *Nature* 396:17–18.

Okada, K. and A. J. Fischer. 2001. Adaptation mechanisms of upland rice genotypes to highly weathered acid soils of South American Savannah's. In: *Plant Nutrient Acquisition: New Perspectives*, eds. N. Ae, J. Arihara, K. Okada, and A. Srinivasan, pp. 185–200. Tokyo, Japan: Springer.

Ortas, I. and C. Akpinar. 2011. Response of maize genotypes to several mycorrhizal inoculums in terms of plant growth, nutrient uptake and spore production. *J. Plant Nutr.* 34:970–987.

Ozanne, P. G. 1980. Phosphate nutrition of plants: A general treatise. In: *The Role of Phosphorus in Agriculture*, ed. R. C. Dinauer, pp. 559–589. Madison, WI: ASA, CSSA, and SSSA.

Parniske, M. 2008. Arbuscular mycorrhiza: The mother of plant root endosymbiosis. *Nat. Rev. Microbiol.* 6:763–775.

Payne, T. S., D. D. Stuthman, R. L. McGraw, and P. P. Bregitzer. 1986. Physiological changes associated with three cycles of recurrent selection for grain yield improvement in oats. *Crop Sci.* 26:734–736.

Peng, S., R. C. Laza, R. M. Visperas, A. L. Sanico, K. G. Cassman, and G. S. Khush. 2000. Grain yield of rice cultivars and lines developed in the Philippines since 1966. *Crop Sci.* 40:307–314.

Pettersson, S. and P. Jensen. 1983. Variation among species and varieties in uptake and utilization of K. *Plant Soil* 72:231–237.

Puget, P. and L. E. Drinkwater. 2001. Short-term dynamics of root and shoot derived carbon from a leguminous green manure. *Soil Sci. Soc. Am. J.* 65:771–779.

Qian, J. H. and J. W. Doran. 1996. Available carbon released from crop roots during growth as determined by carbon-13 natural abundance. *Soil Sci. Soc. Am. J.* 60:828–831.

Raaijmakers, J. M., T. C. Paulitz, C. Steinberg, C. Alabouvette, and Y. Moenne-Loccoz. 2009. The rhizosphere: A playground and battlefield for soilborne pathogens and beneficial microorganisms. *Plant Soil* 321:341–361.

Radin, J. W. and J. Lynch. 1994. Nutritional limitations to yield: Alternatives to fertilization. In: *Physiology and Determination of Crop Yield*, eds. K. J. Boote, J. M. Bennett, T. R. Sinclair, and G. M. Paulsen, pp. 277–283. Madison, WI: ASA, CSSA, SSSA.

Raghu, K. and I. C. MacRae. 1966. Occurrence of phosphate-dissolving micro-organisms in the rhizosphere of rice plants and in submerged soils. *J. Appl. Bacteriol.* 29:582–586.

Ramaekers, L., R. Remans, I. M. Rao, M. W. Blair, and J. Vanderleyden. 2010. Strategies for improving phosphorus acquisition efficiency of crop plants. *Field Crops Res.* 117:169–176.

Richardson, A. E. 2001. Prospects for using soil microorganisms to improve the acquisition of phosphorus by plants. *Aust. J. Plant Physiol.* 28:897–906.

Riley, D. and S. A. Barber. 1971. Effect of ammonium and nitrate fertilization on phosphorus uptake as related to root-induced pH changes at the root soil-interface. *Soil Sci. Soc. Am. Proc.* 35:301–306.

Sahrawat, K. L. and M. Sika. 2002. Direct and residual phosphorus effects on soil test values and their relationships with grain yield and phosphorus uptake of upland rice on an Ultisol. *Commun. Soil Sci. Plant Anal.* 33:321–332.

Sainju, U. M., B. P. Singh, and W. F. Whitehead. 2005. Tillage, cover crops, and nitrogen fertilization effects on cotton and sorghum root biomass, carbon, and nitrogen. *Agron. J.* 97:1279–1290.

Schachtman, D. P., R. J. Reid, and S. M. Ayling. 1998. Update on phosphorus uptake. Phosphorus uptake by plants: From soil to cell. *Plant Physiol.* 116:447–453.

Schmidt, J. W. 1984. Genetic contributions to yield grains in wheat. In: *Genetic Contributions to Yield Grains of Five Major Crop Plants*, ed. W. R. Fehr, pp. 89–110. Madison, WI: ASA.

Scott-Russell, R. 1977. *Plant Root Systems: Their Function and Interaction with the Soil.* London, U.K.: McGraw-Hill Book Company.

Shen, H., H. Yang, and T. Guo. 2011. Influence of arbuscular mycorrhizal fungi and ammonium:nitrate ratios on growth and pungency of spring onion plants. *J. Plant Nutr.* 34:743–752.

Shibata, R. and K. Yano. 2003. Phosphorus acquisition from non-labile sources in peanut and pigeon pea with mycorrhizal interaction. *Appl. Soil Ecol.* 24:133–141.

Singh, Y., A. Dobermann, B. Singh, K. F. Bronson, and C. S. Khind. 2000. Optimal phosphorus management strategies for wheat-rice cropping on a loamy sand. *Soil Sci. Soc. Am. J.* 64:1413–1422.

Smith, F. W., A. L. Rae, and M. J. Hawkesford. 2000. Molecular mechanisms of phosphate and sulphate transport in plants. *Biochim. Biophys. Acta* 1465:236–245.

Soil Science Society of America. 2008. *Glossary of Soil Science Terms.* Madison, WI: Soil Science Society of America.

Stone, L. R. and A. J. Schlegel. 2006. Yield-water supply relationships of grain sorghum and winter wheat. *Agron. J.* 98:1359–1366.

Sumner, M. E., M. V. Fey, and A. D. Noble. 1991. Nutrient status and toxicity problem in acid soils. In: *Soil Acidity*, eds. B. Ulrich and M. E. Sumner. Berlin, Germany: Springer-Verlag.

Tinker, P. B. 1984. The role of microorganisms in mediating and facilitating the uptake of plant nutrients from soil. *Plant Soil* 76:77–91.

Tollenaar, M. 1989. Genetic improvement in grain yield of commercial maize hybrids grown in Ontario from 1959 to 1988. *Crop Sci.* 29:1365–1371.

Troeh, Z. I. and T. E. Loynachan. 2009. Diversity of arbuscular mycorrhizal fungal species in soils of cultivated soybean fields. *Agron. J.* 101:1453–1462.

Vance, C. P. 2010. Quantitative trait loci, epigenetics, sugars, and microRNAs: Quaternaries in phosphate acquisition and use. *Plant Physiol.* 154:582–588.

Veneklaas, E., H. Lambers, J. Bragg, P. M. Finnegan, C. E. Lovelock, W. C. Plaxton, C. A. Price et al. 2012. Opportunities for improving phosphorus-use efficiency in crop plants. *New Phytol.* 195:306–320.

Vose, P. B. 1984. Effects of genetic factors on nutritional requirement of plants. In: *Crop Breeding: A Contemporary Basis*, eds. P. B. Vose and S. G. Blixt, pp. 67–114. Oxford, U.K.: Pergamon Press.

Wallace, D. H., J. L. Ozbun, and H. M. Munger. 1972. Physiological genetics of crop yield. *Adv. Agron.* 24:97–146.

Wang, B., D. M. Funakoshi, Y. Dalpe, and C. Hamel. 2002. Phosphorus 32 absorption and translocation to host plants by arbuscular mycorrhizal fungi at low root-zone temperature. *Mycorrhiza* 12:93–96.

Wang, J., S. Kang, F. Li, F. Zhang, Z. Li, and J. Zhang. 2008. Effects of alternate partial root-zone irrigation on soil microorganisms and maize root. *Plant Soil* 302:45–52.

Whipps, J. M. and J. M. Lynch. 1986. The influence of the rhizosphere on crop productivity. *Adv. Microb. Ecol.* 9:187–244.

Whitelaw, M. A. 2000. Growth promotion of plants inoculated with phosphate-solubilizing fungi. *Adv. Agron.* 69:99–151.

Wissuwa, M., K. Gatdula, and A. Ismail. 2005. Candidate gene characterization at the Pup 1 locus, a major QTL increasing tolerance to phosphorus deficiency. In: *Rice is Life, Scientific Perspectives for the 21st Century*, eds. K. Toriyama, K. L. Heong, and B. Hardy, pp. 83–85. Los Banõs, Philippines: IRRI.

Witt, C., A. Dobermann, S. Abulrachman, H. C. Gines, G. H. Wang, R. Nagrajan, S. Satawathananont et al. 1999. Internal nutrient efficiencies of irrigated lowland rice in tropical and subtropical Asia. *Field Crops Res.* 63:113–138.

Wych, R. D. and D. C. Rasmusson. 1983. Genetic improvement in malting barley cultivars since 1920. *Crop Sci.* 23:1037–1040.

Wych, R. D. and D. D. Stuthman. 1983. Genetic improvement in Minnesota adapted oat cultivars since 1923. *Crop Sci.* 23:879–881.

Yang, X., W. Wang, Z. Ye, Z. He, and V. C. Baligar. 2004. Physiological and genetic aspects of crop plant adaptation to elemental stresses in acid soils. In: *The Red Soils of China: Their Nature, Management and Utilization*, eds. M. J. Wilson, Z. He, and X. Yang, pp. 171–218. Dordrecht, the Netherlands: Kluwer Academic Publishers.

Zhang, H. and B. G. Forde. 1998. An *Arabidopsis* MADS box gene that controls nutrient induced changes in root architecture. *Science* 279:407–409.

Zhang, Q. F. 2007. Strategies for developing green super rice. *Proc. Natl. Acad. Sci. USA* 104:16402–16409.

5 Phosphorus Interactions with Other Nutrients

5.1 INTRODUCTION

An interaction occurs when the level of one production factor differentially influences the response to another factor. The Soil Science Society of America (2008) defined nutrient interaction as the response from two or more nutrients applied together that deviate from additive individual responses when applied separately. This term may also be used to describe a metabolic or ion uptake phenomenon. Nutrient interaction in crop plants is probably one of the most important factors affecting yields of annual crops. Nutrient interaction can be positive, negative, or result in no interaction (Fageria et al., 2011a, 2014b). A positive interaction occurs when the influence of the combined practices exceeds the sum of the influence of an individual practice. Positive interactions have served as the scientifically based justification for the development of a balanced plant nutrition program. When factors in combination result in a growth response that is less than the sum of their individual effects, the interaction is negative (Sumner and Farina, 1986).

When crop yield reaches an early plateau in a fertilizer experiment, it may be associated with the deficiency of another essential nutrient, known as Liebig's law of the minimum (Aulakh and Malhi, 2005). When climatic factors such as solar radiation, precipitation, and temperature are at an optimum level, plant nutrient requirements will be higher due to higher yields. Wallace (1990) proposed the law of the maximum in contrast to the law of the minimum. Law of the maximum states that when the need is fully satisfied for every factor involved in the process, the rate of the process can be at its maximum potential, which is greater than the sum of its parts because of a sequentially additive interaction (Wallace, 1990; Aulakh and Malhi, 2005).

Nutrient interactions occur when the supply of one nutrient affects the absorption, distribution, or function of another nutrient (Robson and Pitman, 1983). Induced deficiencies, toxicities, modified growth responses, and/or modified nutrient composition may occur (Wilkinson et al., 2000). Interactions may be specific or nonspecific. Nonspecific interactions become important when the contents of both nutrients are near the deficiency range or are excessive in total or proportion. When the supply of one nutrient is increased, dilution may induce a deficiency of the other nutrient, especially when the supply of the other nutrient is limiting. Such nonspecific interactions are theoretically possible for any mineral nutrient combination. Specific nutrient interactions may occur when (1) competition occurs between ions that have similar physicochemical properties (valence or diameter) or that form chemical bonds and (2) ions with sufficiently similar chemical properties compete for adsorption or absorption sites or transport within xylem and phloem or metabolic functions (Robinson and Pitman, 1983; Wilkinson et al., 2000).

Pan (2012) reported that nutrient interactions will be delineated as to their specificity, specific or primary interactions being those in which two nutrients directly react in a chemical or biological process. Furthermore, nonspecific or secondary nutrient interactions occur when the uptake of one nutrient is indirectly affected by the activity of another nutrient through a series of intermediate plant processes. Pan (2012) listed the examples of primary or specific interactions as cation–anion exchange, cation–anion precipitation, ion pairing, and ion uptake synergism or antagonism. Secondary or nonspecific interaction examples include increased yield potential and nutrient demand, altered nutrient uptake or utilization efficiency, modification of rhizosphere chemistry, biology, and modification of soil solution ionic strength (Pan, 2012).

Wilkinson et al. (2000) reported that interactions may also be associated with absorption, adsorption, translocation, and/or precipitation at any sites at the soil–root interface, which can affect uptake by roots and subsequent translocation. Zhang et al. (2006) and Pan (2012) reported that nutrients interact during numerous physical, chemical, and biological processes along the soil–root continuum such that the level of one nutrient alters the availability, uptake, or plant response to another nutrient.

Nutrient interaction can be measured in terms of crop growth and nutrient concentrations in plant tissue. Soil, plant, and climatic factors can influence interaction. In nutrient interaction studies, all other factors should be at an optimum level except the variation in level of the nutrient under investigation. Nutrient interaction can occur at the root surface or within the plant. Interactions at the root surface are due to the formation of chemical bonds by ions and precipitation or complexes. One example of this type of interaction is liming of acid soils that decreases the concentration of iron (Fe), zinc (Zn), copper (Cu), and manganese (Mn) (Fageria and Zimmermann, 1998). A second type of interaction is between ions whose chemical properties are sufficiently similar that they compete for site of absorption, transport, and function on plant root surface or within plant tissues. Such interactions are more common between nutrients of similar size, charge, geometry of coordination, and electronic configuration (Robinson and Pitman, 1983).

Interactions between phosphorus (P) and other essential plant nutrients in the soil are the manifestation of specific chemical reactions, few of which have been quantitatively defined (Adams, 1980). Qualitatively, however, these reactions are useful expressions in terms of nutrient availability and the efficiency of P fertilizers. Interaction varies from nutrient to nutrient and among crop species and even among cultivars of same species. P interactions with other essential nutrients can occur in soil as well as in plant. Therefore, nutrient interaction is a complex issue in mineral nutrition and not well understood in annual crops. The objective of this chapter is to discuss P interaction with other essential plant nutrients, which may be useful in optimizing management of P for crop production.

5.2 PHOSPHORUS INTERACTIONS WITH MACRONUTRIENTS

Micronutrients that have significant interactions with P are nitrogen (N), potassium (K), calcium (Ca), magnesium (Mg), and sulfur (S). Deficiencies of these macronutrients are widely reported in the literature in crop plants in most agroecological regions (Fageria, 2009, 2014; Fageria et al., 2011a). Deficiency of N, P, and K in upland rice, lowland rice, and dry bean grown on Brazilian Oxisols and Inceptisols is presented (Figures 5.1 through 5.3).

5.2.1 Phosphorus versus Nitrogen

N and P are the most yield-limiting nutrients in crop production worldwide. Adequate supply of these two nutrients along with K accounts for a major share of the current annual fertilizer use. P × N interaction can be the single most important nutrient interaction of practical significance. Sumner and Farina (1986) reported that because of the dominant role of N and P as fertilizer nutrients in most cropping systems, P × N interactions are probably economically the most important of all interactions involving P. This interaction is often synergistic, is occasionally additive, and, in rare cases, may be antagonistic (Aulakh and Malhi, 2005). Single application of N or P has not improved the yield of annual crops to a desired level (Singh, 1991; Dwivedi et al., 2003; Aulakh and Malhi, 2005). Therefore, N and P should be applied in adequate rates and proper proportions to obtain maximum economic yield of crops.

Aulakh and Malhi (2005) reported that growers that cannot afford to apply both N and P in optimum amounts may apply smaller or suboptimum amounts of both N and P instead of a large amount of N or P alone. For example, corn (*Zea mays*) grown in red soils produced 370 kg more grain with the application of 75 kg N + 30 kg P_2O_5 ha^{-1} as compared to 100 kg N ha^{-1} without P

FIGURE 5.1 Two lowland rice genotypes growth at two nitrogen (N) levels (top) and upland rice growth at three N levels and two dry bean genotypes growth at two N levels (bottom). (From Fageria, N.K., *Mineral Nutrition of Rice*, CRC Press, Boca Raton, FL, 2014; Fageria, N.K. et al., *Mineral Nutrition of Dry Bean*, EMBRAPA/Rice and Bean Research Center, Brasilia, Brazil, 2014b.)

(Satyanarayana et al., 1978). Sharma and Tandon (1992) reported that sorghum's (*Sorghum bicolor* L. Moench) yield in different soils (Alfisol, Vertisol, and Inceptisol) was higher when 60 kg N ha^{-1} + 60 kg P_2O_5 ha^{-1} were added, as compared to 60 kg N ha^{-1} + 0 kg P ha^{-1}. Srinivas and Rao (1984) reported that dry bean (*Phaseolus vulgaris* L.), while N alone was beneficial only up to 30 kg N ha^{-1}, made an effective use of 60 kg N ha^{-1} when this was combined with 100 kg P_2O_5 ha^{-1}. They reported that dry bean yield could be increased more than five times by a judicious N + P combination, of which 59% was due to the interaction effect.

P application can create more favorable conditions for biological N fixation. While application of N alone, particularly beyond 20 kg N ha^{-1}, reduced nitrogenase activity, a balance between N and P application maintained nitrogenase activity at a high level in field peas (*Pisum sativum*) (Aulakh and Malhi, 2005). P × N interaction is favorable for biological N fixation in legumes enhancing N_2 fixation in faba beans (*Vicia faba*) (Muller et al., 1993; Amanuel et al., 2000). Nodulation at the late flowering stage and consequent total N yield of faba beans were significantly improved by the application of P in all the three locations studied in Ethiopia (Amanuel et al., 2000).

P × N interactions related to yield increases are associated with N-induced increase in P uptake by plants (Terman et al., 1977; Sumner and Farina, 1986). Mechanisms involved are not well understood, but a number of both soil- and plant-related mechanisms have been proposed (Adams, 1980). Sumner and Farina (1986) reported that increasing P and N rates increased leaf P content of corn at flowering and this was related to grain yield. N applications resulted in a 50% increase in leaf P content at tasseling, even though the mean yield of dryland corn exceeded 11 Mg ha^{-1} (Sumner and Farina, 1986).

Fageria and Santos (2008) studied the effects of P on the uptake of N in the straw and grain of lowland rice (*Oryza glaberrima*) (Table 5.1). N uptake or accumulation (shoot or grain weight × concentration in shoot or grain) in shoot and grain at harvest was significantly influenced by

FIGURE 5.2 Upland rice growth at vegetative and maturity growth stage without and with phosphorus (P) at the top and dry bean growth at two P levels at the bottom. (From Fageria, N.K., *Mineral Nutrition of Rice*, CRC Press, Boca Raton, FL, 2014; Fageria, N.K. et al., *Mineral Nutrition of Dry Bean*, EMBRAPA/Rice and Bean Research Center, Brasilia, Brazil, 2014b.)

P rates (Table 5.1). A quadratic uptake of N in the straw and grain occurred as P rates increased (Table 5.2). Quadratic response of N accumulation in shoot and grain is related to a quadratic increase in shoot and grain yield of rice crop with increasing P rates (Figure 5.4). P has a significant positive relationship with N in lowland rice production. Positive interaction of P with N has been reported (Terman et al., 1977; Wilkinson et al., 2000; Fageria and Baligar, 2005). Schulthess et al. (1997) reported that accumulation of N and P in the shoot and grain of wheat (*Triticum aestivum* L.) was positively associated. Similarly, Fageria and Baligar (2005) reported positive association between N and P in dry bean production. Positive interaction of N with P may be associated with improved yield with the addition of N (Fageria and Baligar, 2005). Pederson et al. (2002) reported that N concentration was highly correlated with P concentration in aboveground plant parts of ryegrass (*Lolium multiflora* Lam.). Improvement in the uptake of P with the addition of N may also be related to an increase in root hairs, chemical changes in the rhizosphere, and physiological changes stimulated by N, which influence transport of P (Marschner, 1995; Baligar et al., 2001). N has a synergistic relationship with P in crop plants (Black, 1993). Kaiser and Kim (2013) also reported synergistic interaction of N and P in soybean (*Glycine max*) production. Abbasi et al. (2012) reported that application of P significantly increased uptake of N in soybean.

Fageria et al. (2014a) studied the uptake of N under different P levels in the roots of five tropical legume cover crops (Table 5.3). Uptake (concentration × dry matter) was significantly affected by

FIGURE 5.3 Upland rice growth without and with potassium (K) at the top and dry bean growth without and with K at the bottom. (From Fageria, N.K., *Mineral Nutrition of Rice*, CRC Press, Boca Raton, FL, 2014; Fageria, N.K. et al., *Mineral Nutrition of Dry Bean*, EMBRAPA/Rice and Bean Research Center, Brasilia, Brazil, 2014b.)

P levels as well as cover crop treatments (Table 5.3). Similarly, P × cover crop interaction was also significant for N uptake, indicating variable responses of cover crops in N uptake with changing P levels. Uptake of N increased with the increasing P levels, indicating a positive interaction between these two elements (Fageria et al., 2014a). Positive interaction of P × N may be related to an increase in root growth due to P fertilization, which may have been responsible for higher amounts of N uptake in the cover crop species (Tables 5.4 and 5.5).

Increased P absorption by plants is a common consequence of adding N fertilizers. In reviewing the effect of N on P uptake by plants, Grunes et al. (1958) summarized that N fertilization increased shoot and root growth, altered metabolisms, and enhanced solubility of soil P. Adams (1980) reported that yields of most nonlegumes are increased by N fertilizers, usually resulting in expected increase in demand for inorganic nutrients, that is, P. Effect of N on P uptake in such cases may be best explained by physiological stimulation occurring within the plant as a consequence of the greater N supply (Grunes et al., 1958; Bennett et al., 1962; Soltanpour, 1969).

TABLE 5.1
Influence of P on the Uptake of N in the Straw and Grain and Straw and Grain Yield of Lowland Rice

	N Uptake in Straw	N Uptake in Grain	Straw Yield
P Rate (kg ha^{-1})		kg ha^{-1}	
0	30.5	13.8	3930.3
131	61.0	39.4	7088.7
262	56.0	52.2	7753.5
393	59.5	63.5	7664.3
524	69.8	56.2	8093.3
655	46.3	59.3	7021.0
F-test			
Year (Y)	**	**	**
P rate (P)	**	**	**
Y × P	NS	NS	NS
CV(%)	38	21	22

Source: Fageria, N.K. and Santos, A.B., *Commun. Soil Sci. Plant Anal.*, 39, 873, 2008. Data are averages of 2 years' field experimentation.
**, and NSSignificant at the 1% probability levels and nonsignificant, respectively.

TABLE 5.2
Regression Equations Showing the Relationship between Phosphorus Rate and Nitrogen Uptake in Straw, Grain, and Straw Yield of Lowland Rice

Variables	Regression Equations	R^2
P rate vs. N uptake in straw	$Y = 16.80 + 0.16X - 0.00013X^2$ 0.3137*	0.32*
P rate vs. N uptake in grain	$Y = -21.07 + 0.05X - 0.00028X^2$	0.90**
P rate vs. straw yield	$Y = 4297.54 + 19.07X - 0.02X^2$	0.63**

Source: Fageria, N.K. and Santos, A.B., *Commun. Soil Sci. Plant Anal.*, 39, 873, 2008.
*,**Significant at the 5% and 1% probability levels, respectively.

5.2.2 Phosphorus versus Potassium

P and K interaction is important because K uptake is maximum or equal to N in most crop plants (Fageria et al., 2011a). In addition, K is absorbed as K^+ ion and P uptake is mostly as anions ($H_2PO_4^-$ and HPO_4^{2-}) by plants. Theoretically there should not be negative interactions. Fageria (2009) reported that P has positive interaction with K. Positive interaction of P with K may be associated with improvement in growth and yield of crop plants with P fertilization (Fageria, 2009).

Fageria et al. (2013) studied K concentration in 14 tropical legume crops under three P levels (Table 5.6). K concentration varied from 8.91 to 31.57 g kg^{-1} at a low P level, with a mean value of 20.6 g kg^{-1} (Table 5.6). At a medium P level, K concentration varied from 10.5 to 40.4 g kg^{-1}, with a mean value of 29.5 g kg^{-1}. Similarly, at a high P level, K concentration among cover crop species varied from 18.1 to 30.1 g kg^{-1}, with a mean value of 25.7 g kg^{-1}. Overall, K concentration in cover crop species shoots increased with the addition of P fertilizer in the soil. Hence, P was having synergetic effect in K uptake by cover crops. Positive interaction of K with P has been reported

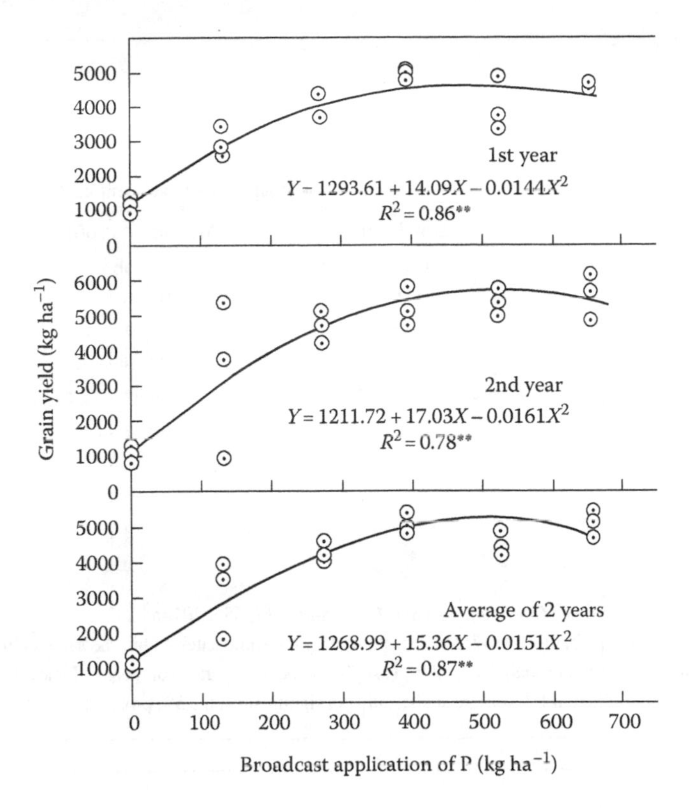

FIGURE 5.4 Relationship between broadcast application of phosphorus and grain yield of lowland rice grown on an Inceptisol. (From Fageria, N.K. and Santos, A.B., *Commun. Soil Sci. Plant Anal.*, 39, 873, 2008.)

TABLE 5.3
Influence of P on Ni Uptake (mg Plant^{-1}) in Five Tropical Legume Cover Crops

	P Fertilizer Rate (mg P kg^{-1})		
Cover Crop Species	**0**	**100**	**200**
Sunn hemp	2.90a	9.16a	10.15c
Pigeon pea	6.92a	6.63a	2.09d
Lablab	2.48a	20.81a	20.13b
Gray mucuna bean	12.79a	21.90a	37.14a
White jack bean	13.96a	18.67a	20.15b
Average	7.81b	15.43a	17.93a
F-test			
P level (P)	**		
Cover crops (CC)	**		
P × CC	**		

Source: Fageria, N.K. et al., *Commun. Soil Sci. Plant Anal.*, 45, 555, 2014a.

**Significant at the 1% probability level. Means followed by the same letter within the same column are statistically not significant at the 5% probability level by Tukey's test. For comparison of means of P levels, the same letter in the same line under different P levels are statistically not significant at the 5% probability level by Tukey's test.

TABLE 5.4
Root Dry Weight (g $Plant^{-1}$) of Five Tropical Legume Cover Crops as Influenced by P Fertilization

	Phosphorus Level (mg kg^{-1})		
Cover Crops	**Low P (0)**	**Medium P (100)**	**High P (200)**
Sunn hemp	0.17c	0.83ab	0.64c
Pigeon pea	0.16c	0.51b	0.13d
Lablab	0.13c	1.14a	0.96b
Gray mucuna bean	0.53b	0.83ab	1.42a
White jack bean	0.77a	1.12a	0.93b
Average	0.35b	0.89a	0.82a
F-test			
P level (P)	**		
Cover crops (CC)	**		
P × CC	**		

Source: Fageria, N.K. et al., *Commun. Soil Sci. Plant Anal.*, 45, 555, 2014a.
**Significant at the 1% probability level. Means followed by the same letter within the same column are statistically not significant at the 5% probability level by Tukey's test. For comparison of means of P levels, the same letter in the same line under different P levels are statistically not significant at the 5% probability level by Tukey's test.

TABLE 5.5
Maximum Root Length (cm) of Five Tropical Legume Cover Crops as Influenced by P Fertilization

	Phosphorus Level (mg kg^{-1})		
Cover Crops	**Low P (0)**	**Medium P (100)**	**High P (200)**
Sunn hemp	31.50ab	24.50d	22.00b
Pigeon pea	21.33bc	30.00cd	23.00b
Lablab	19.67c	32.00c	29.00b
Gray mucuna bean	31.50abc	39.00b	51.00a
White jack bean	36.00a	50.33a	52.33a
Average	27.40b	35.17a	35.47a
F-test			
P level (P)	**		
Cover crops (CC)	**		
P × CC	**		

Source: Fageria, N.K. et al., *Commun. Soil Sci. Plant Anal.*, 45, 555, 2014a.
**Significant at the 1% probability level. Means followed by the same letter within the same column are statistically not significant at the 5% probability level by Tukey's test. For comparison of means of P levels, the same letter in the same line under different P levels are statistically not significant at the 5% probability level by Tukey's test.

(Dibb and Thompson, 1985; Fageria, 2009). Fageria (2009) reported that the positive interaction of P with macronutrients may be associated with improvement in growth and yield of crop plants with the P fertilization. Increased growth and yield required more nutrients, as compared to low growth and yield. Wilkinson et al. (2000) also reported that increased growth requires more nutrients to maintain tissue composition within acceptable limits.

TABLE 5.6
Potassium Concentration (g kg^{-1}) in the Shoots of 14 Tropical Cover Crops under Three P Levels

	P Fertilizer Rate (mg P kg^{-1})		
Cover Crops	**0**	**100**	**200**
Crotalaria breviflora	22.62abc	25.89abcd	24.47a
Crotalaria juncea L.	22.45abc	23.34bcd	23.05a
Crotalaria mucronata	22.12abc	32.26abc	26.43a
Crotalaria spectabilis Roth	31.32a	39.32ab	29.53a
Crotalaria ochroleuca G. Don	31.57a	40.35a	30.08a
Calopogonium mucunoides	26.63ab	34.70ab	23.18a
Pueraria phaseoloides Roxb.	27.23ab	37.89ab	29.82a
Cajanus cajan L. Millspaugh	18.38bcde	33.14abc	26.19a
Cajanus cajan L. Millspaugh	20.49bcd	34.03abc	30.02a
Dolichos lablab L.	15.20cde	29.13abcd	28.23a
Mucuna deeringiana (Bort) Merr.	11.54de	24.94abcd	29.55a
Mucuna aterrima (Piper & Tracy) Holland	9.77e	18.38cd	18.11a
Mucuna cinereum L.	8.91e	10.53d	18.64a
Canavalia ensiformis L. DC.	20.13bcd	28.97abcd	22.76a
Average	20.60	29.49	25.72
F-test			
P levels (P)	**		
Cover crops (C)	**		
P × C	**		
CV(%)	17.77		

Source: Fageria, N.K. et al., *Commun. Soil Sci. Plant Anal.*, 44, 3340, 2013.
**Significant at the 1% probability level. Mean followed by the same letters within the same column is not significant at the 5% probability level by Tukey's test.

Fageria et al. (2014a) studied the influence of P on the uptake of K in the roots of five tropical legume cover crops (Table 5.7). K uptake in the roots of cover crops significantly increased with the addition of P in the soil, indicating positive influence of P on the uptake of K. Positive effect of P on the uptake of K in legumes has been reported by Fageria (2009). Studies conducted by Potash and Phosphate Institute (1999) indicated P × K interactions in soybean, wheat, and corn production (Tables 5.8 through 5.10). Soybean yield increased 8% when 34 kg P_2O_5 ha^{-1} was added as compared to the control treatment. Similarly, the increase in grain yield was 54% as compared to the control when 134 kg K_2O ha^{-1} was added. However, the yield increase was 87%, as compared to the control when 34 kg P_2O_5 ha^{-1} and 134 kg K_2O ha^{-1} were added. Improvement in grain yield of wheat occurred with the addition of P and K fertilizers (Table 5.9). Similarly, corn yield was also increased when NPK was added together, as compared to omission of PK (Table 5.10). Therefore, P × K interaction was positive in soybean, wheat, and corn production. Similarly, Darst and Wallingford (1985) reported a positive interaction between P and K in corn production. Balanced plant nutrition facilitated all production inputs to operate more efficiently (Darst and Wallingford, 1985).

Fageria and Oliveira (2014) studied the interaction among N, P, and K in upland rice production. Significant N × P × K interaction occurred for plan height, shoot dry weight, grain yield, and grain harvest index (GHI) (Table 5.11). Response of these plant characteristics is associated with an adequate rate of N, P, and K fertilization. Plant height varied from 24.3 cm in the $N_0P_0K_0$ treatment to 111 cm in the $N_1P_1K_2$ (N = 150 mg kg^{-1}, P = 100 mg kg^{-1}, and K = 200 mg kg^{-1}) treatment, with a

TABLE 5.7
Potassium Uptake (mg Plant^{-1}) in Five Tropical Legume Cover Crops as Influenced by P Fertilization

	P Fertilizer Rate (mg P kg^{-1})		
Cover Crops	**0**	**100**	**200**
Sunn hemp	1.06a	4.39a	4.87ab
Pigeon pea	1.11a	1.80a	0.94b
Lablab	0.58a	5.95a	9.02ab
Gray mucuna bean	1.21a	3.29a	16.37a
White jack bean	6.11a	8.36a	12.72ab
Average	2.02c	4.76b	8.78a
F-test			
P level (P)	**		
Cover crops (CC)	**		
P × CC	**		

Source: Fageria, N. K. et al., *Commun. Soil Sci. Plant Anal.*, 45, 555, 2014a.
**Significant at the 1% probability level. Means followed by the same letter within the same column are statistically not significant at the 5% probability level by Tukey's test. For comparison of means of P levels, the same letter in the same line under different P levels are statistically not significant at the 5% probability level by Tukey's test.

TABLE 5.8
Interaction between P and K in Soybean Production in Virginia, USA

P_2O_5 (kg ha^{-1})	K_2O (kg ha^{-1})	Grain Yield (kg ha^{-1})	Yield Increase (%)
0	0	1613	—
34	0	1747	8
0	134	2486	54
34	134	3024	87

Source: Potash and Phosphate Institute, *Better Crops*, 83, 11–13, 1999.

TABLE 5.9
Interaction of P and K in Wheat Production

	K Fertilizer Rate (kg K_2O ha^{-1})		
	0	**45**	**90**
P_2O_5 (kg ha^{-1})		**Grain Yield (kg ha^{-1})**	
0	3494	4301	4301
34	5242	5645	5846
68	5174	5914	6115

Source: Potash and Phosphate Institute, *Better Crops*, 83, 11–13, 1999.

TABLE 5.10
Interaction between N, P, and K in Corn Production

NPK Treatments	Grain Yield (kg ha^{-1})	Increase in Yield (%)
N (–PK)	7,087	—
NK (–P)	8,530	20
NP (–K)	8,718	23
NPK	11,101	57

Source: Potash and Phosphate Institute, *Better Crops*, 83, 11–13, 1999.

mean value of 74.9 cm. Shoot dry weight varied from 0.30 g plant^{-1} in the $N_0P_0K_0$ treatment to 28.7 g plant^{-1} in the $N_2P_2K_2$ (N = 300 mg kg^{-1}, P = 200 mg kg^{-1}, and K = 200 mg kg^{-1}) treatment, with a mean value of 10.7 g plant^{-1}. Fageria and Baligar (1997) reported significant increase in rice plant height and shoot dry weight with the addition of N, P, and K fertilization in the Brazilian Oxisol.

Grain yield varied from 0 to 23.01 g plant^{-1} in the $N_0P_0K_0$ and $N_2P_2K_2$ treatments, respectively, with a mean value of 6.78 g plant^{-1}. Plants that did not receive P fertilization but received adequate rate of N and K did not produce panicle or grain. Therefore, P is the most yield-limiting nutrient in highly weathered Brazilian Oxisol. Fageria and Baligar (1997, 2001) have reported similar results. A positive interaction occurred among N, P, and K fertilization in grain yield of upland production. Increasing N rate increases the demand for other nutrients, especially P and K, and higher yields were obtained at the highest rates of N, P, and K (Wilkinson et al., 2000). Wilson (1993) also confirmed the generalization that the response to one nutrient depends on the sufficiency level of other nutrients. Yield reductions occurred when high levels of one nutrient were combined with low levels of the other nutrients (Wilkinson et al., 2000). Alleviating the yield by depressing the effect of excessive macronutrient supply involved removing the limitation of a low supply of other nutrients.

GHI is an important index in determining the partitioning of dry matter between shoot and grain. GHI varied from 0 in the treatment that did not receive P to 0.45 in the treatment $N_2P_2K_2$, with a mean value of 0.25 (Table 5.12). Fageria (2009) reported that rice GHI was influenced by environmental factors, including mineral nutrition. Fageria and Baligar (2005) reported that variation in rice GHI is from 0.23 to 0.50. However, Kiniry et al. (2001) reported that rice GHI values varied greatly among cultivars, locations, seasons, and ecosystems and ranged from 0.35 to 0.62. GHI values of modern crop cultivars are commonly higher than traditional cultivars for major field crops (Ludlow and Muchlow, 1990). Limits to which GHI can be increased are considered to be about 0.60 (Austin et al., 1980). Cultivar with low harvest indices would indicate that further improvement in partitioning of biomass would be possible (Fageria and Baligar, 2005). On the other hand, cultivars with harvest indices between 0.50 and 0.60 would probably not benefit by increasing harvest index (Sharma and Smith, 1986).

Mean response values of plant height, shoot dry weight, grain yield, and GHI with the application of N, P, and K nutrients are presented (Table 5.12). Maximum plant height was achieved at the N_1 level (150 mg N kg^{-1} of soil), P_2 level (100 mg P kg^{-1} of soil), and K_2 level (200 mg K kg^{-1} of soil). Similarly, shoot dry weight, grain yield, and grain harvest values were maximum at the highest levels of N_2 (300 mg kg^{-1} of soil), P_2 (200 mg P kg^{-1} of soil), and K_2 (200 mg kg^{-1} of soil). Increase in plant height was 25% at N_2 level, as compared to the N_0 level. Similarly, increase in plant height at the P_2 level was 179%, as compared to the P_0 level and 25% at the K_2 level, as compared to the K_0 level. Shoot dry weight increase was 191% at highest N level, as compared to the lowest N level; in case of P the increase at the highest level was 1550%, as compared to the lowest P level. Similarly, the increase in shoot dry weight at highest K level was 55%, as compared to zero K level. Grain yield increase was 372% at 300 mg N kg^{-1}, as compared to control treatment. At highest K level

TABLE 5.11
Influence of N, P, and K Treatments on Plant Height, Shoot Dry Weight, Grain Yield, and GHI of Upland Rice

N, P, and K Treatments	Plant Height (cm)	Shoot Dry Weight (g Plants^{-1})	Grain Yield (g Plant^{-1})	GHI
N0P0K0	24.33i	0.30i	0.00j	0.00g
N0P0K1	26.00i	0.40hi	0.00j	0.00g
N0P0K2	46.83f	0.86hi	0.00j	0.00g
N0P1K0	75.00e	3.82gh	2.32ij	0.38abcde
N0P1K1	75.17de	8.55ef	3.05ij	0.26f
N0P1K2	91.92bc	7.30ef	3.75i	0.34bcdef
N0P2K0	78.42de	8.56ef	3.20ij	0.27f
N0P2K1	79.92de	8.17ef	3.34ij	0.29ef
N0P2K2	83.33cde	8.86e	4.05hi	0.31def
N1P0K0	39.21fgh	0.52hi	0.00j	0.00g
N1P0K1	43.33fg	0.49hi	0.00j	0.00g
N1P0K2	35.08ghi	0.54hi	0.00j	0.00g
N1P1K0	85.25cde	12.75d	5.93ghi	0.32cdef
N1P1K1	106.83a	16.32c	10.71ef	0.39abcde
N1P1K2	111.58a	17.69c	12.64de	0.42abc
N1P2K0	93.67bc	15.35cd	10.21ef	0.40abcd
N1P2K1	111.08a	18.66c	12.35cd	0.43ab
N1P2K2	109.08a	22.19b	16.64bc	0.43ab
N2P0K0	27.75hi	0.35hi	0.00j	0.00g
N2P0K1	30.75hi	0.44hi	0.00j	0.00g
N2P0K2	35.83fghi	5.24fg	0.00j	0.00g
N2P1K0	86.67cd	15.29cd	7.56fgh	0.33bcdef
N2P1K1	110.67a	22.38b	16.20bcd	0.42abc
N2P1K2	110.58a	22.89b	18.63b	0.45a
N2P2K0	92.75bc	16.73c	8.82fg	0.35abcdef
N2P2K1	103.33ab	24.28b	18.75b	0.44ab
N2P2K2	108.42a	28.66a	23.01a	0.45a
Average	74.92	10.65	6.78	0.25
F-Test				
N	**	**	**	**
P	**	**	**	**
K	**	**	**	**
N × P	**	**	**	**
N × K	**	**	**	**
P × K	**	**	**	**
N × P × K	**	**	**	**
CV(%)	4.69	9.98	16.31	9.66

Source: Fageria, N.K. and Oliveira, J.P., *J. Plant Nutr.*, 31, 1586, 2014.
Means within the same column followed by the same letter do not differ significantly at the 5% probability level by Tukey's test.

TABLE 5.12
Average Values of Plant Height, Shoot Dry Weight, Grain Yield, and GHI across N, P, and K Levels

N, P, and K Treatments	Plant Height (cm)	Shoot Dry Weight (g Plant^{-1})	Grain Yield (g Plant^{-1})	GHI
N_0	65.55b	5.20c	2.19c	0.21b
N_1	81.68a	11.61b	7.83b	0.27a
N_2	78.53a	15.12a	10.33a	0.27a
Average	74.92	10.65	6.78	0.25
P_0	34.35b	1.02c	0.00c	0.00b
P_1	94.85a	12.11b	8.98b	0.37a
P_2	95.56a	16.83a	11.37a	0.37a
Average	74.92	10.65	6.78	0.25
K_0	67.00b	8.18b	4.23b	0.23a
K_1	76.34a	11.08a	7.38a	0.25a
K_2	81.41a	12.69a	8.75a	0.27a
Average	74.92	10.65	6.78	0.25

Source: Fageria, N.K. and Oliveira, J.P., *J. Plant Nutr.*, 31, 1586, 2014.
Means within the same column and same nutrient levels followed by the same letter do not differ significantly at the 5% probability level by Tukey's test.

(200 mg K kg^{-1}), the increase was 107%, as compared to control treatment. In case of P at zero P level, plants did not produce grain. An increase in harvest index at the highest N, P, and K level followed the same pattern as the plant height, shoot dry weight, and grain yield. P was the most yield-limiting nutrient, followed by N and K, in upland rice production in Brazilian Oxisol. Fageria and Baligar (1997); Fageria et al. (2011a) reported similar conclusions. The low availability of P in Brazilian Oxisol is associated with natural low levels of this element and higher P immobilization capacity (Fageria, 1989; Fageria and Baligar, 2008).

Panicles per plant were significantly influenced by N, P, and K treatments and their interactions (Table 5.13). Grain weight (1000) was also influenced by N, P, and K treatments and N × P and N × K interactions (Table 5.13). Root dry weight was influenced by N, P, and K treatments and N × P, N × K, and P × K interactions (Table 5.13). This means that to obtain maximum panicle number, 1000 grain weight and root dry weight, adequate levels of N, P, and K in the growth medium are required. Average analysis of N, P, and K showed maximum effect of P on panicle number, followed by N and K (Table 5.14). Root dry weight and maximum root length were influenced by N and P treatments. K application improved these parameters, but the effect was not significant. Spikelet sterility was reduced with the application of K, as compared to control treatment. Improvement in panicle number, 1000 grain weight, and root growth with the application of N, P, and K have been reported by Fageria et al. (2011a) and Fageria (2009).

5.2.3 Phosphorus versus Calcium

Ca is mostly supplied by liming in acid soils. Liming is the most effective practice to improve crop yield on acid soils. Use of dolomitic lime not only supplies Ca and Mg but also improves soil pH and soil biological properties of acid soils. Soil pH significantly influenced the availability of most essential nutrients, including P (Fageria and Baligar, 2008). Fageria and Zimmermann (1995) studied the interaction between lime and P in dry bean and corn (Table 5.15). There were significant effects of lime and P treatments on shoot dry weight of dry bean and corn. Lime × P interactions were also

TABLE 5.13
Influence of N, P, and K levels on Panicle Number, 1000 Grain Weight, and Root Dry Weight

N, P, and K Treatments	Panicle Number (Plant^{-1})	1000 Grain Weight (g)	Root Dry Weight (g Plant^{-1})
1. N_0P_0	0.00e	0.00e	0.18d
2. N_0P_1	2.42d	29.67ab	1.59cd
3. N_0P_2	2.66d	30.29a	1.53cd
4. N_1P_0	0.0e	0.00e	0.22d
5. N_1P_1	3.89c	27.92bc	2.77bc
6. N_1P_2	5.58b	26.45cd	4.32ab
7. N_2P_0	0.0e	0.00e	0.28b
8. N_2P_1	5.66b	26.27cd	4.09ab
9. N_2P_2	7.27a	25.65d	5.11a
1. N_0K_0	1.44a	20.13a	1.04c
2. N_0K_1	1.44a	20.12a	1.03c
3. N_0K_2	2.19a	19.71ab	1.23c
4. N_1K_0	2.55a	17.73bcd	2.15bc
5. N_1K_1	3.11a	18.79abc	2.04bc
6. N_1K_2	3.81a	17.89bcd	3.12ab
7. N_2K_0	3.25a	16.11d	1.68bc
8. N_2K_1	4.53a	17.41cd	3.21ab
9. N_2K_2	5.17a	18.39abc	4.58a
1. P_0K_0	0.00d	0.00a	0.14c
2. P_0K_1	0.00d	0.00a	0.26c
3. P_0K_2	0.00d	0.00a	0.29c
4. P_1K_0	2.58c	27.66a	2.31b
5. P_1K_1	4.42b	28.11a	2.43b
6. P_1K_2	4.97b	28.10a	3.73ab
7. P_2K_0	4.67b	26.31a	2.44b
8. P_2K_1	4.66b	28.22a	3.61ab
9. P_2K_2	6.19a	27.86a	4.91a
F-test			
N	**	**	**
P	**	**	**
K	**	**	**
N × P	**	**	**
N × K	NS	**	**
P × K	**	NS	**
N × P × K	**	NS	NS
CV(%)	22.40	7.20	47.36

Source: Fageria, N.K. and Oliveira, J.P., *J. Plant Nutr.*, 31, 1586, 2014.

Means within the same column and interactions N × P, N × K, and P × K nutrient levels followed by the same letter do not differ significantly at the 5% probability level by Tukey's test.

significant for these traits in both the crops. Influence of lime on the growth of dry bean and corn varied with the variation in P levels in the soil. Dry bean and corn yield were higher at higher lime and P rates, as compared to lower lime and P rates, indicating significant positive interactions between lime and P in the dry weight of shoot of dry bean and corn. Similarly, uptake of P and Ca in the shoot of these two crops was also significantly influenced by lime × P interactions, indicating differential

TABLE 5.14
Average Values of Panicle Number, 1000 Grain Weight, and Root Dry Weight across N, P, and K Levels

N, P, and K Treatments	Panicle Number ($Plant^{-1}$)	1000 Grain Weight (g)	Root Dry Weight (g $Plant^{-1}$)	Maximum Root Length (cm)	Spikelet Sterility (%)
N_0	1.69c	19.99a	1.10a	29.56b	9.17b
N_1	3.16b	18.13b	2.44ab	32.93a	11.75a
N_2	4.31a	17.31b	3.16a	34.63a	9.98ab
Average	3.06	18.48	2.23	32.37	10.30
P_0	0.00c	0.00b	0.23b	24.70c	0.00b
P_1	3.99b	27.96a	2.82a	33.89b	14.25a
P_2	5.18a	27.47a	3.66a	38.52a	16.64a
Average	3.06	18.48	2.23	32.37	10.30
K_0	2.42a	17.99a	1.63a	32.37a	12.27a
K_1	3.03ab	18.78a	2.10a	32.33a	8.92b
K_2	3.72a	18.65a	2.98a	32.41a	9.70b
Average	3.06	18.48	2.23	32.37	10.30

Source: Fageria, N.K. and Oliveira, J.P., *J. Plant Nutr.*, 31, 1586, 2014.
Means within the same column and same nutrient levels followed by the same letter do not differ significantly at the 5% probability level Tukey's test.

TABLE 5.15
Influence of Lime and P on Dry Matter Yield of Dry Bean and Corn

Lime Rate (g kg^{-1})	P Rate (mg kg^{-1})	Dry Bean SDW (g Pot^{-1})	Corn SDW (g Pot^{-1})
0	0	1.25	1.10
0	50	8.30	4.93
0	175	10.60	8.73
2	0	1.30	1.43
2	50	9.00	7.13
2	175	10.60	11.47
4	0	1.70	1.10
4	50	10.50	6.60
4	175	12.00	9.93
F-test			
Lime (L)		**	**
Phosphorus (P)		**	**
L × P		*	*
CV(%)		7.47	11.16
Average across the lime (L) and P level			
L_0		6.68b	4.92c
L_2		6.93b	6.68a
L_3		8.07a	5.88b
P_0		1.14c	1.21c
P_{50}		9.22b	6.22b
P_{175}		11.06a	10.04a

Source: Fageria, N.K. and Zimmerman, F.J.P., *J. Plant Nutr.*, 18, 2519, 1995.
SDW, shoot dry weight.
*,**Significant at the 5% and 1% probability levels, respectively. Means followed by the same letter within the same column are not significant at the 5% probability level by Tukey's test.

TABLE 5.16
Influence of Lime and Phosphorus on Uptake of P and Ca in the Shoot of Dry Bean and Corn

		Dry Bean		Corn	
Lime (g kg^{-1})	P (mg kg^{-1})	P mg Pot^{-1}	Ca mg Pot^{-1}	P mg Pot^{-1}	Ca mg Pot^{-1}
0	0	2.7	24	1.1	5.6
0	50	10.8	97	8.1	19.4
0	175	25.9	114	20.6	38.8
2	0	1.2	34	1.4	11.2
2	50	18.3	150	10.7	42.0
2	175	25.9	195	27.7	63.2
4	0	1.2	58	0.9	11.4
4	50	21.4	199	9.5	52.7
4	175	28.8	274	20.8	71.8
F-test					
Lime (L)		**	**	NS	**
Phosphorus (P)		**	**	**	**
L × P		**	**	*	**
CV(%)		14.5	17	21.1	9.9
Average across the lime (L) and P level					
L_0		13.1b	78c	11.1a	21.3b
L_2		15.1ab	126b	13.3a	42.2a
L_3		17.2a	177a	10.4a	45.3a
P_0		1.7c	38c	2.4c	9.4c
P_{50}		16.9b	149b	9.4b	41.4b
P_{175}		26.9a	194a	23.1a	57.9a

Source: Fageria, N.K. and Zimmerman, F.J.P., *J. Plant Nutr.*, 18, 2519, 1995.
*,**, and NS Significant at the 5% and 1% probability levels and not significant, respectively. Means followed by the same letter within the same column are not significant at the 5% probability level by Tukey's test.

responses of lime and P for uptake of P and Ca at different P levels (Table 5.16). Uptake of P and Ca was higher at higher lime and P levels, as compared to lower lime and P levels. Lime and P were having positive interactions in the uptake of P and Ca in shoot of dry bean and corn.

Ca has long been recognized for its special synergistic role in facilitating ion uptake (Viets, 1944), which is commonly attributed to its positive maintenance of membrane integrity and subsequent membrane transport selectivity of other ions (Marschner, 1995; Epstein and Bloom, 2005). Membrane stabilization is achieved by Ca bridging of carboxylate and phosphate groups of the membrane phospholipids (Caldwell and Huag, 1981).

Robson et al. (1970) studied Ca and P uptake by alfalfa (*Medicago sativa*) and *Trifolium* species in solution culture using the concentrations of these ions generally found in the soil solution. They reported that Ca enhanced P uptake by these plant species. Effect of Ca on uptake of P was more pronounced at lower P concentrations, as compared to higher P concentrations. Adams (1980) reported that Ca increased the transport rate of P because of its effect on P carriers. Other favored mechanisms included a screening action of Ca of electronegative sites, resulting in greater accessibility to absorption sites by $H_2PO_4^-$ ions (Adams, 1980).

Fageria et al. (2013) studied the influence of P on the concentration and uptake of Ca in 14 tropical legume cover crops. Ca concentration varied from 5.60 to 13.1 g kg^{-1} at the low P rate, with a

TABLE 5.17
Influence of P on Ca Concentration (g kg^{-1}) in the Shoots of 14 Tropical Legume Cover Crops

	P Fertilizer Rate (mg P kg^{-1})		
Cover Crops	**0**	**100**	**200**
Crotalaria breviflora	11.95abcd	12.95bcd	14.70ab
Crotalaria juncea L.	11.41abcd	11.80cde	14.21ab
Crotalaria mucronata	8.47cde	9.65de	10.26c
Crotalaria spectabilis Roth	15.25a	17.10ab	16.16a
Crotalaria ochroleuca G. Don	5.60e	7.84e	9.20c
Calopogonium mucunoides	12.74abc	14.86abc	12.42bc
Pueraria phaseoloides Roxb.	12.37abc	12.82bcde	9.77c
Cajanus cajan L. Millspaugh	11.15abcd	17.73ab	14.33ab
Cajanus cajan L. Millspaugh	12.89abc	17.76ab	15.88ab
Dolichos lablab L.	9.46bcde	15.64abc	15.70ab
Mucuna deeringiana (Bort) Merr.	11.94abcd	18.82a	17.59a
Mucuna aterrima (Piper & Tracy) Holland	10.31bcd	13.79abcd	14.06ab
Mucuna cinereum L.	7.74de	12.74bcde	14.03ab
Canavalia ensiformis L. DC.	13.05ab	16.68ab	17.54a
Average	11.02	14.23	13.99
F-test			
P levels (P)	**		
Cover crops (CC)	**		
P × CC	**		
CV(%)	10.90		

Source: Fageria, N.K. et al., *Commun. Soil Sci. Plant Anal.*, 44, 3340, 2013.
**Significant at the 1% probability level. Mean followed by the same letters within the same column is not significant at the 5% probability level by Tukey's test.

mean value of 11.0 g kg^{-1} (Table 5.17). At medium P levels, Ca concentration in the tops of cover crop species varied from 7.84 to 18.8 g kg^{-1}, with a mean value of 14.2 g kg^{-1}. At the higher P rates, the Ca concentration varied from 9.20 to 17.6 g kg^{-1}, with a mean value of 14.0 g kg^{-1}. Overall, Ca concentration increased with the addition of P fertilizer. *Crotalaria ochroleuca* was having a minimum concentration of Ca at three P levels as compared with other cover crop species. Ca concentration increase was 29% at the medium P level and 27% at the high P level as compared with control treatment. Fageria (2009) reported positive association between P and Ca uptake in crop plants.

Uptake (concentration × dry matter yield) of Ca was significantly affected by P level as well as cover crop treatments (Table 5.18). Similarly, P × cover crop interaction was significant for Ca uptake, indicating variable responses of cover crops with changing P levels on Ca uptake. Overall, Ca uptake increased with increasing P levels. Fageria (2009) reported that generally P has positive significant interactions with most macronutrients, including Ca. Positive effect of P on the uptake of Ca was related to an increase in dry weight of cover crops with the addition of P (Fageria et al., 2013).

5.2.4 Phosphorus versus Magnesium

P improves Mg uptake and Mg improves P uptake in crop plants. Uptake (concentration × dry matter yield) of Mg was significantly affected by P level as well as cover crop treatments (Table 5.19). Similarly, P × cover crop interaction was significant for Mg uptake, indicating differential responses

TABLE 5.18
Influence of P on Ca Uptake (mg Plant^{-1}) in the Shoots of 14 Tropical Legume Cover Crops

	P Fertilizer Rate (mg P kg^{-1})		
Cover Crops	**0**	**100**	**200**
Crotalaria breviflora	1.58g	6.99e	10.17d
Crotalaria juncea L.	14.16de	40.95c	66.53c
Crotalaria mucronata	1.66g	7.43de	7.93d
Crotalaria spectabilis Roth	4.71fg	17.11cde	22.21d
Crotalaria ochroleuca G. Don	1.70g	10.60de	21.77d
Calopogonium mucunoides	3.38fg	13.95cde	3.18d
Pueraria phaseoloides Roxb.	1.98g	9.51de	10.02d
Cajanus cajan L. Millspaugh	7.88efg	36.25cd	28.95d
Cajanus cajan L. Millspaugh	5.12fg	28.36cde	27.93d
Dolichos lablab L.	9.15ef	72.47b	96.70b
Mucuna deeringiana (Bort) Merr.	26.77c	85.52b	94.13bc
Mucuna aterrima (Piper & Tracy) Holland	16.78d	78.90b	97.37b
Mucuna cinereum L.	36.56b	73.35b	110.63b
Canavalia ensiformis L. DC.	79.48a	154.22a	169.60a
Average	15.06	45.40	54.79
F-test			
P levels (P)	**		
Cover crops (C)	**		
P × C	**		
CV(%)	19.79		

Source: Fageria, N.K. et al., *Commun. Soil Sci. Plant Anal.*, 44, 3340, 2013.
**Significant at the 1% probability level. Mean followed by the same letters in the same column is not significant at the 5% probability level by Tukey's test.

of cover crops with changing P levels on Mg uptake. Overall, uptake of Fageria (2009) reported that generally P has positive significant interactions with most macronutrients, including Mg. Positive effect of P on uptake of these macronutrients was related to an increase in dry weight of cover crops with the addition of P (Fageria et al., 2013).

Increased levels of Mg improved P uptake in crop plants (Edwards, 1968; Franklin, 1969; Agbim, 1981; Sumner and Farina, 1986). These authors further reported that positive effect of Mg on P uptake was related to the fact that Mg is an activator for almost all reactions involving phosphate transfer within the plants. A positive relation between the P and Mg contents of plants was previously reported by Truog et al. (1947).

5.2.5 Phosphorus versus Sulfur

Fageria et al. (2013) studied the influence of P on S concentration and uptake in 14 tropical legume cover crops (Tables 5.20 and 5.21). S concentration varied from 1.11 to 2.92 g kg^{-1}, with a mean value of 2.13 g kg^{-1} at the low P level (Table 5.20). At the medium P level, S concentration varied from 1.86 to 3.41 g kg^{-1}, with a mean value of 2.77 g kg^{-1}. At 200 mg P kg^{-1} level, S concentration varied from 1.81 to 3.13 g kg^{-1}, with a mean value of 2.54 g kg^{-1}. Overall, S concentration in the shoots of cover crops increased with the addition of P in the soil. Positive effects of P on S uptake were reported by Fageria (2009) in crop plants.

TABLE 5.19
Influence of P on Mg Uptake (mg Plant^{-1}) in the Shoots of 14 Tropical Legume Cover Crops

	P Fertilizer Rate (mg P kg^{-1})		
Cover Crops	**0**	**100**	**200**
Crotalaria breviflora	0.54g	2.46d	2.86b
Crotalaria juncea L.	5.70de	15.20bc	27.20a
Crotalaria mucronata	0.97fg	3.04d	3.32b
Crotalaria spectabilis Roth	1.20fg	3.87d	5.46b
Crotalaria ochroleuca G. Don	1.40fg	6.48cd	12.35b
Calopogonium mucunoides	1.24fg	4.23d	1.09b
Pueraria phaseoloides Roxb.	1.13fg	3.91d	4.69b
Cajanus cajan L. Millspaugh	2.78efg	11.38cd	7.88b
Cajanus cajan L. Millspaugh	1.51fg	7.42cd	6.64b
Dolichos lablab L.	4.02def	21.60b	32.09a
Mucuna deeringiana (Bort) Merr.	10.85c	23.17b	26.62a
Mucuna aterrima (Piper & Tracy) Holland	7.14d	22.05b	28.25a
Mucuna cinereum L.	17.33b	23.86b	36.11a
Canavalia ensiformis L. DC.	21.34a	33.26a	38.51a
Average	5.51	12.99	16.65a
F-test			
P levels (P)	**		
Cover crops (CC)	**		
P × CC	**		
CV(%)	24.92		

Source: Fageria, N.K. et al., *Commun. Soil Sci. Plant Anal.*, 44, 3340, 2013.
**Significant at the 1% probability level. Mean followed by the same letters within the same column is not significant at the 5% probability level by Tukey's test.

Uptake (concentration × dry matter yield) of S was significantly affected by P level as well as cover crop treatments (Table 5.21). Similarly, P × cover crop interaction was significant for S uptake, indicating differential responses of cover crops with changing P levels on S uptake. Overall, S uptake increased with increasing P levels. Fageria (2009) reported that generally P has positive significant interaction with most of the macronutrients, including S. Positive effect of P on the uptake of S was related to an increase in dry weight of cover crops with the addition of P (Fageria et al., 2013).

The purpose of adding P fertilizer to soils is to reduce the amount of adsorbed sulfate (SO_4^{2-}) (Kamprath et al., 1956; Chao et al., 1962) and thereby increase its availability. This concept has been applied to correlation experiments for soil testing, and P-containing solutions have been reported to be a highly effective SO_4^{2-} extractant (Hoeft et al., 1973; Sumner and Farina, 1986).

5.3 PHOSPHORUS VERSUS MICRONUTRIENTS

Essential micronutrients for crop plants are Zn, Fe, Mn, Cu, boron (B), molybdenum (Mo), chlorine (Cl), and nickel (Ni). In some publications, Co is also cited as an essential micronutrient; however, essentiality of this element is not proven for crop plants (Fageria et al., 2002). Cl is classified as a micronutrient, even though its concentration in plant tissue is often equivalent to those of macronutrients. Micronutrients are required by plants in small amounts, as compared to macronutrients. But their influence is as important as macronutrients in crop production. Based on physiological properties, the

TABLE 5.20
Influence of P on S Concentration (g kg^{-1}) in the Shoots of 14 Tropical Legume Cover Crop

	P Fertilizer Rate (mg P kg^{-1})		
Cover Crops	**0**	**100**	**200**
Crotalaria breviflora	2.22abcde	2.67abcd	2.93ab
Crotalaria juncea L.	2.40abcd	2.32bcd	2.73abc
Crotalaria mucronata	2.76ab	2.70abcd	2.70abc
Crotalaria spectabilis Roth	2.92a	3.27a	3.01ab
Crotalaria ochroleuca G. Don	2.50abc	3.15ab	3.13a
Calopogonium mucunoides	2.56ab	3.11ab	2.21bcd
Pueraria phaseoloides Roxb.	2.56ab	3.01abc	2.15bcd
Cajanus cajan L. Millspaugh	2.14bcde	3.41a	2.57abcd
Cajanus cajan L. Millspaugh	2.22abcde	3.41a	2.97ab
Dolichos lablab L.	1.78cdef	3.15ab	2.74abc
Mucuna deeringiana (Bort) Merr.	1.61ef	2.73abcd	2.33abcd
Mucuna aterrima (Piper & Tracy) Holland	1.28f	1.91d	1.81d
Mucuna cinereum L.	1.11f	1.86d	2.04cd
Canavalia ensiformis L. DC.	1.77def	2.15cd	2.33abcd
Average	2.13	2.77	2.54
F-test			
P levels (P)	**		
Cover crops (Cc)	**		
P × CC	**		
CV(%)	10.70		

Source: Fageria, N.K. et al., *Commun. Soil Sci. Plant Anal.*, 44, 3340, 2013.
**Significant at the 1% probability level. Mean followed by the same letters within the same column is not significant at the 5% probability level by Tukey's test.

essential plant micronutrients are all metals except for B and Cl. For micronutrients the deficient and toxic concentration range is very narrow, as compared with macronutrients. For macronutrients, the sufficiency range is very broad and toxicity rarely occurs. Among the micronutrients, deficiency and toxicity range is narrower for Mo and B than for any of the other micronutrients. Most of the micronutrients are immobile in plants. Therefore, deficiency symptoms first appear in the younger leaves. Micronutrient deficiency symptoms in rice leaves are presented (Figure 5.5). The accumulation of micronutrients by plants generally follows the order of Cl > Fe > Mn > Zn > B > Cu > Mo (Marschner, 1995; Fageria et al., 2002; Fageria and Stone, 2008). This order may change among plant species and with growth conditions (e.g., flooded rice). P interactions with micronutrients (positive and negative) have been reported on a wide variety of crops (Potash and Phosphate Institute, 1999).

5.3.1 Phosphorus versus Zinc

P interaction with Zn is common in crop plants and widely reported in the literature (Loneragan et al., 1979; Cakmak and Marschner, 1987; Fageria et al., 2002). Increasing P levels in the soil can reduce Zn uptake, and Zn deficiency symptoms have been reported by Fageria (2014) in upland rice grown on a Brazilian Oxisol. This disorder, commonly known as "P-induced Zn deficiency," is the most widely studied P-trace element interaction in the soil–plant system (Bolan et al., 2005). The decrease in Zn availability at higher levels of P fertilization has been reported to be related to

TABLE 5.21
Influence of P on S Uptake (mg Plant^{-1}) in the Shoots of 14 Tropical Legume Cover Crops

	P Fertilizer Rate (mg P kg^{-1})		
Cover Crops	**0**	**100**	**200**
Crotalaria breviflora	0.30h	1.45g	2.01de
Crotalaria juncea L.	3.05cd	8.90cde	13.30bc
Crotalaria mucronata	0.56gh	2.08fg	2.03de
Crotalaria spectabilis Roth	0.90fgh	3.27efg	4.10de
Crotalaria ochroleuca G. Don	0.76fgh	4.26fg	7.34cd
Calopogonium mucunoides	0.67gh	2.90g	0.57e
Pueraria phaseoloides Roxb.	0.44h	2.29def	2.21de
Cajanus cajan L. Millspaugh	1.62efg	7.13efg	5.31de
Cajanus cajan L. Millspaugh	0.89fgh	5.50b	5.19de
Dolichos lablab L.	1.83ef	14.94bc	18.22ab
Mucuna deeringiana (Bort) Merr.	3.55c	12.46bcd	12.50bc
Mucuna aterrima (Piper & Tracy) Holland	2.06de	10.92bcd	12.56bc
Mucuna cinereum L.	5.47b	11.53a	19.80a
Canavalia ensiformis L. DC.	10.87a	20.25a	23.53a
Average	2.35	7.71	9.19
F-test			
P levels (P)	**		
Cover crops (CC)	**		
P × CC	**		
CV(%)	22.25		

Source: Fageria, N.K. et al., *Commun. Soil Sci. Plant Anal.*, 44, 3340, 2013.
**Significant at the 1% probability level. Mean followed by the same letters within the same column is not significant at the 5% probability level by Tukey's test.

five mechanisms (Loneragan et al., 1979; Bolan et al., 2005). These mechanism are as follows: (1) dilution of Zn in plant issue by promotion of plant growth with the P fertilizer, (2) inhibition of Zn absorption by plant roots through the cations added with P compounds, (3) P-induced Zn adsorption by soil rich in variable Fe and Al oxides and hydroxide in soils, (4) P-induced higher physiological requirement for Zn in plant shoots, and (5) inhibition of Zn translocation from root to shoot due to physiological inactivations of Zn within the root in the presence of higher P levels (Bolan et al., 2005).

Previous studies indicated a decrease in Zn uptake with increasing P levels, while others have reported the reverse (Sumner and Farina, 1986). Boawn et al. (1954) and Ellis et al. (1964) reported that Zn deficiency did not occur in corn and dry bean when high rates of P were applied, despite the fact that P content of plant tissue was more than doubled. Fageria et al. (2013) studied Zn uptake in the shoots of 14 tropical legume cover crops (Table 5.22). Uptake of Zn was significantly influenced by P level and cover crop treatments. P × cover crop interaction was significant for Zn uptake, indicating that differential response occurred in Zn uptake among cover crops with varying P levels. Similarly, a positive interaction of P with Zn has been reported in corn (Table 5.23).

Nutrient solution experiments are often used to clarify concept of mineral nutrition (Adams, 1980). In the case of P–Zn interactions, however, they have not been particularly enlightening. For example, P may have no effect on Zn uptake (Bingham, 1963), P may increase Zn uptake (Wallace et al., 1973), and P may decrease Zn uptake (Racz and Haluschak, 1974). P–Zn interactions are not conclusive in literature: positive, negative, or no interactions.

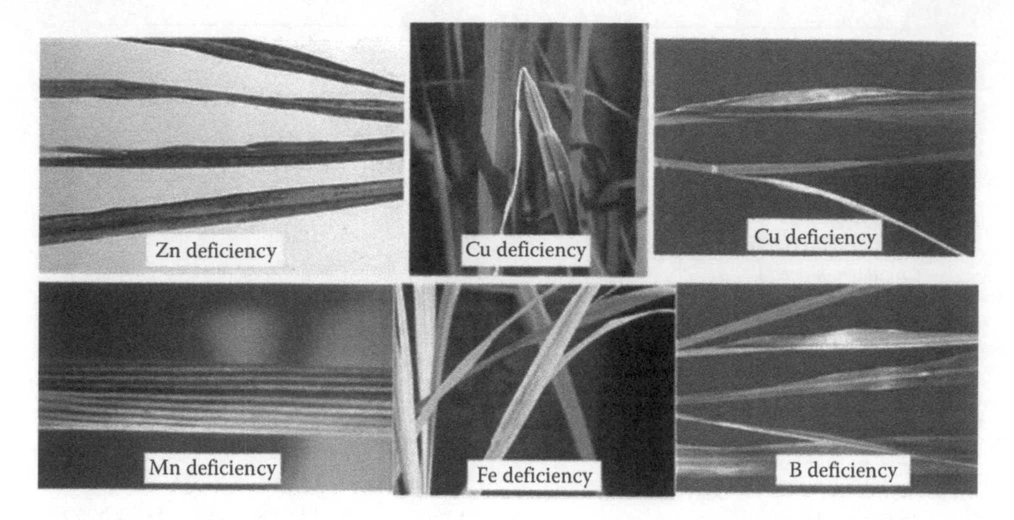

FIGURE 5.5 Micronutrient deficiency symptoms in rice leaves. (From Fageria, N.K., *Fertilization and Mineral Nutrition of Rice*, CNPAF, EMBRAPA, Santo Antonio de Goias, Brazil, 1984; Fageria, N.K. and Barbosa Filho, M.P., Nutritional deficiency in rice and their correction, EMBRAPA/CNPAF Document No. 42, 1994.)

TABLE 5.22
Influence of P on Zn Uptake (μg Plant⁻¹) in the Shoots of 14 Tropical Legume Cover Crops

	P Fertilizer Rate (mg P kg^{-1})		
Cover Crops	**0**	**100**	**200**
Crotalaria breviflora	3.43d	13.07e	15.75fg
Crotalaria juncea L.	31.46cd	82.92bcde	88.26bcd
Crotalaria mucronata	8.51d	28.69de	24.22efg
Crotalaria spectabilis Roth	11.01d	31.29cde	39.25defg
Crotalaria ochroleuca G. Don	11.07d	41.01cde	61.63defg
Calopogonium mucunoides	7.79d	24.89de	6.79g
Pueraria phaseoloides Roxb.	5.49d	29.96de	25.20efg
Cajanus cajan L. Millspaugh	33.55cd	102.49bcd	79.33cde
Cajanus cajan L. Millspaugh	18.83d	83.98bcde	66.52def
Dolichos lablab L.	26.91cd	108.31bc	134.74abc
Mucuna deeringiana (Bort) Merr.	57.83bc	92.11bcd	150.87a
Mucuna aterrima (Piper & Tracy) Holland	33.12cd	126.52ab	144.81ab
Mucuna cinereum L.	85.48ab	136.36ab	177.20a
Canavalia ensiformis L. DC.	89.70a	191.64a	167.69a
Average	30.30	78.09	84.45
F-test			
P levels (P)	**		
Cover crops (CC)	**		
P × CC	**		
CV(%)	27.17		

Source: Fageria, N.K. et al., *Commun. Soil Sci. Plant Anal.*, 44, 3340, 2013.
**Significant at the 1% probability level. Mean followed by the same letters within the same column is not significant at the 5% probability level by Tukey's test.

TABLE 5.23
Phosphorus and Zn Interaction in Corn Production

Phosphorus Rate (P_2O_5 kg ha^{-1})	Zinc Rate (kg Zn ha^{-1})	Grain Yield (kg ha^{-1})
0	0	8,216
90	0	7,464
0	22	6,836
90	22	10,976

Source: Potash and Phosphate Institute, *Better Crops*, 83, 11–13, 1999.

5.3.2 Phosphorus versus Copper

P–Cu interactions are positive or negative (Adams, 1980). P-induced Zn deficiency has been reported in citrus (Bingham and Martin, 1956; Bingham and Garber, 1960). In contrast to most citrus experiments, the Cu–P interaction in subterranean clover (*Trifolium subterraneum* L.) was reported to be synergistic instead of antagonistic (Greenwood and Hallsworth, 1960).

Fageria et al. (2013) studied the interaction between P and Zn in tropical legume cover crops. Cu concentration in the shoots of 14 cover crops was significantly affected by cover crops and P × cover crop interactions (Table 5.24). Cu concentration varied from 3.76 to 14.52 mg kg^{-1}, with a mean value of 9.68 mg kg^{-1} at the low P level. At the medium P level, Cu concentration varied from 5.86 to 16.32 mg kg^{-1}, with a mean value of 10.47 mg kg^{-1}. At the high P level, Cu concentration varied from 5.34 to 13.60 mg kg^{-1}, with a mean value of 9.75 mg kg^{-1}. Overall, Cu concentration increased with the addition of P fertilization. Fageria (2009) reported positive effect of P on Cu uptake in crop plants. Cu concentration was minimal in the cover crop shoots as compared to other micronutrients. Fageria (2009) reported that Cu is taken up by the plants in only very small quantities. The Cu concentration of most plants is generally between 2 and 20 mg kg^{-1} in the dry plant tissues (Mengel et al., 2001).

Similarly, Cu uptake was significantly influenced by P and cover crop treatments, and P × cover crop interaction was significant for Cu uptake (Table 5.25). Uptake of Cu increased with the increasing P levels. Cu uptake was 2.8 and 3.2 times higher at the 100 mg P kg^{-1} and 200 mg P kg^{-1} P levels, respectively, as compared to control treatments. P and Cu interaction is believed to occur at the site of absorption, possibly with Cu precipitation at the root surface. Application of P reduced the effect of toxic levels of Cu. Excess Cu can decrease P absorption (Potash and Phosphate Institute, 1999).

5.3.3 Phosphorus versus Manganese

Chemistry of Mn in soils is complex because three oxidation states are involved, that is, Mn^{2+}, Mn^{3+}, and Mn^{4+}. Mn forms hydrated oxides with mixed valence states (Lindsay, 1979). Soil solution concentrations of Mn^{2+}, as well as exchangeable Mn^{2+}, are governed by the solubility of Mn oxides, such as Mn_2O_3 and MnO_2 (Adams, 1980). Soil pH decrease associated with high P rates can account for numerous reports that P fertilizer increased Mn uptake by plants (Bingham and Garber, 1960; Larsen, 1964; Smilde, 1973). Page et al. (1963) reported a significant correlation between the decreased soil pH and increased Mn uptake of oats associated with application of P fertilizer.

Fageria et al. (2013) studied the influence of P on concentration and uptake of Mn in the shoots of 14 tropical legume cover crops. Mn concentration varied from 29.5 to 117.9 mg kg^{-1}, with a mean value of 76.6 mg kg^{-1} at the low P level (Table 5.26). At the medium P level, Mn concentration in shoots of cover crop plants varied from 45.70 to 172.01 mg kg^{-1}, with a mean value of 101.5 mg kg^{-1}. At the higher P level, Mn concentration varied from 46.6 to 158.1 mg kg^{-1}, with a mean value of

TABLE 5.24
Influence of P on Cu Concentration (mg kg^{-1}) in the Shoots of 14 Tropical Legume Cover Crops

	P Fertilizer Rate (mg P kg^{-1})		
Cover Crops	**0**	**100**	**200**
Crotalaria breviflora	13.30ab	10.34abc	11.97abcd
Crotalaria juncea L.	8.28bcde	8.11c	7.36ef
Crotalaria mucronata	14.51a	11.52abc	9.84abcde
Crotalaria spectabilis Roth	9.72abcd	9.22bc	9.07cdef
Crotalaria ochroleuca G. Don	9.01abcde	9.30bc	9.70bcde
Calopogonium mucunoides	8.83abcde	10.49abc	8.46def
Pueraria phaseoloides Roxb.	12.48abc	9.88bc	8.22def
Cajanus cajan L. Millspaugh	14.07ab	16.32a	13.60a
Cajanus cajan L. Millspaugh	14.52a	16.26a	12.59abc
Dolichos lablab L.	3.76e	6.32c	7.25ef
Mucuna deeringiana (Bort) Merr.	11.12abc	14.64ab	13.48ab
Mucuna aterrima (Piper & Tracy) Holland	4.77de	8.06c	8.84cdef
Mucuna cinereum L.	6.82cde	10.28abc	10.85abcde
Canavalia ensiformis L. DC.	4.36de	5.86c	5.34f
Average	9.68	10.47	9.75
F-test			
P levels (P)	NS		
Cover crops (CC)	**		
P × CC	**		
CV(%)	17.46		

Source: Fageria, N.K. et al., *Commun. Soil Sci. Plant Anal.*, 44, 3340, 2013.
**Significant at the 1% probability level. Mean followed by the same letters within the same column is not significant at the 5% probability level by Tukey's test.

94.0 mg kg^{-1}. With the increasing level of P in the growth medium, there is an improved Mn uptake by legume cover crops. Similarly, uptake of Mn at three P levels was also significantly increased (Table 5.27). Fageria et al. (2013) reported that a significant increase in Mn uptake by tropical legume crops may be related to an increase in dry weight of shoots of these crops with the addition of P. On some soils, the increase in Mn uptake is believed to be partially due to increased soil acidity from high rates of P (Potash and Phosphate Institute, 1999).

5.3.4 Phosphorus versus Iron

P and Fe interactions are reported as positive and negative (Adams, 1980; Sumner and Farina, 1986). Ajakaiye (1979) reported negative interactions between P × Fe for millet (*Pancilin menhouse*) and sorghum in nutrient solution. Similarly, Wallace et al. (1973) reported negative interaction between P and Fe in dry bean in nutrient solution. Sumner and Farina (1986) reported that in general, Fe deficiency or stress that may occur in a field crop is much more likely to result from elevated pH rather than from high P levels.

Fageria et al. (2013) studied the influence of P on the concentration and uptake of Fe in the shoots of 14 tropical legume cover crops (Tables 5.28 and 5.29). Fe concentration varied from 136 to 1384 mg kg^{-1}, with a mean value of 371 mg kg^{-1} at the low P level. At the medium P level, Fe concentration in plant tissue of cover crop species varied from 106 to 716 mg kg^{-1}, with a mean

TABLE 5.25
Influence of P on Cu Uptake (μg Plant^{-1}) in the Shoots of 14 Tropical Legume Cover Crops

	P Fertilizer Rate (mg P kg^{-1})		
Cover Crops	**0**	**100**	**200**
Crotalaria breviflora	1.77d	5.64f	8.33ef
Crotalaria juncea L.	13.19cd	39.92bcd	40.74cd
Crotalaria mucronata	2.96d	8.85ef	7.07ef
Crotalaria spectabilis Roth	2.99d	9.21ef	12.42ef
Crotalaria ochroleuca G. Don	2.73d	12.54def	23.02def
Calopogonium mucunoides	2.32d	9.63ef	2.17f
Pueraria phaseoloides Roxb.	2.24d	7.75f	8.50ef
Cajanus cajan L. Millspaugh	11.64cd	38.61bcde	29.14de
Cajanus cajan L. Millspaugh	5.95d	25.79cdef	22.18def
Dolichos lablab L.	5.16d	44.24abc	57.73bc
Mucuna deeringiana (Bort) Merr.	25.49bc	66.88ab	72.35b
Mucuna aterrima (Piper & Tracy) Holland	7.63d	46.41abc	61.19bc
Mucuna cinereum L.	45.88a	73.81a	114.81a
Canavalia ensiformis L. DC.	33.67ab	65.44ab	62.04bc
Average	11.69	32.48	37.26
F-test			
P levels (P)	**		
Cover crops (CC)	**		
P × CC	**		
CV(%)	30.81		

Source: Fageria, N.K. et al., *Commun. Soil Sci. Plant Anal.*, 44, 3340, 2013.
**Significant at the 1% probability level. Mean followed by the same letters within the same column is not significant at the 5% probability level by Tukey's test.

value of 241 mg kg^{-1}. At the high P level, Fe concentration varied from 118 to 352 mg kg^{-1}, with a mean value of 192 mg kg^{-1}. Fe concentration was maximal in the plant tissue of cover crops as compared with other micronutrients. This may be related to higher Fe content in the Brazilian Oxisols. Fageria and Baligar (2005) reported that average values of Fe were 116 mg kg^{-1} of 200 soil samples collected from six states covering Cerrado region of Brazil with Oxisols. Overall, Fe concentration decreased with increasing P levels in the soil. Decrease was 54% at the medium P level and 93% at the high P level. Fe uptake is reported to be decreased with the addition of P in the growth medium (Follett et al., 1981; Fageria, 2009). Specific absorption rate of Fe decreased with increasing P supply due to physiological interaction of P and Fe (Fageria, 2009). The inhibition of Fe uptake by P may be related to its competing with the roots for Fe^{2+} and interfering with the reduction of Fe^{3+} in solution (Chaney and Coulombe, 1982). However, Fe uptake in the shoots of 14 cover crops increased with increasing P levels. This increase may be related to an increase in dry matter of shoots of 14 cover crop species (Fageria et al., 2013).

5.3.5 Phosphorus versus Boron

There are limited data on the interactions between P and B in crop plants. However, Bingham and Garber (1960) and Bingham and Martin (1956) reported that the application of P to soils in southern California resulted in a lower availability of B, especially in acid soils, as measured by plant uptake

TABLE 5.26
Influence of P on Mn Concentration (mg kg⁻¹) in the Shoots of 14 Tropical Legume Cover Crops

	P Fertilizer Rate (mg P kg^{-1})		
Cover Crops	**0**	**100**	**200**
Crotalaria breviflora	63.96abc	45.70d	54.36e
Crotalaria juncea L.	80.95abc	87.41bcd	110.59bc
Crotalaria mucronata	114.20a	127.87abc	126.45ab
Crotalaria spectabilis Roth	73.38abc	84.38bcd	89.26bcde
Crotalaria ochroleuca G. Don	55.02bc	67.38cd	66.73de
Calopogonium mucunoides	52.40c	73.38cd	54.00e
Pueraria phaseoloides Roxb.	112.64ab	111.38abcd	68.71cde
Cajanus cajan L. Millspaugh	86.66abc	130.93abc	98.23bcd
Cajanus cajan L. Millspaugh	87.32abc	126.68abc	82.53cde
Dolichos lablab L.	64.57abc	91.46bcd	104.95bcd
Mucuna deeringiana (Bort) Merr.	117.86a	172.01a	155.06a
Mucuna aterrima (Piper & Tracy) Holland	61.94abc	103.22abcd	100.83bcd
Mucuna cinereum L.	72.67abc	151.88ab	158.11a
Canavalia ensiformis L. DC.	29.54a	47.45d	46.58e
Average	76.65	101.51	94.03
F-test			
P levels (P)	*		
Cover crops (CC)	**		
P × CC	**		
CV(%)	20.88		

Source: Fageria, N.K. et al., *Commun. Soil Sci. Plant Anal.*, 44, 3340, 2013.
**Significant at the 1% probability level. Mean followed by the same letters within the same column is not significant at the 5% probability level by Tukey's test.

and water extractability. However, Gupta (1993) reported that there is no information on P × B interaction. P may reduce, enhance, or have no effect on B uptake.

5.3.6 Phosphorus versus Molybdenum

P has been reported to strongly compete with molybdate anion (MoO_4^{2-}) for adsorption sites, thereby resulting in increased desorption of the latter (Barrow, 1973; Xie et al., 1993; Bolan et al., 2005). Neunhauserer et al. (2001) reported that the value of P fertilizers in the phytoremediation of Mo^- contaminated soils through the greatly enhanced solubilization of Mo, and this facilitated its removal through plant uptake. However, Potash and Phosphate Institute (1999) reported that P application improved Mo uptake in acid soils but decreased in alkaline soils. Increase in Mo with acidic soils is believed to be the result of enhanced absorption and translocation due to the $H_2PO_4^-$ ion (Potash and Phosphate Institute, 1999).

5.3.7 Phosphorus versus Chlorine

Interaction between P and Cl^- was reported to be positive, negative, or not affected (Fixen, 1993). However, Gausman et al. (1958) reported that an optimum or critical level of Cl^- existed for maximum P uptake to occur, with uptake decreasing on either side of this level. In their studies, the optimum

TABLE 5.27
Influence of P on Mn Uptake (μg Plant^{-1}) in the Shoots of 14 Tropical Legume Cover Crops

	P Fertilizer Rate (mg P kg^{-1})		
Cover Crops	**0**	**100**	**200**
Crotalaria breviflora	8.19e	24.62f	38.19e
Crotalaria juncea L.	103.72c	353.83cdef	536.69bcde
Crotalaria mucronata	23.29cde	98.67def	96.64cde
Crotalaria spectabilis Roth	22.88cde	84.53def	122.26cde
Crotalaria ochroleuca G. Don	16.71de	91.47def	157.37cde
Calopogonium mucunoides	13.69e	67.41ef	13.86e
Pueraria phaseoloides Roxb.	17.66de	80.31def	70.20de
Cajanus cajan L. Millspaugh	78.30cde	281.71cdef	215.65bcde
Cajanus cajan L. Millspaugh	34.91cde	203.55cdef	146.47cde
Dolichos lablab L.	62.77cde	466.84bcde	665.95bcd
Mucuna deeringiana (Bort) Merr.	268.96b	783.39ab	830.77ab
Mucuna aterrima (Piper & Tracy) Holland	97.82cd	588.75abc	698.08bc
Mucuna cinereum L.	360.21a	910.05a	1440.76a
Canavalia ensiformis L. DC.	205.16b	482.31bcd	481.60bcde
Average	93.88	322.67	393.89
F-test			
P levels (P)	**		
Cover crops (CC)	**		
P × CC	**		
CV(%)	43.21		

Source: Fageria, N.K. et al., *Commun. Soil Sci. Plant Anal.*, 44, 3340, 2013.
**Significant at the 1% probability level. Mean followed by the same letters within the same column is not significant at the 5% probability level by Tukey's test.

level is reported to be 100 mg Cl^- kg^{-1} soil for potato (*Solanum tuberosum*). Reduction in P uptake at the Cl^- levels exceeding the optimum was attributed to anion competition (Fixen, 1993).

5.3.8 Phosphorus versus Nickel

Limited data are available on P × Ni interactions in crop plants. However, corn grown in calcareous soil with Ni application decreased P concentration (Karimian, 1995). In wheat, application of P fertilizer decreased the accumulation of Ni (Mishra and Kar, 1971).

5.3.9 Phosphorus versus Soil Salinity

Salt-affected soils can be defined as those soils that have been adversely modified for the growth of most crop plants by the presence of soluble salts, with or without high amounts of exchangeable sodium (Soil Science Society of America, 2008). Common ions contributing to this problem are Ca^{2+}, Mg^{2+}, Cl^-, Na^+, SO_4^{2-}, and HCO_3^- and in some cases K^+ and NO_3^- (Bernstein, 1975). Salt-affected soils limit crop production around the world. Civilizations have been destroyed by the encroachment of salinity on the soils; as a result, vast areas of the land are rendered unfit for agriculture. Salt-affected soils are found in many regions of the world (Fageria et al., 2011b). Salt-affected soils normally occur in arid and semiarid regions where rainfall is insufficient to

TABLE 5.28
Influence of P on Fe Concentration (mg kg⁻¹) in the Shoots of 14 Tropical Legume Cover Crops

	P Fertilizer Rate (mg P kg^{-1})		
Cover Crops	**0**	**100**	**200**
Crotalaria breviflora	1384.06a	116.65c	175.42ab
Crotalaria juncea L.	218.17b	120.25c	119.37b
Crotalaria mucronata	136.76b	118.60c	134.09ab
Crotalaria spectabilis Roth	186.03b	106.76c	296.71ab
Crotalaria ochroleuca G. Don	341.42b	140.21c	118.17b
Calopogonium mucunoides	642.50ab	716.65a	352.14a
Pueraria phaseoloides Roxb.	715.88ab	633.39ab	239.54ab
Cajanus cajan L. Millspaugh	178.18b	214.20c	125.16b
Cajanus cajan L. Millspaugh	189.16b	181.48c	163.75ab
Dolichos lablab L.	464.36b	206.30c	232.02ab
Mucuna deeringiana (Bort) Merr.	178.03b	248.86bc	150.55ab
Mucuna aterrima (Piper & Tracy) Holland	197.59b	167.11c	134.52ab
Mucuna cinereum L.	200.55b	201.48c	197.02ab
Canavalia ensiformis L. DC.	170.56b	209.01	255.45ab
Average	371.66	241.50	192.42
F-test			
P levels (P)	*		
Cover crops (CC)	**		
P × CC	**		
CV(%)	64.76		

Source: Fageria, N.K. et al., *Commun. Soil Sci. Plant Anal.*, 44, 3340, 2013.
**Significant at the 1% probability level. Mean followed by the same letters within the same column is not significant at the 5% probability level by Tukey's test.

leach salts from the root zone. Salt problems, however, are not restricted to arid or semiarid regions. They can develop even in subhumid and humid regions under appropriate conditions (Bohn et al., 1979). In addition, these soils may also occur in coastal areas subject to tides. Salts generally originate from native soil and irrigation water. Roughly 263 million ha is the irrigated area worldwide, and in most of that area, salinity is a growing threat (Epstein and Bloom, 2005). Irrigated area represents about 20% of the total land used for crop production (Fageria, 1992). This represents about 19% of the total area of the world under crop production. Use of inappropriate levels of fertilizers with inadequate management practices can create saline conditions even in humid conditions.

In the salt-affected environment, there is a preponderance of nonessential elements over essential elements. In the salt-affected soils, plants must absorb the essential nutrients from a diluted source in the presence of highly concentrated nonessential nutrients. This requires extra energy, and plants sometimes are unable to fulfill their nutritional requirements. There are two main stresses imposed by salinity on plant growth. One is water stress imposed by the increase in osmotic potential of the rhizosphere as a result of high salt concentration. Another stress is toxic effect of high concentration of ions. Hale and Orcutt (1987) reported that if the salt concentration is high enough to lower the water potential by 0.05–0.1 MPa, then plant is under salt stress. If the salt concentration is not this high, the stress is ion stress and may be caused by one particular species of ion (Hale and Orcutt, 1987).

TABLE 5.29
Influence of P on Fe Uptake (μg Plant⁻¹) in the Shoots of 14 Tropical Legume Cover Crops

	P Fertilizer Rate (mg P kg⁻¹)		
Cover Crops	**0**	**100**	**200**
Crotalaria breviflora	172.12b	63.63b	112.87a
Crotalaria juncea L.	1662.66b	13,813.22a	6,823.21a
Crotalaria mucronata	27.94b	91.26b	105.54a
Crotalaria spectabilis Roth	55.47b	106.84b	420.70a
Crotalaria ochroleuca G. Don	103.53b	190.15b	280.01a
Calopogonium mucunoides	141.11b	662.48b	90.42a
Pueraria phaseoloides Roxb.	114.30b	428.48b	245.79a
Cajanus cajan L. Millspaugh	1410.29b	5,285.04ab	1,587.00a
Cajanus cajan L. Millspaugh	74.05b	285.74b	286.58a
Dolichos lablab L.	1200.81b	16,899.73a	11,782.02a
Mucuna deeringiana (Bort) Merr.	387.08b	1,126.44b	805.65a
Mucuna aterrima (Piper & Tracy) Holland	309.55b	949.58b	930.88a
Mucuna cinereum L.	7515.34a	14,525.86a	27,640.48a
Canavalia ensiformis L. DC.	6961.62a	14,362.68a	9,863.09a
Average	1438.28	4,913.65	4,355.30
F-test			
P levels (P)	**		
Cover crops (CC)	**		
P × CC	**		
CV(%)	35.54		

Source: Fageria, N.K. et al., *Commun. Soil Sci. Plant Anal.*, 44, 3340, 2013.
**Significant at the 1% probability level. Mean followed by the same letters within the same column is not significant at the 5% probability level by Tukey's test.

Interaction between salinity and mineral nutrition is very complex because it is influenced by plant species and genotypes within species, plant age, composition and level of salinity, concentration of nutrients in the substrate, and climatic conditions. Salt stress reduces plant growth including leaf area, which reduces photosynthetic process and nutrient use efficiency. Grattan and Grieve (1992) reported that nutrient acquisition by plants can be disrupted by excessive ions in solution either via direct ionic competition between ions or by the decreased osmotic potential of the solution reducing the mass flow of mineral nutrients to the root. Munns (2002) and Bernstein et al. (1995) reported that specific ion effect on plant growth in salt-affected soils induces mineral deficiency. Uptake and utilization of mineral nutrients by plants are adversely affected under high salt concentration of soil or nutrient solution (Pessarrakali and Tucker, 1985).

Nutrient imbalances or interactions in salt-affected soils may occur in various ways. Imbalances may result from the effect of salinity on nutrient availability, competitive uptake, transport or partitioning within the plant, or by physiological inactivation of a given nutrient resulting in an increase in the plant's internal requirement for that essential element (Grattan and Grieve, 1994). Two or more of these processes may occur simultaneously, but whether they ultimately affect crop yield or quality depends upon the salinity level, composition of salts, crop species, nutrient in question, and a number of environmental factors (Grattan and Grieve, 1999).

Champagnol (1979) reviewed 17 publications and reported that P, added to saline soils, increased crop growth and yield in 34 of the 37 crops studied. Similar to the effect of added N, added P did not necessarily increase crop salt tolerance (Grattan and Grieve, 1999). In most cases, salinity decreases

TABLE 5.30
Influence of Soil Salinity on the Concentration of P and K in Two Lowland Rice Genotypes

	Genotype CNA 810098		Genotype CNA 810162[a]	
Salinity Level (ds m^{-1} at 25°C)	P conc. (g kg^{-1})	K conc. (g kg^{-1})	P conc. (g kg^{-1})	K conc. (g kg^{-1})
0.29 (control)	2.9	35.5	2.5	34.0
5	2.7	32.5	2.0	32.8
10	2.4	25.7	2.1	24.3
15	1.5	22.5	—	—

Source: Fageria, N.K., *Plant Soil*, 88, 237, 1985.
[a] Genotype CNA 810162 did not produce sufficient dry matter at the highest salinity level.

the concentration of P in plant tissue (Sharpley et al., 1992), but other studies indicate salinity either increased or had no effect on P uptake (Grattan and Grieve, 1999). Plant environmental conditions, crop species, and genotypes within species have a large role in P accumulation (Grattan and Grieve, 1994). P uptake in plants decreased with increasing salt concentration in lowland rice (Table 5.30). In genotype CNA 810098, the decrease was 93% at 15 ds m^{-1} salinity level as compared to the control treatment. Similarly, in genotype CNA 810162, the decrease was only 19% at the 10 ds m^{-1} salinity level as compared to control. Effect of salt stress on P uptake varies with plant genotypes and external salinity concentration in the growth medium.

Phosphate availability is reduced in saline soils not only because of ionic strength effects that reduce the activity of P but also because of P concentrations in soil solution that are highly controlled by sorption processes and by the low solubility of Ca–P minerals. Phosphate concentration in field-grown agronomic crops decreased as salinity ($NaCl + CaCl_2$) increased (Sharpley et al., 1992). In many cases, tissue P concentration was reduced between 20% and 50%; however, there was no evidence of P deficiency in the crop (Grattan and Grieve, 1999). Due to reduction in P uptake in salt-affected soils, P requirement of certain crops may increase.

Physiological and biochemical reasons were presented for P uptake reduction in salt-affected soils. Hale and Orcutt (1987) reported that the decrease in P uptake with the addition of salts in the plant growth medium is associated with a decrease in activity of adenosine triphosphate (ATP) coenzyme that supplies energy for P uptake. Hale and Orcutt (1987) reported that salinity damages mechanisms controlling intracellular P concentrations. In kidney bean, a decrease occurred in P, ATP, and energy available for the young leaves with the addition of salts in the rooting medium (Maas and Nieman, 1978). In addition, salt-stressed plants often appeared as P-deficient plants with small, dark green leaves, decreased shoot to root ratios, decreased tillering in cereals, delayed flowering, and reduced grain size in cereals and legumes (Hewitt, 1963).

5.4 CONCLUSIONS

Balanced supply of essential plant nutrients is important for crop production in most cropping systems. Recognition of the nutrient balance in crop production is an indirect reflection of the contribution of interactions to yield. Due to a large use of N and P in crop production, the P × N interactions are probably economically the most important of all interactions involving P. Nutrient interactions in crop plants can be measured by yield response or nutrient uptake. Nutrient interactions may be positive, negative, or neutral. When yield increase with the addition of two nutrients is higher as compared to individual one, the interaction is known as positive or synergistic. However, when the combined effects of two nutrients on yield are less than the sum of their individual, the interaction is negative or antagonistic. When there is no significant change in the yield with the addition of two nutrients or their individual values, the interaction is neutral or there is no interaction.

Phosphorus interaction with other nutrients is more important when concentrations of P and other nutrients are at low levels or at critical levels in the growth medium because this situation may create deficiency of one or another due to low availability in the growth medium. Nutrient interactions may occur in the rhizosphere and/or in the plant uptake process involving adsorption, precipitation, absorption, or translocation. Major interactions of P with macro- and micronutrients are positive or synergistic. However, there may be some exception in both the group of nutrients. In most cases, salinity decreases the concentration of P in plant tissue, but results of some studies indicate salinity either increased or had no effect on P uptake. Plant environmental conditions, crop species, and genotypes within species have a large role in P accumulation. P interactions with other nutrients were conducted under controlled conditions in most studies. Field data on nutrient interactions are still not sufficient for many crop species, and more research is needed to reach a definitive conclusion.

REFERENCES

Abbasi, M. K., M. M. Tahir, W. Azam, Z. Abbas, and N. Rahim. 2012. Soybean yield and chemical composition in response to phosphorus-potassium nutrition in Kashmir. *Agron. J.* 104:1476–1484.

Adams, F. 1980. Interactions of phosphorus with other elements in soil and plants. In: *The Role of Phosphorus in Agriculture*, ed. R. C. Dinauer, pp. 655–680. Madison, WI: ASA, CSSA, and SSSA.

Agbim, N. N. 1981. Interactions of phosphorus, magnesium and zinc on the yield and nutrient content of maize. *J. Agric. Sci. Camb.* 96:509–514.

Ajakaiye, C. O. 1979. Effect of phosphorus on growth and iron nutrition of millet and sorghum. *Plant Soil* 51:551–561.

Amanuel, G., R. F. Kuhne, D. G. Tanner, and P. L. G. Vlek. 2000. Biological nitrogen fixation in faba bean (*Vicia faba* L.) in the Ethiopian highlands as affected by P fertilization and inoculation. *Biol. Fertil. Soils* 32:353–359.

Aulakh, M. S. and S. S. Malhi. 2005. Interactions of nitrogen with other nutrients and water: Effect on crop yield and quality, nutrient use efficiency, carbon sequestration, and environmental pollution. *Adv. Agron.* 86:341–409.

Austin, R. B., J. Bingham, R. D. Blackwell, L. T. Evans, M. A. Ford, C. L. Morgan, and M. Taylo. 1980. Genetic improvements in winter wheat yields since 1900 and associated physiological changes. *J. Agric. Sci.* 94:675–689.

Baligar, V. C., N. K. Fageria, and Z. L. He. 2001. Nutrient use efficiency in plants. *Commun. Soil Sci. Plant Anal.* 32:921–950.

Barrow, N. J. 1973. On the displacement of adsorbed anions from soil; Displacement of molybdenum by phosphate and by hydroxide. *Soil Sci.* 116:423–451.

Bennett, W. F., J. Pesek, and J. Hanway. 1962. Effect of nitrogen on phosphorus absorption by corn. *Agron. J.* 54:437–442.

Bernstein, L. 1975. Effects of salinity and sodicity on plant growth. *Annu. Rev. Phytopathol.* 13:295–312.

Bernstein, N., W. K. Silk, and A. Lauchli. 1995. Growth and development of sorghum leaves under conditions of NaCl stress. *Planta* 191:433–439.

Bingham, F. T. 1963. Relation between phosphorus and micronutrients in plants. *Soil Sci. Soc. Am. Proc.* 27:389–391.

Bingham, F. T. and M. J. Garber. 1960. Solubility and availability of micronutrients in relation to phosphorus fertilization. *Soil Sci. Soc. Am. Proc.* 24:209–213.

Bingham, F. T. and J. P. Martin. 1956. Effects of soil phosphorus on growth and minor element nutrition of citrus. *Soil Sci. Soc. Am. Proc.* 20:382–385.

Black, C. A. 1993. *Soil Fertility Evaluation and Control.* Boca Raton, FL: CRC Press.

Boawn, L. C., F. G. Viets, and C. L. Crawford. 1954. Effect of phosphate fertilizer on zinc nutrition of field beans. *Soil Sci.* 78:1–7.

Bohn, H., B. McNeal, and G. O'Connor. 1979. *Soil Chemistry.* New York: John Wiley & Sons.

Bolan, N. S., D. C. Adriano, R. Naidu, M. L. Mora, and M. Santiago. 2005. Phosphorus-trace element interactions in soil-plant systems. In: *Phosphorus: Agriculture and the Environment*, ed. L. K. Al-Amoodi, pp. 317–352. Madison, WI: ASA, CSSA, and SSSA.

Cakmak, I. and H. Marschner. 1987. Mechanisms of phosphate-induced zinc deficiency in cotton III. Changes in physiological availability of zinc in plants. *Physiol. Plant* 70:13–20.

Caldwell, C. R. and A. Huag. 1981. Temperature dependence of the barley root plasma membrane-bound Ca^{2+} and Mg^{2+} dependent ATPase. *Physiol. Plant.* 53:117–125.

Champagnol, F. 1979. Relationships between phosphate nutrition of plants and salt toxicity. *Phosphorus Agric.* 76:35–43.

Chaney, R. L. and B. A. Coulombe. 1982. Effect of phosphate on regulation of Fe-stress response in soybean and peanut. *J. Plant Nutr.* 5:469–487.

Chao, T. T., M. E. Harward, and S. C. Fang. 1962. Movement of S tagged sulfate through soil columns. *Soil Sci. Soc. Am. Proc.* 26:27–32.

Darst, B. C. and G. W. Wallingford. 1985. Interrelationships of potassium with cultural and management practices. In: *Potassium in Agriculture*, ed. R. D. Munson, pp. 559–573. Madison, WI: ASA, CSSA, and SSSA.

Dibb, D. W. and W. R. Thompson, Jr. 1985. Interaction of potassium with other nutrients. In: *Potassium in Agriculture*, ed. R. D. Munson, pp. 515–533. Madison, WI: ASA, CSSA, and SSSA.

Dwivedi, B. S., A. K. Shukla, V. K. Singh, and R. L. Yadav. 2003. Improving nitrogen and phosphorus use efficiencies through inclusion of forage cowpea in the rice-wheat systems in the Indo-Gangetic plains of India. *Field Crops Res.* 80:167–193.

Edwards, D. G. 1968. Cation effects on phosphate absorption from solution by *Trifolium subterraneum. Aust. J. Biol. Sci.* 21:1–11.

Ellis, R., J. F. Davis, and D. L. Thurlow. 1964. Zinc availability in calcareous Michigan soils as influenced by phosphorus level and temperature. *Soil Sci. Soc. Am. Proc.* 28:83–86.

Epstein, E. and A. J. Bloom. 2005. *Mineral Nutrition of Plants: Principles and Perspectives*, 2nd edn. Sunderland, MA: Sinauer Associates Inc. Publishers.

Fageria, N. K. 1984. *Fertilization and Mineral Nutrition of Rice*. Brasilia, Brazil: EMBRAPA/Editora Campus.

Fageria, N. K. 1985. Salt tolerance of rice cultivars. *Plant Soil* 88:237–243.

Fageria, N. K. 1989. *Tropical Soils and Physiological Aspects of Crops*. Brasilia, Brazil: EMBRAPA.

Fageria, N. K. 1992. *Maximizing Crop Yields*. New York: Marcel Dekker.

Fageria, N. K. 2009. *The Use of Nutrients in Crop Plants*. Boca Raton, FL: CRC Press.

Fageria, N. K. 2014. *Mineral Nutrition of Rice*. Boca Raton, FL: CRC Press.

Fageria, N. K. and V. C. Baligar. 1997. Response of common bean, upland rice, corn, wheat, and soybean to soil fertility of an Oxisol. *J. Plant Nutr.* 20:1279–1289.

Fageria, N. K. and V. C. Baligar. 2001. Improving nutrient use efficiency of annual crops in Brazilian acid soils for sustainable crop production. *Commun. Soil Sci. Plant Anal.* 32:1303–1319.

Fageria, N. K. and V. C. Baligar. 2005. Enhancing nitrogen use efficiency in crop plants. *Adv. Agron.* 88:97–185.

Fageria, N. K. and V. C. Baligar. 2008. Ameliorating soil acidity of tropical Oxisols by liming for sustainable crop production. *Adv. Agron.* 99:345–399.

Fageria, N. K., V. C. Baligar, and R. B. Clark. 2002. Micronutrients in crop production. *Adv. Agron.* 77:189–272.

Fageria, N. K., V. C. Baligar, and C. A. Jones. 2011a. *Growth and Mineral Nutrition of Field Crops*, 3rd edn. Boca Raton, FL: CRC Press.

Fageria, N. K., V. C. Baligar, A. Moreira, and L. C. A. Moraes. 2013. Soil phosphorus influence on growth and nutrition of tropical legume crops in acid soil. *Commun. Soil Sci. Plant Anal.* 44:3340–3361.

Fageria, N. K. and M. P. Barbosa Filho. 1994. Nutritional deficiency in rice and their correction. EMBRAPA/CNPAF Document No. 42.

Fageria, N. K., H. R. Gheyi, and A. Moreira. 2011b. Nutrient bioavailability in salt affected soils. *J. Plant Nutr.* 34:945–962.

Fageria, N. K., A. Moreira, L. C. A. Moraes, and M. F. Moraes. 2014a. Root growth, nutrient uptake, and nutrient use efficiency by roots of tropical legume cover crops as influenced by phosphorus fertilization. *Commun. Soil Sci. Plant Anal.* 45:555–569.

Fageria, N. K. and J. P. Oliveira. 2014. Nitrogen, phosphorus and potassium interactions in upland rice. *J. Plant Nutr.*37:1586–1600.

Fageria, N. K. and A. B. Santos. 2008. Lowland rice response to thermophosphate fertilization. *Commun. Soil Sci. Plant Anal.* 39:873–889.

Fageria, N. K. and L. F. Stone. 2008. Micronutrient deficiency problems in South America. In: *Micronutrient Deficiencies in Global Crop Production*, ed. B. J. Allooway, pp. 245–266. New York: Springer.

Fageria, N. K., L. F. Stone, A. B. Santos, and M. C. S. Carvalho. 2014b. *Mineral Nutrition of Dry Bean*. Brasilia, Brazil: EMBRAPA/Rice and Bean Research Center.

Fageria, N. K. and F. J. P. Zimmermann. 1995. Lime and phosphorus interactions on growth and nutrient uptake by upland rice, wheat, common bean, and corn in an Oxisol. *J. Plant Nutr.* 18:2519–2532.

Fageria, N. K. and F. J. P. Zimmermann. 1998. Influence of pH on growth and nutrient uptake by crop species in an Oxisol. *Commun. Soil Sci. Plant Anal.* 29:2675–2682.

Fixen, P. E. 1993. Crop responses to chloride. *Adv. Agron.* 50:107–150.

Follett, R. H., L. S. Murphy, and R. L. Donahue. 1981. *Fertilizers and Soil Amendments.* Englewood Cliffs, NJ: Prentice-Hall.

Franklin, R. E. 1969. Effect of adsorbed cations on phosphorus uptake by excised roots. *Plant Physiol.* 44:697–700.

Gausman, H. W., C. E. Cummingham, and R. A. Struchtemeyer. 1958. Effects of chloride and sulfate on 32P uptake by potatoes. *Agron. J.* 50:90–91.

Grattan, S. R. and C. M. Grieve. 1992. Mineral element acquisition and growth response of plants grown in saline environment. *Agric. Ecosyst. Environ.* 38:275–300.

Grattan, S. R. and C. M. Grieve. 1994. Mineral nutrient acquisition and response by plants grown in saline environments. In: *Handbook of Plant and Crop Stress*, ed. M. Pessarkli, pp. 203–226. New York: Marcel Dekker.

Grattan, S. R. and C. M. Grieve. 1999. Salinity-mineral nutrient relations in horticulture crops. *Sci. Horticult.* 78:127–157.

Greenwood, E. H. N. and E. G. Hallsworth. 1960. Studies on the nutrition of forage legumes. II. Some interactions of calcium, phosphorus, copper, and molybdenum on the growth and chemical composition of *Trifolium subterraneum* L. *Plant Soil* 12:97–127.

Grunes, D. L., F. G. Viet, Jr., and S. H. Shih. 1958. Proportionate uptake of soil and fertilizer phosphorus by plants as affected by nitrogen fertilization. I. Growth chamber experiment. *Soil Sci. Soc. Am. Proc.* 22:43–48.

Gupta, U. C. 1993. Factors affecting boron uptake by plants. In: *Boron and Its Role in Crop Production*, ed. U. C. Gupta, pp. 87–123. Boca Raton, FL: CRC Press.

Hale, M. G. and D. M. Orcutt. 1987. *The Physiology of Plants under Stress.* New York: John Wiley & Sons.

Hewitt, E. J. 1963. The essential nutrient elements: Requirements and interactions in plants. In: *Plant Physiology: A Treatise*, Vol. 3, ed. F. C. Steward, pp. 137–360. New York: Academic Press.

Hoeft, R. G., L. M. Walsh, and D. R. Keeney. 1973. Evaluation of various extractants for available soil sulfur. *Soil Sci. Soc. Am. Proc.* 37:401–404.

Kaiser, D. E. and K. I. Kim. 2013. Soybean response to sulfur fertilizer applied as a broadcast or starter using replicated strip trials. *Agron. J.* 105:1189–1198.

Kamprath, E. J., W. L. Nelson, and J. W. Fitts. 1956. The effect of ph, sulfate, and phosphate concentrations on the adsorption of sulfate by soils. *Soil Sci. Soc. Am. Proc.* 20;463–466.

Karimian, N. 1995. Effect of nitrogen and phosphorus on zinc nutrition of corn in a calcareous soil. *J. Plant Nutr.* 18:2261–2271.

Kiniry, J. R., G. McCauley, Y. Xie, and J. G. Arnold. 2001. Rice parameters describing crop performance of four U.S. cultivars. *Agron. J.* 93:1354–1361.

Larsen, S. 1964. The effect of phosphate applications on manganese content of plants grown on neutral and alkaline soils. *Plant Soil* 21:37–42.

Lindsay, W. L. 1979. *Chemical Equilibrium in Soils.* New York: John Wiley & Sons.

Loneragan, J. F., T. S. Grove, A. D. Robson, and K. Snowball. 1979. Phosphorus toxicity as a factor in zinc-phosphorus interactions in plants. *Soil Sci. Soc. Am. J.* 43:966–972.

Ludlow, M. M. and R. C. Muchlow. 1990. A critical evaluation of traits for improving crop yields in water-limited environments. *Adv. Agron.* 43:107–153.

Maas, E. V. and R. H. Nieman. 1978. Physiology of plant tolerance to salinity. In: *Crop Tolerance to Suboptimal Land Conditions*, ed. G. A. Jung, pp. 277–299. Madison, WI: American Society of Agronomy.

Marschner, H. 1995. *Mineral Nutrition of Higher Plants*, 2nd edn. New York: Academic Press.

Mengel, K., E. A. Kirkby, H. Kosegarten, and T. Appel. 2001. *Principles of Plant Nutrition,* 5th edn. Dordrecht, the Netherlands: Kluwer Academic Publishers.

Mishra, D. and M. Kar. 1971. Nickel in plant growth and metabolism. *Bot. Rev.* 40:395–452.

Muller, S., L. Heinrich, and I. Weigert. 1993. Influence of differentiated phosphorus and nitrogen fertilizer application on nutrient uptake and seed yield of faba bean. *Bodenkultur* 44:127–133.

Munns, R. 2002. Comparative physiology of salt and water stress. *Plant Cell Environ.* 25:239–250.

Neunhauserer, C., M. Berreck, and H. Insam. 2001. Remediation of soils contaminated with molybdenum using soil amendments and phytoremediation. *Water Air Soil Pollut.* 128:85–96.

Page, E. R., E. K. Schofield-Palmer, and A. J. MacGregor. 1963. Studies in soil and plant manganese. IV. Superphosphate fertilization and manganese content of young oat plants. *Plant Soil* 19:255–264.

Pan, W. L. 2012. Nutrient interactions in soil fertility and plant nutrition. In: *Handbook of Soil Sciences: Resource Management and Environmental Impacts*, 2nd edn., eds. P. M. Huang, Y. Li, and M. E. Sumner, pp. 16-1–16-11. Boca Raton, FL: CRC Press.

Pederson, G. A., G. F. Brink, and T. E. Fairbrother. 2002. Nutrient uptake in plant parts of sixteen forages fertilized with poultry litter: Nitrogen, phosphorus, potassium, copper, and zinc. *Agron. J.* 94:895–904.

Pessarrakali, M. and T. C. Tucker. 1985. Uptake of nitrogen-15 by cotton under salt stress. *Soil Sci. Soc. Am. J.* 49:149–152.

Potash and Phosphate Institute. 1999. Phosphorus interactions with other nutrients. *Better Crops* 83:11–13.

Racz, G. J. and Haluschak. 1974. Effects of phosphorus on Cu, Zn, Fe, and Mn utilization by wheat. *Can. J. Soil Sci.* 54:357–367.

Robinson, A. D. and J. B. Pitman. 1983. Interactions between nutrients in higher plants. In: *Inorganic Plant Nutrition: Encyclopedia of Plant Physiology*, vol. 1A, eds. A. Lauchli and R. L. Bieleski, pp. 147–180. New York: Springer-Verlag.

Robson, A. D., D. G. Edwards, and J. F. Loneragan. 1970. Calcium stimulation of phosphate absorption by annual legumes. *Aust. J. Agric. Res.* 21:601–612.

Satyanarayana, T., V. P. Badanur, and G. V. Havanagi. 1978. Nitrogen and phosphorus fertilization of ragi on red soils of Bangalore. *Indian J. Agron.* 23:37–39.

Schulthess, U., B. Feil, and S. C. Jutzi. 1997. Yield-independent variation in grain nitrogen and phosphorus concentration among Ethiopian wheats. *Agron. J.* 89:497–506.

Sharma, P. K. and H. L. S. Tandon. 1992. The interaction between nitrogen and phosphorus in crop production. In: *Management of Nutrient Interactions in Agriculture*, ed. H. L. S. Tandon, pp. 1–20. New Delhi, India: Fertilizer Development and Consultation Organization.

Sharma, R. C. and E. L. Smith. 1986. Selection for high and low harvest index in three winter wheat populations. *Crop Sci.* 26:1147–1150.

Sharpley, A. N., J. J. Meisinger, J. F. Power, and D. L. Suarez. 1992. Root extraction of nutrients associated with long-term soil management. *Adv. Soil Sci.* 19:151–217.

Singh, A. K. 1991. Response of pre-flood, early season maize to graded levels of nitrogen and phosphorus in *Ganga diara* tract of Bihar. *Indian J. Agron.* 36:508–510.

Smilde, K. W. 1973. Phosphorus and micronutrient metal uptake by some tree species as affected by phosphate and lime applied to an acid sandy soil. *Plant Soil* 39:131–148.

Soil Science Society of America. 2008. *Glossary of Soil Science Terms*. Madison, WI: SSSA.

Soltanpour, P. N. 1969. Effect of nitrogen, phosphorus, and zinc placement on yield and composition of potatoes. *Agron. J.* 61:288–289.

Srinivas, K. and J. V. Rao. 1984. Response of French bean to N and P fertilization. *Indian J. Agron.* 29:146–149.

Sumner, M. E. and M. P. W. Farina. 1986. Phosphorus interactions with other nutrients and lime in field cropping systems. *Adv. Soil Sci.* 5:201–236.

Terman, G. L., J. C. Noggle, and C. M. Hunt. 1977. Growth rate-nutrient concentration relationships during early growth of corn as affected by applied N, P, and K. *Soil Sci. Soc. Am. J.* 41:363–368.

Truog, E., R. J. Goates, C. G. Gerloff, and K. C. Berger. 1947. Magnesium-phosphorus relationships in plant nutrition. *Soil Sci.* 63:19–25.

Viets, F. Jr. 1944. Calcium and other polyvalent cations as accelerators of ion accumulation by excised barley roots. *Plant Physiol.* 19:466–480.

Wallace, A. 1990. Crop improvement through multi-disciplinary approaches to different types of stresses: Law of maximum. *J. Plant Nutr.* 13:313–325.

Wallace, A., A. Elgazzar, and G. V. Alexander. 1973. Phosphorus levels on zinc and other heavy metal concentrations in Hawkeye and P154619-5-1 soybeans. *Commun. Soil Sci. Plant Anal.* 4:343–345.

Wilkinson, S. R., D. L. Grunes, and M. E. Sumner. 2000. Nutrient interactions in soil and plant nutrition. In: *Handbook of Soil Science*, ed. M. E. Sumner, pp. D-89–D-111. Boca Raton, FL: CRC Press.

Wilson, J. B. 1993. Macronutrient (NPK) toxicity and interactions in the grass *Festuca ovina*. *J. Plant Nutr.* 16:1151–1159.

Xie, R. J., A. F. Mackezie, and Z. J. Lou. 1993. Casual-modeling ph and phosphate effects on molybdate sorption in 3 temperature soils. *Soil Sci.* 155:385–397.

Zhang, F. S., J. Shen, and Y. G. Zhu. 2006. Nutrient interactions in soil-plant system. In: *Encyclopedia of Soil Science*, ed. R. Lal, pp. 1153–1156. New York: John Wiley & Sons.

6 Phosphorus and the Environment

6.1 INTRODUCTION

Food crop yield needs to be increased not only to feed the growing world population but also to avoid hunger and malnutrition. In addition, an increase in crop yields in the future also demands that good quality be produced at a sustainable rate with lower costs and clean environment. Improvement in crop yields in the future will be challenging for the world's farming community and agricultural scientists. An increase in crop yield can be accomplished by increasing land area, cropping intensity, and yield per unit area. Increase in land area is possible only in Africa and South America due to the availability of new land areas. Furthermore, all the three options require more water and fertilizer inputs to achieve the desirable goal.

Phosphorus is essential for all organisms (Cade-Menum and Liu, 2014). Commercial P fertilizers are primarily derived or produced from rock phosphate, a resource with finite stocks (Cordell et al., 2009). Soils that are low in available P require the addition of P fertilizers to produce maximum economic yields (Richardson et al., 2011). In contrast, intensive agriculture and overfertilization, particularly with manure, have produced high soil P concentrations in some regions, beyond what is needed for plant growth and exceeding the soil P storage capacity (Jarvie et al., 2013). This excess P may be lost from soils to water bodies in runoff and erosion, where it can enhance growth of nuisance algae (Elser and Bennett, 2011; Jarvie et al., 2013; Cade-Menum and Liu, 2014). Managing P under conditions from deficiency to excess requires detailed information about P concentration and P speciation because the chemical forms of P determine their bioavailability and environmental reactivity (Contron et al., 2005; Pierzynski et al., 2005; Condron and Newman, 2011). Levels of P in surface water greater than 10 ppb (10 microgram P L^{-1}) have been associated with enhanced algae growth in streams and lakes (Foth and Ellis, 1988).

Phosphorus originating from agricultural lands has long been recognized as a surface water pollutant (Sims and Kleinman, 2005). Phosphorus loss from agricultural lands via hydrological processes, that is, erosion, interflow, overland flow, matrix flow, preferential flow to water bodies, has been reported (Haygarth and Sharpley, 2000; Heathwaite et al., 2000). As P controls eutrophication of most freshwater systems (Foy, 2005), and even some estuarine systems (National Research Council, 2000), preventing agricultural nonpoint P pollution is now a worldwide environmental priority. Sims and Kleinman (2005) reported that currently eutrophication persists as the most pervasive surface water impairment in the United States, with agriculture identified as a major source of P to eutrophic waters.

Agricultural P loss is a global concern due to nutrient enrichment and eutrophication in water bodies (Correll, 1998; Liu et al., 2014). For years, descriptions and predictions of soil P transport for effective environment risk management have been documented. However, most of these studies have focused on particulate and dissolved P operationally discriminated by membrane filtration (typically 0.45 µm pore size; Heckrath et al., 1995; Anderson et al., 2013). This fractionation neglects mobile colloids, particles ranging from 1 nm to 1 µm (Baalousha et al., 2005), by which P could be transported across a long distance (Liu et al., 2014). Accumulated evidence indicates that colloid-facilitated P transport is an important mechanism for P transfer from agricultural land to aquatic ecosystems (Heathwaite et al., 2005a; Ilg et al., 2005; Siemens et al., 2008; Regelink et al., 2013). Colloidal P has been reported to contribute >50% of total P in soil water samples (Hens and Merckx, 2001). Due to their large specific surface area, soil colloids tend to have high P concentrations as compared with bulk soils (Schoumans and Chardon, 2003). Colloidal P also represents a major source that contributes to algae available P in water bodies (Van Moorleghem et al., 2013).

Although there are no direct detrimental effects of P on the terrestrial environment, the continued application of fertilizer P to agricultural land can result in the buildup of natural trace contaminants contained in the fertilizer (Sharpley and Menzel, 1987). All fields are linked hydrologically by either surface or groundwater pathways to a water course. Transfer of P along rapid preferential surface flow (overland flow) and subsurface flow (drain flow) pathways influence the risk of P loss (Withers et al., 2005). Transport of P from terrestrial to aquatic environment in surface and subsurface runoff can result in a deterioration in water quality from accelerated eutrophication. Soil nutrient surveys in many countries indicate that many agricultural soils have accumulated soil P to a level that a large amount of P application is no longer necessary (Withers et al., 2005). Since fertilizer P use has been linked to nutrient enrichment of streams draining agricultural watersheds (Sharpley and Syers, 1979b; Calhoun et al., 2002), the possible environmental consequences of their indiscriminate use are significant (Sharpley et al., 1987). In addition, in recent years, there has been increasing concern over the management of P inputs in fertilizers and manures, and the excessive buildup of P fertility in soils in relation to potential adverse effects on water quality and biological diversity (Withers et al., 2005). The objective of this chapter is to discuss transport of P from the terrestrial to aquatic environments, the impact of P on the terrestrial environment, and suggesting appropriate management practices to reduce P losses from agricultural lands to the environment.

6.2 TRANSPORT OF PHOSPHORUS FROM AGRICULTURAL LANDS TO AQUATIC ENVIRONMENTS

Some terms that are commonly used in the loss of P from landscapes to water bodies are "transport," "transfer," and "loss." These terms should be defined before discussing the P loss from soil–plant systems to water bodies or environment. "Transport" is the term more traditionally used to describe the nonsource aspects of P loss from agriculture—technically, though, it refers more to P movement by flowing water after it has been mobilized. "Transfer" is a more generic term, recently proposed and accepted, that refers to the integration of P mobilization with spatial and temporal dynamics of hydrology, resulting in transport at the soil and hillslope scale, to ultimately the receiving surface waters (Gburek et al., 2005). Leaching is one example of transfer commonly used to describe the eluviation of P through soils (Weaver et al., 1988a,b; Wagenet, 1990). Finally, the term "loss," as in P loss, is often used in the context of a discipline, and typically refers to the transfer of P from one compartment to another, such as soil to water bodies (Robinson et al., 1994; Gburek and Sharpley, 1998). "Soil erosion" is a term that refers to actual transfer of soil material from one place to another and should not be confused with P loss (Gburek et al., 2005).

Haygarth and Sharpley (2000) also defined terminology for P transfer from catchment to water bodies. According to these authors, the terminology for P transfer can be defined as (1) *processes* (or modes) (i.e., erosion, leaching, incidental), (2) *pathways* (i.e., overland flow, subsurface flow, drainage flow), and (3) *form* terms (i.e., those that can be described in soil or water samples). They suggested a method of classifying pathways by scale, plane, and time, and a particular caution is noted for leaching, which is a process, not a pathway, and runoff (a vaguely defined pathway).

Phosphorus can exist in either inorganic or organic forms and can be mobilized both in soluble forms and in association with particles and colloids (Espinosa et al., 1999; Gburek et al., 2005). Transport of P from landscape to water bodies is governed by soil P status and leaching intensity (Heckrath et al., 1995), while particulate and colloid P transport is most commonly associated with soil erosion intensity (Owens and Walling, 2002; McDowell et al., 2003; Gburek et al., 2005). Mobilization or movement of P in soil–plant systems occurs by three primary processes (Haygarth and Jarvis, 1999). These processes are (1) solubilization, operationally defined as all P from analysis after filtered through a <0.45 μm membrane, where the driving mechanism is chemical nonequilibrium; (2) physical detachment of soil particles and colloids with attached P, where the driving mechanism is the force exerted by moving water; and (3) mobilization where anthropogenic sources of P (i.e., manures or fertilizers) as originally applied coincide with large flows. This type of mobilization, a more direct movement of the P source itself, has been termed incidental (Preedy et al., 2001).

Loss of P from landscape to water bodies can occur through runoff loss as either soluble or particulate P. Particulate P include P sorbed by soil particles and organic matter eroded during runoff (Sharpley and Menzel, 1987). Soil erosion is a selective process in which runoff sediment becomes enriched in finer-sized particles and lighter organic matter. Because P is strongly adsorbed on clay particles and organic matter contains relatively high levels of P, the major proportion of P transported to the aquatic environment from cultivated land is usually in the particulate form (Sharpley and Menzel, 1987). In runoff from grassland, forest, and/or sandy soils, which carry minimal suspended soil, most of the P may be transported in soluble form (Sharpley and Menzel, 1987; He et al., 2006).

In the soil environment P is subject to several soil processes that control its availability to plants and potential movement to surface waters. In the soil solution, P is present as either a monovalent ($H_2PO_4^-$, acid soils) or a divalent (HPO_4^{2-}, pH higher than 7) anion. Phosphorus enters the soil solution via (1) dissolution of primary minerals; (2) dissolution of secondary minerals; (3) desorption of P from clays, oxides, and minerals; and (4) biological conversion of P in organic materials to inorganic forms (mineralization). It should be noted that all of these processes are reversible (Wood, 1998). In most soils, soil solution P ranges between <0.01 and 1 g L^{-1}, and a value of 0.2 mg L^{-1} is commonly accepted as the solution P concentration needed to meet the nutritional needs of most agronomic crops (Wood, 1998).

After application, fertilizers and manures react with soil particles, and P sorption and desorption can occur. Processes bringing fertilizer and manure into contact with the soil can be biological (microorganisms or earthworms), physical (by water), or man-induced (incorporation by tillage operation). Once the P from fertilizers or manures has been brought into contact with soil materials, its fate in the context of potential for subsequent loss from the soil–plant system depends, to a large degree, on the chemical processes governing P concentration in the soil solution. Two important processes that govern P concentration in soil solution are designated as sorption and desorption.

6.2.1 Phosphorus Sorption

Sorption is defined as the removal of an ion or molecule from solution by adsorption and absorption (Soil Science Society of America, 2008). Sorption of P is an important phenomenon in controlling P concentration in the soil solution. Sorption determines P availability to plants as well as loss from soil–plant systems. Soils are divided into two groups, that is, having low P sorption capacity such as sandy soils and high P sorption capacity such as clay soils. Sandy soils have low surface area as compared to clay soils. Gburek et al. (2005) reviewed the literature on P sorption and reported that P saturated sandy soils in the Netherlands lost a significant amount of P by leaching. These authors reported that during the winter, about 40% of shallow groundwater samples in the catchment exceeded 1 mg P L^{-1} and leaching through the soil toward the ground water accounted for about 85%–90% of the P losses from the agricultural land in the area. Histosols are also prone to high losses in subsurface drainage due to their limited sorption capacity (Gburek et al., 2005). Izuno et al. (1991) reported that for the Everglades agricultural area in Florida, United States, high P losses occurred by leaching. In fine-textured soils P is strongly sorbed on soil colloids, a characteristic discovered as early as 1850 (Way, 1850). Main P sorbers are Fe, Al oxides, and clays.

6.2.2 Phosphorus Desorption

Desorption is defined by the Soil Science Society of America (2008) as the migration of adsorbate off the adsorption sites. It is the reverse process of adsorption. Desorption or dissolution is the first step in the transport of P from croplands to water bodies (lakes, river, streams, and ocean). Desorption of P from the soil in relation to availability and water quality has been studied using

TABLE 6.1
Phosphorus Extraction Methods to Determine Desorbed Phosphorus

Extracting Solution	Solution:Soil Ratio	Reference
0.1 M NaCl	100:1	Li et al. (1972)
	50:1	Ryden et al. (1972)
	50:1	Romkens and Nelson (1974)
0.01 M $CaCl_2$	10:1	White and Beckett (1964)
	10:1	Taylor and Kunishi (1971)
	10:1	Gardner and Jones (1973)
	5:1	Elrashidi and Larsen (1978)
	25:1	Green et al. (1978)
	50:1	Oloya and Logan (1980).
Anion exchange		
Resin (Dowex I-X4)	30:1	Ballaux and Peaslee (1975)
Resin (Dowex I-XB)	100:1	Evans and Jurinak (1976)
Distilled water	10:1–1000:1	Sharpley et al. (1981)
Filter lake water	2000:1–4000:1	Bahnick (1977)

various extracting solutions and solution: soil ratios (Table 6.1). Desorption of P from the soil colloids occurs in a short time period. Sharpley and Menzel (1987) in a review of literature reported that that 50% of P desorbed from a desert soil in 50 h occurred in the first hour of reaction, and reported that approximately 75% of the P desorbed in 4 h from several soils occurred in the initial 30 min. Consequently, P can be desorbed from surface soil by short rainfall and runoff events (Sharpley and Menzel, 1987). Desorption of soil P by rainfall runoff water is brought about by interaction with a thin layer of surface soil (1–3 mm). If the surface water percolates through the soil profile, P sorption by P-deficient subsoils generally results in low concentrations of soluble P in subsurface flow. Exceptions may occur in organic or peat soils, where organic matter may accelerate the downward movement of P together with organic acids and Fe and Al (Sharpley and Menzel, 1987). Phosphorus is more mobile or desorbed more in sandy soils that have low cation exchange capacity as compared to soils with high clay content.

6.2.3 Phosphorus Leaching

Leaching is defined as the removal of soluble materials from one zone in soil to another via water movement in the profile (Soil Science Society of America, 2008). "Leaching" is a particularly ambiguous term that is associated with the soil profile scale and does not describe a pathway, although it has occasionally been used in such context to describe an amalgam of all pathways of water drainage through soil (Bromfield and Jones, 1972; Jordan and Smith, 1985; Heckrath et al., 1995; Haygarth and Jarvis, 1999). "Leaching" is therefore a process term, describing the eluviation of solutes, such as P, down through soil, especially in freely draining coarse textured soils (Weaver et al., 1988a,b; Wagenet, 1990).

Transport of P from the growing croplands to water bodies can occur through leaching. This type of transport varies from crop to crop and watershed to watershed. Rainfall intensity and soil properties also determine the quantity of P leached from the soil profile and transported to water streams. Increased losses of soluble P in runoff from alfalfa (*Medicago sativa*) plots were 33 g P ha^{-1}, as compared to forests 4 g P ha^{-1}, oats (*Avena fatua*) 16 g P ha^{-1}, and corn (*Zea mays*) plots 11 g P ha^{-1} (Sharpley and Menzel, 1987).

Phosphorus leaching losses are significant and larger than originally believed (Heckrath et al., 1995; Sims et al., 1998b), particularly in soil where continuous application of organic wastes or manure has raised the soil to an excess P content and decreased its P sorption capacity (Mozaffari and Sims, 1996). Long-term application of inorganic fertilizers, manures, poultry litters, swine effluents, or municipal wastes results in P accumulation in the upper soil layer and can induce significant P leaching in subplot layer horizons as indicated by increases in soil P tests from these locations (Sims et al., 1998a). Leinweber et al. (1999) measured soil test P, P saturation capacity, and degree of P sorption saturation for 20 differently managed soils and compared the results with leaching losses with lysimeters. Sandy soils under grass receiving organic fertilizers or farmyard manures had the highest P losses. Elliott et al. (2002) measured P leaching in three acidic soils low in organic matter and with a gradient in Fe and Al oxides and reported that increasing quantities of P in column leachates occurred with decreasing Fe and Al oxides. Clay soils that are susceptible to cracking and preferential flow can have significant P leaching (Stamm et al., 1998; Jensen et al., 2000), especially during rain after a dry period (Beauchemin et al., 1998).

Phosphorus concentration in fertilizers or organic manures can influence P leaching. Sharpley and Moyer (2000) measured the amounts of inorganic and organic P in leachates obtained with simulated rainfall on several manures (dairy, poultry, and swine) and composts (dairy and poultry). Proportion of P lost in leachates followed the order of dairy manure > poultry manure > poultry compost > dairy compost = swine slurry, and most P in all leachate was present in inorganic forms. Potential for P to be leached was closely related to water extractable inorganic P concentration of the respective material. Siddique et al. (2000), in leaching trials on soil columns, reported higher amounts of P to be leached from inorganic P fertilizers than from sludge-treated soils. The amount of P leached was related by a curvilinear relation to the degree of P sorption saturation values at all depths in all leached soil than did sludge. Elliott et al. (2002) obtained similar results using triple superphosphate and eight biosolids.

6.2.4 Detachment and Transport of Particulate Phosphorus

Detachment and transport of particulate P mainly occur through wind and water erosion of the surface soil. Primary source of sediment in watersheds with a permanent vegetative cover, such as forest or pasture, is from streambank erosion (Sharpley and Menzel, 1987). The degree of enrichment of P in runoff sediment due to the preferential transport of fine-sized particles and lighter organic matters is expressed as an enrichment ratio (Soil Science Society of America, 2008). For P, this is calculated as the ratio of the concentration of P in the runoff sediment to that in the source soil. Enrichment ratio values of 1.3 for total P and 3.3 for 0.001 M H_2SO_4 extractable P for a silt loam situated on a 20%–25% slope were reported by Sharpley and Menzel (1987), whereas values between 1.9 and 2.2 occurred for water-soluble plus acid-extractable P for silt loam in Wisconsin.

Detachment and transport of P particulates also depend on solubilization and ionic strength of the soil solution. Solubilization indicates the transfer of P from a solid phase to a water phase due to chemical nonequilibrium between the two phases. Desorption from sorbing surfaces, as well as dissolution from minerals, contributes to solubilization. In contrast to these chemical processes, particles may become suspended due to the mechanical forces exerted by water flowing over or through the soil; this is the major mobilization step for detachment, a physical process (Gburek et al., 2005). However, resistance of a particle to detachment may also depend on chemical properties of the solution, with its ionic strength being of major importance.

Losses of dissolved and particulate P from agricultural cropland have been implicated as a major factor responsible for accelerated eutrophication of surface waters in a number of locations throughout the United States (Carpenter et al., 1998; Daniel et al., 1998; Parry, 1998; Mullins et al., 2005) and the potential for high P soils to negatively affect surface water quality is of concern in the United States as well as some European countries and Canada (Sims, 1993; Sharpley et al., 1994; Breeuwsma et al., 1995). These environmental concerns have resulted in legislation in several states in the United States and policy changes that will require more intensive management of P in agricultural systems, especially in areas with concentrated animal feeding operations (Sharpley et al., 2000).

6.2.5 Changes in Phosphate Forms during Transport

Interchange between soluble and particulate P can occur during transport process in the soil–plant system or stream flow. Fine materials will have more impact on the transportation due to their greater capacity to sorb or desorb P and will be important in determining the short-term potential of runoff to increase algal growth (Sharpley and Menzel, 1987). In addition, soluble P may be removed by stream macrophytes and particulate P deposited or eroded from the stream bed with a change in stream velocity. Quantity of soluble and particulate P may be different entering water streams from those entering stream flow (Sharpley and Menzel, 1987). Input of sediment from heavily P-fertilized soils may increase the soluble P concentration of stream flow significantly (Taylor and Kunishi, 1971). During phosphorus transport processes, available forms of P may be converted to unavailable forms (Sharpley and Menzel, 1987).

6.2.6 Amount of Phosphorus Loss from Landscape to Aquatic Environments

Phosphorus has many physiological and biochemical function in plants but it does not have rapid cycling like carbon (C) or Nitrogen (N) in the soil–plant systems. Phosphorus's lower mobility and higher immobilization characteristics make it one of the most growth-limiting elements for crop production, especially in highly weathered tropical as well as temperate soils (Fageria and Baligar, 2008). By the year 2000 the global mobilization of the nutrient had roughly tripled as compared to its natural flows because of (1) increased soil erosion and runoff from fields, (2) recycling of crop residues and manures, (3) discharges of urban and industrial wastes, and (4) applications of inorganic fertilizers (15 million Mg P year^{-1}) (Smil, 2000). Global food production is now highly dependent on the continuing use of P, which accounts for 50%–60% of all P supply, although crops use the nutrient with relatively high efficiency; lost P that reaches water is commonly the main cause of environmental pollution (Smil, 2000). This undesirable process affects fresh and ocean waters in many parts of the world.

Loss of P from soil–plant systems to environment depends on cropping systems, rate of P applied to the crop, soil types, crop yield level, methods of P application, form of fertilizers, amount of precipitation after P fertilizer application, vegetative cover, and tillage system. Examples of P losses from different cropping systems are presented in Table 6.2. Data in Table 6.2 show that a major part of P loss from soil–plant system is by loss through particulate P or erosion of the soil. Sharpley and Menzel (1987) reported that it is difficult to distinguish between losses of fertilizer P and native soil P; the losses of fertilizer P are generally less than 1% of the applied. Losses of P in subsurface drainage are small, with application of fertilizers at recommended rates normally having no significant effect on P loss (Sharpley and Menzel, 1987).

In addition to P loss from agricultural lands (Table 6.2), U.S. national surveys indicated that combined totals of water erosion (sheet and rill) and wind erosion in the United States range mostly between 10 and 25 Mg ha^{-1}, with the mean just below 15 Mg ha^{-1}(Lee, 1990; Bloodworth and Berc, 1998; Smil, 2000). Global mean is higher, at least 20 Mg ha^{-1} (Smil, 1999b), implying an annual loss of 10 kg P ha^{-1} and 15 Mt P year^{-1} from the world's crop fields. Erosion has been increased by overgrazing, which now affects more than half (i.e., at least 1.7 billion ha) of the world's permanent pastures; an erosion rate of at least 15 Mg ha^{-1} would release about 13 Mt P annually from overgrazed land (Smil, 2000). Adding more than 2 Mt P eroded annually from undisturbed land brings the global total to over 30 Mt P year^{-1} (Smil, 2000).

Subtracting about 3 Mt P year^{-1} eroded by wind would leave 27 Mt of waterborne P; not all of this nutrient reaches the ocean, as at least 25% of it is redeposited on adjacent cropland, grassland, or on more distant places (Smil, 1999a). Consequently, river-borne input of particulate organic and inorganic P into the ocean is appropriately 20 Mt year^{-1} (Smil, 2000). Conversion of roughly 1.5 billion ha of forest and grasslands to crop fields and development, accompanied by an increase of 0.2 kg P ha^{-1} in solution (from 0.1 to 0.3 kg P ha^{-1}), would have added about 0.3 Mt P year^{-1}; a similar loss

TABLE 6.2
Soluble and Particulate Phosphorus Losses from Different Cropping Systems and P Rates

Land Use	Fertilizer P Rate (kg P ha^{-1} Year^{-1})	Soluble P (kg ha^{-1} Year^{-1})	Particulate P (kg ha^{-1} Year^{-1})	Reference
Fallow	0	0.10	33.15	Burwell et al. (1975)
Hay	0	0.39	0.02	Burwell et al. (1975)
Contour corn	29	0.25	18.19	Burwell et al. (1975)
Rotation corn	29	0.15	8.43	Burwell et al. (1975)
Rotation oats	30	0.14	5.01	Burwell et al. (1975)
Contour corn	66	0.15	0.76	Burwell et al. (1977)
Contour corn	40	0.12	0.45	Burwell et al. (1977)
Grazed bromegrass	41	0.16	0.08	Burwell et al. (1977)
Terraced corn	67	0.10	0.20	Burwell et al. (1977)
Native forest	0	0.01	0.20	McColl et al. (1977)
Pasture	75	0.04	0.29	McColl et al. (1977)
Corn grain	30	4.3	0.02	McColl et al. (1977)
Rotation grazing	0	0.04	0.50	Menzel et al. (1978)
Wheat	6.5	0.30	1.90	Menzel et al. (1978)
Cotton	25	1.10	5.60	Menzel et al. (1978)
Wheat/summer fallow	54	1.20	2.90	Nicholaichuk and Read (1978)
Pasture	0	0.50	0.67	Sharpley and Syers (1979b)

from 1.7 billion ha of overgrazed pastures would have doubled that loss. Even if inorganic fertilizers were to lose 2% of their P owing to leaching, the additional burden would be less than 0.4 Mt P year^{-1}. Enhanced urban loss owing to the leaching of lawn and garden fertilizers would increase the total to just over 1 Mt P year^{-1}, doubling the pre-agricultural rate to over 2 Mt P year^{-1}. Grand total of particulate and dissolved P transfer to the ocean would then be 22 Mt P year^{-1} (Smil, 2000).

6.2.7 Impact of Phosphorus on the Environment

Phosphorus has a negative impact on environment when present in soil–plant system in low as well as in high amounts. When P is present in the soil in small amount, land degradation occurs due to lower growth of vegetation. In Oxisols and Ultisols of temperate as well as tropical regions, P deficiency is very common for most crop plants. Phosphorus deficiency in these soils is related to low level of P in these highly weathered soils. In addition, these soils also have high immobilization capacity of P due to the presence of high amount of Al and Fe oxides (Fageria and Baligar, 2008; Fageria, 2009). Brady and Weil (2002) reported that there are probably 1–2 billion ha of land in the world where P deficiency limits growth of both crops and native vegetation.

In intensive agriculture, especially in Europe and North America, higher amount of P has been applied during the past decades than that has been removed by the crops (Brady and Weil, 2002). Excess amount of P moved from croplands to water bodies have occurred by runoff, leaching, and erosion, thereby creating environmental issues. Impact of P accumulation in excess amount in water bodies is attributed to eutrophication of aquatic ecosystems and, in particular, lakes, where it is commonly regarded as the limiting nutrient governing production (Foy, 2005). Eutrophication is the condition in an aquatic ecosystem where excessive nutrient concentrations result in high biological productivity, typically associated with algae blooms, that causes oxygen depletion to be detrimental to other organisms, especially to fish and shrimp (Soil Science Society of America, 2008).

Eutrophication can transform clear, oxygenated, good tasting water into cloudy, oxygen poor, foul smelling, bad tasting, and possibly toxic water (Brady and Weil, 2002). Eutrophic conditions favor growth of cyanobacteria, blue-green algae that are mostly undesirable food for zooplankton, a major food source for fish. This cyanobacteria produces toxins and bad tasting and smelling compounds that result in water unsuitable for human or animal consumption (Brady and Weil, 2002). In addition, eutrophic waters reduce biological diversity, especially fish and shrimp.

Transparency and color are the most obvious indicators of the nutrient conditions of a water body. Transparent oligotrophic waters support low plant productivity and appear either blue or brown; eutrophic waters have high primary productivity as large amounts of phytoplankton make them turbid and limit their transparency to less than 50 cm (Smil, 2000). Advanced eutrophication is marked by blooms of cyanobacteria and siliceous algae, scum-forming algae, and potentially toxic algae such as *Dinophysis* and *Gonyaulax* (Smil, 2000). Eventual decomposition of this phytomass creates hypoxic or anoxic conditions near the bottom or throughout a shallow water column (Smil, 2000). Water may be nonpotable and may have different odor or taste. Kotak et al. (1994) and Martin and Cooke (1994) reported that eutrophic water requires expensive treatments before consumption. Formation of trihalomethanes during chlorination can cause serious health hazards to livestock and people by ingestion of soluble neuro- and hepatotoxins released by decomposing algal blooms. Phosphorus is toxic to human and animals in high concentration. Eutrophication seriously disrupts coastal ecosystems in regions receiving high P inputs.

Wood (1998) and Foy (2005) reported that there are several adverse ecological effects of eutrophication with a high level of P in the water bodies. These are (1) replacement of high-quality edible fish, submerged macrophytic vegetation, and benthic organisms with coarse, rapid growing fish and algae and noxious aquatic plants; (2) increased sedimentation with eutrophication impairs navigational and recreational use; (3) lake depths are reduced; (4) enhanced vegetative growth blocks navigable waterways; (5) decaying algal biomass produces surface scums; (6) undesirable odors occur (hydrogen sulfide, methane, etc.); (7) populations of insect pests such as mosquitos increase; (8) increased potential for fish kills; (9) declining fish species that are intolerant of low oxygen; (10) production and release of algal toxins; (11) threatening public and animal health; (12) release of manganese (Mn) and iron (Fe) from lake sediments, reflecting low oxygen supply; and (13) reduced valuation of shoreline properties. These implications clearly demonstrate that enrichment of surface waters with P is undesirable.

Concentrations above 0.01 mg L^{-1} of dissolved P are likely to cause eutrophication, but nutrient supply (loading), rather than P or N concentration in water, is the key anthropogenic factor in the process (Smil, 2000). Brady and Weil (2002) reported that critical levels of P in water, above which eutrophication is likely to be triggered, are approximately 0.03 mg L^{-1} of dissolved P and 0.1 mg L^{-1} of total P. Mullins et al. (2005) and Sharpley et al. (1996) reported that P concentrations that cause eutrophication can range from 0.01 to 0.03 mg L^{-1}. Comparisons of polluted lakes and estuaries have shown that excessive eutrophication can be generally prevented if annual loading is lower than 1 g P m^{-2} (10 kg P ha^{-1}) of water surface (Smil, 2000). Algal growth in surface waters is usually limited by P availability, although no clear guidelines exist regarding concentrations of total or dissolved P in runoff that will induce eutrophication (Wood, 1998). However, recommendations have been made in regard to critical P concentrations that are expected to cause noxious aquatic growth in downstream waters (Table 6.3).

Phosphorus concentrations that cause eutrophication can range from 0.01 to 0.03 mg L^{-1} (Sharpley et al., 1996). To determine the threshold level of soil P accumulation, Dutch regulators have set a critical limit of 0.10 mg L^{-1} as dissolved P tolerance in ground water at a given soil depth (mean highest water level) (Daniel et al., 1998). Pote et al. (1996) reported that a Mehlich 3 P concentration of 50 mg kg^{-1} (optimum for many crops) had a dissolved reactive P (DRP) concentration in overland flow from grassland in Arkansas of 0.5 mg L^{-1}. Soils with soil test P (STP) concentrations similar to those recommended for optimum crop growth may sustain DRP concentrations in surface runoff above levels accelerating eutrophication in surface water bodies (McDowell and Sharpley, 2001).

TABLE 6.3
Critical Phosphorus Concentrations Reported for Surface Waters

P Concentration (mg L^{-1})	Type of P/Water Bodies	References
0.01	Dissolved P, critical concentration for lakes	Sawyer (1947); Vollenweider (1968)
0.10	Total P, critical concentration for streams	USEPA (1986)
0.05	Total P, critical concentration for lakes	USEPA (1986)
0.05	Dissolved P, critical concentration allowed to enter Florida Everglades	South Florida Water Management District (1994)
1.0	Flow-weighted annual dissolved P, proposed allowable limit for agricultural runoff	USEPA (1986)

Appropriate management of P is an important aspect not only for higher crop yields but also for environmental protection.

6.2.8 Reclamation of Aquatic/Soil Environment Contaminated with Excess Phosphorus

Excess P inputs in the water sources should be avoided to keep water quality at the desirable level. However, once a water source is contaminated, chemical methods are available to reduce the P levels in lakes and streams. Commonly used chemicals to reduce P-contaminated water bodies are aluminum sulfate and sodium aluminate, due to the stability of flocculated Al hydroxides with redox changes (Sharpley and Menzel, 1987). In this chemical process, P is removed by precipitation of $AlPO_4$, by coagulation or entrapment of P-containing particulates, or by sorption of P on the surfaces of Al hydroxide polymers (Sharpley and Menzel, 1987).

When using chemicals for P reduction, caution should be taken about the rate to be used since higher rates may be toxic to fish or shrimp. Experimental data are limited on the Al concentration toxic to aquatic biota. However, Sharpley and Menzel (1987) reported that an Al concentration 0.05 mg L^{-1} was reported to be nontoxic to fish. A predetermined amount of $Al_2(SO_4)_3$ is applied as a slurry from lake surface if P removal from the epilimnion is required. If control of P release from sediments is required, then application to the hypolimnion is necessary (Sharpley and Menzel, 1987). As $Al_2(SO_4)_3$ removes dissolved organic P inefficiently, applications should be made in early spring when the major proportion of P in lake water is in the inorganic form. The continued presence of organic P may be significant, as certain nuisance blue algae can produce a phosphatase enzyme under P-limiting conditions that is capable of mineralizing organic to inorganic P at rates sufficient to support algal blooms. Application time will not be critical for treatment of P desorption from lake sediments. However, the relative importance of lake sediments as a P source should be assessed prior to $Al_2(SO_4)_3$ application. For example, lakes receiving substantial inputs of clay in addition to P may contain sediment with high sorption capacities from P (Sharpley and Menzel, 1987).

Sharply and Menzel (1987) reviewed the literature on artificial removal of P from lakes and concluded that application of $Al_2(SO_4)_3$ just below the surface of Horseshoe Lake, Wisconsin, resulted in a significant decrease in the P content of both the epilimnion and hypolimnion. They reported that prior to application of aluminum sulfate the lake had experienced algal blooms and fish kills that were partially attributed to agricultural inputs of P. Sharpley and Menzel (1987) reported that the hypolimnetic application of $Al_2(SO_4)_3$ to the eutrophic West Twin Lake, Ohio, resulted in an 88% reduction in total P concentration, indicating that the layer of $Al_2(SO_4)_3$ deposited on the sediments reduced P release to overlying waters by 98%. Three years later the lake was mesotrophic (Sharpley and Menzel, 1987). Excess P in soil can also be removed by phytoremediation that is removing plant biomass after a period of P accumulation, and this process has the potential to reduce P losses (Kovar and Claassen, 2005; Lu et al., 2010).

6.3 MANAGEMENT PRACTICES TO REDUCE PHOSPHORUS LOSS FROM SOIL–PLANT SYSTEM

Phosphorus and nitrogen are both nutrients often associated with accelerated eutrophication of lakes and streams (Levine and Schindler, 1989). However, P is most often the element limiting accelerated eutrophication because many blue-green algae are able to utilize atmospheric N_2. Therefore, minimizing lake eutrophication from agricultural nonpoint source pollution often requires controlling P inputs to surface water. The International Joint Commission between the United States and Canada recommended this approach for managing nonpoint source pollution in the Great Lakes Basin (Pote et al., 1996). Similarly, Florida nonpoint source programs have focused on P (Little, 1988). In the Netherlands, the national strategy for minimizing nonpoint source pollution is to limit entry of P into both surface and groundwater (Pote et al., 1996). Sharpley et al. (1994) and Daniel et al. (1994) and Pote et al. (1996) identified the importance of developing P management strategies to limit surface water eutrophication from agricultural nonpoint source pollution.

Phosphorus is a key element for modern crop production. Fertilization of crops comprises the largest proportion of P used in agriculture, and P fertilizer use has increased steadily since 1960 in an effort to balance the gradual depletion of soil P caused by removal of P in harvested biomass (Wood, 1998). However, excess use of P in crop production can leach to water bodies, such as lakes, streams, rivers, and oceans, and can result in water pollution. Sites with high runoff potential include sloping fields with impermeable or compacted soils, intensively tile-drained clay soils, and land bordering watercourses, which are subjected to flooding (Withers et al., 2005). These sites require careful management to minimize runoff, and soil loss and incidental P loss may arise directly after application. There are a large number of best management practices (BMPs) that have been suggested to control nonpoint P loss (Withers and Jarvis, 1998; Sims and Kleinman, 2005; Withers et al., 2005).

Management practices that can be adopted to prevent such adverse impact on water bodies may be use of adequate P rate, use of conservation tillage, restrictions on use of P-contaminated detergents, and limits on the number of animals in problem areas. Inputs of P fertilizers can be lowered by relying on a variety of BMPs. In addition, nutrient management strategies for animal manure and other by-products will need to include concentrated efforts to reduce P losses in surface runoff and subsurface drainage. Sims and Kleinman (2005) reported that there is no single BMP to control a particular pollutant in all situations and suggested that the best management system should be designed on (1) type, source, and cause of the pollutants; (2) agricultural, climatic, and environmental conditions; (3) economic situation of the farm operators; (4) experience of the system designers; and (5) acceptability of the practices by the farmers. BMPs to control loss of P from the croplands to water bodies are discussed in this section.

6.3.1 Control of Soil Erosion

Soil erosion is one of the most important processes in transporting P from agricultural lands to the water bodies. Soil erosion transports particulate P bound to soil and in vegetative matter; only that portion of particulate P in equilibrium with dissolved P is available for aquatic biota. Thus, bioavailable P includes dissolved P and a portion of particulate P. Bioavailable P moves from agricultural fields into receiving waters it contributes to eutrophication. In the United States, most P indices use the Universal Soil Loss Equation (USLE) (Soil Science Society of America, 2008) to predict the risk of erosion. USLE is an equation for predicting A, the mean annual soil loss in mass per unit area per year, and is defined as $A = RKLSCP$, where R is the rainfall factor, K is the soil erodibility factor, L is the length of slope, S is the percent slope, C is the cropping and management factors, and P is the conservation practice factor. Many European countries have national erosion mapping systems; for example, the Norwegian erosion-risk map is based on soil texture and slope, and the U.K. erosion risk system is based on slope, soil texture, and crop type (Heathwaite et al., 2005b).

6.3.2 Adequate Phosphorus Rate

Use of adequate rate of P for crop production is fundamental to avoid loss of this element in the soil–plant system. Adequate rate is defined as the quantity of P removed by a crop that is mainly determined by yield goals. This practice may reduce the chances of excess P accumulation in the soil and reduces chances of P transport from agricultural lands to water bodies.

6.3.3 Agronomic and Environmental Soil P Tests

Soil testing to determine the nutrient requirements for optimum plant growth began in the nineteenth century. Throughout twentieth century soil scientists and agronomists have conducted countless experiments to develop soil P tests for different crops and soils (Sims et al., 1998a). The objective of this research is to identify or establish optimum P level in the soil and quantity of P required to obtain maximum economic yield of crops. Soil testing programs have components such as collecting soil samples, preparation, analysis, and interpretation of analysis results. All these components are important in identifying P fertilizer requirements of a crop in a given soil and making accurate P fertilizer recommendations. Sims (1993, 2000) and Fageria et al. (2011) reported detailed discussion of these soil testing components.

In recent years, there has been increased interest in using existing soil tests, or new soil testing methods for environmental as well as agronomic purposes (Sims, 1993; Sharpley et al., 1994). A major reason for the increased interest in environmental soil testing for P has been that a considerable body of research suggests that the extractable P content of soils influences the amount of P in runoff water and subsurface drainage, particularly if soil test P values exceed those needed for optimum crop growth (Sharpley et al., 1977, 1978, 1985, 1996; Sims, 1993; Heckrath et al., 1995; Pote et al., 1996; Moore et al., 1998; Sims et al., 1998b). This has pointed to the need for soil testing methods that can not only predict the probability of crop response to inputs of P, but also accurately quantify the likelihood that environmental problems will be caused by agricultural P (Moore et al., 1998).

Agronomic soil test for P is to ensure optimum P level in the soil for the growth and development of a crop and to produce maximum economic yield. Use of agronomic soil test is to ensure that a crop is not supplied with excess P, which can be subsequently transported to the water bodies and can cause eutrophication. Values of agronomic soil P test are generally lower than environmental soil P test values. Agronomic soil P test values for different crops are discussed in Chapter 4. In addition, agronomic and environmental soil P test values are presented in Table 6.4.

TABLE 6.4
Threshold Soil Test P Values for Agronomic and Environmental Purposes

P Extraction Method	Agronomic (mg kg^{-1})[a]	Environmental (mg kg^{-1})[a]
Mehlich 1	13–25	>55
Mehlich 3	25–50	>50
Bray 1	20–40	>75
Olsen	12	>50

Sources: Compiled from Sharpley, A.N. et al., *J. Soil Water Conserv.*, 51, 160, 1996; Fageria, N.K. et al., *Agron. J.*, 89, 737, 1997; Shober, A.L. and Sims, J.T., *J. Environ. Qual.*, 32, 1955, 2003; Fageria, N.K., *The Use of Nutrients in Crop Plants*, CRC Press, Boca Raton, FL, 2009.

[a] Threshold value is defined as the soil test concentration above which the soil test level is considered optimum for plant growth and responses to the addition of the nutrient are unlikely to occur. Threshold values cited in this table are approximate and can be affected by soil type, crop, and management practices. Environmental-threshold value is defined as the soil test concentration above which risk of environmental contamination is very high.

6.3.4 Conservation Tillage

Conservation tillage is an important management practice to reduce soil erosion and consequently nutrient loss from soil–plant systems, including P. Conservation tillage is defined as any tillage sequence, the object of which is to minimize or reduce loss of soil and water, operationally, a tillage or tillage and planting combination that leaves 30% or greater cover of crop residue on the soil surface (Soil Science Society of America, 2008). The terms "minimum tillage," "no-tillage," or "zero tillage" are also used in the literature. Minimum tillage or zero tillage is defined as a procedure whereby a crop is planted directly into the soil with no primary or secondary tillage since harvest of the previous crop. In this process, usually a special planter is necessary to prepare a narrow shallow seedbed immediately surrounding the seed being planted. No-tillage is sometimes practiced in combination with subsoiling to facilitate seeding and early root growth, whereby the surface residue is left undisturbed except for a small slot in the path of the subsoil shank. Conservation, minimum tillage, or no-tillage has been widely adopted in developed as well as developing countries in recent years for crop production. Until the 1970s, most arable cropland in the United States and other areas in the world relied upon regular tillage practices (moldboard plowing, disking, rototilling) to prepare the soil for planting and to control weeds and pests (Sims and Kleinman, 2005). For a variety of reasons, especially erosion control, conservation tillage systems have become widespread in the United States since 1970 (Sims and Kleinman, 2005). It is projected that conservation tillage will be practiced on 75% of cropland in the United States by 2020 (Lal, 1997). Kern and Johnson (1993) reported that increasing conservation tillage to 76% of planted cropland would change agricultural systems from C sources to C sink.

There is a general concept that tillage decreases aggregate stability by increasing mineralization of organic matter and exposing aggregates to additional raindrop impact energies (Tisdall and Oades, 1982; Elliott, 1986; Angers et al., 1992; Amezketa, 1999; Balesdent et al., 2000; Park and Smucker, 2005). Tillage promotes soil organic matter (SOM) loss through crop residue incorporation into soil, physical breakdown of residues, and disruption of macroaggregates (Six et al., 2000a; Wright and Hons, 2004). In contrast, conservation or no-tillage reduces soil mixing and soil disturbance, which allows SOM accumulation (Blevins and Frye, 1993). Conservation tillage has been reported to improve soil aggregation and aggregate stability (Beare et al., 1994; Six et al., 2000b). Conservation or minimum tillage promotes soil aggregation through enhanced binding of soil particles as a result of greater SOM content (Six et al., 2002). Microaggregates often form around particles of undecomposed SOM, providing protection from decomposition (Six et al., 2002; Wright and Hons, 2004). Microaggregates are more stable than macroaggregates, and thus tillage is more disruptive of larger aggregates than smaller aggregates, making SOM from large aggregates more susceptible to mineralization (Cambardella and Elloitt, 1993; Six et al., 2002a; Wright and Hons, 2004). Since tillage often increases the proportion of microaggregates to macroaggregates, there may be less crop-derived SOM in conventional tillage than conservation or no-tillage (Six et al., 2000b; Wright and Hons, 2004). Fungal growth and mycorrhizal fungi, which are promoted by no-tillage, contribute to the formation and stabilization of macroaggregates (Tisdall and Oades, 1982; Beare and Bruce, 1993).

Larger SOM accumulation in conservation tillage had been observed in intensive cropping systems, where multiple crops are grown yearly (Ortega et al., 2002; Wright and Hons, 2004). Use of conservation tillage, including no-till, is being considered as part of a strategy to reduce C loss from agricultural soils (Kern and Johnson, 1993; Paustian et al., 1997; Denef et al., 2004). Crop species also influence SOM accumulation in the soil. Residue quality often plays an important role in regulating long-term SOM storage (Lynch and Bragg, 1985). Crop residues having low N concentration, such as wheat (*Triticum aestivum* L.), generally decompose at slower rates than residues with higher N, such as sorghum (*Sorghum bicolor* L. *Moench*) and soybean (*Glycine max*) (Franzluebbers et al., 1995; Wright and Hons, 2005). Since wheat residues often persist longer and increase SOM more than sorghum or soybean (Wright and Hons, 2005). All these beneficial effects of conservation or no-tillage reduce soil erosion and conserve more nutrients and avoid loss of P in the soil–plant system to water bodies.

6.3.5 Cover Crops

Increasing crop productivity and maintaining a clean environment are major challenges to agricultural scientists in the twenty-first century. To meet these challenges, crop production practices need to be modified in favor of higher yields and minimized environmental pollution. Management of crop residues is a key component of sustainable cropping systems (Fageria et al., 2005). Historically, crop residues have had an important role as mulch for soil and water conservation, an input for maintaining soil organic matter and returning nutrients to soil. To achieve these objectives the use of cover crops in cropping systems is an important strategy. Before discussing cover crops and their role in crop production, it is important to define the term "cover crop." Cover crop is defined as the close-growing crop that provides soil protection, seeding protection, and soil improvement between periods of normal crop production, or between trees in orchards and vines in vineyards. When plowed under and incorporated into the soil, cover crops may be referred to as green manure crops (Soil Science Society of America, 2008).

Planting cover crops before or between main crops as well as between trees or shrubs of plantation crops can improve soil physical, chemical, and biological properties, and consequently lead to improved soil health and yield of principal crops. Leaving cover crops as surface mulches in no-till crop production systems has the advantage of increasing N economy, conserving soil moisture, reducing soil erosion, improving soil physical properties, increasing nutrient retention, increasing soil fertility, suppressing weeds, reducing diseases and insects, reducing global warming potential, and increasing crop yields (Fageria et al., 2005).

Loss of topsoil by wind and water erosion caused by poor soil management is by far the largest single factor contributing to deterioration of soil physical, chemical, and biological properties and to the further decline in productivity of most croplands (Fageria et al., 2005). Soil erosion removes the top soil layer, which generally contains large amounts of soil organic matter and immobile nutrients. Loss of such top soil layers ultimately reduces crop production. Magnitude of the effect of erosion on yields also varies among soils, crops, and management practices (Fageria et al., 2005, 2011).

When the soil surface is exposed to raindrop impact, the permeability of the soils is reduced by seal formation (Fageria et al., 2011). Seal formation reduces the infiltration rate, thus increasing runoff and subsequent soil loss. Reduction in soil erosion by cover crops is associated with increasing organic matter content, which improves soil water infiltration and holding capacity. With more infiltration and less runoff from each rainfall event, soil erosion is significantly reduced. Cover crops growing after soybean increase surface cover, anchor residues, and reduce rill erosion (Fageria et al., 2005). Cover crops include grasses, legumes, or other herbaceous plants that are established for seasonal cover, to protect against soil erosion, to scavenge excess nutrients remaining in the soil profile after harvest of the previous crop, to loosen root-restricting layers, to improve soil structure, to increase soil organic matter, and/or other conservation purposes to provide soil protection during the periods when primary row crops or cash crops are not being grown (Meisinger et al., 1991; Sims and Kleinman, 2005). Kleinman et al. (2001) reported that cover crops reduced total P concentration in springtime runoff to 36% of that from conventional corn. Important cover crops of tropical and temperate regions are presented in Table 6.5.

6.3.6 Diet Manipulation to Reduce Phosphorus in Animal Wastes

Phosphorus is an important nutrient for animal health. If P is present in lower amounts than necessary, the effect on animal health is negative. However, if P is present in higher amounts than necessary, P pollution may occur when animal manures of high dietary values are applied to

TABLE 6.5
Major Cover Crops of Tropical and Temperate Regions

Tropical Region		Temperate Region	
Common Name	**Scientific Name**	**Common Name**	**Scientific Name**
Sunn hemp	*Crotalaria juncea* L.	Hairy vetch	*Vicia villosa* Roth
Sesbania	*Sesbania aculeata* Retz Poir	Barrel medic	*Medicago truncatula* Gaertn
Sesbania	*Sesbania rostrata* Bremek & Oberm	Alfalfa	*Medicago sativa* L.
Cowpea	*Vigna unguiculata* L. Walp.	Black lentil	*Lens culinaris* Medikus
Soybean	*Glycine max* L. Merr.	Red clover	*Trifolium pratense* L.
Cluster bean	*Cyamopsis tetragonoloba*	Soybean	*Glycine max* L. Merr.
Alfalfa	*Medicago sativa* L.	Faba bean	*Vicia faba* L.
Egyptian clover	*Trifolium alexandrinum* L.	Crimson clover	*Trifolium incarnatum* L.
Wild indigo	*Indigofera tinctoria* L.	Ladino clover	*Trifolium repens* L.
Pigeon pea	*Cajanus cajan* L. Millspaugh	Subterranean clover	*Trifolium subterraneum* L.
Mungbean	*Vigna radiata* L. Wilczek	Common vetch	*Vicia sativa* L.
Lablab	*Lablab purpureus* L.	Purple vetch	*Vicia benghalensis* L.
Graybean	*Mucuna cinerecum* L.	Cura clover	*Trifolium ambiguum* Bieb.
Buffalo bean	*Mucuna aterrima* L. Piper & Tracy	Sweet lover	*Melilotus officinalis* L.
Crotalaria breviflora	*Crotalaria breviflora*	Winter pea	*Pisum sativum* L.
White lupin	*Lupinus albus* L.	Narrowleaf vetch	*Vicia angustifolia* L.
Milk vetch	*Astragalus sinicus* L.	Milk vetch	*Astragalus sinicus* L.
Crotalaria	*Crotalaria striata*		
Zornia	*Zornia latifolia*		
Jack bean	*Canavalia ensiformis* L. DC.		
Tropical kudzu	*Pueraria phaseoloides* (Roxb.) Benth.		
Velvet bean	*Mucuna deeringiana* Bort. Merr.		
Adzuki bean	*Vigna angularis*		
Brazilian stylo	*Stylosanthes guianensis*		
Jumbie bean	*Leucaena leucocephala* Lam. De Wit		
Desmodium	*Desmodium ovalifolium* Guillemin & Perrottet		
Pueraria	*Pueraria phaseoloides* Roxb.		

Source: Fageria, N.K. et al., *Commun. Soil Sci. Plant Anal.*, 36, 2733, 2005.

croplands (Chang et al., 1991). In a long-term study in Canada, Dormaar and Chang (1995) applied beef cattle feedlot manure for 14 years to an irrigated and nonirrigated loam soil, followed by 6 years with no further additions of manures. Concentrations of bicarbonate extractable P in the residual treatments were lower as compared to those continuing to receive annual applications of manures, but they were still above levels considered adequate for optimum plant growth.

Studies conducted in the United States indicated that P level in the animal diet is frequently above the level considered adequate (Poulsen, 2000; Valk et al., 2000). A survey conducted in Mid-South regions of the United States indicated that P was fed at levels above the National Research Council guidelines (Mullins et al., 2005). Wu et al. (2000) reported that reducing dietary P from 4.9 to 4.0 g kg^{-1} was sufficient to maintain P balance and the level of milk production, but fecal P excretion was reduced by 23%. Mullins et al. (2005) reviewed the literature and reported that P in dairy diets could be reduced from approximately 4.8 to 3.8 g kg^{-1} with corresponding reduction in P excretion by 25%–30%. These authors further reported that a 20% decrease in dietary P could be achieved without decreasing animal performance, and this decrease would result in a 25%–30% decrease in the P content of manure. Effects of reduced dietary P may have even greater environmental impacts due to reduced P losses following land application (Mullins et al., 2005).

6.3.7 Manure Management

Phosphorus usually is the limiting nutrient in many freshwater ecosystems. Excess P concentration leads to eutrophication. Large P losses from agricultural fields impair water resources in many regions, mainly in areas of developed countries with intensive animal production (Mallarino and Scheperes, 2005). Often animal manures are applied to fields at frequencies and rates that exceed P amount required for optimum crop yield or P removed in harvested crop parts. Because of the relative N and P content of manure and larger N than P losses before the manure is applied to fields, continued use of manure at rates that is based on N needs of crops usually results in P accumulation in soils. Upper limit for P fertilizers or manures that could be applied to fields with minimal nutrient loss could be ultimately determined by soil P test, the application method, potential for P delivery from fields through soil erosion, surface water runoff, and subsurface drainage (Mallarino and Schepers, 2005). Watershed level studies have indicated that small areas usually deliver disproportionately large amounts of P to streams or lakes (Gburek and Sharpley, 1998; Gburek et al., 2000, 2005; Klatt et al., 2003; Mallarino and Schepers, 2005).

Use of animal manures in adequate amounts and properly mixed into the soil for crop production is important in reducing P losses from the croplands. In addition, advancements in manure management include the addition of amendments to swine and poultry manures to chemically stabilize P in less soluble forms, therefore decreasing the potential for losses via runoff and subsurface transport following land application (Sims and Luka-McCafferty, 2002). Amendments that can be used to reduce P losses are metal salts or by-products containing Al, Fe, or Ca to solid or liquid manures, which is similar to the procedures used by wastewater treatment plants to remove P (Dao, 1999, 2001; Codling et al., 2000; Moore et al., 2000). To date, the most effective and accepted on-farm use of this BMP has been the use of alum $[Al_2(SO_4)_3{\cdot}14H_2O]$ as an amendment to poultry litter in poultry houses (Mullins et al., 2005). Alum is a dry acid and has been reported to decrease ammonia volatilization from manure, improve animal health and weight gains, and decrease the solubility of P in poultry litter (Moore et al., 1999, 2000; Worley et al., 2000; Mullins et al., 2005). Moore et al. (2000) reported that amending broiler with alum (aluminum sulfate) could reduce P solubility and decrease the risk of soluble P losses in runoff. Alum as a litter treatment is now a recommended BMP in some U.S. states (Sims and Kleinman, 2005). Similarly, Penn and Sims (2002) reported amending biosolids with $FeCl_3$ decreased soluble P in soils and dissolved P losses in runoff as compared to biosolids derived from biological nutrient removal processes.

Elrashidi et al. (1999) reported that fluidized bed combustion (FBC) material reduced P leaching from dairy manure by 88% as compared to untreated manure. They suggested that the reduction in P leaching may have been due to one of four mechanisms, including (1) retention of soluble P organic complexes by FBC minerals surfaces, (2) P sorption by Al and Fe, (3) precipitation of calcium phosphate minerals, and (4) occlusion of P in precipitated carbonate minerals. According to Mullins et al. (2005) more research data are required for use of amendments in reducing P losses from animal and bird manures to control P losses from soil–plant systems.

In addition, adequate storage of animal manures on the farm is also important in reducing surface water contamination with P. As livestock constantly generate manure, storage facilities provide growers with flexibility in manure management particularly as to when manure must be land-applied. Inadequate storage is a common problem on many animal feeding operations and often promotes land application of manures during periods when crops are not growing or when the ground is frozen. Proper storage also provides opportunities for treatment of manures, ranging from separation of liquid and solid fractions, to precipitation of soluble P into forms that are conducive to recovery, such as struvite (Greaves et al., 1999; Burns et al., 2001). Specific storage options vary with livestock type and individual farm characteristics, ranging from cement storage pads to anaerobic and aerobic lagoons to oxidation ponds and ditches (Day and Funk, 1998). Giasson (2002) evaluated manure storage options for New York dairy farms and reported that installation of manure lagoons with a 3-month storage capacity resulted in the most cost-effective control of nonpoint source P losses.

6.3.8 Recovery of Precipitated Phosphorus from Wastes

Direct discharge of P from sewage treatment works into water courses has been largely responsible for the widespread and undesirable P contamination of rivers, lakes, and water streams (Muscutt and Withers, 1996; Withers et al., 2005). Excess P should be removed from wastewaters at sewage treatment works. A number of technologies exist for P removal and recovery, but one option that has been increasingly considered in recent years is the recovery of ammonium magnesium phosphate (struvites) from wastewaters and livestock wastes (Greaves et al., 1999; Doyle and Parsons, 2002; Burns et al., 2003; Withers et al., 2005). In granular form, struvite is a potentially useful slow-release fertilizer that can substitute for existing fertilizers either in isolation, or when blended with other fertilizers (Gaterell et al., 2000).

6.3.9 Thermal Drying of Sewage Sludge

Sewage sludge is an important source of P and organic matter for crop production, and can be used to improve soil physical, chemical, and biological properties (Withers et al., 2005). However, if not applied properly, it may add P to excess level and this may be transported to water bodies (Shober and Sims, 2003). Heating sewage sludge at high temperature (100°C–800°C) and subsequent conversion to pellets or granules provide an opportunity to improve the handling and acceptability of biosolids-based products to farmers and to expand the land based upon which biosolids can be applied and can reduce the rapid P buildup of the soil (Withers et al., 2005). Withers et al. (2005) reported that that thermal dried sludge applied to soil reduced P concentration in the soil and in the sludge. However, such materials may not be suitable for low-P soils but are quite effective in medium- to high-P soils to obtain optimum crop yields (Withers et al., 2005).

6.3.10 Use of Buffers

There are several types of buffers to control P transportation from agricultural land to surface waters. These buffers are known as riparian buffers, filter strips, contour grass strips, field borders, alley cropping, and vegetative barriers (Sims and Kleinman, 2005). Buffers are used to trap sediment leaving agricultural fields and hillslopes, and are primarily viewed as a means to control particulate P transport. However, buffers can promote infiltration of runoff water and attenuate soluble contaminants in runoff (Sims and Kleinman, 2005). Plant uptake in filter strips and riparian buffers may sequester dissolved form of nutrients (Lawrence et al., 1985; Muscutt and Withers, 1996; Uusi-Kamppa et al., 2000; Sims and Kleinman, 2005).

Buffers are effective in reducing nutrient loss from surface and subsurface transport. Buffers are usually recommended to be designed with at least two or three zones (USEPA, 1998). Zone 1 is directly adjacent to the stream or water body, contains permanent woody vegetation, and extends a minimum of 5 m. Its main function is to enhance ecosystem stability and assist in controlling the physical, chemical, and trophic status of the stream. For optimum effectiveness vegetation should remain undisturbed (Sims and Kleinman, 2005). Zone 2 adjoins and may directly contribute to the function of zone 1. It is the primary location for biological processes that remove pollutants via surface and subsurface flow. Natural Resource Conservation (NRCS) recommends that zone 2 extend a minimum of 6–7 m from the edge of zone 1 and should consist of woody vegetation that may be harvested on a regular basis provided the process does not compromise buffer functions (Sims and Kleinman, 2005). Zone 3 contains herbaceous vegetation (i.e., conservation grasses) and is added to the buffer when adjacent to cropland or highly erosive areas to filter sediments and other

TABLE 6.6
Effects of Different Buffer Width and Types on Sediment and P Loss Reduction in Surface Runoff from Agricultural Fields

Buffer Width (m)	Buffer Type	Sediment Reduction (%)	Phosphorus Loss Reduction (%)
4.6	Grass	61	29
9.2	Grass	75	24
19	Forest	90	70
23.6	Grass/Forest[a]	96	79
28.2	Grass/Forest[b]	97	77

Source: Adapted from USEPA, Water quality functions of riparian forest buffer systems in the Chesapeake Bay watershed, Report no. EPA 903-R-95-004, Nutrient subcommittee, Chesapeake Bay Program, USEPA, Annapolis, MD, 1998.

[a] 4.6 m of grass buffer and 19 m of forest.

[b] 9.2 m of grass buffer and 19 m of forest.

pollutants from concentrated flow and promote sheet flow entering the buffer system (Sims and Kleinman, 2005). Buffer can be an effective means to reduce sediment and P loss from agricultural lands and reduce nonpoint P pollution (Table 6.6). However, buffers should be well designed and maintained to be effective in erosion control (Tate et al., 2000). Chaubey et al. (1994) reported a correlation between vegetative filter strip width and removal of P from runoff water in filter strips <9 m wide. Filter strips >9 m in width did not differ in removal of dissolved and total P. Heathwaite et al. (1998) reported 10 m grassed buffer strips removed 98% of P from grassland receiving mineral fertilizers, but only 10% P from grassland amended with cattle slurry.

6.3.11 Grass Waterways

Grass waterways are designed to convey concentrated runoff from agricultural fields without causing channel erosion (Sims and Kleinman, 2005). In some cases, they may be constructed as cross-slope diversions to intercept runoff and break up effective slope length. Chow et al. (1999) reported that installation of a grassed waterway/terrace combination resulted in a 20-fold reduction in annual erosion from a potato (*Solanum tuberosum*) field in New Brunswick, Canada.

6.3.12 Irrigation Management

Soil erosion can be greater under artificial irrigation as compared to natural rainfall if irrigation system is not well planned (Lentz et al., 1998). In general, furrow irrigation is associated with the largest potential for erosion, and subsequently particulate P losses, followed by sprinkler and then drip irrigation (Koluvek et al., 1993). Irrigation management, especially use of mulches to reduce soil erosion, and reducing P losses from agricultural lands to water bodies have been reported (Smith et al., 1990; Levy et al., 1995; Sojka and Bjorneberg, 2002).

6.3.13 Streambank Protection

Streambank protection is an important strategy to reduce soil erosion and pollution of water source from P pollution. Exclusion of livestock from streams reduces streambank erosion and direct deposit of manure into surface waters (Sims and Kleinman, 2005). In addition to streambank fencing, construction of stable crossings such as bridges and gravel travel lanes may reduce erosion while facilitating livestock to cross waterways. In a watershed study of streambank fencing, Meals (2000) reported reduction in stream total P concentration, with nearly a 50% decrease in P loads exposed from the watershed. Similarly, Line et al. (2000) reported a 76% reduction in total P loads and an 82% reduction in suspended sediments in a small stream draining a 15 ha pasture after implementation of stream bank fencing.

6.3.14 Phosphorus Index

In the early 1990s the USDA began to develop assessment tools for areas with water quality problems. A group of scientists from universities and governmental agencies met in 1990 to discuss the P issue and later formed a work group known as phosphorus index core team (PICT) to more formally address this problem. Members of PICT soon realized that despite the many scientists conducting independent research on soil P, there was a lack of integrated research that could be used to develop the field-scale assessment tool for P. Consequently, the first priority of PICT was a simple, field-based planning tool that could integrate, through a multi-parameter matrix, the soil properties, hydrology, and agricultural management practices within a defined geographic area, thus assessing, in a relative way, the risk of P movement from soil to water. This planning tool, now referred to as *the phosphorus index*, has since been used in several U.S. states to enhance efforts to prevent nonpoint source pollution of surface waters by agricultural P (Lemunyon and Gilbert, 1993; Sharpley, 1995; Lemunyon and Daniel, 1998).

Phosphorus index uses eight characteristics to obtain an overall rating for a site (Table 6.7). The general interpretation of the P index is reported in Table 6.8. Each characteristic is assigned

TABLE 6.7
Phosphorus Index Relating Site Characteristics and P Loss Rating

Site Characteristic (Weighing Factor)	None (0)	Low (<20)	Medium (20–40)	High (40–80)	Very High (>80)
Soil erosion (1.5)	N/A	<11.2 Mg ha^{-1}	11.2–22.4 Mg ha^{-1}	22.4–33.6 Mg ha^{-1}	>33.6 Mg ha^{-1}
Irrigation erosion (1.5)	N/A	Infrequent irrigation on well-drained soils	Moderate irrigation on soils with slopes <5%	Frequent irrigation on soils with slopes of 2%–5%	Frequent irrigation on soils with slopes >5%
Soil runoff class (0.5)	N/A	Very low or low	Medium	High	Very high
Soil test P (1.0)	N/A	Low	Medium	High	Excessive
P fertilizer rate (kg ha^{-1}) (0.75)	None applied	<15	15–45	46–75	>75
P fertilizer application method (0.5)	None applied	Placed with planter deeper than 5 cm	Incorporate immediately before crop	Incorporated surface applied 3 months before crop	Surface applied >3 months before crop
Organic P source application rate (kg ha^{-1}) (1.0)	None applied	<15	15–45	46–75	>75
Organic P source application method (1.0)	None	Injected deeper than 5 cm	Incorporate immediately before crop	Incorporated >3 months before crop or surface applied <3 months before crop	Surface applied to pasture or >3 months before crop

Source: Beegle, D. et al., Interpretation of soil test phosphorus for environmental purposes, in: *Soil Testing for Phosphorus: Environmental Uses and Implications*, Sims, J.T., ed., University of Delaware, Newark, DE, 1998, pp. 41–43.

TABLE 6.8
Generalized Interpretation of the P Index

P Index	General Vulnerability to P Loss
<8	Low potential for P movement from the site. If current farming practices are maintained, there is a low probability of an adverse impacts to surface waters from P losses at this site.
8–14	Medium potential for P movement from the site. The chance for an adverse impact to surface waters exists. Some remedial action should be taken to lessen the probability of P loss.
15–32	High potential for P movement from the site and for an adverse impact on surface waters to occur unless remedial action is taken. Soil and water conservation as well as P management practices are necessary to reduce the risk of P movement and water quality degradation.
>32	Very high potential for P movement from the site and for an adverse impact on surface waters. Remedial action is required to reduce the risk for P loss. All necessary soil and water conservation practices, plus a P management plan, must be put in place to avoid the potential for water quality degradation.

Source: Beegle, D. et al., Interpretation soil test phosphorus for environmental purposes, in: *Soil Testing for Phosphorus: Environmental Uses and Implications*, Sims, J. T., ed., University of Delaware, Newark, DE, 1998, pp. 41–43.

an interpretive rating with a corresponding numerical value such as none (0), low (10), medium (20), high (4), or very high (8), based on the relationship between the characteristics and the potential for P loss from a site. Suggested ranges appropriate to each rating for a site characteristic are then assigned (Beegle et al., 1998). Each of the characteristics in the P index is also given a weighting factor that reflects its relative importance to P loss. For example, erosion (weighting factor = 1.5) is generally more important to P loss than P fertilizer application methods (weighing factor = 0.5). These weighing factors used are arbitrarily selected based on the professional experience of the scientists, not on the basis of experimental results (Beegle et al., 1998). According to Beegle et al. (1998) individual states or regions should modify the weighting factors as appropriate, based on local soil properties, hydrologic conditions, and agricultural management practices.

Framework and rationale for the P index approach are based on observations that most P from agricultural watersheds appears to come from only a small but well-defined area of the landscape, where surface runoff generation coincides with zones of high P concentration (Gburek and Sharpley, 1998). Phosphorus indices are designated to rank site or field vulnerability to P loss in surface and subsurface pathways from agricultural landscapes (Heathwaite et al., 2005b). They identify specific areas within a watershed that are likely to contribute most to P loss from land to surface waters and provided site-specific, yet flexible management options to minimize this loss (Gburek et al., 2000).

Use of P indices in the United States is largely in response to USDA-EPA (United States Department of Agriculture-Environmental Protection Agency) proposals that all animal feeding operations (AFOs) have a comprehensive nutrient (P and N) management plan in place by 2008 to address water quality concerns related to nutrient management. Similar approaches are being developed in some European countries (Djodjic et al., 2002). In Denmark and Norway, the P indices have been developed on the basis of the principles used in the Pennsylvania P index (Heathwaite et al., 2005b).

Mallarino and Schepers (2005) reported that P indices have been developed to improve the risk estimation of P loss from agricultural fields as compared with estimates provided solely by soil test P and planned P application methods or rates. Phosphorus index can be used to classify fields or field areas into classes according to the risk of P loss through various transport mechanisms and, therefore, can provide guidelines for improved soil conservation and P management practices. A P index does not directly provide management recommendations, but partial ratings for the various index components can provide clues as to the major causes of a certain risk of P loss (Mallarino and Schepers, 2005).

6.3.15 Fertilizer and Manure Application Methods and Timing

Methods by which fertilizers and manures are applied significantly influence P loss in runoff (Zhao et al., 2001; Sims and Kleinman, 2005). Fertilizer P is often applied as broadcast and/or band application. However, animal manures (solids) are generally applied as broadcast on the soil surface. Liquid manures can be surface-applied as sprays, injected, or knifed into the soil. Where possible, incorporation of fertilizers and manures in the soil, in conjunction with soil erosion control practices, should be used to prevent loss of particulate P, and subsequently reduce the risk of P loss to water (Sims and Kleinman, 2005). Timing of P application to soils relative to when runoff occurs is a key BMP for preventing runoff losses (Sharpley, 1997). Immediately following fertilizer or manure application, the potential for P loss peaks declines over time, as water-soluble P applied in the fertilizers or manures gradually interacts with soils and is converted to increasingly recalcitrant forms (Edwards and Daniel, 1993). Sharpley and Syers (1979a) reported declining DRP from >0.25 to <0.1 mg L^{-1} and total P (TP) concentration from >0.7 to <0.1 mg L^{-1} in tile drainage over one month following temporary intensive grazing of paddocks by dairy cattle. Similarly, Gascho et al. (1998) reported exponential declines in DRP concentrations in surface runoff from >5 to <1 mg L^{-1} over roughly the same time period after mineral fertilizer application.

6.4 CONCLUSIONS

Phosphorus is an important nutrient in crop production under most agroecological regions. Besides its role in many physiological and biochemical processes, adequate P rate in the soil also increases the response to N and K fertilizers. However, its judicious use in crop production is important to avoid water pollution. Important sources of P loss from soil–plant systems are soil erosion and leaching from soil profiles. Impacts of P loss from soil–plant systems are enormous in terms of water pollution. One such impact is known as eutrophication. Phosphorus in agricultural runoff can cause accelerated lake and stream eutrophication. Where producers have applied P at rates exceeding crop uptake, soil P has sometimes become the main source of P in runoff. Eutrophication is the condition in an aquatic ecosystem where excessive nutrient concentrations result in high biological productivity, typically associated with algae blooms that cause sufficient oxygen depletion to be detrimental to other organisms, especially to fish and shrimp. Eutrophic water may be unfit for human and animal consumption and requires a high cost to overcome the impact. Eutrophication restricts water use for fisheries, recreation, industry, and drinking due to the increased growth of undesirable algae and aquatic weeds and to oxygen shortages caused by their death and decomposition. Critical levels of P in water, above which eutrophication is likely to be triggered, are approximately 0.03 mg L^{-1} of dissolved P and 0.1 mg L^{-1} of total P. These values are an order of magnitude lower than P concentrations in soil solution critical for plant growth (0.2–0.3 mg L^{-1}), emphasizing the disparity between critical water, soil P concentrations, and the importance of controlling P losses to limit eutrophication.

Perhaps the most significant change in agricultural P management in the past few decades has been the rapid increase in the number of countries or states taking a more regulatory approach to nonpoint P pollution. Losses of P from land areas to water bodies can be significantly reduced by adopting appropriate soil and crop management practices. Soil and crop management practices that can reduce P losses from soil–plant systems are control of soil erosion, use of P rates based on soil test, use of organic manures at an adequate rate, irrigation management to minimize runoff and erosion, and adopting appropriate methods of P fertilization of crop plants. Reducing P loss in agricultural runoff can occur through source and transport control strategies. These include refining feed rations, using feed additives to increase P absorption by animals, moving manures from surplus to deficit areas, finding alternative uses for manures, and targeting conservation practices, such as reduced tillage, buffer strips, and cover crops, to critical areas of P export from a watershed. In most cases the only permanent solution to reduce P losses is balancing farm and watershed P inputs and outputs.

More research data are needed to improve the partitioning models for soluble and particulate P transport in runoff, in lakes and impoundments. Investigation should focus on mechanisms of exchange between labile P and solution P, and method to routinely quantify the amounts of desirable or bioavailable P in various materials.

REFERENCES

Amezketa, E. 1999. Soil aggregate stability: A review. *J. Sustain. Agric.* 14:83–151.

Anderson, H., L. Bergstrom, F. Djodjie, B. Ulen, and H. Kirchmann. 2013. Topsoil and subsoil properties influence phosphorus leaching from four agricultural soils. *J. Environ. Qual.* 42:455–463.

Angers, D. A., A. Pesant, and J. Vigneux. 1992. Early cropping induced changes in soil aggregation, organic matter, and microbial biomass. *Soil Sci. Soc. Am. J.* 56:115–119.

Baalousha, M., F. V. Kammer, M. Motelica-Heino, and P. L. Coustumer. 2005. 3D Characterization of natural colloids by FIFFF-MALLS-TEM. *Anal. Bional. Chem.* 383:549–556.

Bahnick, D. A. 1977. Contribution of red clay erosion to orthophosphate loadings into Southwestern Lake-Superior. *J. Environ. Qual.* 6:217–222.

Balesdent, J., C. Chenu, and M. Balabane. 2000. Relationship of soil organic matter dynamics to physical protection and tillage. *Soil Tillage Res.* 53:215–230.

Ballaux, J. C. and D. E. Peaslee. 1975. Relationships between sorption and desorption of phosphorus by soils. *Soil Sci. Soc. Am. Proc.* 39:275–278.

Beare, M. H. and R. R. Bruce. 1993. A comparison of methods for measuring water-stable aggregates: Implications for determining environmental effects on soil structure. *Geoderma* 56:87–104.

Beare, M. H., M. L. Cabrera, P. F. Hendrix, and D. C. Coleman. 1994. Aggregate protected and unprotected organic matter pools in conventional and no-tillage soils. *Soil Sci. Soc. Am. J.* 58:787–795.

Beauchemin, S., R. R. Simard, and D. Cluis. 1998. Forms and concentrations of phosphorus in drainage water of twenty-seven tile drained soils. *J. Environ. Qual.* 27:721–728.

Beegle, D., A. N. Sharpley, and D. Graetz. 1998. Interpretation soil test phosphorus for environmental purposes. In: *Soil Testing for Phosphorus: Environmental Uses and Implications*, ed. J. T. Sims, pp. 41–43. Newark, DE: University of Delaware.

Blevins, R. L. and W. W. Frye. 1993. Conservation tillage: An ecological approach to soil management. *Adv. Agron.* 51:33–78.

Bloodworth, H. and J. L. Berc. 1998. *Cropland Acreage, Soil Erosion, and Installation of Conservation Buffer Strips: Preliminary Estimates of the 1997 National Resources Inventory*. Washington, DC: USDA.

Brady, N. C. and R. R. Weil. 2002. *The Nature and Properties of Soils*, 13th edn. Upper Saddle River, NJ: Prentice Hall.

Breeuwsma, A., J. G. A. Reijerink, and O. F. Schoumans. 1995. Impact of manure on accumulation and leaching of phosphate in areas of intensive livestock farming. In: *Animal Wastes and the Land Water Interface*, ed. K. Steele, pp. 239–250. Boca Raton, FL: Lewis Publishers.

Bromfield, S. M. and O. L. Jones. 1972. The initial leaching of hayed-off pasture plants in relation to the recycling of phosphorus. *Aust. J. Agric. Res.* 23:811–824.

Burns, R. T., L. B. Moody, I. Celen, and J. R. Buchanan. 2003. Optimizing phosphorus precipitation from swine manure slurries to enhance recovery. *Water Sci. Technol.* 48:139–146.

Burns, R. T., L. B. Moody, F. R. Walker, and D. R. Raman. 2001. Laboratory and in situ reductions in soluble phosphorus in liquid swine waste slurries. *Environ. Technol.* 22:1273–1278.

Burwell, R. E., G. E. Shuman, H. G. Heinemann, and R. G. Spomer. 1977. Nitrogen and phosphorus movement from agricultural watersheds. *Soil Water Conserv.* 32:266–230.

Burwell, R. E., D. R. Timmons, and R. F. Holt. 1975. Nutrient transport in surface runoff as influenced by soil cover and seasonal periods. *Soil Sci. Soc. Am. Proc.* 39:523–528.

Cade-Menum, B. and C. W. Liu. 2014. Solution phosphorus-31 nuclear magnetic resonance spectroscopy of soils from 2005 to 2013: A review of sample preparation and experimental parameters. *Soil Sci. Soc. Am. J.* 78:19–37.

Calhoun, F. G., J. M. Bigham, and B. K. Slater. 2002. Relationships between plant available phosphorus, fertilizer sales, and water quality in Northeastern Ohio. *J. Environ. Qual.* 31:38–46.

Cambardella, C. A. and E. T. Elliott. 1993. Carbon and nitrogen distribution in aggregates from cultivated and native grassland soils. *Soil Sci. Soc. Am. J.* 57:1071–1076.

Carpenter, S. R., N. F. Caraco, D. L. Correll, R. W. Horwarth, A. N. Sharpley, and V. H. Smith. 1998. Nonpoint pollution of surface waters with phosphorus and nitrogen. *Ecol. Appl.* 8:559–568.

Chang, C., T. G. Sommerfeldt, and T. Entz. 1991. Soil chemistry after eleven annual applications of cattle feedlot manure. *J. Environ. Qual.* 20:475–480.

Chaubey, I., D. R. Edwards, T. C. Daniel, P. A. Moore, and D. J. Nichols. 1994. Effectiveness of vegetative filter strips in retaining surface-applied swine manure constituents. *Trans. ASAE* 37:845–850.

Chow, T. L., H. W. Rees, and H. Daigle. 1999. Effectiveness of terrace/grassed waterways systems for soil and water conservation: A field evaluation. *J. Soil Water Conserv.* 654:577–583.

Codling, E. E., R. L. Chaney, and C. L. Mulchi. 2000. Use of aluminum and iron rich residues to immobilize phosphorus in poultry litter and litter amended soils. *J. Environ. Qual.* 29:1924–1931.

Condron, L. M. and S. Newman. 2011. Revising the fundamentals of phosphorus fractionation of sediments and soils. *J. Soils Sediments* 11:830–840.

Contron, L. M., B. L. Turner, and B. J. Cade-Menum. 2005. Chemistry and dynamics of soil organic phosphorus. In: *Phosphorus: Agriculture and the Environment*, ed. L. K. Al-Amoodi, pp. 87–121. Madison, WI: ASA, CSSA, and SSSA.

Cordell, D., J. O. Drangert, and S. White. 2009. The story of phosphorus: Global food security and food for thought. *Glob. Environ. Change* 19:292–305.

Correll, D. L. 1998. The role of phosphorus in the eutrophication of receiving waters: A review. *J. Environ. Qual.* 27:261–266.

Daniel, T. C., A. N. Sharpley, D. R. Edwards, R. Wedepohl, and J. L. Lemunyon. 1994. Minimizing surface water eutrophication from agricultural by phosphorus management. *J. Soil Water Conserv.* 49:30–38.

Daniel, T. C., A. N. Sharpley, and J. L. Lemunyon. 1998. Agricultural phosphorus and eutrophication: A symposium overview. *J. Environ. Qual.* 27:251–257.

Dao, T. H. 1999. Co-amendments to modify phosphorus extractability and nitrogen/phosphorus ration in feedlot manure and composted manure. *J. Environ. Qual.* 28:1114–1121.

Dao, T. H., T. H. Sikora, A. Hamasaki, and R. L. Chaney. 2001. Manure phosphorus extractability affected by aluminum and iron by products and aerobic composting. *J. Environ. Qual.* 30:1693–1698.

Day, D. L. and T. L. Funk. 1998. Processing manure: Physical, chemical, and biological treatment. In: *Animal Waste Utilization: Effective Use of Manure as a Soil Resource*, eds. J. L. Hatfield and B. A. Stewart, pp. 243–282. Chelsea, MI: Anna Arbor Press.

Denef, K., J. Six, R. Merckx, and K. Paustian. 2004. Carbon sequestration in microaggregates of no-tillage soils with different clay mineralogy. *Soil Sci. Soc. Am. J.* 68:1935–1944.

Djodjic, F., H. Months, A. Shirmohammadi, L. Bergstrom, and B. Ulen. 2002. A decision support system for phosphorus management at a watershed scale. *J. Environ. Qual.* 28:1273–1282.

Dormaar, J. F. and C. Chang. 1995. Effect of 20 annual applications of excess feedlot manure on labile soil phosphorus. *Can. J. Soil Sci.* 75:507–512.

Doyle, J. and S. A. Parsons. 2002. Struvite formation, control and recovery. *Water Res.* 36:3925–3940.

Edwards, D. R. and T. C. Daniel. 1993. Runoff quality impacts of swine manure applied to fescue plots. *Trans. ASAE* 36:81–86.

Elliott, E. T. 1986. Aggregate structure and carbon, nitrogen, and phosphorus in native and cultivated soils. *Soil Sci. Soc. Am. J.* 50:627–633.

Elliott, H. A., G. A. O'Connor, and S. Brinton. 2002. Phosphorus leaching from biosolids-amended sandy soils. *J. Environ. Qual.* 31:681–689.

Elrashidi, M. A., V. C. Baligar, R. F. Korcak, N. Persaud, and K. D. Richey. 1999. Chemical composition of leachate of dairy manure mixed with fluidized bed combustion residue. *J. Environ. Qual.* 28:1243–1251.

Elrashidi, M. A. and S. Larsen. 1978. Effect of phosphate addition on the solubility of phosphate in soil. *Plant Soil* 50:585–596.

Elser, J. and E. Bennett. 2011. A broken biogeochemical cycle. *Nature* 478:29–31.

Espinosa, M., B. L. Turner, and P. M. Haygarth. 1999. Preconcentration and separation of trace phosphorus compounds in soil leachate. *J. Environ. Qual.* 28:1497–1504.

Evans, R. L. and J. J. Jurinak. 1976. Kinetics of phosphate release from a desert soil. *Soil Sci.* 121:205–211.

Fageria, N. K. 2009. *The Use of Nutrients by Crop Plants*. Boca Raton, FL: CRC Press.

Fageria, N. K. and V. C. Baligar. 2008. Ameliorating soil acidity of tropical Oxisols by liming for sustainable crop production. *Adv. Agron.* 99:345–399.

Fageria, N. K., V. C. Baligar, and B. A. Bailey. 2005. Role of cover crops in improving soil and row crop productivity. *Commun. Soil Sci. Plant Anal.* 36:2733–2757.

Fageria, N. K., V. C. Baligar, and C. A. Jones. 2011. *Growth and Mineral Nutrition of Field Crops*, 3rd edn. Boca Raton, FL: CRC Press.

Fageria, N. K., A. B. Santos, and V. C. Baligar. 1997. Phosphorus soil test calibration for lowland rice on an Inceptisol. *Agron. J.* 89:737–742.

Foth, H. D. and B. G. Ellis. 1988. *Soil Fertility*. New York: John Wiley & Sons.

Foy, R. H. 2005. The return of the phosphorus paradigm: Agricultural phosphorus and eutrophication. In: *Phosphorus: Agriculture and the Environment*, pp. 911–939. Madison, WI: ASA, CSSA, and SSSA.

Franzluebbers, A. J., F. M. Hons, and D. A. Zuberer. 1995. Soil organic carbon, microbial biomass, and mineralizable carbon and nitrogen in sorghum. *Soil Sci. Soc. Am. J.* 59:460–466.

Gardner, B. R. and P. J. Jones. 1973. Effects of temperature on phosphate sorption isotherms and phosphate desorption. *Commun. Soil Sci. Plant Anal.* 4:83–93.

Gascho, G. J., R. D. Wauchope, J. G. Davis, C. C. Truman, C. C. Dowler, J. E. Hook, H. R. Sumner, and A. W. Johnson. 1998. Nitrate-nitrogen, soluble, and bioavailable phosphorus runoff from simulated rainfall after fertilizer application. *Soil Sci. Soc. Am. J.* 62:1711–1718.

Gaterell, M. R., R. Gay, R. Wilson, R. J. Gochin, and J. N. Lester. 2000. An economic and environmental evaluation of the opportunities for substituting phosphorus recovered from waste water treatment works in existing U.K. fertilizer markets. *Environ. Technol.* 21:1067–1084.

Gburek, W. J., E. Barberis, P. M. Haygarth, B. Kronvang, and C. Stamm. 2005. Phosphorus in the landscape. In: *Phosphorus: Agriculture and the Environment*, pp. 941–979. Madison, WI: ASA, CSSA, and SSSA.

Gburek, W. J. and A. N. Sharpley. 1998. Hydraulic controls on phosphorus loss from upland agricultural watersheds. *J. Environ. Qual.* 27:267–277.

Gburek, W. J., A. N. Sharpley, A. L. Heathwaite, and G. J. Folmar. 2000. Phosphorus management at the watershed scale: A modification of the phosphorus index. *J. Environ. Qual.* 29:130–144.

Giasson, E. 2002. *Application of Decision Analysis Techniques for Land Use Planning and Management*. Ithaca, NY: Cornell University.

Greaves, J., P. Hobbes, D. Chadwick, and P. Haygarth. 1999. Prospects for the recovery of phosphorus from animal manures: A review. *Environ. Technol.* 20:697–708.

Green, D. B., T. J. Logan, and N. E. Smeck. 1978. Phosphate adsorption-desorption characteristics of suspensed sediments in Maumee river basin of Ohio. *J. Environ. Qual.* 7:208–212.

Haygarth, P. M. and S. C. Jarvis. 1999. Transfer of phosphorus from agricultural soils. *Adv. Agron.* 66:195–249.

Haygarth, P. M. and A. N. Sharpley. 2000. Terminology for phosphorus transfer. *J. Environ. Qual.* 29:10–15.

He, Z. L., M. Zhang, P. J. Stoffella, X. E. Yang, D. J. Banks, and D. V. Calvert. 2006. Phosphorus concentrations and loads in surface runoff from sandy soils under crop production. *Soil Sci. Soc. Am. J.* 70:1807–1816.

Heathwaite, A. L., P. Griffiths, and R. J. Parkinson. 1998. Nitrogen and phosphorus in runoff from grassland with buffer strips following application of fertilizers and manures. *Soil Use Manage.* 14:142–148.

Heathwaite, A. L., P. Haygarth, R. Matthews, N. Preedy, and P. Butler. 2005a. Evaluating colloidal phosphorus delivery to surface waters from diffuse agricultural sources. *J. Environ. Qual.* 34:287–298.

Heathwaite, A. L., P. M. Hygarth, and R. Dils. 2000. Pathway of phosphorus transport. In: *Agriculture and Phosphorus Management: The Chesapeake Bay*, ed. A. N. Sharpley, pp. 107–130. Boca Raton, FL: Lewis Publishers.

Heathwaite, A. L., A. N. Sharpley, M. Bechmann, and S. Rekolainen. 2005b. Assessing the risk and magnitude of agricultural nonpoint source phosphorus pollution. In: *Phosphorus: Agriculture and the Environment*, ed. L. K. Al-Amoodi, pp. 981–1020. Madison, WI: ASA, CSSA, and SSSA.

Heckrath, G., P. Bookes, P. Pouloton, and K. Goulding. 1995. Phosphorus leaching from soils containing different phosphorus concentrations in the Broadbalk Experiment. *J. Environ. Qual.* 24:904–910.

Hens, M. and R. Merckx. 2001. Functional characterization of colloidal phosphorus species in the soil solution of sandy soils. *Environ. Sci. Technol.* 35:493–500.

Ilg, K., J. Siemens, and M. Kaupenjohnn. 2005. Colloidal and dissolved phosphorus in sandy soils affected by phosphorus saturation. *J. Environ. Qual.* 34:926–935.

Izuno, F. T., C. A. Sanchez, F. J. Cole, A. B. Bottcher, and D. B. Jones. 1991. Phosphorus concentrations in drainage waters in the Everglades agricultural areas. *J. Environ. Qual.* 20:608–619.

Jarvie, H. P., A. N. Sharply, P. J. A. Withers, J. T. Scott, B. E. Haggard, and C. Neal. 2013. Phosphorus mitigation to control river eutrophication: Murky waters, inconvenient truths and postnormal science. *J. Environ. Qual.* 42:295–304.

Jensen, M. B., T. B. Olsen, H. C. B. Hansen, and J. Magid. 2000. Dissolved and particulate phosphorus in leachate from structure soil amended with fresh cattle faeces. *Nutr. Cycl. Agroecosyst.* 56:253–261.

Jordan, C. and R. V. Smith. 1985. Factors affecting leaching of nutrients from an intensively managed grassland in County Knuuttila, Northern Ireland. *J. Environ. Manag.* 20:1–15.

Kern, J. S. and M. G. Johnson. 1993. Conservation tillage impacts on national soil and atmospheric carbon levels. *Soil Sci. Soc. Am. J.* 57:200–210.

Klatt, J. G., A. P. Mallarino, J. A. Downing, J. A. Kopaska, and D. J. Wittry. 2003. Soil phosphorus management practices and their relationship to phosphorus delivery in the Iowa Clear Lake agricultural watershed. *J. Environ. Qual.* 32:2140–2149.

Kleinman, P. J. A., A. N. Sharpley, K. Gartley, W. M. Jarrell, S. Kuo, R. G. Menon, R. Myers, K. R. Reddy, and E. O. Skogley. 2001. Interlaboratory comparison of soil phosphorus extracted by various soil test methods. *Commun. Soil Sci. Plant Anal.* 32:2325–2345.

Koluvek, P. K., K. K. Tanji, and T. J. Trout. 1993. Overview of erosion from irrigation. *J. Irrig. Drain. Div. Am. Soc. Civ. Eng.* 119:929–946.

Kotak, B. G., E. E. Prepas, and S. E. Hrudey. 1994. Blue green algal toxins in drinking water supplies: Research in Alberta. *Lake Line* 14:37–40.

Kovar, J. L. and N. Claassen. 2005. Soil–root interactions and phosphorus nutrition of plants. In: *Phosphorus: Agriculture and the Environment*, ed. L. K. Al-Amoodi, pp. 379–414. Madison, WI: ASA, CSSA, and SSSA.

Lal, R. 1997. Residue management, conservation tillage and soil restoration for mitigating greenhouse effects by CO_2-enrichment. *Soil Tillage Res.* 43:81–107.

Lawrence, R. R., R. A. Leonard, and J. M. Sheridan. 1985. Managing riparian ecosystems to control non-point pollution. *J. Soil Water Conserv.* 40:87–91.

Lee, L. K. 1990. The dynamics of declining soil erosion rates. *J. Soil Water Conserv.* 45:622–624.

Leinweber, P., R. Meissner, K. U. Eckhardt, and J. Seeger. 1999. Management effects on forms of phosphorus in soil and leaching potential. *Eur. J. Soil Sci.* 50:413–424.

Lemunyon, J. and T. C. Daniel. 1998. Phosphorus management for water quality protection: A national effort. In: *Soil Testing for Phosphorus: Environmental Uses and Implications*, ed. J. T. Sims, pp. 1–4. Newark, DE: University of Delaware.

Lemunyon, J. L. and R. G. Gilbert. 1993. Concept and need for a phosphorus assessment tool. *J. Prod. Agric.* 6:483–486.

Lentz, R. D., R. E. Sojka, and C. W. Robbins. 1998. Reducing phosphorus losses from surface irrigated fields: Emerging polyacrylamide technology. *J. Environ. Qual.* 27:305–312.

Levine, S. L. and D. W. Schindler. 1989. Phosphorus, nitrogen and carbon dynamics of Experimnetal Lake. 303 during recovery from eutrophication. *Can. J. Fish Aquat. Sci.* 46:2–10.

Levy, G. J., J. Levin, and I. Shainberg. 1995. Polymer effects on runoff and soil erosion from sodic soils. *Irrig. Sci.* 16:9–14.

Li, W. C., D. E. Armstrong, J. D. H. Williams, F. F. Harris, and J. K. Syers. 1972. Rate and extent of inorganic-phosphate exchange in lake sediments. *Soil Sci. Soc. Am. Proc.* 36:279–285.

Line, D. E., W. A. Harman, G. D. Jennings, E. J. Thompson, and D. L. Osmond. 2000. Nonpoint source pollutant load reduction associated with livestock exclusion. *J. Environ. Qual.* 29:1882–1890.

Little, C. E. 1988. Rural clean water: The Okeechobee story. *J. Soil Water Conserv.* 43:386–390.

Liu, J., J. Yang, X. Liang, Y. Zhao, B. J. Cade-Menum, and Y. Hu. 2014. Molecular speciation of phosphorus present in readily dispersible colloids from agricultural soils. *Soil Sci. Soc. Am. J.* 78:47–53.

Lu, Q., Z. L. He, D. A. Graetz, P. J. Stoffella, and X. E. Yang. 2010. Phytoremediation to remove nutrients and improve eutrophic stormwaters using water lettuce (*Pistia stratiotes* L.). *Environ. Sci. Pollut. Res.* 17:84–96.

Lynch, J. M. and E. Bragg. 1985. Microorganisms and soil aggregate stability. *Adv. Soil Sci.* 2:133–171.

Mallarino, A. P. and J. S. Schepers. 2005. Role of precision agriculture in phosphorus management practices. In: *Phosphorus : Agriculture and Environment*, ed. L. K. Al-Amoodi, pp. 881–908. Madison, WI: ASA, CSSA, and SSSA.

Martin, A. and G. D. Cooke. 1994. Health risks in eutrophication water supplies. *Lake Line* 14:24–26.

McColl, R. H. S., E. White, and A. R. Gibson. 1977. Phosphorus and nitrate runoff in hill pasture and forest catchments. *N. Z. J. Mar. Freshwater Res.* 11:729–744.

McDowell, R. W., J. J. Drewery, R. W. Muirhead, and R. J. Paton. 2003. Cattle treading and phosphorus and sediment loss in overland flow from grazed cropland. *Aust. J. Soil Res.* 41:1521–1532.

McDowell, R. W. and A. N. Sharpley. 2001. Approximating phosphorus release from soils to surface runoff and subsurface drainage. *J. Environ. Qual.* 30:508–520.

Meals, D. W. 2000. Lake Champlain basin agricultural watershed section 319 national monitoring program project, Year 6 Annual Report: May 1998–September 1999. Waterbury, CT: Vermont Department of Environmental Conservation.

Meisinger, J. J., W. L. Hargrove, R. B. Mikkelsen Jr, J. R. Williams, and V. W. Benson. 1991. Effects of cover crops on groundwater quality. In: *Cover Crops for Clean Water. Proceedings of the International Conference on Soil Water Conservation Society of America*, ed. W. L. Hargrove, pp. 57–68. Ankeny, IA: Soil Conservation Society of America.

Menzel, R. G., E. D. Rhoades, A. E. Olness, and S. J. Smith. 1978. Variability of annual nutrient and sediment discharges in runoff from Oklahoma cropland and rangeland. *J. Environ. Qual.* 7:401–406.

Moore, P. A. Jr, T. C. Daniel, and D. R. Edwards. 1999. Reducing phosphorus runoff and improving poultry production with alum. *Poult. Sci.* 78:692–698.

Moore, P. A. Jr, T. C. Daniel, and D. R. Edwards. 2000. Reducing phosphorus runoff and inhibiting ammonium loss from poultry manure with aluminum sulfate. *J. Environ. Qual.* 29:37–49.

Moore, P. A. Jr, B. C. Joern, and T. L. Provin. 1998. Improvements needed in environmental soil testing for phosphorus. In: *Soil Testing for Phosphorus: Environmental Uses and Implications*, ed. J. T. Sims, pp. 21–29. Newark, DE: University of Delaware.

Mozaffari, M. and J. T. Sims. 1996. Phosphorus transformations in poultry litter-amended soils of the Atlantic Coastal Plain. *J. Environ. Qual.* 25:1357–1365.

Mullins, G., B. Joern, and P. Moore. 2005. By-product phosphorus: Sources, characteristics, and management. In: *Phosphorus: Agriculture and the Environment*, ed. L. K. Al-Amoodi, pp. 829–879. Madison, WI: ASA, CSSA, and SSSA.

Muscutt, A. D. and P. J. A. Withers. 1996. The phosphorus content of rivers in England and Wales. *Water Res.* 30:1258–1268.

National Research Council. 2000. *Clean Coastal Waters: Understanding and Reducing the Effects of Nutrient Pollution*. Washington, DC: National Academic Press.

Nicholaichuk, W. and D. W. L. Read. 1978. Nutrient runoff from fertilized and unfertilized fields in Western Canada. *J. Environ. Qual.* 7:542–544.

Oloya, T. O. and T. J. Logan. 1980. Phosphate desorption from soils and sediments with varying levels of extractable phosphate. *J. Environ. Qual.* 9:526–531.

Ortega, R. A., G. A. Peterson, and D. G. Westfall. 2002. Residue accumulation and changes in soil organic matter as affected by cropping intensity in no-till dryland agroecosystems. *Agron. J.* 94:944–954.

Owens, P. N. and D. E. Walling. 2002. The phosphorus content of fluvial sediment in rural and industrialized river basins. *Water Res.* 36:685–701.

Park, E. J. and A. J. M. Smucker. 2005. Saturated hydraulic conductivity and porosity within macroaggregates modified by tillage. *Soil Sci. Soc. Am. J.* 69:38–45.

Parry, R. 1998. Agricultural phosphorus and water quality: A US Environmental protection agency perspective. *J. Environ. Qual.* 27:258–261.

Paustian, K., O. Andren, H. Janzen, R. Lal, P. Smith, G. Tian, H. Tiessen, M. V. Noordwijk, and P. Woomer. 1997. Agricultural soil as a C sink to offset CO_2 emissions. *Soil Use Manage.* 13:230–244.

Penn, C. J. and J. T. Sims. 2002. Phosphorus forms in biosolids amended soils and losses in runoff: Effects of wastewater treatment process. *J. Environ. Qual.* 31:1349–1361.

Pierzynski, G. M., R. W. McDowell, and J. T. Sims. 2005. Chemistry, cycling and potential movement of inorganic phosphorus in soils. In: *Phosphorus: Agriculture and the Environment*, eds. J. T. Sims and A. N. Sharpley, pp. 53–86. Madison, WI: ASA, CSSA, and SSSA.

Pote, D. H., T. C. Daniel, A. N. Sharpley, P. A. Moore Jr, D. R. Edwards, and D. J. Nichols. 1996. Relating extractable soil phosphorus to phosphorus losses in runoff. *Soil Sci. Soc. Am. J.* 60:855–859.

Poulsen, H. D. 2000. Phosphorus utilization and excretion in pig production. *J. Environ. Qual.* 29:24–27.

Preedy, N., K. B. McTiernan, R. Matthews, L. Heathwaite, and P. M. Haygarth. 2001. Rapid incidental phosphorus transfers from grassland. *J. Environ. Qual.* 30:2105–2112.

Regelink, I. C., G. F. Koopmans, C. V. D. Salm, L. Weng, and W. H. V. Riemsdijk. 2013. Characterization of colloidal phosphorus species in drainage waters from a clay soil using asymmetric flow field flow fractionation. *J. Environ. Qual.* 42:464–473.

Richardson, A. E., J. P. Lynch, P. R. Ryan, E. Delhaize, E. A. Smith, and S. E. Smith. 2011. Plant and microbial strategies to improve the phosphorus efficiency of agriculture. *Plant Soil* 349:121–156.

Robinson, J. S., A. N. Sharpley, and S. J. Smith. 1994. Development of a method to determine bioavailable phosphorus loss in agricultural runoff. *Agric. Ecosyst. Environ.* 47:287–297.

Romkens, M. J. and D. W. Nelson. 1974. Phosphorus relationships in runoff from fertilized soils. *J. Environ. Qual.* 3:10–13.

Ryden, J. C., J. K. Syers, and R. F. 1972. Harris. Sorption of inorganic-phosphate by laboratory ware—Implications in environmental phosphorus techniques. *J. Environ. Qual.* 1:431–434.

Sawyer, C. N. 1947. Fertilization of lakes by agricultural and urban drainage. *N. Engl. Water Works Assoc. J.* 61:109–127.

Schoumans, O. F. and W. J. Chardon. 2003. Risk assessment methodologies for predicting phosphorus losses. *J. Plant Nutr. Soil Sci.* 166:403–408.

Sharpley, A. N. 1995. Identifying sites vulnerable to phosphorus loss on agricultural runoff. *J. Environ. Qual.* 24:947–951.

Sharpley, A. N. 1997. Rainfall frequency and nitrogen and phosphorus in runoff from soil amended with poultry litter. *J. Environ. Qual.* 26:1127–1132.

Sharpley, A. N., L. R. Ahuja, M. Yamamoto, and R. G. Menzel. 1981. The kinetics of phosphorus desorption from soil. *Soil Sci. Soc. Am. J.* 45:493–496.

Sharpley, A. N., S. C. Chapra, R. Wedepohl, J. T. Sims, T. C. Daniel, and K. R. Reddy. 1994. Managing agricultural phosphorus for protection of surface waters: Issues and options. *J. Environ. Qual.* 23:437–451.

Sharpley, A. N., T. C. Daniel, J. T. Sims, and D. H. Pote. 1996. Determination environmentally sound soil phosphorus levels. *J. Soil Water Conserv.* 51:160–166.

Sharpley, A. N., B. Foy, and P. Withers. 2000. Practical and innovative measures for the control of agricultural phosphorus losses to water: An overview. *J. Environ. Qual.* 29:1–9.

Sharpley, A. N. and R. G. Menzel. 1987. The impact of soil and fertilizer phosphorus on the environment. *Adv. Agron.* 41:297–324.

Sharpley, A. N. and B. Moyer. 2000. Phosphorus forms in manure and compost and their release during simulated rainfall. *J. Environ. Qual.* 29:1462–1469.

Sharpley, A. N., R. W. Rillman, and J. K. Syers. 1977. Use of laboratory extraction data to predict losses of dissolved inorganic phosphate in surface runoff and tile drainage. *J. Environ. Qual.* 6:33–36.

Sharpley, A. N., S. J. Smith, W. A. Berg, and J. R. Williams. 1985. Nutrient runoff losses as predicted by annual and monthly soil sampling. *J. Environ. Qual.* 14:354–360.

Sharpley, A. N., S. J. Smith, and J. W. Naney. 1987. Environmental impact of agricultural nitrogen and phosphorus use. *J. Agric. Food Chem.* 35:812–817.

Sharpley, A. N. and J. K. Syers. 1979a. Loss of nitrogen and phosphorus in tile drainage as influenced by urea application and grazing animals. *N. Z. J. Agric. Res.* 22:127–131.

Sharpley, A. N. and J. K. Syers. 1979b. Phosphorus inputs into a stream draining an agricultural watershed. 2. Amounts contributed and relative significance of runoff types. *Water Air Soil Pollut.* 11:417–428.

Sharpley, A. N., J. K. Syers, and R. W. Tillman. 1978. An improved soil-sampling procedure for the prediction of dissolved inorganic phosphate concentrations in surface runoff from pasture. *J. Environ. Qual.* 7:455–456.

Shober, A. L. and J. T. Sims. 2003. Phosphorus restrictions for land application of biosolids: Current status and future trends. *J. Environ. Qual.* 32:1955–1964.

Siddique, M. T., J. S. Robinson, and B. J. Alloway. 2000. Phosphorus reactions and leaching potential in soils amended with a sewage sludge. *J. Environ. Qual.* 29:1931–1938.

Siemens, J., K. Ilg, H. Pagel, and M. Kaupenjohann. 2008. Is colloid-faciliated phosphorus leaching triggered by phosphorus accumulation in sandy soils? *J. Environ. Qual.* 37:2100–2107.

Sims, J. T. 1993. Environmental soil testing for phosphorus. *J. Prod. Agric.* 6:501–507.

Sims, J. T. 2000. The role of soil testing in environmental risk assessment for phosphorus. In: *Agriculture and Phosphorus Management: The Chesapeake Bay*, ed. A. N. Sharpley, pp. 57–81. Boca Raton, FL: CRC Press.

Sims, J. T., S. Hodges, and J. Davis. 1998a. Soil testing for phosphorus: Current status and uses in nutrient management programs. In: *Soil Testing for Phosphorus: Environmental Uses and Implications*, ed. J. T. Sims, pp. 13–20. Newark, DE: University of Delaware.

Sims, J. T. and P. J. A. Kleinman. 2005. Managing agricultural phosphorus for environmental protection. In: *Phosphorus: Agriculture and the Environment*, ed. L. K. Al-Amoodi, pp. 1021–1068. Madison, WI: ASA, CSSA, and SSSA.

Sims, J. T. and N. J. Luka-McCafferty. 2002. On-farm evaluation of aluminum sulfate (alum) as a poultry litter amendment: Effects on litter properties. *J. Environ. Qual.* 31:2066–2073.

Sims, J. T., R. R. Simard, and B. C. Joern. 1998b. Phosphorus losses in agricultural drainage: Historical perspectives and current research. *J. Environ. Qual.* 27:277–293.

Six, J., E. T. Elliott, and K. Paustain. 2000a. Soil microaggregate turnover and microaggregate formation: A mechanism for c sequestration under no-tillage agriculture. *Soil Biol. Biochem.* 32:2099–2013.

Six, J., C. Feller, K. Denef, S. M. Ogle, J. C. Moraes, and A. Albrecht. 2002a. Soil organic matter, biota and aggregation in temperate and tropical soils-effects of no-tillage. *Agronomie* 22:755–775.

Six, J., K. Paustian, E. T. Elliott, and C. Combrink. 2000b. Soil structure and organic matter: I. Distribution of aggregate size classes and aggregate-associated carbon. *Soil Sci. Soc. Am. J.* 64:681–689.

Smil, V. 1999a. Nitrogen in crop production: An account of glob. flows. *Glob. Biogeochem. Cycles* 13:647–662.

Smil, V. 1999b. Crop residues: Agricultures largest harvest. *BioScience* 49:299–308.

Smil, V. 2000. Phosphorus in the environment: Natural flows and human interferences. *Annu. Rev. Energy Environ.* 25:53–88.

Smith, H. J. C., G. J. Levy, and I. Shainberg. 1990. Water-droplet energy and soil amendments: Effect on infiltration and erosion. *Soil Sci. Soc. Am. J.* 54:1084–1087.

Soil Science Society of America. 2008. *Glossary of Soil Science Terms*. Madison, WI: SSSA.

Sojka, R. E. and D. L. Bjorneberg. 2002. Erosion, controlling irrigation-induced. In: *Encyclopedia of Soil Science*, 1st edn., ed. R. Lal, pp. 411–414. New York: Marcel Dekker.

South Florida Water Management District. 1994. U.S. district court/southern district, Case number 88-1880-CIV.

Stamm, C., H. Fluhler, R. Gachter, J. Leuenberger, and H. Wunderli. 1998. Preferential transport of phosphorus in drained grassland soils. *J. Environ. Qual.* 27:515–522.

Tate, K. W., G. A. Nader, D. J. Lewis, E. R. Atwill, and J. M. Conor. 2000. Evaluation of buffers to improve the quality of runoff from irrigated pastures. *J. Soil Water Conserv.* 55:473–478.

Taylor, A. W. and H. M. Kunishi. 1971. Phosphate equilibria on stream sediment and soil in a watershed draining an agricultural region. *J. Agric. Food Chem.* 19:827–831.

Tisdall, J. M. and J. M. Oades. 1982. Organic matter and water-stable aggregates in soils. *J. Soil Sci.* 33:141–163.

USEPA (U.S. Environmental Protection Agency). 1986. Quality criteria for water, Report no. EPA-440/5-86-001. Washington, DC: Office of Water Regulation and Standards.

USEPA (U.S. Environmental Protection Agency). 1998. Water quality functions of riparian forest buffer systems in the Chesapeake Bay watershed, Report no. EPA 903-R-95-004. Nutrient subcommittee, Chesapeake Bay Program. Annapolis, MD: USEPA.

Uusi-Kamppa, J., B. Braskerud, H. Jansson, N. Syverson, and R. Uusitalo. 2000. Buffer zones and constructed wetlands as filters for agricultural phosphorus. *J. Environ. Qual.* 29:151–158.

Valk, H., H. A. Metcalf, and P. J. A. Withers. 2000. Prospects for minimizing phosphorus excretion in ruminants by dietary manipulation. *J. Environ. Qual.* 29:167–175.

Van Moorleghem, C., N. D. Schutter, E. Smolders, and R. Merckx. 2013. The bioavailability of colloidal and dissolved organic phosphorus to the alga *Pseudokichnerlella subcapitata* in relation to analytical phosphorus measurements. *Hydribiologia* 709:41–53.

Vollenweider, R. A. 1968. Scientific fundamentals of the eutrophication of lakes and flowing waters, with particular reference to nitrogen and phosphorus. Publication No. DAS/Sai/68.27. Paris, France: Organization for eEconomic Cooperation and Development, Directorate for Scientific Affairs.

Wagenet, R. J. 1990. Quantative prediction of the leaching of organic and inorganic solutes in soil. *Philos. Trans. R. Soc. Lond. B* 329:321–330.

Way, J. T. 1850. On the power of soils to absorb manure. *J. R. Agric. Soc. Engl.* 11:313–379.

Weaver, D. M., G. S. P. Ritchie, and G. C. Andrson. 1988a. Phosphorus leaching in sandy soils II. Laboratory studies of the long-term effects of the phosphorus source. *Aust. J. Soil Res.* 26:191–200.

Weaver, D. M., G. S. P. Ritchie, G. C. Andrson, and D. M. Deeley. 1988b. Phosphorus leaching in sandy soils I. Short term effects of fertilizer applications and environmental conditions. *Aust. J. Soil Res.* 26:177–190.

White, R. E. and P. H. T. Beckett. 1964. Studies on the phosphate potential of soils. Part I—The measurement of phosphate potential. *Plant Soil* 20:1–16.

Withers, P. J. A. and S. C. Jarvis. 1998. Mitigation options for diffuse phosphorus loss to water. *Soil Use Mange.* 14:186–192.

Withers, P. J. A., D. M. Nash, and C. A. M. Laboski. 2005. Environmental management of phosphorus fertilizers. In: *Phosphorus: Agriculture and the Environment*, eds. J. T. Sims and A. N. Sharpley, pp. 781–827. Madison, WI: ASA, CSSA, and SSSA.

Wood, C. W. 1998. Agricultural phosphorus and water quality: An overview. In: *Soil Testing for Phosphorus: Environmental Uses and Implications*, ed. J. T. Sims, pp. 5–12. Newark, DE: University of Delaware.

Worley, J. W., M. L. Cabrera, and L. M. Risse. 2000. Reduced levels of alum to amend broiler litter. *Appl. Eng. Agric.* 16:441–444.

Wright, A. L. and F. M. Hons. 2004. Soil aggregation and carbon and nitrogen storage under soybean cropping sequences. *Soil Sci. Soc. Am. J.* 68:507–513.

Wright, A. L., and F. M. Hons. 2005. Carbon and nitrogen sequestration and soil aggregation under sorghum cropping sequences. *Biol Fertil Soils* 41: 95–100.

Wu, Z., L. D. Satter, and R. Sojo. 2000. Milk production, reproductive performance, and fecal excretion of phosphorus by dairy cows fed three amounts of phosphorus. *J. Dairy Sci.* 83:1028–1041.

Zhao, S. L., S. C. Gupta, D. R. Huggins, and J. F. Moncrief. 2001. Tillage and nutrient source effects on surface and subsurface water quality at corn planting. *J. Environ. Qual.* 30:998–1008.

7 Management Practices for Optimizing Phosphorus Availability to Crop Plants

7.1 INTRODUCTION

Phosphorus is essential for growth and development of crop plants. P has been used liberally during the last few decades to raise soil P content and improve yield in developed as well as developing countries. A 40% increase in P fertilizer use occurred in the world during the 2010 cropping season as compared to the 2000 cropping season. During the same period, an increase in nitrogen (N) was about 31% and potassium (K) about 25% (Table 7.1). However, in recent years, economic and environmental concerns are encouraging growers in developed as well as developing countries to use P more rationally (Withers et al., 2005). In addition, the use of chemical fertilizers should be combined with the use of organic manures and other management practices to produce higher crop yields and reduce risk of P loss from the soil–plant system (Fageria, 2009).

Use of essential plant nutrients in adequate amount and proper balance is key to successful crop production. The best strategy to use nutrients in adequate amount and balance, including P, is to adopt an integrated nutrient management system. Integrated nutrient management includes the use of chemical fertilizers in adequate rate and effective methods of application at appropriate timing during crop growth. In addition, the use of appropriate crop rotation, conservation tillage, improving water use efficiency (WUE), and controlling diseases, insects, and weeds are also part of an integrated nutrient management system. Fertilizers that are applied in a balanced method and used efficiently for optimum crop production will avoid unnecessary wastage, minimize surplus P and potential soil P accumulation, and provide the optimum economic returns to growers (Steen, 1996; Withers et al., 2005; Fageria, 2009, 2014).

The objective of this chapter is to provide an overview of a variety of efficient management practices and associated concepts that are currently applied or can potentially be applied to agronomic and environmental P management.

7.2 SOIL MANAGEMENT

Soil management practices that can optimize P uptake and use efficiency by crop plants are liming acid soils, use of gypsum in acid soils, and use of organic manures. These soil management practices improve soil environment that in turn improve plant root growth as well as P availability to crop plants. However, these soil management practices should be used in combination with fertilizers and crop management practices to achieve optimum results in relation to crop yields and environmental benefits.

7.2.1 LIMING ACID SOILS

Soil acidity affects a large land area in various parts of the world. Liming is a dominant and effective practice to overcome these constraints and improve crop production on acid soils. Lime is called the foundation of crop production or "workhorse" in acid soils. Lime requirement for crops grown on acid soils is determined by the quality of liming material, soil fertility, crop species, cultivars within species, crop management practices, and economic considerations. Soil pH, base saturation, and aluminum (Al) saturation are important acidity indices that can be used as a basis for liming acid soils. In addition, crop

TABLE 7.1
Use of N, P, and K (10^6 Tons) in Different Regions of the World

	N		P		K	
Region	**2000**	**2010**	**2000**	**2010**	**2000**	**2010**
Africa	2.46	3.01	0.41	0.50	0.39	0.32
Eastern	0.33	0.58	0.11	0.14	0.07	0.07
Middle	0.03	0.02	0.00	0.00	0.02	0.01
Northern	1.45	1.72	0.16	0.23	0.10	0.06
Southern	0.43	0.39	0.08	0.07	0.12	0.09
Western	0.22	0.29	0.05	0.06	0.09	0.08
America	17.19	20.86	3.55	4.18	6.83	7.99
Northern	12.03	13.47	1.94	1.97	3.97	3.80
Central	1.75	1.56	0.19	0.13	0.28	0.27
Southern	3.25	5.65	1.40	2.06	2.51	3.88
Caribbean	0.17	0.18	0.03	0.02	0.07	0.03
Asia	46.72	67.03	7.73	13.02	6.52	10.48
Central	0.76	0.83	0.06	0.06	0.02	0.02
Eastern	23.21	35.75	4.11	7.65	3.40	4.79
Southern	15.26	21.34	2.45	4.20	1.53	3.27
Western	2.28	1.99	0.46	0.32	0.14	0.12
Southeastern	5.22	7.12	0.66	0.79	1.43	2.28
Europe	13.21	13.47	1.79	1.57	3.97	3.51
Northern	2.35	2.22	0.25	0.21	0.69	0.53
Southern	2.69	2.03	0.59	0.38	0.91	0.66
Eastern	3.54	5.00	0.39	0.69	0.94	1.51
Western	4.63	4.21	0.55	0.28	1.43	0.81
Oceania	1.19	1.51	0.69	0.58	0.29	0.20
World	80.79	105.89	14.17	19.86	18.00	22.50

Source: FAO, FAOSTAT data base-Agricultural production, Food and Agricultural Organization of the United Nations, Rome, Italy, http://faostat3.fao.org/faostat-gateway/go/to/home/E, accessed November 4, 2013.

responses to lime rate are vital tools for making liming recommendations for crops grown on acid soils (Fageria and Baligar, 2008). Liming acid soil influences the soil's physical, chemical, and biological properties in favor of higher crop yields and consequently higher P use efficiency (PUE).

7.2.1.1 Beneficial Effects of Liming

Liming improves crop yields and changes soils' physical and chemical properties. In addition, it reduces P immobilization in soils, improves growth of beneficial microorganisms, controls plant diseases, reduces leaching of heavy metals, and mitigates nitrous oxide emission from soils. These effects are synthesized in the succeeding section.

7.2.1.2 Improves Crop Yields

Liming improves crop yields in acid soils. Soybean (*Glycine max*) grain yield (GY) quadratically increased by the addition of lime (0–18 Mg ha^{-1}) (Figure 7.1). GY of soybean was about 2 Mg ha^{-1} with no lime and increased to 3.2 Mg ha^{-1} with the addition of 12 Mg lime ha^{-1}. Variation in GY was 86% with the use of dolomitic lime. Improvement in GY of soybean, common bean (*Phaseolus vulgaris* L.), and corn (*Zea mays*) grown on a Brazilian Oxisol soil is presented (Figure 7.2).

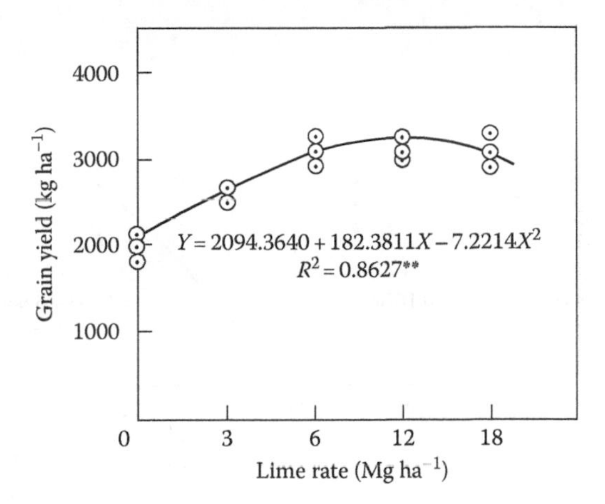

FIGURE 7.1 Relationship between lime rate and GY of soybean grown on Brazilian Oxisol. Values are averages of 3 years' field trial. (From Fageria, N.K., *The Use of Nutrients in Crop Plants*, CRC Press, Boca Raton, FL, 2009.)

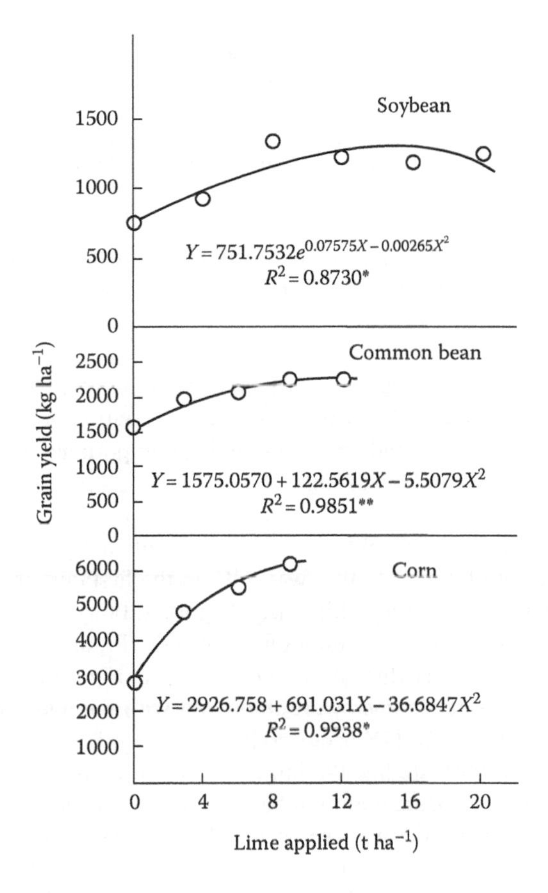

FIGURE 7.2 Influence of lime on GY of soybean, common bean, and corn. (From Fageria, N.K. and Stone, L.F., Managing soil acidity of Cerrado and Varzea of Brazil, Document no. 42, National Rice and Bean Research Center of EMBRAPA, Santo Antônio de Goiás, Brazil, 1999.)

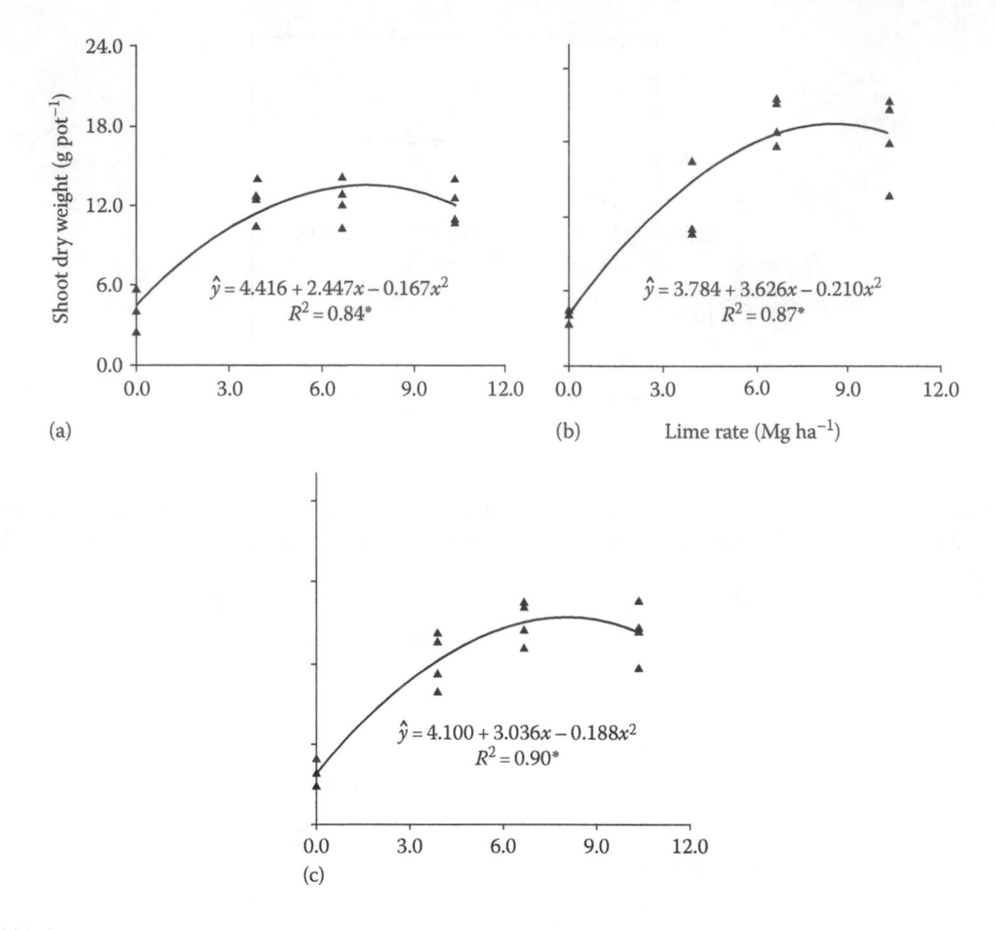

FIGURE 7.3 Influence of lime application on shoot dry weight of alfalfa at (a) first and (b) second cuttings and (c) average of two cuttings. (From Moreira, A. and Fageria, N.K., *Revista brasileira de ciência do solo*, 34, 1231, 2010.)

GY of three crops quadratically increased with increasing lime rate: 0–20 Mg ha^{-1} for soybean, 0–12 Mg ha^{-1} for common bean, and 0–9 Mg ha^{-1} for corn. Variation in GY due to liming was 87% for soybean, 99% for common bean, and corn. Liming is an important soil amendment for improving yield of cereals and legumes on Oxisol soils.

Moreira and Fageria (2010) studied the influence of liming on dry matter yield of alfalfa (*Medicago sativa*) (Figure 7.3). Dry weight of alfalfa was significantly increased in the two cuttings, and variation in yield increased due to liming was 84% in the first cutting, 87% in the second cutting, and 90% in the mean of two cuttings. Similarly, Fageria (2008) studied the influence of liming on the GY of dry bean (Figure 7.4), base saturation versus GY (Figure 7.5), calcium (Ca) saturation versus GY (Figure 7.6), magnesium (Mg) saturation versus GY (Figure 7.7), and K saturation versus GY (Figure 7.8). All these acidity indices were improved with the addition of lime, and GY was quadratically increased, except for K. GY occurred linearly when K content was increased from 1% to 3.5%. Fageria and Stone (1999) studied the influence of pH on the dry weight of shoot of wheat (*Triticum aestivum* L.), corn, soybean, common bean, and lowland rice (*Oryza glaberrima*) grown on a Brazilian Inceptisol. A quadratic increase occurred in the dry weight of wheat, corn, soybean, and common bean (Figure 7.9). The variation in shoot dry weight due to the change in soil pH was 66% for wheat, 50% for corn, 84% for soybean, and 46% for common bean. However, lowland rice dry matter was significantly reduced with increasing soil pH. This may be related to an increase in soil pH with flooding of lowland rice and/or acidity tolerance of rice plants (Fageria, 2014).

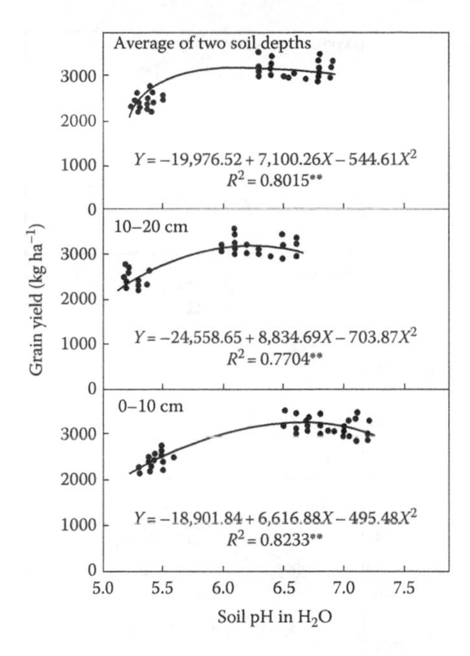

FIGURE 7.4 Influence of pH on GY of dry bean. (From Fageria, N.K., *Commun. Soil Sci. Plant Anal.*, 39, 845, 2008.)

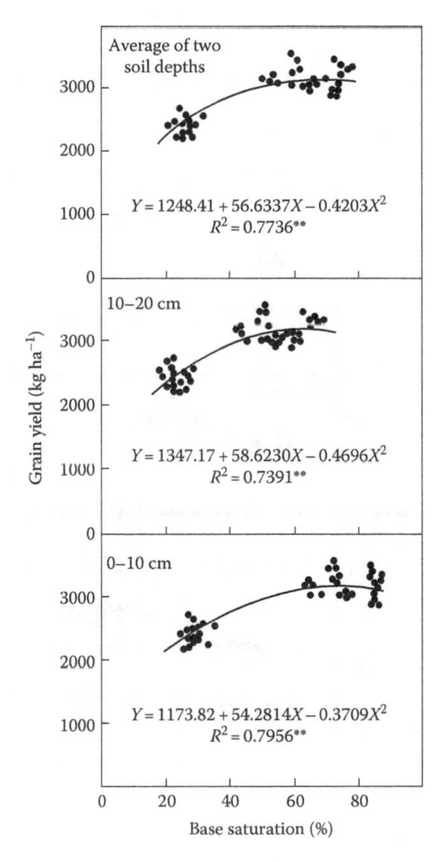

FIGURE 7.5 Influence of base saturation on GY of dry bean. (From Fageria, N.K., *Commun. Soil Sci. Plant Anal.*, 39, 845, 2008.)

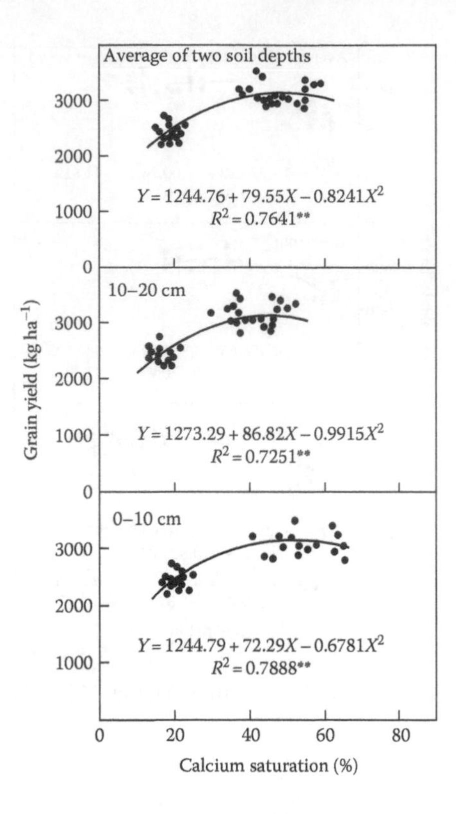

FIGURE 7.6 Influence of Ca saturation on GY of dry bean. (From Fageria, N.K., *Commun. Soil Sci. Plant Anal.*, 39, 845, 2008.)

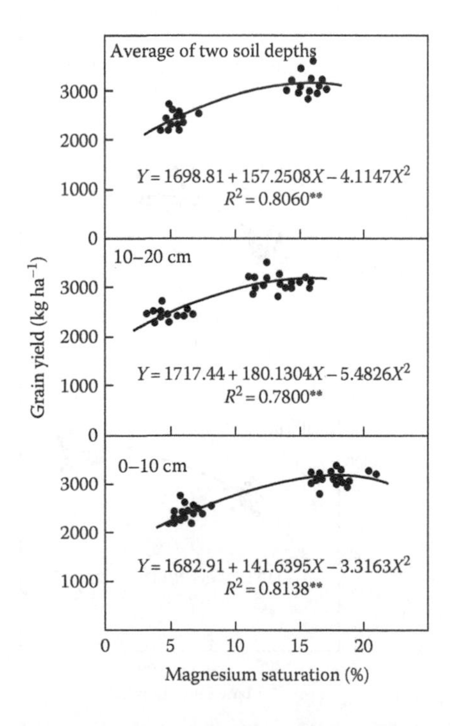

FIGURE 7.7 Influence of Mg saturation on GY of dry bean. (From Fageria, N.K., *Commun. Soil Sci. Plant Anal.*, 39, 845, 2008.)

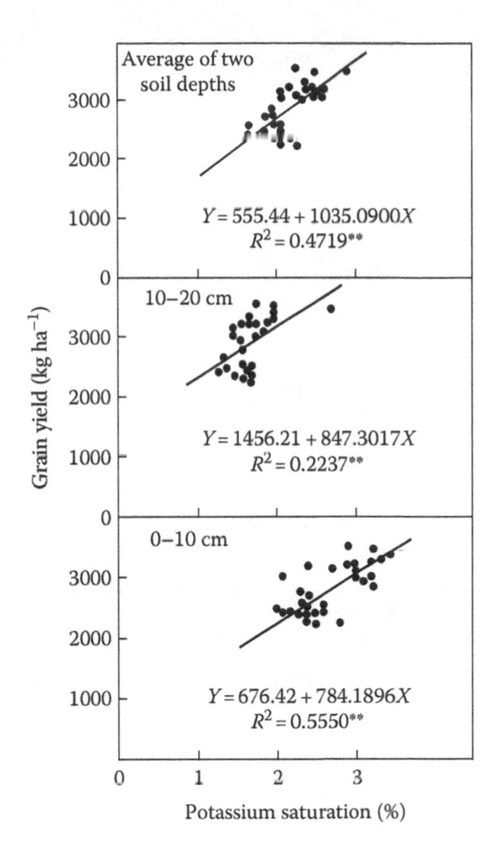

FIGURE 7.8 Influence of K saturation on GY of dry bean. (From Fageria, N.K., *Commun. Soil Sci. Plant Anal.*, 39, 845, 2008.)

7.2.1.3 Improves Soil Chemical Properties

Lime application significantly increases soil pH, Ca and Mg contents, base saturation, and cation exchange capacity (CEC) and reduces acidity (H + Al) in acid soils (Table 7.2). Analysis of variance indicated year × lime rate interactions were significant for all the soil chemical properties analyzed. Values for 3 years of these chemical properties are presented (Table 7.2). In the first year, pH increases from 5.3 to 7.3 with the lime rate of 0–18 Mg ha^{-1}. In the second year, the pH increase was 5.2–7.2 in the same lime rate range, and in the third year, the increase in pH was 4.7–7.7. Overall, a pH increase was 5.1–7.4 in the lime rate range of 0–18 Mg ha^{-1}. Increase in pH with lime application was associated with neutralization of Al + H ions and an increase in Ca and Mg concentration in the soil solution. Fageria and Stone (2004) and Fageria (2006) reported similar increases in pH with the application of lime in the range of 0–24 Mg ha^{-1} in Brazilian Oxisol soils. Overall, an increase in base saturation was 16.4%–91.5%, the H + Al decrease was 3.88–0.43 $cmol_c$ kg^{-1}, the Al decrease was 1.52–0 $cmol_c$ kg^{-1}, Ca increase was 0.45–3.14 $cmol_c$ kg^{-1}, Mg increase was 0.19–1.32 $cmol_c$ kg^{-1}, and the CEC increase was 4.64–5.01 $cmol_c$ kg^{-1} with the application of 0–18 Mg lime ha^{-1}. Fageria and Stone (2004) reported similar increases or decreases in acidity indices of Brazilian Oxisol with the application of lime in the range of 0–24 Mg ha^{-1}. Fageria (2001a) reported similar increases in Ca and Mg concentration in Brazilian Oxisol soils with the application of lime in the range of 0–20 Mg ha^{-1}.

Application of 3 Mg lime ha^{-1} neutralized all the Al^{3+}. This suggests that acidity that requires higher lime rate represents H^+ ions in the soil solution. Fageria and Morais (1987) reported similar results with the application of lime in Brazilian Oxisol soils. Soil acidity indices (pH, Ca, Mg,

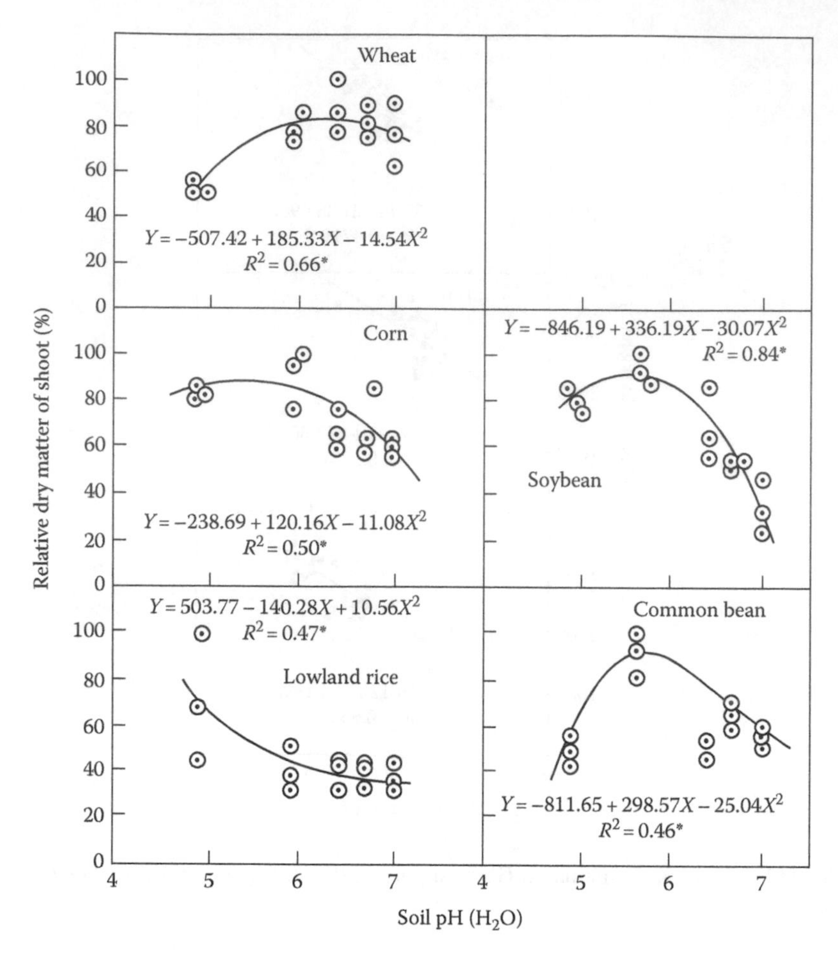

FIGURE 7.9 Influence of soil pH on relative dry matter yield of wheat, corn, soybean, common bean, and lowland rice. (From Fageria, N.K. and Stone, L.F., Managing soil acidity of Cerrado and Varzea of Brazil, Document no. 42, National Rice and Bean Research Center of EMBRAPA, Santo Antônio de Goiás, Brazil, 1999.)

base saturation, H + Al, acidity saturation, Ca/K, and Mg/K) are quadratically associated with GY (Table 7.3). Variability in GY was 93% due to soil pH, 96% due to soil Ca content, 94% due to soil Mg content, 97% due to base saturation, 91% due to H + Al content, 94% due to acidity saturation, 89% due to Ca/Mg ratio, 90% due to Ca/K ratio, and 91% due to Mg/K ratio (Table 7.3). The importance of acidity indices in increasing soybean yield was in the order of base saturation > Ca > Mg > acidity saturation > pH > Mg/K > H + Al > Ca/K > Ca/Mg. Fageria (2001b) reported more or less similar attribution of these indices in increasing soybean GY in Brazilian Oxisol soils.

Values for maximum GY (3100 kg ha^{-1}) calculated by quadratic regression equations were 7.1 for pH, 2.7 comol$_c$ kg^{-1} for Ca, 1.6 comol$_c$ kg^{-1} for Mg, 88% for base saturation, 0.49 comol$_c$ kg^{-1} for H + Al, 5.2 cmol$_c$ kg^{-1} for CEC, 1.92 for Ca/Mg ratio, 9.5 for Ca/K ratio, and 5.4 for Mg/K ratio. These values for 90% of the maximum GY (2900 kg ha^{-1}) were lowered as compared with maximum economic values (Table 7.3). Fageria (2001b) reported that maximum GY of soybean in Brazilian Oxisol soils was obtained with 63% base saturation, and at pH of 6.8, Ca and Mg values for maximum GY of soybean were 4 and 1.4 comol$_c$ kg^{-1}. According to EMBRAPA (1995), base saturation for soybean GY in the central region of Brazil should be near 70%. Variation in the results of acidity indices for maximum and economic yield of the present study with those reported earlier in literature may be due to use of different cultivars, variation in yield level, and other soil and crop management practices.

TABLE 7.2
Selected Soil Chemical Properties after Harvest of Three Soybean Crops as Influenced by Liming Treatments

	Lime Rate (Mg ha^{-1})				
Soil Property	**0**	**3**	**6**	**12**	**18**
First crop					
pH	5.3	6.0	6.5	7.0	7.3
Base saturation (%)	17.1	54.1	68.2	80.6	91.1
H + Al (cmol$_c$ kg^{-1})	3.90	2.53	1.85	1.23	0.63
Al (cmol$_c$ kg^{-1})	3.75	0	0	0	0
Ca (cmol$_c$ kg^{-1})	0.47	1.78	2.84	3.80	4.77
Mg (cmol$_c$ kg^{-1})	0.25	1.11	1.06	1.26	1.41
CEC (cmol$_c$ kg^{-1})	4.71	5.51	5.84	6.39	6.90
Second crop					
pH	5.2	5.7	6.4	6.9	7.2
Base saturation (%)	16.9	37.2	56.8	72.8	83.3
H + Al (cmol$_c$ kg^{-1})	3.85	2.75	1.99	1.22	0.67
Al (cmol$_c$ kg^{-1})	0.38	0.1	0	0	0
Ca (cmol$_c$ kg^{-1})	0.49	1.03	1.64	2.05	2.13
Mg (cmol$_c$ kg^{-1})	0.13	0.44	0.85	1.09	1.09
CEC (cmol$_c$ kg^{-1})	4.64	4.38	4.64	4.53	4.06
Third crop					
pH	4.7	6.1	6.7	7.5	7.7
Base saturation (%)	15.2	52.5	70.4	95.5	100
H + Al (cmol$_c$ kg^{-1})	3.88	2.04	1.26	0.19	0
Al (cmol$_c$ kg^{-1})	0.43	0	0	0	0
Ca (cmol$_c$ kg^{-1})	0.38	1.33	1.69	2.34	2.50
Mg (cmol$_c$ kg^{-1})	0.18	0.76	1.10	1.41	1.46
CEC (cmol$_c$ kg^{-1})	4.57	4.30	4.22	4.1	4.08
Average of three crops					
pH	5.1	5.9	6.5	7.1	7.4
BS (%)	16.4	47.9	65.1	83.0	91.5
H + Al (cmol$_c$ kg^{-1})	3.88	2.44	1.70	0.88	0.43
Al (cmol$_c$ kg^{-1})	1.52	0.03	0	0	0
Ca (cmol$_c$ kg^{-1})	0.45	1.38	2.06	2.73	3.14
Mg (cmol$_c$ kg^{-1})	0.19	0.77	1.00	1.25	1.32
CEC (cmol$_c$ kg^{-1})	4.64	4.73	4.9	5.01	5.01
Statistical analysis					
Year (Y)		**			
Lime rate (L)		**			
Y × L		**			

Source: Fageria, N.K. et al., *Commun. Soil Sci. Plant Anal.*, 44, 2941, 2013b.
**Significance at 1% probability level.

TABLE 7.3
Relationship between Soil Chemical Property (*X*) and Soybean GY

Soil Property	Regression Equation	R^2	VMY[a]	VMEY[b]
pH in H_2O	$Y = -9{,}884.7040 + 3{,}636.8190X - 254.6528X^2$	0.9260**	7.1	6.0
Ca (comol$_c$ kg^{-1})	$Y = 1{,}484.3560 + 1{,}189.5530X - 216.6682X^2$	0.9577**	2.7	1.6
Mg (comol$_c$ kg^{-1})	$Y = 1{,}650.7640 + 1{,}881.7360X - 584.0436X^2$	0.9362**	1.6	0.9
Base sat. (%)	$Y = 1{,}397.4520 + 38.7096X - 0.2203X^2$	0.9713**	88	51.0
H + Al (comol$_c$ kg^{-1})	$Y = 3{,}080.3400 + 93.4309X - 95.7709X^2$	0.9076**	0.49	0
Acidity sat. (%)	$Y = 3{,}041.1380 + 11.3545X - 0.5417X^2$	0.9409**	10.5	0
CEC (cmol$_c$ kg^{-1})	$Y = -42{,}520.15 + 17{,}455.66X - 1{,}670.3430X^2$	0.5101**	5.2	4.8
Ca/Mg ratio	$Y = 5{,}359.008 - 2{,}288.174X + 281.131X^2$	0.8903**	1.92	1.9
Ca/K ratio	$Y = 1{,}277.9740 + 397.1924X - 20.9609X^2$	0.9006**	9.5	5.6
Mg/K	$Y = 1{,}599.9570 + 573.1361X - 52.9977X^2$	0.9124**	5.4	3.0

Source: Fageria, N.K. et al., *Commun. Soil Sci. Plant Anal.*, 44, 2941, 2013b.
Note: Values are averages of three crops.
[a] VMY, Value of maximum yield was calculated by quadratic regression equation.
[b] VMEY, Value of maximum economic yield was calculated by regression equation on the basis of 90% of maximum yield.
**Significance at 1% probability level.

7.2.1.4 Reducing Phosphorus Immobilization

Oxisols are naturally deficient in total and plant-available P, and significant portions of applied P are immobilized due to either precipitation of P as insoluble Fe/Al phosphates or chemisorption to Fe/Al oxides and clay minerals (Nurlaeny et al., 1996). Smyth and Cravo (1992) reported that Oxisols are notorious for P immobilization since they have higher iron (Fe) oxide contents in their surface horizons than any other type of soils. P fixation capacity in Oxisols is directly related to the surface area and clay contents of the soil material and inversely related to SiO_2/R_2O_3 ratios (Curi and Camargo, 1988).

Bolan et al. (1999) reported that in variable charge soils, a decrease in pH increases the anion exchange capacity, thereby increasing the retention of P. Hence, improving crop yields on these soils requires high rates of P application (Sanchez and Salinas, 1981; Fageria, 1989). Reports regarding the effects of liming on P availability in highly weathered acid soils are in conflict (Friesen et al., 1980a; Haynes, 1984). Liming can increase, decrease, or have no effect on P availability (Haynes, 1982; Fageria, 1984; Mahler and McDole, 1985; Anjos and Rowell, 1987; Curtin and Syers, 2001). However, in a recent study, Fageria and Santos (2005) reported a linear increase in Mehlich 1 extractable P with increasing soil pH in the range of 5.3–6.9 (mean of 0–10 and 10–20 cm soil depth) in Brazilian Oxisol soils (Fageria, 2008). Mansell et al. (1984) and Edmeades and Perrott (2004) reported that in acid soils of New Zealand, primary benefit of liming occurs through an increase in the availability of P by decreasing P adsorption and stimulating the mineralization of organic P. Fageria (1984) reported that in Brazilian Oxisol soils, there was a quadratic increase in the Mehlich 1 extractable P in the pH range of 5–6.5, which was thereafter decreased. An increase in P availability in the pH range of 5–6.5 was associated with release of P ions from Al and Fe oxides, which were responsible for P fixation (Fageria, 1989). At higher pH (>6.5), the reduction of extractable P was associated with precipitation of P as Ca phosphate (Naidu et al., 1990). The increase in extractable P or liberation of P in the pH range of 5–6.5 and reduction in the higher pH range (>6.5) can be explained by the following equations:

$$AlPO_4(\text{P fixed}) + 3OH^- \Leftrightarrow Al(OH)_3 + PO_4^{3-}(\text{P released})$$

$$Ca(H_2PO_4)_2(\text{soluble P}) + 2Ca^{2+} \Leftrightarrow Ca_3(PO_4)_2(\text{insoluble P}) + 4H^+$$

Liming of acid soils results in the release of P for plant uptake; this effect is often referred to as "P priming effect" of lime (Bolan et al., 2003). Bolan et al. (2003) reported that in soils high in exchangeable and soluble Al, liming may increase plant P uptake by decreasing Al, rather than by increasing P availability per se. This may be due to improved root growth where Al toxicity is alleviated, facilitating a greater volume of soil to be explored (Friesen et al., 1980b).

7.2.1.5 Improving Activities of Beneficial Microorganisms

Soil microbiological properties can serve as soil quality indicators since soil microorganisms are the second most important (after plants) biological agents in the agricultural ecosystem (Yakovchenko et al., 1996; Fageria, 2002). Soil microorganisms provide the primary driving force for many chemical and biochemical processes and therefore affect nutrient cycling, soil fertility, and carbon (C) cycling (He et al., 2003). Plant roots and rhizosphere are colonized by many plant beneficial microorganisms such as symbiotic and nonsymbiotic dinitrogen (N_2)-fixing bacteria, plant growth-promoting rhizobacteria, saprophytic microorganisms, biocontrol agents, and mycorrhizae and free-living fungi. Soil acidity restricts the activities of these beneficial microorganisms, except fungi, which grow well over a wide range of soil pH (Brady and Weil, 2002). Enhancing the activities of beneficial microbes such as rhizobia, diazotropic bacteria, and mycorrhizae in the rhizosphere has improved root growth by the fixation of atmospheric N, suppressing pathogens, producing phytohormones, enhancing root surface area to facilitate the uptake of less mobile nutrients such as P and micronutrients, and mobilizing and solubilizing unavailable nutrients (Fageria and Baligar, 2008).

Low soil pH adversely affects the activities of rhizobium, including a loss of its ability to fix N (Angle, 1998). Mulder et al. (1977) reported that low soil pH reduced the activity and their ability to multiply. Holdings and Lowe (1971) demonstrated that low soil pH increased the number of ineffective rhizobia in soil. Angle (1998) reported that soil pH below 5.5 reduced rhizobial populations and rhizobia that survive such a pH lack the capacity to fix atmospheric N. Ibekwe et al. (1995) reported that plants grown in an unamended control soil with low pH often exhibited low rates of N fixation. Ibekwe et al. (1995) reported high rates of N fixation when high concentrations of heavy metals were present, but soil pH was near neutral. Franco and Munns (1982) reported that decreasing the pH of nutrient solutions from 5.5 to 5.0 decreased the number of nodules formed by common bean.

Bacteria are divided into three groups based on their tolerance to acidity. First group is known as acidophiles (grow well under acidic conditions), the second group is neutrophiles (grow well under neutral pH), and the third group is called as alkaliphiles (grow well under alkaline conditions) (Fageria and Baligar, 2008). Soils contain all these groups of bacteria. Most soil bacteria, however, including the N-fixing rhizobium, belong to the neutrophiles group. Therefore, acidic pH ranges are detrimental to bacterial activities.

Lime ameliorates harmful effects of soil acidity (Cregan et al., 1989). Studies on bacteria suggest that the success of liming may be due not only to an effect on soil pH but also to a direct effect on the bacteria themselves (Reeve et al., 1993). Like most neutrophilic bacteria, rhizobia appear to maintain an intracellular pH between 7.2 and 7.5, even when the external environment is acidic (Kashket, 1985). However, differences exist between and among acid-tolerant and acid-sensitive strains (Bhandhari and Nicholas, 1985). Strain tolerance to lower pH has been reported for rhizobia (Brockwell et al., 1995) and arbuscular mycorrhizal fungi (Habte, 1995). Muchovej et al. (1986) reported that liming of a Brazilian Oxisol soils improved nodule formation in soybean. Haynes and Swift (1988) reported that liming increased microbial biomass and enzyme activity in acid soils.

Nurlaeny et al. (1996) reported that liming increased shoot dry weight, total root length, and mycorrhizal colonization of roots in soybean and corn grown on tropical acid soils. They reported that mycorrhizal colonization improved P uptake and plant growth. Furthermore, colonization of the roots with arbuscular mycorrhizal fungi can increase plant uptake of P and other nutrients with low mobility, such as zinc (Zn) and copper (Cu) (Marschner and Dell, 1994). Uptake of P by mycorrhizal plants is usually from the same labile soil P pool from which the roots of nonmycorrhizal plants absorb P (Morel and Plenchette, 1994). However, the external hyphae can absorb and translocate P to the host from soil outside the root depletion zone of nonmycorrhizal roots (Johansen et al., 1993). Thereby, under conditions where P and other nutrient diffusions in soil are slow and root density is not high, more immobile nutrients are spatially available to mycorrhizal plants than to nonmycorrhizal plants (Nurlaeny et al., 1996). Furthermore, mycorrhizal plants may utilize organic soil P due to surface phosphatase activity of hyphae (Tarafdar and Marschner, 1995), enhanced activity of P-solubilizing bacteria in the mycorrhizosphere (Linderman, 1992; Tarafdar and Marschner, 1995), or reduced immobilization of labile P in organic matter (Joner and Jakobsen, 1994). Additionally, soil pH affects arbuscular mycorrhizal colonization, species distribution, and effectiveness of the mycorrhizal symbiosis with plant species (Hayman and Tavares, 1985).

7.2.1.6 Reducing Solubility and Leaching of Heavy Metals

Heavy metals are those metals having densities >5.0 Mg m^{-3} (Soil Science Society of America, 2008). In soils, these include the elements Cu, Zn, Fe, manganese (Mn), cadmium (Cd), cobalt (Co), chromium (Cr), mercury (Hg), nickel (Ni), and lead (Pb). Higher concentrations of these heavy metals in soil solution can lead to uptake by crop plants in quantities that are harmful for human and animal health; leaching can occur and contaminate groundwater (Hall, 2002; Epstein and Bloom, 2005). Soil properties such as organic matter content, clay type, redox status, and soil pH are considered the major factors in determining the bioavailability of heavy metals in soil (Huang and Chen, 2003).

Increasing pH by application of lime to acid soils reduces the solubility of most heavy metals (Lindsay, 1979; Mortvedt, 2000; Fageria et al., 2002). In addition, higher soil pH increases the adsorption affinity of Fe oxides, organic matter, and other adsorptive surfaces (Sauve et al., 2000). This practice can reduce the leaching of heavy metals to groundwater as well as their absorption by plants and consequently improve soil and water quality and subsequently human health.

Increasing soil pH with lime can significantly affect the adsorption of heavy metals in soils (Adriano, 1986). Adsorption of Pb, Cd, Ni, Cu, and Zn is significantly decreased with increasing soil pH (Basta and Tabatabai, 1992). Ribeiro et al. (2001) reported that among several soil amendments (dolomitic lime, gypsum, vermicompost, sawdust, and solomax), dolomitic lime was the most effective in reducing bioavailabilities of Zn, Cd, Cu, and Pb. Most metals are relatively more mobile under acid and oxidizing conditions and are strongly retained under alkaline and reducing conditions (Huang and Chen, 2003). Brummer and Herms (1983) reported that Pb, Cd, Hg, Co, Cu, and Zn are more soluble at pH range of 4–5 than in a pH range of 5–7. However, under acidic conditions, arsenic (As), selenium (Se), and molybdenum (Mo) are less soluble because of the formation of these elements into anionic forms (Huang and Chen, 2003). Fageria and Baligar (2008) reported the influence of increasing base saturation on soil-extractable Mn, Fe, Zn, and Cu after the harvest of common beans grown on Brazilian Oxisol soils. According to these authors, with the exception of Cu, concentrations of these microelements in the extractant were significantly reduced with increasing soil base saturation due to liming.

Risk of food chain contamination by toxic substances and elements has been a major concern of both producers and consumers (Treder and Cieslinski, 2005). Concentrations of heavy metals in plants and their distribution to various parts of the plant are the result of the combined influence of soil properties and biological factors. However, transfer of heavy metals from the soil solids to the soil solution, and thus their availability to plants, depends upon several factors. Soil properties such as organic matter content, clay type, redox potential, and soil pH are considered the major factors that determine the bioavailability of heavy metals in soil (Treder and Cieslinski, 2005). Hence, liming reduces the availability of heavy metals to crop plants.

7.2.1.7 Improve Soil Structure

Clustering of soil particles (sand, silt, and clay) into aggregates or peds and their arrangement into various patterns resulted in what is termed soil structure. From the agronomic standpoint, soil structure affects plant growth through its influence on infiltration, percolation, and retention of water, soil aeration, and mechanical impedance to root growth. Its general role in soil–water relations can be evaluated in terms of the extent of soil aggregation, aggregate stability, and pore size distribution.

These soil characteristics change with tillage practices and cropping systems. Major binding agents responsible for aggregate formation are the silicate clays, oxides of Fe and Al, and organic matter and its biological decomposition products. Oxisols and Ultisols, characterized by dominant quantities of Fe and Al oxides, have a high degree of aggregation, and the aggregates are quite stable. Red color of these soils is attributed to Fe oxide minerals (Fageria and Baligar, 2008).

Ca in liming materials increases the formation of soil aggregates, thereby improving soil structure (Chan and Heenan, 1998). Lime-induced improvement in aggregate stability manifests through the effect of liming on dispersion and flocculation of soil particles (Bolan et al., 2003). Liming is often recommended for the successful colonization of earthworm in pasture soils. Lime-induced increase in earthworm (*Lumbricus terrestris*) activity may influence soil structure and macroporosity through the release of polysaccharide and the burrowing activity of earthworm (Springett and Syers, 1984).

7.2.1.8 Improving Nutrient Use Efficiency

Improving nutrient use efficiency is becoming increasingly important in modern crop production due to rising costs associated with fertilizer inputs and growing concern about environmental pollution. Nutrient use efficiency is defined in several ways in the literature. The most common definitions are known as agronomic efficiency, physiological efficiency, agrophysiological efficiency, apparent recovery efficiency, and utilization efficiency (Baligar et al., 2001; Fageria and Baligar, 2005). Most of these definitions refer to nutrient uptake and utilization by plants in dry matter production. In a simple explanation, efficiency is output of economic produce divided by fertilizer input. Crop species or genotypes of the same species producing higher dry matter yield with low nutrient application rate or accumulation are called efficient plant species or genotypes. According to Soil Science Society of America (2008), a nutrient-efficient plant is one that absorbs, translocates, or utilizes more of a specific nutrient than another plant under conditions of relatively low nutrient availability in the soil or growth medium. Fischer (1998) reported that rice cultivars released in 1965 produced less than 40 kg grain kg^{-1} N absorbed by plants. However, cultivars released in 1995 produced almost 55 kg of grain kg^{-1} N absorbed, resulting in a more than 35% increases in N efficiency (Fischer, 1998).

Nutrient use efficiency (recovery efficiency) seldom exceeds 50% in most grain production systems (Fageria and Baligar, 2005). Worldwide, N recovery efficiency for cereal production (rice, wheat, sorghum [*Sorghum bicolor* L. Moench]), millet (*Pancilin menhouse*), barley (*Hordeum vulgare*), corn, oat (*Avena fatua*), and rye (*Elymus* sp.) is approximately 33% (Raun and Johnson, 1999). Low nutrient use efficiency is associated with the use of low input crop management technology. Such management practices include the use of low rate of fertilizers and lime; water deficiency; inadequate control of insects, diseases, and weeds; and planting of inefficient plant species or genotypes of the same species (Fageria, 1992). Low nutrient use efficiency in crop production systems is undesirable, both economically and environmentally.

Soil acidity is responsible for low nutrient use efficiency by crop plants. Fageria et al. (2004) reported that liming of Oxisols improved the use efficiency of P, Zn, Cu, Fe, and Mn by upland rice genotypes. Efficiency of these nutrients was higher under a pH of 6.4 than with pH of 4.5. Improvement in efficiency of these nutrients was associated with decreasing soil acidity, improving their availability, and enhancing root growth (Fageria et al., 2004).

7.2.1.9 Control of Plant Diseases

Mineral nutrition has an important role in the control of plant diseases. Healthy plants provided with adequate essential nutrients in appropriate balance generally have fewer diseases as compared with nutrient-deficient plants (Fageria et al., 2011a). When evaluating the effects of mineral nutrition on plant diseases, in addition to optimum rate of nutrient application, an appropriate nutrient balance is important. Indiscriminate excess supply of a nutrient will create imbalance with other nutrients, and chances of plant infestation with diseases will increase. Furthermore, there is lack of systematic research to determine effects of mineral nutrition on plant diseases. Plant pathologists without collaboration with soil scientists have conducted the majority of studies in this discipline and, invariably, with inappropriate fertilizer rates (excess or deficient). In addition, most of the data related to nutrient disease interactions are collected under controlled growth conditions, and field studies are minimal.

However, liming has been reported to increase or decrease plant diseases. It has been known for nearly 100 years that amending the soil with $CaCO_3$ would provide a significant measure of club-root (*Plasmodiophora brassicae* Wor.) control in crucifers (Engelhard, 1989). Conversely, many diseases of potato (*Solanum tuberosum* L.), such as common scab (*Streptomyces scabies*), powdery scab (*Spongospora subterranea*), black scurf (*Rhizoctonia solani*), and tuber blight (*Phytophthora infestans*), are favored at higher pH as compared to lower pH (Haynes, 1984).

Ca has been implicated in plant resistance to several plant pathogens, including *Erwinia phytophthora*, *R. solani*, *Sclerotium rolfsii*, and *Fusarium oxysporum* (Kiraly, 1976). Haynes (1984) reported that Ca forms rigid linkages with pectic chains and therefore promotes the resistance of plant cell walls to enzymatic degradation by pathogens.

Minimal information is available on the effects of lime on damaging of specific insects to plants. Variable levels of macro- and micronutrients in plants have positive to negative or no effects on insect damage to crop plants (Fageria and Scriber, 2002). Lime improves the availability of Ca, Mg, Mo, and P and reduces the availability of Mn, Zn, Cu, and Fe (Fageria and Baligar, 2008). Evaluations are needed on how the availability of these nutrients in soil affects insect damage in plant.

7.2.1.10 Mitigation of Nitrous Oxide Emission from Soils

Nitrous oxide (N_2O) is globally important, due to its role as a greenhouse gas, and once oxidized to NO_x, it can catalyze stratospheric ozone destruction. N_2O is a potent greenhouse gas with greater global warming potential than carbon dioxide (CO_2) (Izaurralde et al., 2004). N_2O concentration in the atmosphere has increased since preindustrial times, and agricultural lands are the main anthropogenic sources (Clough, 2004). Concentration of N_2O in the atmosphere, estimated at 2.68×10^{-2} mL L^{-1} around 1750, has increased by about 17% as a result of human alterations of the global N cycle (IPCC, 2001). Global annual N_2O emissions from agricultural soils have been estimated to range between 1.9 and 4.2 Tg N, with approximately half arising from anthropogenic sources (IPCC, 2001).

Since soil pH has a potential effect on N_2O production pathways, and the reduction of N_2O to N_2, it has been suggested that liming may provide an option for the mitigation of N_2O emission from agricultural soils (Stevens et al., 1998). Clough et al. (2003, 2004) reported that liming has been promoted as a mitigation option for lowering soil N_2O emissions when soil moisture content is maintained at field capacity. N_2O forms in soils primarily occur during the process of denitrification (Robertson and Tiedje, 1987) and to a lesser extent during nitrification (Tortoso and Hutchinson, 1990).

7.2.1.11 Acid-Tolerant Crop Species and Genotypes within Species

Use of acid-tolerant crop species and/or genotypes within species is another important strategy in optimizing P availability to crop plants. Crop species and genotypes within species differ significantly in relation to their tolerance to soil acidity (Sanchez and Salinas, 1981; Yang et al., 2000; Fageria and Baligar, 2003, 2008; Fageria et al., 2004). Lime requirements vary from species to species and among cultivars within species. Many of the plant species tolerant to acidity have their center of origin in acid soil regions, suggesting that adaptation to soil constraints is part of the evolution processes (Sanchez and Salinas, 1981; Foy, 1984). A typical example of this evolution is the acid soil tolerance of Brazilian upland rice cultivars. In Brazilian Oxisol soils, upland rice grows very well without liming, when other essential nutrients are supplied in adequate amount, and water is not a limiting factor (Fageria, 2001a).

Upland rice production in Brazilian Oxisol soils is an example of crop acidity tolerance evaluation. Fageria et al. (2004) reported that GY and yield components of 20 upland rice genotypes were significantly decreased at low soil acidity (limed to pH 6.4) as compared with high soil acidity (without lime, pH 4.5), demonstrating the tolerance of upland rice genotypes. Variation in upland rice genotypes to acidity tolerance is presented in Figure 7.10. Out of 10 genotypes, 7 genotypes produced higher GY at

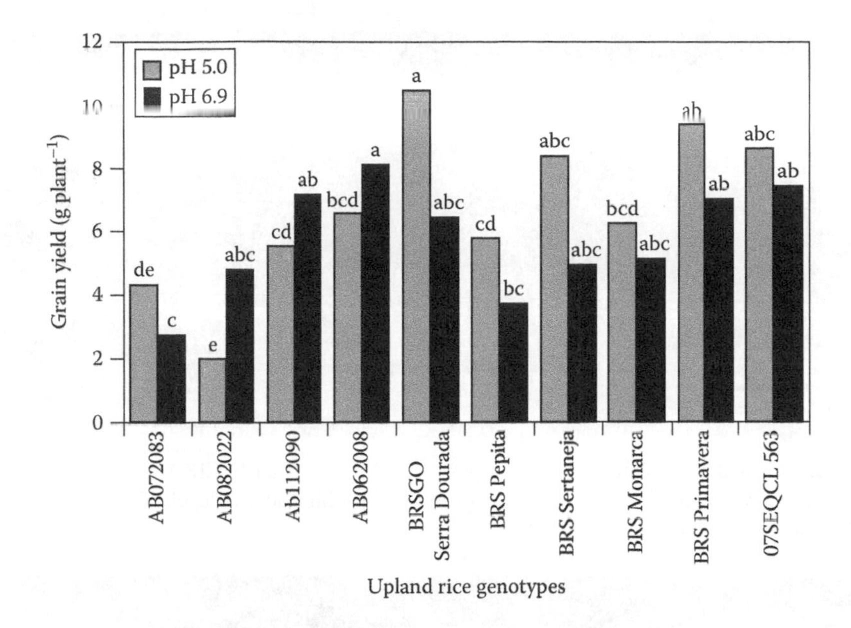

FIGURE 7.10 Upland rice genotype GY at two pH levels. (From Fageria, N.K. et al., *Commun. Soil Sci. Plant Anal.*, 46, 1076, 2015a.)

low pH (5.0) as compared to high pH (6.9). Variations in growth of upland rice genotypes at two lime levels are presented in Figures 7.11 through 7.16. Variations in root growth of 30 upland rice at two lime levels are presented (Figure 7.4). Maximum root length (MRL) and root dry weight (RDW) were having significant lime × genotype interactions, indicating variation in these traits with the variation in acidity level of the soil under investigation (Table 7.4). MRL varied from 22 to 32 cm, with a mean value of 26.1 cm at high acidity level (without lime addition). Similarly, at low acidity level, MRL varied from 17.3 to 23.3 cm, with a mean value of 20.0 cm. Variation in MRL among upland rice genotype has been reported by Fageria (2013) and Fageria and Moreira (2011). Overall, MRL decreased at low acidity level as compared to high acidity level. RDW varied from 0.97 to 7.39 g plant^{-1}, with a mean value of 2.08 g plant^{-1} at high acidity level. Similarly, at low acidity level, RDW varied from 0.46 to 3.37 g plant^{-1}, with a

FIGURE 7.11 Growth of upland rice genotypes BRS Primavera, AB072083, and BRSGO Serra Dourada at two acidity levels. Better growth and yield at high acidity levels. (From Fageria, N.K. et al., *Commun. Soil Sci. Plant Anal.*, 46, 1076, 2015a.)

FIGURE 7.12 Growth of upland rice genotypes BRS Monarca, BRS CIRAD 302, and Primavera CL (07SEQCL 441) at two acidity levels. Better growth and yield at high acidity levels. (From Fageria, N.K. et al., *Commun. Soil Sci. Plant Anal.*, 46, 1076, 2015a.)

FIGURE 7.13 Growth of upland rice genotypes Curinga CL (07SEQCL 563), BRS Pepita, and BRS Sertaneja at two acidity levels. Better growth and yield at high acidity levels. (From Fageria, N.K. et al., *Commun. Soil Sci. Plant Anal.*, 46, 1076, 2015a.)

FIGURE 7.14 Growth of upland rice genotypes AB072085, AB072007, and AB062041 at two acidity levels. Growth and yield almost equal at two acidity levels. (From Fageria, N.K. et al., *Commun. Soil Sci. Plant Anal.*, 46, 1076, 2015a.)

FIGURE 7.15 Growth of upland rice genotypes AB062037, AB062008, and BRA 032048 at two acidity levels. Growth and yield better at high lime or low acidity levels compared to low lime or high acidity levels. (From Fageria, N.K. et al., *Commun. Soil Sci. Plant Anal.*, 46, 1076, 2015a.)

FIGURE 7.16 Growth of upland rice genotypes AB112092, AB072044, and AB112092 at two acidity levels. Growth and yield better at high lime or low acidity levels compared to low lime or high acidity levels. (From Fageria, N.K. et al., *Commun. Soil Sci. Plant Anal.*, 46, 1076, 2015a.)

mean value of 1.31 g plant^{-1}. Variation in upland rice RDW has been reported by Fageria (2013). Overall, RDW decreased by 59% at low acidity level as compared to high acidity level. Fageria (2013) reported that increasing level of Mg in the soil decreased RDW of upland rice. Similarly, Fageria (2013) reported that MRL and RDW of upland rice decreased when lime rate was increased from 1.42 to 2.14 g kg^{-1} in Brazilian Oxisol soils. In the present study, lime rate at low acidity level was 2.5 g kg^{-1} soil. Variation in root growth of upland rice genotypes at two lime levels is presented in Figures 7.17 through 7.19.

Fageria et al. (2014a) studied the acidity tolerance of tropical legume cover crops grown on a Brazilian Oxisol soil (Figure 7.20). Shoot dry weight was significantly influenced by soil pH and cover crop species. Shoot dry weights at three pH levels are presented in Figure 7.20. Shoot dry weight at low soil pH level varied from 3.9 to 21.4 g plant^{-1}, with a mean value of 9.9 g plant^{-1}. At all soil pH levels, white jack bean (*Canavalia ensiformis*) produced the highest shoot dry weight and Brazilian lucerne (*Stylosanthes guianensis*) produced the lowest shoot weight. Interspecies variability in shoot dry weight of tropical legume cover crops has been widely reported (Fageria et al., 2005, 2009; Baligar and Fageria, 2007). Tropical legume cover crops have a high degree of tolerance

TABLE 7.4
MRL and RDW of 30 Upland Rice Genotypes as Influenced by Lime and Genotype Treatments

	MRL (cm)		Root Dry Wt. (g Plant^{-1})	
Genotype	**Without Lime**	**With Lime**	**Without Lime**	**With Lime**
1. AB072083	23.66a–d	19.00a	2.24bc	1.27c–g
2. AB112172	24.00a–d	18.00a	1.46bc	0.51g
3. AB072041	23.00b–d	19.00a	1.71bc	0.78e–g
4. AB072001	23.66a–d	19.66a	1.02c	0.69fg
5. AB072063	26.33a–d	21.66a	1.17bc	0.52g
6. AB082022	29.00a–d	20.66a	2.99bc	1.60b–g
7. AB082021	22.00d	22.66a	2.47bc	2.11a–d
8. AB072047	22.66b–d	21.66a	1.26bc	1.00d–g
9. AB112092	26.00a–d	21.33a	1.11c	0.89d–g
10. AB072085	25.66a–d	19.66a	1.93bc	0.84d–g
11. AB072007	22.00d	19.66a	1.08c	1.01d–g
12. AB072044	27.00a–d	18.66a	2.08bc	1.47b–g
13. AB112090	28.00a–d	21.66a	1.82bc	1.21c–g
14. AB072035	23.33b–d	17.33a	1.51bc	1.19c–g
15. BRS Esmeralda	27.00a–d	18.33a	1.46bc	2.00b–e
16. AB052033	30.00a–d	19.66a	2.33bc	2.44a–c
17. AB062008	25.66a–d	21.00a	1.79bc	1.28c–g
18. AB062037	27.33a–d	21.00a	2.06bc	1.94b–f
19. AB062041	29.00a–d	20.66a	1.92bc	2.64ab
20. AB062045	26.33a–d	17.66a	1.36bc	0.64g
21. BRSMG Caravera	25.00a–d	17.66a	0.97c	0.52g
22. BRSGO Serra Dourada	24.33a–d	21.00a	2.67bc	1.19c–g
23. BRS Pepita	22.33cd	22.00a	1.31bc	0.46g
24. BRS Sertaneja	30.66a–c	21.66a	2.32bc	1.18c–g
25. BRS Monarca	32.00a	23.33a	3.41b	1.71b–g
26. BRS Primavera	26.00a–d	18.66a	2.33bc	1.10d–g
27. BRSMG Caçula (CMG 1152)	24.00a–d	17.33a	1.40bc	0.48g
28. BRS CIRAD 302	30.00a–d	20.00a	2.83bc	2.39a–c
29. 07SEQCL 441	27.33a–d	18.66a	3.04bc	0.85d–g
30. 07SEQCL 563	31.00ab	21.66a	7.39a	3.37a
Average	26.14a	20.03b	2.08a	1.31a
F-test				
Lime level (L)	**		NS	
Genotype (G)	**		**	
L × G	**		**	
CVL (%)	9.98		14.01	
CVG (%)	10.61		15.20	

Source: Fageria, N.K. et al., *Commun. Soil Sci. Plant Anal.*, 46, 1076, 2015a.

** and NS Significance at 1% probability level and nonsignificant, respectively. Means followed by the same letter within the same column are not significantly different at the 5% probability level by Tukey's test. Average values were compared in the same line.

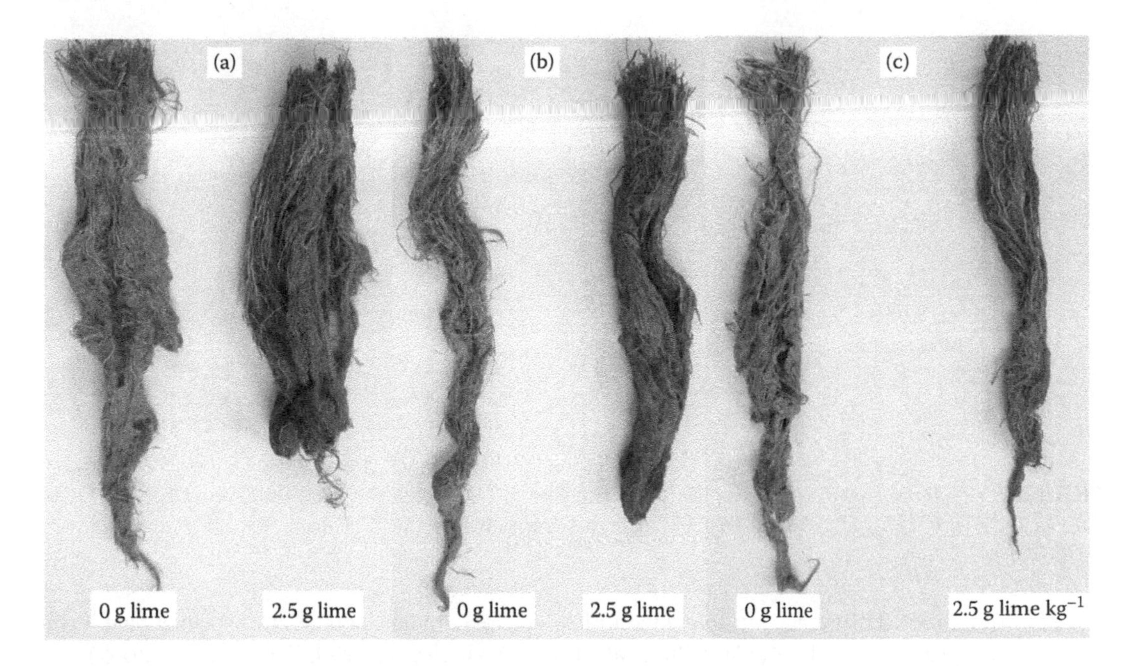

FIGURE 7.17 Root growth of three upland rice genotypes at two acidity levels. (a) B052033, (b) B062041, and (c) BRSMG Caravera. (From Fageria, N.K. et al., *Commun. Soil Sci. Plant Anal.*, 46, 1076, 2015a.)

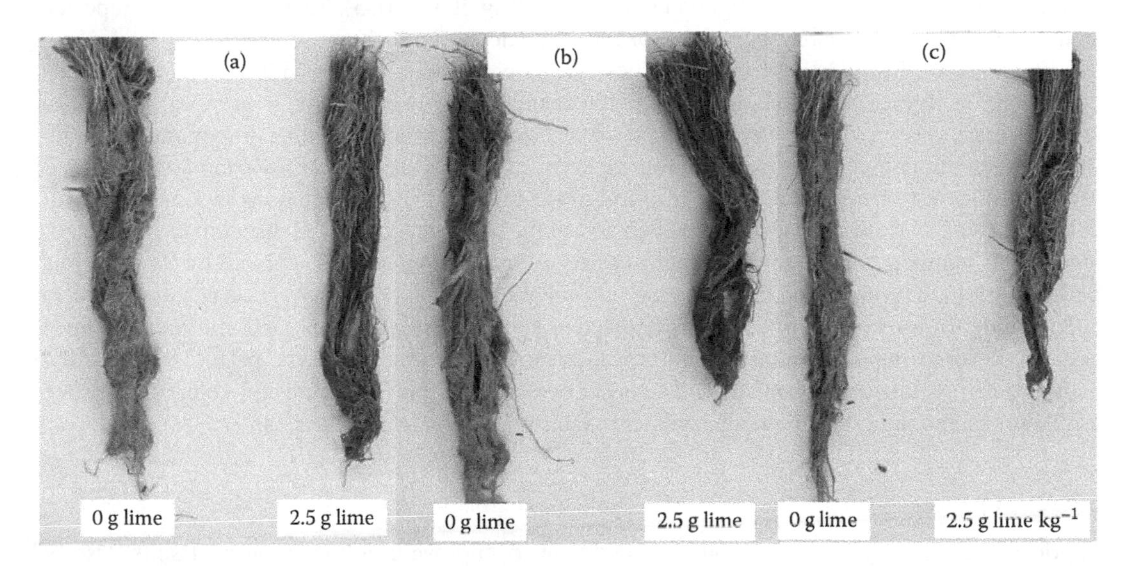

FIGURE 7.18 Root growth of three upland rice genotypes at two acidity levels. (a) BRS Pepita, (b) BRS Primavera, and (c) BRSMG Açula (CMG 1152). (From Fageria, N.K. et al., *Commun. Soil Sci. Plant Anal.*, 46, 1076, 2015a.)

to low soil pH. Tolerance to low soil pH is an important trait of cover crop legumes since they can be planted on strongly acidic soils without lime input. Many tropical legume cover crops have been reported to be tolerant to low soil pH (Fageria et al., 2011a).

Based on shoot dry weight, the tolerance of cover crops to low soil pH was in the order of white jack bean (most tolerant) > black mucuna > gray mucuna > lablab > ochroleuca crotalaria > mucuna bean ana > sunn hemp > bicolor pigeon pea > mulato pigeon pea > black jack bean > black pigeon pea > showy crotalaria > calopo > smooth crotalaria > short-flowered crotalaria > Brazilian lucerne

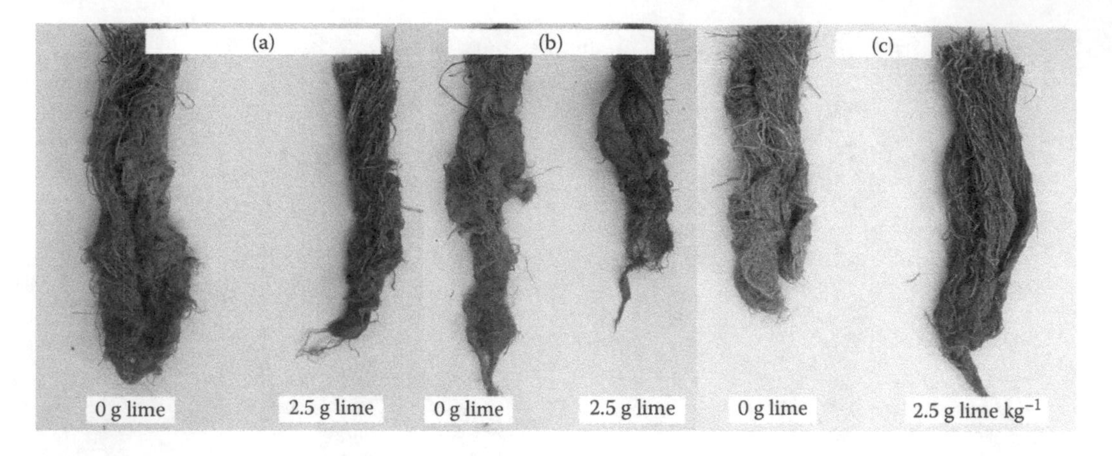

FIGURE 7.19 Root growth of three upland rice genotypes at two acidity levels (a) Primavera CL (07SEQCI), (b) Curinga CL (07SEQCI), and (c) B082021. (From Fageria, N.K. et al., *Commun. Soil Sci. Plant Anal.*, 46, 1076, 2015a.)

(most susceptible). Differences in tolerance to low soil pH among legume crop species have been previously reported (Foy, 1984; Carvalho et al., 1985; Devine et al., 1990; Fageria et al., 2009).

A substantial number of plant species of economic importance are generally regarded as tolerant to acid soil conditions of the tropics (Sanchez and Salinas, 1981). In addition, there are cultivars that are tolerant to soil acidity (Fageria et al., 2004; Yang et al., 2005). Yang et al. (2005) reported significant differences among genotypes of rye, triticale (× Triticosecale), wheat, and buckwheat (*Fagopyrum esculentum*) to Al toxicity. These crop species or cultivars within these species can be planted on tropical acid soils in combination with reduced rates of lime input. Combination of legume–grass pasture and agroforestry system of management are the other important soil acidity management components useful in tropical ecosystems. For example, *Pueraria phaseoloides* is used as understory for rubber, *Gmelina arborea* or *Dalbergia nigra* plantations in Brazil presumably supplying N to the tree crops (Sanchez and Salinas, 1981). A detailed discussion of the combination of legume–grass pasture and agroforestry in tropical America is reported by Sanchez and Salinas (1981). They reported that when an acid-tolerant legume or legume–grass pasture is grown under young tree crops, the soil is more protected, soil erosion is reduced, and nutrient cycling is enhanced. Some important annual food crops, cover or green manure crop, pasture species, and plantation crops tolerant to tropical acid soils are presented (Table 7.5). Acid soil–tolerant crops are useful for establishing low input management systems (Fageria and Baligar, 2008).

7.2.2 Gypsum

In addition to lime, gypsum is another amendment to improve crop yields on acid soils. Use of gypsum can improve Ca and sulfur (S) contents and soil structure. Gypsum may improve water infiltration and enhance the ability of roots to penetrate the soil (Viator et al., 2002; Fageria, 2013). Gypsum application has been reported to increase the yield of many crops in many soils. Thomas et al. (1995) reported a yield increase of 15% in wheat and sorghum with the addition of gypsum. Gypsum applied in irrigation water increased sugar yield and juice extraction percentage of sugarcane (Kumar et al., 1999). Gypsum increased the yield of corn and alfalfa up to 50% (Viator et al., 2002). This yield response was partially attributed to higher exchangeable Ca and S and a complementary reduction in exchangeable Al (Toma et al., 1999). Gypsum improves fertility of subsoil or lower soil depth (>20 cm) (Fageria, 2013).

Fageria et al. (2014c) studied the influence of gypsum on soybean yield and yield components and changes in Oxisol soil chemical properties. Gypsum significantly increased straw yield (SY), GY, pods

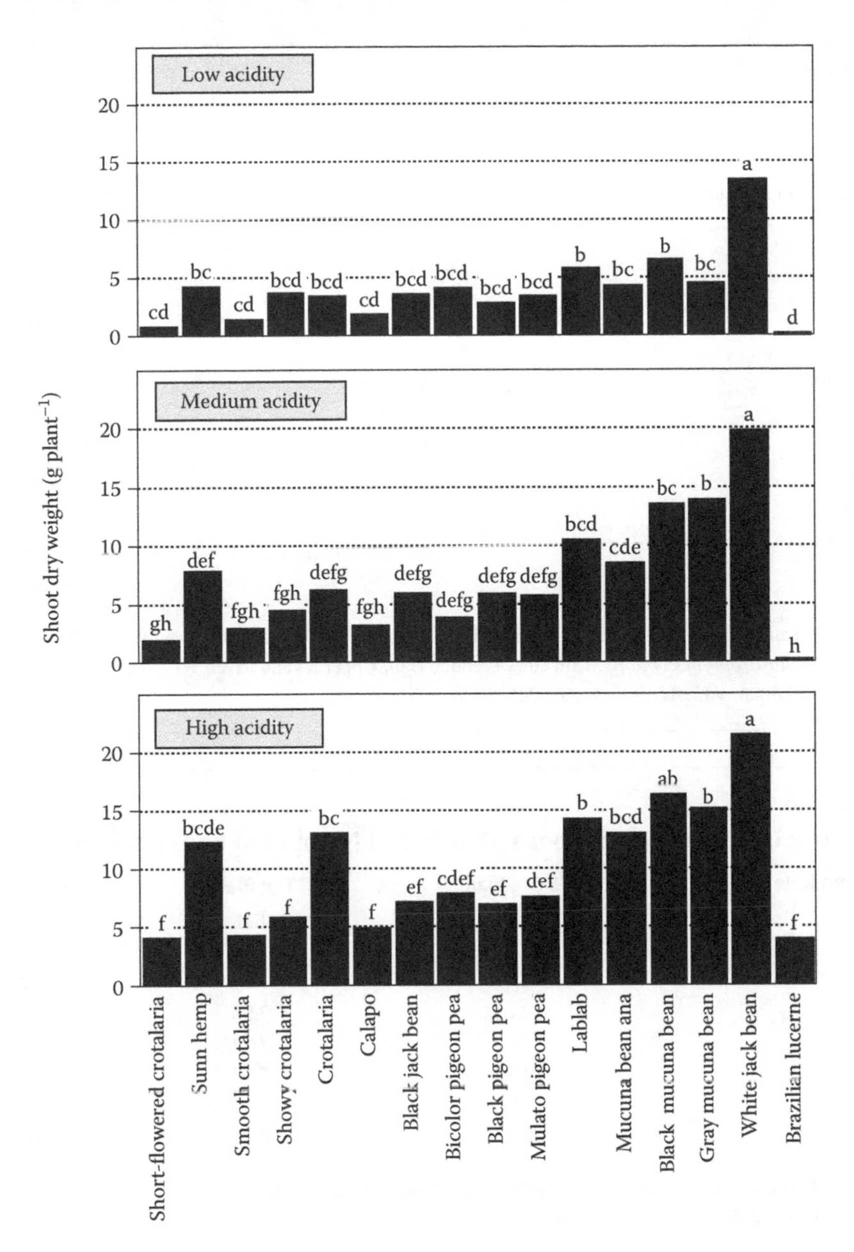

FIGURE 7.20 Shoot dry weight of tropical legume cover crops at different acidity levels. (From Fageria, N.K. et al., *J. Plant Nutr.*, 37, 294, 2014a.)

per plant, grain per pod, thousand grain weight, and grain harvest index (GHI) (Tables 7.6 through 7.9). Quadratic increase in yield and yield components occurred with gypsum rate of 0–2.28 g kg^{-1} soil. Based on a regression equation, maximum SY was obtained with the addition of 1.29 g gypsum kg^{-1} soil; GY was obtained with the addition of 1.43 g gypsum kg^{-1} soil and number of pods with the addition of 1.16 g gypsum kg^{-1} of soil. Maximum grain per pod was obtained with the addition of 1.15 g gypsum kg^{-1} of soil, thousand grain weight with the addition of 2.34 g gypsum kg^{-1} soil, and GHI with the addition of 1.57 g gypsum kg^{-1} soil. Improvement in soybean yield and yield components and related plant characters like SY and GHI may be associated with improvement in soil exchangeable Ca, base saturation, and effective CEC (Table 7.10). The variation in GY with the addition of gypsum was in the order of Ca saturation, exchangeable Ca, base saturation, and effective CEC (Table 7.11). Increase in GY was

TABLE 7.5
Some Important Crop Species, Pasture Species, and Plantation Crops Tolerant to Soil Acidity in the Tropics

Annual Crop Species	Pasture Species	Plantation Crops
Rice	*Brachiaria*	Banana
Peanut	*Andropogon*	Oil palm
Cowpea	*Panicum*	Rubber
Potato	*Digitaria*	Coconut
Cassava	*Napier grass*	Cashew nut
Pigeon pea	*Jaragua grass*	Coffee
Millet	*Centrosema*	Guarana
Kudzu	*Stylosanthes*	Tea
Mucuna		Leucaena
Crotalaria		Brazilian nut
		Eucalyptus
		Papaya

Sources: Sanchez, P.A. and Salinas, J.G., *Adv. Agron.*, 34, 280, 1981; Brady, N.C. and Weil, R.R., *The Nature and Properties of Soils*, 13th edn., Prentice Hall, Upper Saddle River, NJ, 2002; Fageria, N.K. and Baligar, V.C., *Adv. Agron.*, 99, 345, 2008.

TABLE 7.6
Influence of Gypsum on Soybean SY, GY, and Number of Pods per Plant

Gypsum Rate (g kg^{-1})	SY (g Plant^{-1})	GY (g Plant^{-1})	Pods (Plant^{-1})
0	1.48	1.28	23.94
0.28	1.89	6.82	27.44
0.57	2.49	7.89	27.50
1.14	3.06	7.98	31.19
1.71	2.99	7.79	28.81
2.28	1.92	7.60	24.44
F-Test	**	**	*
CV (%)	23.73	8.74	10.86

Source: Fageria, N.K. et al., *Commun. Soil Sci. Plant Anal.*, 45, 271, 2014c.
*,**Significant at the 5% and 1% probability level, respectively.

TABLE 7.7
Relationship between Gypsum Rate and SY, GY, and Number of Pods of Soybean

Variable	Regression Equation	R^2	VMYP (g kg^{-1})
Gypsum rate vs. SY	$Y = 1.35 + 2.71X - 1.06X^2$	0.59**	1.29
Gypsum rate vs. GY	$Y = 2.95 + 8.46X - 2.95X^2$	0.71**	1.43
Gypsum rate vs. no. pods	$Y = 23.97 + 11.11X - 4.78X^2$	0.45**	1.16

Source: Fageria, N.K. et al., *Commun. Soil Sci. Plant Anal.*, 45, 271, 2014c.
Note: Value for maximum yield or pods (VMYP).
**Significant at the 1% probability level.

TABLE 7.8
Influence of Gypsum on Soybean Grain per Pod, 100 Grain Weight, and GHI

Gypsum Rate (g kg^{-1})	Grain per Pod	100 Grain Weight (g)	GHI
0	0.94	5.75	0.46
0.28	1.83	13.75	0.78
0.57	1.63	18.20	0.76
1.14	1.48	17.79	0.73
1.71	1.46	19.21	0.72
2.28	1.33	24.55	0.80
F-test	**	**	**
CV (%)	17.59	26.72	5.36

Source: Fageria, N.K. et al., *Commun. Soil Sci. Plant Anal.*, 45, 271, 2014c.
**Significant at the 1% probability level.

TABLE 7.9
Relationship between Gypsum Rate and Grain per Pod, 100 Grain Weight, and GHI of Soybean

Variable	Regression Equation	R^2	VMYP (g kg^{-1})
Gypsum rate vs. grain per pod (Y)	$Y = 1.24 + 0.71X - 0.31X^2$	0.19^{NS}	1.15
Gypsum rate vs. 100 grain wt. (Y)	$Y = 8.63 + 12.22X - 2.61X^2$	0.61**	2.34
Gypsum rate vs. GHI (Y)	$Y = 0.58 + 0.26X - 0.083X^2$	0.40**	1.57

Source: Fageria, N.K. et al., *Commun. Soil Sci. Plant Anal.*, 45, 271, 2014c.
Note: Value for maximum yield or pods (VMYP).
** and NS Significant at the 1% probability level and nonsignificant, respectively.

TABLE 7.10
Influence of Gypsum on Soil pH, Ca, Mg, H + Al, Base Saturation, and Effective CEC after Harvest of Soybean Crop

Gypsum Rate (g kg^{-1})	pH	Ca (cmol kg^{-1})	Mg (cmol kg^{-1})	H + Al (cmol kg^{-1})	Base Sat. (%)	ECEC (cmol kg^{-1})
0	4.93	0.57	0.27	3.50	24.83	1.19
0.28	4.80	0.90	0.23	2.37	36.54	1.46
0.57	5.07	1.23	0.20	2.67	39.14	1.76
1.14	5.03	1.60	0.20	2.50	44.20	2.08
1.71	5.27	2.27	0.23	2.40	52.64	2.71
2.28	5.07	2.80	0.23	2.40	57.02	3.26
F-test	NS	**	NS	NS	**	**
CV (%)	7.11	12.54	20.69	18.92	12.95	11.84

Source: Fageria, N.K. et al., *Commun. Soil Sci. Plant Anal.*, 45, 271, 2014c.
** and NS Significant at the 1% probability level and nonsignificant, respectively.

TABLE 7.11
Relationship between GY and Soil Chemical Properties as Influenced by Gypsum after Harvest of Soybean Crop

Variable	Regression Equation	R^2	VMY
Ca ($cmol_c\ kg^{-1}$) vs. GY	$Y = -2.21 + 10.40X - 2.44X^2$	0.69**	2.13
BS (%) vs. GY	$Y = -7.27 + 0.56X - 0.005X^2$	0.64**	56
ECEC ($cmol_c\ kg^{-1}$) vs. GY	$Y = -8.22 + 12.55X - 2.31X^2$	0.61**	2.71
Sat. Ca (%)	$Y = -4.59 + 0.63X - 0.0077X^2$	0.80**	41

Source: Fageria, N.K. et al., *Commun. Soil Sci. Plant Anal.*, 45, 271, 2014c.
Note: Value for maximum yield (VMY).
**Significant at the 1% probability level.

associated with the improvement of these soil chemical properties. An increase in soil pH and a decrease in soil acidity (H + Al) were not significant with the addition of gypsum. Fageria (2001b) reported that improvement in these soil chemical properties improved soybean yield in Brazilian Oxisol soils.

7.2.3 Organic Manures

Soil fertility management at an adequate level is one of the main objectives of agricultural research for sustainable crop production. The objective can be achieved if appropriate soil and crop management practices are adopted. The use of organic manures is an important strategy to maintain and/or improve soil fertility for sustainable crop production. Beneficial effects of organic manures in restoring soil productivity were larger than those from inorganic fertilizers (Larney and Janzen, 1996; Fageria and Baligar, 2005). Larney and Janzen (1996) reported that the use of organic manures (livestock and crop residues) may provide an alternative for producers with a desire to restore their eroded soils and simultaneously reduce their inputs of N and P fertilizers. Organic matter cycling is related to the agricultural potential of soils (Tissen et al., 1994), and that green manure production and incorporation represents an alternative source of nutrients to mineral fertilizers (Clement et al., 1998). However, green manuring alone cannot supply sufficient essential plant nutrients for maximum or maximum economic crop yields. The best strategy is to use green manure in conjugation with chemical fertilizers. This combination may reduce the application rate of inorganic fertilizers and risk of environmental pollution and provide sustainability to crop production systems.

Use of organic manures in crop production is an old practice. Organic manures not only supply essential plant nutrients to crop plants but also improve soil's physical and biological properties, which favor better resource availability and improve yields (Fageria and Gheyi, 1999). Brosius et al. (1998) reported that plant- and animal-based wastes may substitute for commercial fertilizers and enhance chemical and biological attributes of soil quality in agricultural production systems. Fageria (2012) reported that soil organic matter (SOM) has a crucial role in maintaining sustainability of cropping systems by improving soil's physical (texture, structure, bulk density, and water-holding capacity), chemical (nutrient availability, CEC, reducing Al toxicity, and allelopathy), and biological (N mineralization bacteria, N_2 fixation, mycorrhizal fungi, and microbial biomass) properties. Preservation of SOM is crucial to ensure long-term sustainability of agricultural ecosystems. Improvement/preservation of SOM can be achieved by adopting appropriate soil and crop management practices (Fageria, 2012). These practices include conservation tillage, crop rotation, application of organic manures, increasing cropping intensity, use of optimum rate of chemical fertilizers, incorporation of crop residues, liming acid soils, and maintaining land under pasture. Organic matter can adsorb heavy metals in the soils, which reduce toxicity of these metals to plants and reduce their leaching into groundwater. Similarly, SOM adsorbs herbicides, which may inhibit the contamination of surface and groundwater. Furthermore, SOM functions as a sink to organic C and mitigates CO_2 gas escape to the environment. Globally, SOM contains approximately three times as much C as found in the world's vegetation. Organic matter has a critical role in the global C balance that is thought to be the major factor affecting global warming. Adequate amount of SOM maintains soil quality and sustainability of cropping systems and reduces environmental pollution (Fageria, 2012).

Addition of organic manures to soil in adequate rate and form improves SOM content (Fageria, 2012). Organic matter increases the soil's abilities to bind and release available essential plant nutrients and to resist the natural tendency of soil to become acid (Cole et al., 1987). There are several forms of organic manures that can be used in crop production. Among the organic manures, biosolids, animal manures, compost, and green manuring are the most common forms of organic manures.

7.2.3.1 Municipal Biosolids

Biosolids is a term introduced by the wastewater treatment industry in the early 1990s. Biosolids is solid, semisolid, or liquid materials generated from the treatment domestic sludge that have been sufficiently processed to permit these materials to be safely used in crop or vegetable production (Mullins et al., 2005). Municipal biosolids contain varying amounts of industrial wastewater, street runoff, human excreta, and residues from household activities (Hue, 1995). Biosolids are treated physically, chemically, or biologically to increase their sustainability for land application and reduce their harmful effects on public health and environmental pollution.

Composition of biosolids can be variable and is dependent on the type of treatment process, composition of the wastewater entering the treatment plant, and method of biosolid handling (Mullins et al., 2005). Nutrient content of biosolids from the United States and Ireland is presented in Table 7.12. Nutrient contents were in the order of N > Ca > P > Mg > K. P contents in the biosolids were in appreciable amounts and can supplement P to soils in an adequate amount. With careful application, biosolids can be a good source of nutrients for agronomic use. Bouwer and Chaney (1974) reported that the liquid effluent portion of sewage treatment may supply plant nutrients at 2% N, 1% P, 1.4% K, 2.4% Ca, and 1.7% Mg (dry weight basis).

Biosolids produced in the United States and Ireland typically contain 6–25 kg P Mg^{-1}, with a mean value of 18 kg P Mg^{-1} (Table 7.12). However, some biosolids are alkaline stabilized by the addition of some type of liming agent, and this process results in biosolids with 7–10 kg P Mg^{-1} (Logan and Harrison, 1995; Christie et al., 2001). Characterization of biosolids has shown that water-soluble orthophosphate concentrations can be relatively low (1–10 mg L^{-1}) as compared to the total P (Sommers et al., 1972), and in general 70%–90% of the total P in biosolids is in inorganic forms (Sommers et al., 1976; Kirkham, 1982). Water-extractable P is dependent on the wastewater treatment processes. Frossard et al. (1996) reported that the water-extractable P in 11 biosolids from

TABLE 7.12
Mean Values of Nutrient Contents of Biosolids from the United States and Ireland

Country/State	N (kg Mg^{-1})	P (kg Mg^{-1})	K (kg Mg^{-1})	Ca (kg Mg^{-1})	Mg (kg Mg^{-1})
North, central, and eastern United States	39	25	4	49	5.4
Michigan	35	22	5	40	7
New York	29	12	2	39	4
Hawaii	38	6	0.6	18	3
Colorado	42	23	4	—	—
Ireland	58.5	19.2	4.9	—	—
Average	40.25	17.9	2.9	36.5	4.9

Source: Mullins, G. et al., Byproduct phosphorus: Sources, characteristics, and management, in: *Phosphorus; Agriculture and the Environment*, L.K. Al-Amoodo, ed., ASA, CSSA, and SSSA, Madison, WI, 2005, pp. 829–879.

TABLE 7.13
Pollutant Concentration Limits of Heavy Metals Applied by Biosolids to Cropland

Heavy Metal	Pollutant Concentration Limit (mg kg^{-1})[a]
As	41
Cd	39
Cr	1200
Cu	1500
Pb	300
Hg	17
Mo	18
Ni	420
Se	36
Zn	2800

Source: Stratton, M.L. and Rechcigl, J.E., Agronomic benefits of agricultural, municipal, and industrial by products and their co-utilization: An overview, in: *Beneficial Co-Utilization of Agricultural, Municipal and Industrial by Products*, S. Brown, J.S. Angle, and L. Jacobs, eds., Kluwer Academic Publishers, Dordrecht, the Netherlands, 1998, pp. 9–34.

[a] The pollution concentration limit is defined as the heavy metal concentration in biosolids in the following text, which can be land applied without restrictive requirements and management practices.

various treatment plants in France varied from 0.2% to 38% of the total P. Mullins et al. (2005) reported that in several European biosolids, the proportion of inorganic P varied with the wastewater treatment system and ranged from 71% to 87% of the total P and concluded that that Fe- and Al-bound P were the dominant forms of inorganic P.

Biosolids should be used for crop production after removal of heavy metals either at the source or by special processing known as autothermal aerobic digestion or liquid composting (Jewell, 1994). Heavy metal concentration limit in the biosolids is presented (Table 7.13). Yield increases of several crops with the addition of biosolids have been reported for corn (Kelling et al., 1977; Parr and Hornick, 1992), sorghum (Hue, 1988), barley (Christie et al., 2001), and wheat (Barbarick et al., 1995). Biosolids increased the yield of crop plants, and their residual effects are observed for several years after application (Mullins et al., 2005). Biosolids have been used for many years as a source of P for agricultural crops (Kirkham, 1982). Potential benefits and problems associated with the use of biosolids for land application have been discussed in several reviews (Kirkham, 1974, 1982; Walker, 1975; Mullins et al., 2005).

7.2.3.2 Animal Manures

There are several types of animal manures or their by-products that can be used as source of P fertilizers in crop production. These consisted of raw feces, urine, waste feed, spilled water, absorptive bedding materials, and any other materials added to the waste stream of a livestock operation (Mullins et al., 2005). Animal manures should be applied in optimum rates; otherwise, it may create a pollution problem, especially N and P, in water bodies (Sharpley et al., 1996; Sims, 1998; Sims et al., 2000). N and P can contribute to eutrophication. P enrichment stimulates the growth of rooted aquatic plants and algae that can lead to the eutrophication of freshwater lakes and streams (Thomann and Mueller, 1987; Mullins et al., 2005).

Composition of animal manures varied with livestock, methods of preparation, and storage and type of manure (solid or liquid). Macronutrient content comparisons among animal manure are presented (Table 7.14). Substantial amount of N and P is present in the cattle and poultry manures. Amount of animal manure produced in different continents is presented (Table 7.15). Mullins et al. (2005) reported that more than 3.6 billion poultry (meat chickens and turkey only), 1 billion cattle, and 750 million hogs are being raised and these animals excrete approximately 13 billion mt of fresh manure each year worldwide. Using typical manure nutrient excretion values (Koelsch, 2007), these animals excrete approximately 67 million mt of N, 15 million mt of P, and 39 million tons of K each

TABLE 7.14
Nutrient Composition of Animal Manures

Nutrient (% Dry Wt.)	Cattle Manure	Poultry Manure	Swine Manure
N	1.2–2.0	1.8–4.1	1.2
P	0.3–0.8	1.5–3.3	0.4
K	1.7	1.5–3.2	0.6
Ca	1.9	1.6	0.3
Mg	0.9	0.4	0.2

Sources: Mills, H.A. and Jones, J.B. Jr., *Plant Analysis Handbook II*, Micro macro Publishing, Inc., Athens, GA, 1996; Elliott, L. F. and Swanson, N. P., Land use of animal wastes, in: *Land Application of Waste Materials*, Soil Conservation Society of America, ed., Soil Conservation Society of America, Ankeny, IA, 1976, pp. 80–90; Stratton, M.L. and Rechcigl, J.E., Agronomic benefits of agricultural, municipal, and industrial by products and their co-utilization: An overview, in: *Beneficial Co-Utilization of Agricultural, Municipal and Industrial by Products*, S. Brown, J.S. Angle, and L. Jacobs, eds., Kluwer Academic Publishers, Dordrecht, the Netherlands, 1998, pp. 9–34.

TABLE 7.15
Animal Manure Produced in Different Continents/Countries

Continent/Country	Manure Produced (Metric Million Tons)		
	Cattle	Hogs	Poultry
Africa	203.08	—	3.71
Australia/New Zealand	314.82	4.16	1.68
Central America	77.44	—	—
Central Europe	124.07	55.92	3.27
East Asia	1,179.10	716.62	41.71
Former Soviet Union	487.87	43.12	1.15
Middle East	116.45	—	2.31
North America	1,145.54	131.42	61.66
South America	2,327.36	50.36	23.07
South Asia	3,740.46	—	2.76
Western Europe	739.21	192.55	29.25
World total	10,455.32	1194.15	170.57

Sources: Adapted from Midwest Plan Service, Manure characteristics, MWPS-18, Sect. 1, Iowa State University, Ames, IA, 2000; Mullins, G. et al., By product phosphorus: Sources, characteristics, and management, in: *Phosphorus: Agriculture and the Environment*, L.K. Al-Amoodi, ed., ASA, CSSA and SSSA, Madison, WI, 2005, pp. 829–879.

year (Mullins et al., 2005). Total quantity of manure nutrients generated is approximately equivalent to 75% of commercial N, 100% of commercial P, and 200% of commercial K (Mullins et al., 2005).

Influence of animal manures on crop can be measured in terms of yield and nutrient uptake. Nutrients are primarily in organic forms and slowly released from the animal manures. Large amounts of animal manures are required to supply nutrients as compared to inorganic fertilizers. Long-term experiment conducted at the Rothamsted experimental station in the United Kingdom evaluated the crop yields on soil receiving farm yard manure (FYM) for nearly 150 years. Results from these classical experiments have demonstrated over the years that the application of FYM (35 Mg ha^{-1}) gives comparable or higher yields as compared to inorganic fertilizers (Mullins et al., 2005). However, in our opinion organic and inorganic fertilizers should be applied in combination to obtain optimum results.

Application of FYM has been reported to improve soil's physical and chemical conditions and to help conserve soil moisture (Gill and Meelu, 1982; Sattar and Gaur, 1989). One-time application of FYM (10–15 t ha^{-1}) increased wheat yields for three successive crop cycles, when applied in conjunction with inorganic N fertilizers under hot and humid conditions in Bangladesh (Mian et al., 1985). Badruddin et al. (1999) reported that application of FYM (10 t ha^{-1}) had the highest wheat yield response (14%) and approximately equivalent levels of NPK had the lowest yield (5.5%), suggesting that organic fertilizers provided growth factors in addition to nutrient content.

7.2.3.3 Compost

Compost is organic residues, or a mixture of organic residues and soil, that has been mixed, piled, and moistened, with or without addition of fertilizers and lime. Compost generally undergoes thermophilic decomposition until the original organic materials have been substantially altered or decomposed, sometimes called artificial manure or synthetic manure. In Europe, compost may refer to a potting mix for container grown plants (Soil Science Society of America, 2008). In addition, composting is a controlled biological process that converts organic constituents, usually wastes, into humus-like material suitable for use as amendment or organic fertilizer (Soil Science Society of America, 2008).

Rynk et al. (1992) defined composting as a controlled aerobic biological process that converts organic materials and nutrients into more stable forms through decomposition and oxidation reactions. Compost has several benefits such as reduction in the volume of the organic manures, which can reduce transportation costs from source or production sites to crop lands. Composts reduces odor and fly breeding potentials and destroys weed seeds and pathogens (Eghball et al., 1997; Mullins et al., 2005). Overall, composting results in significant losses in water, C, and N and an enhancement in the amount of bioavailable P (Barker, 1997). In addition, composting results in enrichment of total P in the finished composted product (Eghball et al., 1997; Dao, 1999; Eghball and Power, 1999). Dao (1999) reported that total P content of cattle manure increased from 2.1 to 5.2 g kg^{-1} following composting. Eghball et al. (1997) reported an increase in total P by 29% when feedlot cattle manure was composed. Eneji et al. (2001) reported a 31% increase for composed livestock manure (mixture of poultry droppings, pig manure, and cattle manure).

The major objective of compost addition is to supply N to crop plants (Hue et al., 1994; Stratton et al., 1995; Sikora and Szmidt, 2001). However, composts are good sources of other plant nutrients including P (Mullins et al., 2005). P concentrations in composts depend on the source feedstocks and can vary from <0.4 to >23 g kg^{-1} (Vogtmann and Turk, 1993; He et al., 1995). Municipal solid waste composts from the United States and European countries have P concentrations ranging from 2 to 6 kg kg^{-1} with a mean of 3.3 g kg^{-1} (Mullins et al., 2005). Plant-available P in compost can range from 20% to 40% of the total P (Vogtmann and Turk, 1993).

Application of compost has been associated with increased crop production and improvement in soil quality (Stratton et al., 1995; Fageria, 2002). During a 4-year study, Eghball and Power (1999) reported increased corn GYs from applying composted feed lot manure as compared to a nonfertilized treatment and similar yields as compared to a chemical fertilizer treatment. Mean N availability was 40% and 18% for the uncomposted manure and compost during the first year after application and 15% and 8% in the second year after application, respectively. Schlegel (1992) reported that composted manure plus additional fertilizer resulted in higher grain sorghum yield than either source applied alone. Alvarez et al. (1995) reported yield increase of tomato (*Solanum lycopersicum*) with the application of three compost materials. Similarly, Roe et al. (1997) reported that combination of low rates of fertilizer with compost from biosolids and yard trimmings resulted in the highest yields of marketable bell peppers (*Capsicum annuum*).

7.2.3.4 Green Manuring

Positive role of green manuring in crop production is known since ancient time. Importance of this soil ameliorating practice is increasing in recent years because of the high cost of chemical fertilizers, increased risk of environmental pollution, and need of sustainable cropping systems (Fageria, 2007a). Green manuring can improve soil's physical, chemical, and biological properties and consequently crop yields. Furthermore, potential benefits of green manuring are reduced NO_3^- leaching risk and lower fertilizer N requirements for succeeding crops. However, its influence may vary from soil to soil, crop to crop, environmental variables, type of green manure crop used, and management system. Beneficial effects of green manuring in crop production should not be evaluated in isolation but in integration with chemical fertilizers (Fageria, 2007a).

Interest in the use of green manures has occurred due to their role in improving soil quality and their beneficial N and non-N rotation effects (Jannink et al., 1996). Addition of organic matter by green manure crops improves soil's physical, chemical, and biological properties (MacRae and Mehuys, 1985; Fageria and Baligar, 2005; Fageria, 2007a). Furthermore, growing green manure crops in rotation with cash crops disrupts life cycle of diseases, insects, or weeds improving cash crop yields (Crookston et al., 1991). Green manuring can increase cropping system sustainability by reducing soil erosion (MacRae and Mehuys, 1985; Smith et al., 1987), by increasing nutrient retention (Dinnes et al., 2002), by improving soil fertility (Fageria and Baligar, 2005), and by reducing global warming potential (Robertson et al., 2000). Use of green manuring in dynamic cropping systems is economically viable, environmentally sustainable, and socially acceptable (Tanaka et al., 2002). Soil scientist during the second half of the twentieth century generated information mostly on sources, methods, and rates of inorganic fertilizers, with less emphasis on organic fertilizers (Pang and Letey, 2000).

7.2.3.4.1 Green Manure Crops

Green manure is defined as a plant material incorporated into soil while green or at maturity, for soil improvement, and green manure crop is any crop grown for the purpose of being turned under while green or soon after maturity for soil improvement (Soil Science Society of America, 2008). Green manure crops can be leguminous as well as nonleguminous and grown in situ or imported into fields as cuttings of trees and shrubs. Latter practice is called green leaf manuring (Singh et al., 1991). The term *green fallow* has been used to describe a green manure farming system that is typically used as partial fallow replacement in a wheat–fallow rotation (Pikul et al., 1997). In this system, a legume is seeded early in the fallow year, grown to about full bloom, and killed by chemicals or tillage. An important aspect of this green fallow system is to balance water use for N_2 fixation with the water and N requirements of the subsequent wheat crop (Pikul et al., 1997).

A vast array of legume species has potential as green manures. There are several hundred species of tropical legumes, but only a fraction of those have been studied for their potential as green manures. In temperate regions, legume crops are numerous, which can be used as green manure crops. Major green manure crops for tropical and temperate regions are presented (Table 7.16). Annual dry matter accumulation by these legumes varies from 1 to over 10 Mg ha^{-1} under ideal growing conditions (Lathwell, 1990). Quantities of N accumulated in the aboveground dry matter range from 20 to over 300 kg ha^{-1} (Lathwell, 1990). Annual legumes have potential as green manure crops in Canadian prairies (Rice et al., 1993). In grain lentil (*Lens culinaris* Medikus)–wheat rotations, there has been a gradual reduction in fertilizer N requirements (Campbell et al., 1992). Vyn et al. (1999) reported that corn GYs were consistently the highest following red clover (*Trifolium pratense*) and often the lowest following annual ryegrass (*Lolium*).

TABLE 7.16
Major Green Manure Crops for Tropical and Temperate Regions

Tropical Region		Temperate Region	
Common Name	**Scientific Name**	**Common Name**	**Scientific Name**
Sunn hemp	*Crotalaria juncea* L.	Hairy vetch	*Vicia villosa* Roth
Sesbania	*Sesbania aculeata* Retz Poir	Barrel medic	*Medicago truncatula* Gaertn
Sesbania	*Sesbania rostrata* Bremek & Oberm	Alfalfa	*Medicago sativa* L.
Cowpea	*Vigna unguiculata* L. Walp.	Black lentil	*L. culinaris* Medikus
Soybean	*G. max* L. Merr.	Red clover	*T. pratense* L.
Cluster bean	*Cyamopsis tetragonoloba*	Soybean	*G. max* L. Merr.
Alfalfa	*M. sativa* L.	Faba bean	*Vicia faba* L.
Egyptian clover	*Trifolium alexandrinum* L.	Crimson clover	*Trifolium incarnatum* L.
Wild indigo	*Indigofera tinctoria* L.	Ladino clover	*Trifolium repens* L.
Pigeon pea	*Cajanus cajan* L. Millspaugh	Subterranean clover	*Trifolium subterraneum* L.
Mung bean	*Vigna radiata* L. Wilczek	Common vetch	*Vicia sativa* L.
Lablab	*Lablab purpureus* L.	Purple vetch	*Vicia benghalensis* L.
Graybean	*Mucuna cinerecum* L.	Cura clover	*Trifolium ambiguum* Bieb.
Buffalo bean	*Mucuna aterrima* L. Piper & Tracy	Sweet clover	*Melilotus officinalis* L.
Shortflower rattlebox	*Crotalaria breviflora*	Winter pea	*Pisum sativum* L.
White lupin	*Lupinus albus* L.	Narrowleaf vetch	*Vicia angustifolia* L.
Milk vetch	*Astragalus sinicus* L.	Milk vetch	*A. sinicus* L.
Crotalaria	*Crotalaria striata*		
Zornia	*Zornia latifolia*		
Jack bean	*C. ensiformis* L. DC.		
Tropical kudzu	*P. phaseoloides* (Roxb.) Benth.		
Velvet bean	*Mucuna deeringiana* Bort. Merr.		
Adzuki bean	*Vigna angularis*		
Brazilian stylo	*S. guianensis*		
Jumbiebean	*Leucaena leucocephala* Lam. De Wit		
Desmodium	*Desmodium ovalifolium* Guillemin & Perrottet		
Pueraria	*P. phaseoloides* Roxb.		

Source: Fageria, N.K., *J. Plant Nutr.*, 30, 691, 2007a.

Legumes are superior green manure crops as compared with nonleguminous crops because they fix atmospheric N. Considerable variation in N fixation can occur, even among legume species (Hesterman et al., 1992). To be agronomically attractive and economically viable, a green manure crop should have certain characteristics. These properties are fast growing for easy adjustment in the cropping system, to produce sufficient dry matter to ameliorate soil's physical, chemical, and biological properties, to fix adequate N, and to require minimum cultural practices during growth period to be relatively more economical.

7.2.3.4.2 Nitrogen Fixation by Green Manure Legumes

Legumes have historically been used to maintain soil N fertility (Fauci and Dick, 1994). Legume rhizobial symbiosis is estimated to account for 40% of the world's fixed N (Ladha et al., 1992). Symbiotic N_2 fixation in legumes is determined by the formation of effective nodules on the roots. Formation of effective nodules depends on plant, soil, and climatic factors and their interactions. Hence, green manure legumes have different N_2 fixation capabilities depending on the environmental conditions, management practices adopted, and legume species (Fageria and Baligar, 2005).

TABLE 7.17
N Accumulation in Major Legume Green Manure Crops

Crop Species	Growth Duration (Days)	N Accumulation (kg ha^{-1})	Reference
G. max	45	115	Meelu et al. (1985)
C. juncea	45	169	Meelu et al. (1985)
C. cajan	45	33	Meelu et al. (1985)
S. aculeata	45	225	Meelu et al. (1985)
V. radiata	45	75	Meelu et al. (1985)
Dolichos lablab	45	63	Meelu et al. (1985)
I. tinctoria	45	45	Meelu et al. (1985)
S. rostrata	56	176	Furoc et al. (1985)
S. aculeata	56	144	Furoc et al. (1985)
V. radiata	45	75	Morris et al. (1986)
V. unguiculata	45	75	Furoc et al. (1985)
S. rostrata	60	219	Ladha et al. (1988)
Sesbania cannabina	60	171	Ladha et al. (1988)
Sesbania aegyptiaca	57	39	Ghai et al. (1985)
Sesbania grandiflora	57	24	Ghai et al. (1985)
Cyamopsis tetragonoloba	49	91	Singh et al. (1992)
Astragalus canadensis L.	Flowering	65–131	Watanabe (1984)
Vicia sativa	Flowering	105–210	Singh et al. (1991)
Melilotus albus	Flowering	150–300	Singh et al. (1991)

N supplied by hairy vetch (*V. villosa* Roth) and crimson clover (*T. incarnatum* L.) in cover crop experiments ranged from 72 to 149 kg N ha^{-1} (Ladha et al., 1988; Fageria, 2007a). Ladha et al. (1988) reported that leguminous green manure crop can accumulate about 2.6 kg N ha^{-1} day^{-1}. Incorporating such crops at 45–65 days results in rice yield equivalent to those with 50–100 kg fertilizer N ha^{-1}. Similarly, they have reported that 45–60-day-old stem-nodulating Sesbania species can provide equivalent of over 200 kg N ha^{-1}.

Various methods of N_2 fixation measurement have been proposed. These methods are N accumulation in green manure crop dry biomass, N balance, N difference, acetylene reduction, substrate-labeled isotope dilution, and $\delta^{15}N$ (natural abundance) dilution (Ladha et al., 1988). According to Ladha et al. (1988), the methods used have provided reasonable estimates of N amount; none are entirely satisfactory. However, N accumulation in the dry tissue of green manure crops and ^{15}N dilution are widely reported references for N_2 fixation (Ladha et al., 1988; Fageria, 2007a). The quantity of N accumulated by major green manure crops is presented (Table 7.17). Furthermore, amounts of N and other nutrient accumulated in 91-day-old graybean (*M. cinerecum* L.) green manure crop are presented (Table 7.18). Green manures not only fix N but also incorporate other essential plant nutrients in considerable amounts for succeeding crop (Table 7.18).

Earlier, only the legume roots were commonly known as sites for nodulation. However, Dreyfus and Dommergues (1981) discovered that *S. rostrata* fixed N on the stem. Since then, several species that form N_2-fixing nodules not only on the roots but also on the subepidermal primordial of the adventitious roots on the stems have been described (Alazard, 1985; Fageria, 2007a). These stem-nodulating species develop large number of parenchymatous, building spongy tissues that may store sufficient oxygen for various metabolic functions in water-saturated ecosystems (Ladha et al., 1992).

7.2.3.4.3 Green Manure Decomposition and Mineralization

Humification and mineralization refer to decomposition of organic matter. However, soil N availability is determined by its mineralization, the microbial conversion of organic N to NH_4^+ and further oxidation to nitrate (NO_3^-) (Gil and Fick, 2001). Soil and plant conditions determine green

TABLE 7.18
Dry Matter Yield and Nutrient Accumulation in the 91-Day-Old Graybean Green Manure Crop Grown on an Inceptisol of Central Brazil

Dry Matter Yield/Nutrient	Value[a]
Dry Matter (kg ha^{-1})	7016
N (kg ha^{-1})	185.2
P (kg ha^{-1})	11.7
K (kg ha^{-1})	161.1
Ca (kg ha^{-1})	72.9
Mg (kg ha^{-1})	18.7
Zn (g ha^{-1})	229.4
Cu (g ha^{-1})	170.0
Mn (g ha^{-1})	1667.6
Fe (g ha^{-1})	2527.8

Source: Fageria, N.K., *J. Plant Nutr.*, 30, 691, 2007a.
[a] Values are averages of two field experiments.

manure decomposition and subsequent N release. Among dominant plant factors are quantity and quality of green manure incorporated into the soil. Soil factors, which determine decomposition rate and N release, are texture, structure, acidity, microbial activity, and soil fertility (Thonnissen et al., 2000b). Lynch and Cotnoir (1956), Sorensen (1975), and Verbene et al. (1990) reported that organic residue decomposition was slow in soils with high clay content as compared with light-textured soils. Similarly, microbial activities were determined by soil's physical (compaction, temperature, and porosity) and chemical conditions (acidity, presence of toxic elements and organic C) (Grant et al., 1993). Greater N mineralization in soils under legume or grass–legume mixture occurred due to the more easily decomposable material, low C/N and lignin/N ratios, low (lignin + polyphenol)/N ratio, and higher litter contribution from legumes, as compared with grasses (Fox et al., 1990; Hatch et al., 1991; Gil and Fick, 2001). Becker et al. (1994), studying green manure and green manure–straw mixtures, identified the lignin to N (L/N) ratio as a significant factor controlling N release.

Residue decomposition depends mainly on temperature and soil moisture (Douglas and Rickman, 1992; Fageria, 2007a). Soil temperature in the range of 20°C–30°C and soil moisture in the range of −0.01 to −0.05 MPa are reported for fast release of NO_3^- following green manure incorporation into soil (Cassman and Munns, 1980). Mary and Recous (1994) reported N immobilization–remineralization following organic residue incorporation as a function of the amount and nature of the residues and soil N, whereas basal mineralization was explained as a function of soil texture and long-term C and N inputs. Alexander (1977) reported that liming acid soils accelerated the decay of plant residues and SOM.

Thonnissen et al. (2000b) reported that soybean and indigofera decomposed rapidly, losing 30%–70% of their biomass within 5 weeks after incorporation. Varco et al. (1989) reported rapid green manure N release within 15 days after incorporation of hairy vetch (*V. villosa* Roth). Broder and Wagner (1988) reported that soil-incorporated soybean residues lost 68% of their total biomass within 32 days. Various studies evaluating the fate of ^{15}N from legume residue decomposition under field conditions led to the conclusions that <30% of legume N was recovered by a subsequent nonlegume crop and large amounts of legume N were retained in soil, mostly in organic forms (Ladd et al., 1983; Broder and Wagner, 1988).

7.2.3.4.4 C/N Ratio of Green Manure Crops

C/N ratio is defined as the ratio of the mass of organic C to the mass of organic N in soil, organic material, plants, or microbial cells (Soil Science Society of America, 2008). C/N ratio of green manure or crop residue incorporated into the soil has an important role in the release or immobilization of soil N because plant tissue is a primary source and sink for C and N. When plant residues having C/N ratio greater than 20 are incorporated into the soil, available soil N is immobilized during the first few weeks of decomposition (Doran and Smith, 1991; Green and Blackmer, 1995; Fageria, 2007a). In aerobic soils, C/N ratios <20 for organic matter are required for net mineralization to occur (Islam et al., 1998).

C/N ratio of green manure species influenced rice N uptake, but only early in the season (Clement et al., 1998). C/N ratios of legume crops and field crops are presented (Table 7.19). Legumes have a low C/N ratio, as compared with cereals (Table 7.19). When high C/N ratio residue is incorporated into the soil, immobilization of available inorganic N may occur during the decomposition period. This occurs since the residue decomposing microbial population increased its biomass in response to the C source. If such immobilization occurs when plants require N for growth and development, the availability of NO_3^- for plants may be reduced (Dinnes et al., 2002). However, this immobilization of N is temporary in nature. As the decomposition proceeds, C/N ratio will began to reduce with time and may approach that of SOM (≈10–12), microbial biomass will decrease, and N from

TABLE 7.19
C/N Ratio of Major Legume Green Manure and Cereal Crops

Crop Specie	Growth Stage/Age in Days	C/N Ratio	Reference
Corn residues (*Z. mays* L.)	Physiol. maturity	67	Burgess et al. (2002)
Rice straw (*Oryza sativa* L.)	Physiol. maturity	69	Eagle et al. (2001)
Rice straw (*O. sativa* L.)	Physiol. maturity	56	Davelouis et al. (1991)
Sorghum (*S. bicolor* L. Moench)	Vegetative	22.0	Clement et al. (1998)
Barley straw (*H. vulgare* L.)	Physiol. maturity	99.1	Larney and Janzen (1996)
Ryegrass (*Lolium multiflorum* Lam)	Vegetative	30	Kuo and Jellum (2002)
Rye (*Secale cereale* L.)	Heading	40	Rannells and Wagger (1996)
Alfalfa hay (*M. sativa* L.)	Not given	15.9	Larney and Janzen (1996)
Pea straw (*P. sativum* L.)	Physiol. maturity	21	Fauci and Dick (1994)
Pea hay (*P. sativum* L.)	Not given	15.4	Larney and Janzen (1996)
Red clover (*T. pratense* L.)	101 days	13.7	Kirchmann (1988)
White clover (*T. repens* L.)	101 days	10.7	Kirchmann (1988)
Yellow trefoil (*Medicago lupulina* L.)	101 days	10.1	Kirchmann (1988)
Persian clover (*Trifolium resupinatum* L.)	101 days	15.8	Kirchmann (1988)
Egyptian clover (*T. alexandrinum* L.)	101 days	16.7	Kirchmann (1988)
Subterranean clover (*T. subterraneum* L.)	101 days	11.4	Kirchmann (1988)
Cowpea (*V. unguiculata* L. Walp.)	Green pods	13.9	Clement et al. (1998)
Sunn hemp (*C. juncea* L.)	Mature pods	20.2	Clement et al. (1998)
Soybean (*G. max* L. Merr.)	Vegetative	17.9	Clement et al. (1998)
Pigeon pea (*C. cajan* L. Millspaugh)	Not given	25.9	Clement et al. (1998)
Wild indigo (*I. tinctoria* L.)	Flowering	15.8	Clement et al. (1998)
Sesbania (*S. rostrata* Bremek & Oberm)	Vegetative	27.8	Clement et al. (1998)
Sesbania (*Sesbania emerus* Aubl. Urb.)	Vegetative	26.5	Clement et al. (1998)
Aeschynomene afraspera	Vegetative	23.9	Clement et al. (1998)
Desmanthus virgatus	Green pods	18.9	Clement et al. (1998)
Tropical kudzu (*P. phaseoloides*)	Not given	19	Davelouis et al. (1991)
Hairy vetch (*V. villosa* Roth)	Vegetative	12	Kuo and Jellum (2002)
Hairy vetch (*V. villosa* Roth)	Flowering	18	Sainju et al. (2002)
Hairy vetch (*V. villosa* Roth)	Early bloom	17	Rannells and Wagger (1996)
Crimson clover (*T. incarnatum* L.)	Midbloom	11	Rannells and Wagger (1996)

plant residues that was incorporated in the microbial mass will once again be released into the soil (Dinnes et al., 2002). Green manure crop should be incorporated in advance to avoid N deficiency for the primary crop followed by the green manuring. Many studies have shown that the low C/N ratio of legume residue enhances soil N availability (Beckie and Brandt, 1997; Fageria, 2007a).

Historically, C/N ratio is the most widely used index of crop residue quality and decomposition rate (Henriksen and Breland, 1999; Fageria, 2007a). However, Vigil and Kissel (1995) concluded that C/N estimated N mineralization parameters poorly, especially when C/N ranged from 10 to 28. Furthermore, Ruffo and Bollero (2003) concluded that the availability of C and N, rather than their total concentration in the residue, has a critical role in residue decomposition and nutrient release. In other studies, soluble C (Kuo and Sainju, 1998), cellulose (Bending et al., 1998), or lignin (Muller et al., 1988) is considered to be most closely related to residue decomposition. Furthermore, some ratios, such as lignin to N or polyphenol plus lignin to N, have been used as indices of residue nutrient release (Fageria, 2007a).

Laboratory and field studies have indicated that residue decomposition occurs in several steps involving chemical and physical transformation. Generally, residues with low N contents or high C/N ratios have slow decomposition rates (Fageria, 2007a). Although C/N ratio and N content have been used to describe residue decomposition, these are not always related to decomposition (Fageria, 2007a). Reinertsen et al. (1984) concluded that early-stage residue decomposition was largely dependent on the size of water-soluble C pool and an intermediately available C pool. This finding is in agreement with a study by Gilmour et al. (1998), who reported that only initial 0–2 weeks' decomposition was related to crop residue organic N and C/N ratio.

Grant et al. (2002) reported that N concentration in the crop residue would determine the net balance between mineralization and immobilization. If the N concentration in the residue is below approximately 20–24 g N kg^{-1}, immobilization will exceed mineralization, and the decomposition residues will bind N rather than release it (Fageria, 2007a). Legume residues contain considerable amounts of N and have a relatively low C/N residue, leading to more rapid release of N than lower N-containing cereal residue (Janzen and Kucey, 1988).

7.2.3.4.5 Amelioration of Soil's Physical, Chemical, and Biological Properties

Crop cultivation reduces SOM and adversely affects soil quality. Soil compaction is the main consequence of low SOM, which restricts plant root development and limits plant growth (Fageria, 2007a). Compaction impedes internal soil drainage, and reduced infiltration may lead to increased soil erosion (Radcliffe et al., 1988). Soil erosion may lead to discharge of sediments and agricultural chemicals, thereby resulting in serious environmental degradation (Dao, 1996).

Green manuring had significant positive influences on soil's physical, chemical, and biological properties and consequently on crop yields (Fageria, 2007a). MacRae and Mehuys (1985) reported that all other factors are equal; a soil with a high SOM level has optimum physical conditions. SOM refers to the organic fraction of the soil; it includes plant and animal residues at various stages of decomposition, cells (living and dead) and tissues of microbes, and substances synthesized by the soil population. The addition of organic matter by green manuring improves the soil structure stability, increases the water-holding capacity of the soil, and increases the infiltration of water into the soil and percolation through the soil (Fageria, 2007a). Martens and Frankenberger (1992) reported that bulk density and aggregate stability are the major factors affecting water infiltration rates. This process is responsible for the reduction of soil erosion. Improved soil physical conditions may promote root growth and increased the use of soil, water, and nutrients (Fageria, 2007a). SOM contents have been reported to be a reliable index of crop productivity in semiarid regions because it positively affects soil's water holding capacity (Fageria, 2007a). Diaz-Zorita et al. (1999) reported that wheat yields over 3 years were linearly related to SOM content in the top 20 cm layer when the organic matter content was <30 g kg^{-1} and higher soil–water holding capacity was one of the factors for this positive effect.

Loss of SOM with cultivation, erosion, or inappropriate soil management practices is a common feature of a crop production system. Under these situations, soil degradation process starts and soil is no longer sustainable for profitable long-term agriculture. Wani et al. (1994) reported that adopting green manures and organic amendments in crop rotations provided a measurable increase in SOM quality and other soil quality attributes as compared with continuous monoculture cereal systems. SOM is considered to be a key attribute of soil quality (Fageria, 2012) and environmental quality (Fageria, 2002). SOM is involved in and related to many soil's physical, chemical, and biological properties (Fageria, 2002, 2012). Organic matter fractions like macroorganic matter, light fraction, microbial biomass, and mineralizable C describe the quality of SOM (Carter, 2002). These fractions have biological significance for several soil functions and processes and are sensitive indicators of change in total SOM content. Total SOM influences soil compactibility, friability, and soil–water holding capacity, while aggregated SOM has major implications for the functioning of soil in regulating air and water infiltration, conserving nutrients, and influencing soil permeability and erodibility (Carter, 2002). Biederbeck et al. (1994) reported that the accumulation of crop residues with frequent inclusion of legume crops in a rotation improved the physical and biochemical properties of the soil by increasing labile organic matter.

Hargrove (1986) reported that legume cover crops lowered soil pH, resulted in re-distribution of K^+ to the soil surface from deeper in the soil profile, and lower C/N ratio in SOM. Legume green manures maintain ground cover, usually between cultivated crops, reducing erosion and providing weed control. Crop uptake of soil NO_3^- reduced risk of leaching (Shipley et al., 1992), reduced water runoff and soil losses of N during intense rainfall, enhanced N availability to succeeding crops, and subsequently reduced the need for N fertilizers (Fageria, 2007a). Sainju and Singh (1997) reported that reduction in NO_3^- leaching by nonlegumes ranges from 29% to 94%, as compared with 6% to 48% by legumes. McCracken et al. (1994) reported that rye (*S. cereale* L.) reduced NO_3^- leaching by 94%, as compared with 48% for hairy vetch. Sainju et al. (1998) concluded that nonlegume cover crops, such as rye, may be more effective in reducing residual NO_3^- and potential leaching of NO_3^- from the soil early in the growing season than are legume green manure crops, such as hairy vetch or crimson clover.

Cover crops promote mycorrhizae on the roots of succeeding crops, increasing soil P and micronutrient availability. Cover crops may suppress plant pests such as nematodes (Lathwell, 1990). In addition to improving soil properties, green manure crops can be used in controlling weeds and other pests (Fageria, 2007a). Weed suppression has been reported under seeded biennial sweet clover (*M. officinalis* L. Lam.) (Entz et al., 1995). Similarly, weed suppression effect of perennial alfalfa has been shown to last for 3 years for wild oat (*A. fatua*) and other, but not all, weed species (Entz et al., 1995). Al-Khatib et al. (1997) and Krishnan et al. (1998) reported that species of Brassica family are often used as cover or green manure crops. They can suppress weeds when incorporated into the soil. These authors concluded that weed suppression by these species may be due to secondary plant metabolites. Although the biological activity of these secondary plant metabolites is low, they have a key role in weed suppression as they can be converted to the corresponding isothiocyanates by the plant enzyme myrosinase. Weed emergence suppression by green manure may be associated with reducing light penetration and soil temperature fluctuations (Teasdale and Mohler, 1993).

Green manure can improve P uptake of succeeding crops via converting unavailable native and residual fertilizer P to more available chemical forms (Fageria, 2007a). Furthermore, decomposition of green manure residues can form H_2CO_3, which can solubilize soil mineral P and consequently higher P availability to plants (Fageria, 2007a). In addition, green manure crops generally accumulate a large amount of P, and on decomposition of residues, these crops can provide a larger pool of mineralizable soil organic P for succeeding crops (Tissen et al., 1994). In soils with high P fixation capacity, P availability may be enhanced by organic compounds released during decomposition process by blocking P adsorption sites or via anion exchange (Kafkafi et al., 1988).

7.2.3.4.6 Yield Response of Annual Crops to Green Manuring

Yield increases in annual crops by the application of green manure depend on crop species, environmental conditions, management practices, and succeeding field crops. The role of N from organic sources such as green manure is tied to complex microbial cycling of C and N; therefore, the availability and effects of legume N are more difficult to predict than those of chemical fertilizer N (Fageria, 2007a). Mineralization of N from green manure crop residue can contribute to either a small or a substantial fraction (4%–30%) of the N assimilated by a subsequent crop, depending on the attributes of the green manure crop such as C/N ratio, soil type, and management practices (Jensen, 1992; Jackson, 2000; Fageria, 2007a). Ladha et al. (1996) reported GY increases of 0.5–1.4 Mg ha^{-1} with green manure N ranging from 81 to 162 kg ha^{-1}, as compared with no green manure, but with no consistent relationship between yield and green manure N rates. Yield increases due to green manure in lowland rice are variable (Singh et al., 1991). A yield advantage of 1–2 Mg ha^{-1} is common with green manure as compared with no green manure, which corresponds to approximately 20–40 kg N ha^{-1} additional N uptake by the rice crop (George et al., 1998). Becker et al. (1995) reported that depending on the season and established method, *S. rostrata* substituted for 35–90 kg of split applied urea N in rice.

Rice GY increase varied from 45% to 130% depending on the environmental conditions and management practices adopted (Kalidurai and Kannaiyan, 1989). Experiments conducted in India, Thailand, Nigeria, and Senegal (Fageria, 2007b) reported substantial yield increases of rice after incorporation of *S. rostrata*. In the Philippines, Ladha et al. (1992) reported rice yield increases ranging from 0.8 to 2.8 Mg ha^{-1} with the use of *S. rostrata* and *A. afraspera* as green manures as compared to a control treatment.

Quantity of N that can be substituted by green manure has been reviewed by Becker et al. (1988) and Ladha et al. (1988). They reported that incorporating 45–60-day-old green manure crop resulted in rice yield increases equivalent to those obtained with 50–100 kg N ha^{-1}. Further, if green manure incorporation is combined with fertilizer N application, rice yields are higher than those obtained with an equivalent amount of N alone. Similarly, Fageria and Baligar (1996) and Fageria and Souza (1995) reported yield increases of rice and common bean in Brazilian Inceptisol and Oxisol soils with green manuring as compared with inorganic fertilizers (Table 7.20).

A conceptual framework for interpreting the effects of green manuring on succeeding crop yields has been proposed by Bouldin (1988). Residual effects of green manuring on yield increases of subsequent lowland and upland rice crops have been reported by Buresh and De Datta (1991) and for lowland rice by Diekmann et al. (1993). According to Ladha et al. (1992), residual effects of green manure on succeeding crop are determined by the amount of N added and soil's physical, chemical, and biological properties. When green manure is used prior to wet season, a modest residual effect on a second rice crop during the dry season can be expected (Morris et al., 1989).

Thonnissen et al. (2000a) reported that tomato yields responded to green manuring N in the wet season in Taiwan and in the northern Philippines, comparing favorably with fertilizer at 38–120 kg N ha^{-1}. They concluded that tomato yield response to green manuring N was high on infertile soils and tomato N requirement can be substituted fully or partially by green manure, depending on soil N mineralization. Thonnissen et al. (2000a) reported that the residual effect of soybean green manure applied to tomato on the following maize was similar to that of 120 kg N ha^{-1}. Guldan et al. (1997) reported that hairy vetch and alfalfa had fertilizer replacement value based on cumulative sorghum yield ranging from 78 to 140 kg N ha^{-1}.

7.2.4 Control of Soil Erosion

A large part of nutrients, including P, is lost by soil erosion. Control of soil erosion is important in improving PUE by field crops. Water and wind erosion is important soil quality deterioration process on the agricultural lands worldwide. This deterioration is due to removal of fertile soil layer, decreased rain water infiltration, decreased soil depth, increased sedimentation of lake and dams, accelerated river flooding, fertile land being covered by sand dunes, and decreased soil productivity. Adopting appropriate erosion control measures can reverse or halt this soil quality deterioration process (Fageria, 2002).

TABLE 7.20
Response to Fertilization of Rice and Common Bean Grown in Rotation in Cerrado and Varzea Acid Soils

Fertility Level	Rice GY (Mg ha^{-1})[a]	Common Bean GY (Mg ha^{-1})[a]
Oxisol of Cerrado[b]		
Low	1.7b	1.2c
Medium	2.1a	1.8b
High	2.1a	2.2a
Medium + green manure	2.4a	1.5a
F-test	*	**
Inceptisol of Varzea[c]		
Low	4.3b	2.9b
Medium	5.5a	6.6a
High	5.5a	8.5a
Medium + green manure	6.3a	8.2a
F-test	**	**

Source: Fageria, N.K. and Souza, N.P., *Pesq. Agropec. Bras*., 30, 359, 1995.

[a] Values are averages of three crops grown in rice–bean rotation.

[b] Cerrado soil fertility levels for rice were low (without addition of fertilizers), medium (50 kg N ha^{-1}, 26 kg P ha^{-1}, 33 kg K ha^{-1}, 30 kg ha^{-1} fritted glass material as a source of micronutrients), and high (all the nutrients were applied at the double the medium level). *C. cajan* L. was used as a green manure at the rate of 25.6 t ha^{-1} green matter. For common bean, the fertility levels were low (without addition of fertilizers), medium (35 kg N ha^{-1}, 44 kg P ha^{-1}, 42 kg K ha^{-1}, 30 kg ha^{-1} fritted glass material as a source of micronutrients), and high (all the nutrients were applied at the double the medium level).

[c] Varzea soil fertility levels for rice were low (without addition of fertilizers), medium (100 kg N ha^{-1}, 44 kg P ha^{-1}, 50 kg K ha^{-1}, 40 kg ha^{-1} fritted glass material as a source of micronutrients), and high (all the nutrients were applied at the double the medium level). *C. cajan* L. was used as a green manure at the rate of 28 t ha^{-1} green matter. For common bean, the fertility levels were low (without addition of fertilizers), medium (35 kg N ha^{-1}, 52 kg P ha^{-1}, 50 kg K ha^{-1}, 40 kg fritted glass material as a source of micronutrients), and high (all the nutrients were applied at the double the medium level).

*,**Significant at the 0.05 and 0.01 probability levels, respectively. Within the same column, means followed by the same letter do not differ significantly at the 5% probability levels by Tukey's test.

7.3 FERTILIZER MANAGEMENT

Use of inorganic fertilizers is important for obtaining higher yield of modern crop cultivars. Applying P fertilizer to crops, where it is needed, is an investment that provides positive financial returns. Economic responses to applying P fertilizers depend entirely on soil P levels, soil moisture, the prevailing cost of P fertilizer, and gross returns for grain. Conversely, yield penalty for not using sufficient P fertilizer can be severe. Fertilizers should be used with caution in relations to rate, time, and methods of application. These precautions are necessary not only to improve crop yields but also to lower the cost of crop production and to reduce environmental pollution. In addition, best response to P fertilizer application is only achieved when other management practices are optimized.

7.3.1 Efficient Phosphorus Source

Use of efficient P source is a prerequisite to obtain optimum results of applied P to crops when P level in a soil is low. There are several P sources available in the market (Table 7.21). Major P

TABLE 7.21
Major P Fertilizers, Their P Content, and Solubility

Common Name	Chemical Composition	P_2O_5 Content (%)	Solubility
Simple superphosphate (SSP)	$Ca(H_2PO_4)_2 + CaSO_4$	18–22	Water soluble
TSP	$Ca(H_2PO_4)_2$	46–47	Water soluble
Monoammonium phosphate	$NH_4H_2PO_4$	48–50	Water soluble
Diammonium phosphate	$(NH_4)_2HPO_4$	54	Water soluble
Phosphoric acid	H_3PO_4	55	Water soluble
Thermophosphate (yoorin)	$[3MgO \cdot CaO \cdot P_2O_5 + 3(CaO\ SiO_2)]$	17–18	Citric acid soluble
Rock phosphates	Apatites	24–40	Citric acid soluble
Basic slag	$Ca_3P_2O_8\ CaO + CaO \cdot SiO_2$	10–22	Citric acid soluble

Source: Fageria, N.K., *The Use of Nutrients in Crop Plants*, CRC Press, Boca Raton, FL, 2009.

forms applied to crops are supplied by triple superphosphate (TSP), simple or ordinary superphosphate, and mono- or diammonium phosphate. Historically, the P content of chemical fertilizers is expressed in P_2O_5. However, scientifically P is generally expressed as P, rather than P_2O_5. Values of P_2O_5 can be converted to % P or vice versa by using the following equation:

$$\%P = \%P_2O_5 \times 0.43; \%P_2O_5 = \%P \times 2.29.$$

Soil pH has an influence on P source reaction in the soil. If the soil is highly acidic (pH < 5.0), P-soluble fertilizers react with Al and Fe to form insoluble phosphates and their availability to plants decreased. If pH is higher than 7.0, the soluble fertilizers react with Ca to form insoluble phosphates (Bundy et al., 2005). Soil pH range from 6.0 to 7.0 is considered as an ideal pH value for P availability to plants. In addition to soil pH, soil temperature and moisture content influence P fertilizer reactions and solubility and availability to plants. Fageria et al. (2014b) has compared six P sources for upland rice grown on a Brazilian Oxisol soil. Plant height (PH), SY, and GY were significantly affected by P treatments (Table 7.22). PH varied from 41.33 cm produced at 0 mg P kg^{-1} to 89.33 cm produced at 200 mg P kg^{-1} added by polymer-coated triple superphosphate, with a mean value of 74.6 cm. SY varied from 1.58 g $plant^{-1}$ produced at the control treatment to 17.8 g $plant^{-1}$ produced at the 200 mg P kg^{-1} by SSP, with a mean value of 12.9 g $plant^{-1}$. Similarly, GY varied from 0.64 g $plant^{-1}$ at control treatment to 7.22 g $plant^{-1}$ at 50 mg P kg^{-1} by SSP, with a mean value of 5.04 g $plant^{-1}$. PH, SY, and GY increased in a quadratic fashion with the application of P (0–400 mg kg^{-1}) for all six P sources (Table 7.23).

Based on regression analysis, maximum variation in PH (92%) was due to MAP P source and minimum variation (68%) was due to PSSP. P requirement for maximum PH varied from 224 mg P kg^{-1} supplied by PTSP to 262 mg P kg^{-1} supplied by PMAP, with a mean value of 235 mg P kg^{-1}. SY variation was 37% with the application of PMAP to 84% with the application of MAP source, with a mean value of 76% variation. P requirement for maximum SY varied from 260 to 286 mg kg^{-1}, with a mean value of 265 mg P kg^{-1} (Table 7.23).

Variation in GY varied from 54% to 93%, with a mean value of 86% with all P sources. P requirements for maximum GY varied from 192 to 250 mg kg^{-1}, with a mean value of 227 mg P kg^{-1}. P requirements for SY were higher as compared to GY. This may be related to higher SY as compared to GY obtained with the application of six P sources. Increase in PH, SY, and GY of upland rice with the addition of P in the Brazilian Oxisol soils has been reported by Fageria et al. (1982, 2011a) and Fageria (2009). To classify the P source efficiency in grain production, GY was plotted against P rate (Figure 7.21). Maximum GY was achieved with P source in the order of PMAP > SSP = MAP > PSSP > TSP > PTSP. Estimated values of GY were calculated with regression equations for each P source (Table 7.22). Maximum GY at an adequate P rate (192–250 mg P kg^{-1}) of about 7 g $plant^{-1}$ (Figure 7.21) was obtained with three P sources, that is, SSP, PMAP, and MAP, and

TABLE 7.22
Influence of P Treatments on PH, SY, and GY of Upland Rice

Code of P Treatment	PH (cm)	SY (g Plant^{-1})	GY (g Plant^{-1})
1. P0 (control)	41.33g	1.58h	0.64i
2. SSP50	73.00a–f	17.06abc	7.22a
3. SPS100	81.00a–e	17.43ab	5.84a–d
4. SPS200	82.33a–d	17.77a	6.41ab
5. SPS400	69.66a–f	16.68a–d	2.26hi
6. PSSP50	81.66a–e	15.06a–f	6.61ab
7. PSSP100	88.00ab	15.93a–d	6.32ab
8. PSSP200	86.33abc	15.43a–e	5.96abc
9. PSSP400	74.33a–f	16.01a–d	5.45b–e
10. TSP50	62.33def	11.52d–g	3.06gh
11. TSP100	77.66a–f	13.70a–g	5.67a–d
12. TSP200	81.33a–e	12.71a–g	5.98abc
13. TSP400	61.33efg	13.83a–g	3.16gh
14. PTSP50	67.33b–f	11.73c–g	5.66a–d
15. PTSP100	84.33abc	12.28b–g	6.29ab
16. PTSP200	89.33a	12.06b–g	5.32b–f
17. PTSP400	65.66c–f	12.93a–g	3.88e–h
18. MAP50	59.00fg	9.20g	4.42c–g
19. MAP100	78.00a–f	10.23efg	5.84a–d
20. MAP200	83.00a–d	11.90c–g	6.37ab
21. MAP400	68.66a–f	12.65a–g	4.54c–g
22. PMAP50	70.00a–f	12.88a–g	3.71fgh
23. PMAP100	75.33a–f	9.73fg	4.99b–f
24. PMAP200	84.00abc	9.52g	6.28ab
25. PMAP400	80.00a–e	11.67c–g	4.27d–g
Average	74.60	12.86	5.04
F-test	**	**	**
CV (%)	8.85	13.38	10.45

Source: Fageria, N.K. et al., *Commun. Soil Sci. Plant Anal.*, 45, 1399, 2014b.
1 = 0 mg P kg^{-1} (control); 2 = 50 mg P kg^{-1} with SSP; 3 = 100 mg P kg^{-1} with SSP; 4 = 200 mg P kg^{-1} with SPP; 5 = 400 mg P kg^{-1} with SPP; 6 = 50 mg P kg^{-1} with polymer-coated simple superphosphate (PSSP); 7 = 100 mg P kg^{-1} with PSSP; 8 = 200 mg P kg^{-1} with PSSP; 9 = 400 mg P kg^{-1} with PSSP; 10 = 50 mg P kg^{-1} with TSP; 11 = 100 mg P kg^{-1} with TSP; 12 = 200 mg P kg^{-1} with TSP; 13 = 400 mg P kg^{-1} with TSP; 14 = 50 mg P kg^{-1} with polymer-coated triple superphosphate (PTSP); 15 = 100 mg P kg^{-1} with PTSP; 16 = 200 mg P kg^{-1} with PTSP; 17 = 400 mg P mg P kg^{-1} with PTSP; 18 = 50 mg P kg^{-1} with monoammonium phosphate (MAP); 19 = 100 mg P kg^{-1} with MAP; 20 = 200 mg P kg^{-1} with MAP; 21 = 400 mg P kg^{-1} with MAP; 22 = 50 mg P kg^{-1} with polymer-coated monoammonium phosphate (PMAP); 23 = 100 mg P kg^{-1} with PMAP; 24 = 200 mg P kg^{-1} with PMAP; and 25 = 400 mg P kg^{-1} with PMAP.
**Significant at the 1% probability level.

minimum GY at an adequate P rate (average value of 227 mg P kg^{-1}) of about 6.2 g was obtained by TSP. Difference in GY increase was about 13% at higher GY-producing P sources as compared to lower GY-producing P sources. PMAP source can be considered the optimum P source among the six P sources due to higher yield at the adequate P as well as at higher P levels.

Panicle density (PD) and thousand grain weight were significantly influenced by P treatments (Table 7.24). However, the thousand grain weight means were similar among different P treatments. PD varied from 0.91 plant^{-1} produced by the control treatment to 10.33 panicle plant^{-1} produced by

TABLE 7.23
Relationship between P Rate and PH, SY, and GY of Upland Rice

Variable	Regression Equation	R^2	PMV
PH			
P rate with SSP vs. PH	$Y = 48.53 + 0.36X - 0.00077X^2$	0.70**	234
P rate with PSSP vs. PH	$Y = 51.61 + 0.39X - 0.00086X^2$	0.68**	227
P rate with TSP vs. PH	$Y = 44.13 + 0.37X - 0.00082X^2$	0.85**	226
P rate with PTSP vs. PH	$Y = 44.94 + 0.44X - 0.00098X^2$	0.88**	224
P rate with MAP vs. PH	$Y = 42.88 + 0.37X - 0.00078X^2$	0.92**	237
P rate with PMAP vs. PH	$Y = 46.99 + 0.33X - 0.00063X^2$	0.72**	262
Average	$Y = 46.52 + 0.38X - 0.00081X^2$	0.86**	235
SY			
P rate with SSP vs. SY	$Y = 5.72 + 0.13X - 0.00025X^2$	0.67**	260
P rate with PSSP vs. SY	$Y = 5.35 + 0.11X - 0.00021X^2$	0.64**	262
P rate with TSP vs. SY	$Y = 4.37 + 0.09X - 0.00016X^2$	0.58**	281
P rate with PTSP vs. SY	$Y = 4.42 + 0.08X - 0.00015X^2$	0.65**	267
P rate with MAP vs. SY	$Y = 3.32 + 0.08X - 0.00014X^2$	0.84**	286
P rate with PMAP vs. SY	$Y = 5.14 + 0.05X - 0.000096X^2$	0.37*	260
Average	$Y = 4.13 + 0.09X - 0.00017X^2$	0.76**	265
GY			
P rate with SSP vs. GY	$Y = 2.29 + 0.05X - 0.00013X^2$	0.61**	192
P rate with PSSP vs. GY	$Y = 2.36 + 0.04X - 0.00009X^2$	0.54**	222
P rate with TSP vs. GY	$Y = 0.88 + 0.05X - 0.00011X^2$	0.93**	227
P rate with PTSP vs. GY	$Y = 2.09 + 0.04X - 0.000097X^2$	0.59**	206
P rate with MAP vs. GY	$Y = 1.37 + 0.05X - 0.00011X^2$	0.87**	227
P rate with PMAP vs. GY	$Y = 1.04 + 0.05X - 0.00010X^2$	0.92**	250
Average	$Y = 1.44 + 0.05X - 0.00011X^2$	0.86**	227

Source: Fageria, N.K. et al., *Commun. Soil Sci. Plant Anal.*, 45, 1399, 2014b.
Note: PMV, P rate for maximum PH, SY, and GY.
SSP, simple superphosphate; PSSP, polymer-coated simple superphosphate; TSP, triple superphosphate; PTSP, polymer-coated triple superphosphate; MAP, monoammonium phosphate; PMAP, polymer-coated monoammonium phosphate.
*,**Significant at the 5% and 1% probability levels, respectively.

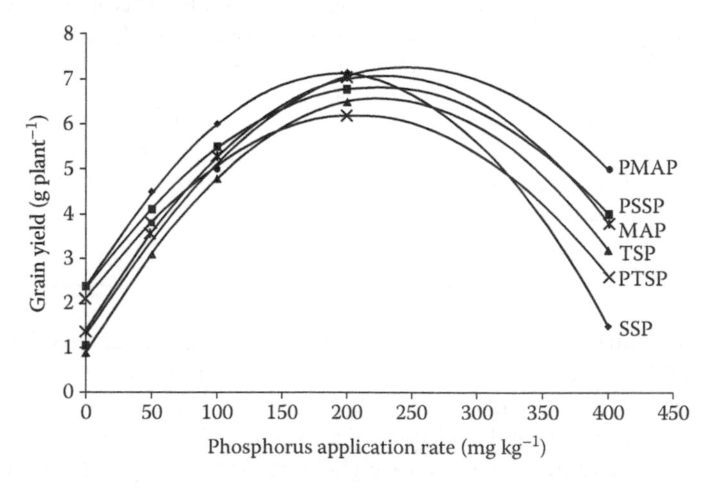

FIGURE 7.21 Relationship between P level of different P sources and GY of upland rice. (From Fageria, N.K. et al., *Commun. Soil Sci. Plant Anal.*, 45(10), 1399, 2014.)

TABLE 7.24
Influence of P Treatments on PD, Spikelet Sterility, and Thousand Grain Weight (TGW) of Upland Rice

Code of P Treatment	PD (Plant^{-1})	Thousand Grain Weight (g)
1. P0 (control)	0.91e	20.59a
2. SSP50	7.66a–d	21.75a
3. SPS100	8.58a–d	25.93a
4. SPS200	10.00ab	25.69a
5. SPS400	8.91abc	21.32a
6. PSSP50	7.41a–d	23.11a
7. PSSP100	10.33a	26.90a
8. PSSP200	8.17a–d	28.64a
9. PSSP400	6.50bcd	21.44a
10. TSP50	5.58cd	25.41a
11. TSP100	6.58bcd	22.78a
12. TSP200	8.08a–d	29.30a
13. TSP400	6.25cd	20.21a
14. PTSP50	6.33bcd	20.37a
15. PTSP100	6.58bcd	19.46a
16. PTSP200	8.17a–d	23.54a
17. PTSP400	7.75a–d	27.26a
18. MAP50	5.41cd	20.52a
19. MAP100	8.25a–d	27.71a
20. MAP200	8.75abc	20.52a
21. MAP400	6.58bcd	23.97a
22. PMAP50	6.25cd	20.86a
23. PMAP100	5.83cd	20.66a
24. PMAP200	6.41bcd	24.27a
25. PMAP400	4.91d	21.65a
Average	7.05	23.35
F-test	**	*
CV (%)	16.63	15.08

Source: Fageria, N.K. et al., *Commun. Soil Sci. Plant Anal.*, 45, 1399, 2014b.

1 = 0 mg P kg^{-1}(control); 2 = 50 mg P kg^{-1} with SSP; 3 = 100 mg P kg^{-1} with SSP; 4 = 200 mg P kg^{-1} with SSP; 5 = 400 mg P kg^{-1} with SSP; 6 = 50 mg P kg^{-1} with PSSP; 7 = 100 mg P kg^{-1} with PSSP; 8 = 200 mg P kg^{-1} with PSSP; 9 = 400 mg P kg^{-1} with PSSP; 10 = 50 mg P kg^{-1} with TSP; 11 = 100 mg P kg^{-1} with TSP; 12 = 200 mg P kg^{-1} with TSP; 13 = 400 mg P kg^{-1} with TSP; 14 = 50 mg P kg^{-1} with PTSP; 15 = 100 mg P kg^{-1} with PTSP; 16 = 200 mg P kg^{-1} with PTSP; 17 = 400 mg P mg P kg^{-1} with PTSP; 18 = 50 mg P kg^{-1} with MAP; 19 = 100 mg P kg^{-1} with MAP; 20 = 200 mg P kg^{-1} with MAP; 21 = 400 mg P kg^{-1} with MAP; 22 = 50 mg P kg^{-1} with PMAP; 23 = 100 mg P kg^{-1} with PMAP; 24 = 200 mg P kg^{-1} with PMAP; and 25 = 400 mg P kg^{-1} with PMAP.

*,**Significant at the 5% and 1% probability levels, respectively.

PSSP, with a mean value of 7.05 panicle per plant. PD increased quadratically with increasing P rate (0–400 mg kg^{-1}) for all six P sources (Table 7.25). Maximum variation in PD was 83% due to the application of TSP, PTSP, and MAP. Minimum variation in PD (55%) was due to the application of PSSP (Table 7.25).

P rate required for maximum PD varied from 208 to 300 mg kg^{-1}, with a mean value of 231 mg kg^{-1}. For PD, maximum amount of P is required by PTSP and minimum by PMAP. Improvement in PD with the application of P in Brazilian Oxisol soils has been reported by Fageria (2009). Fageria and Baligar (1997) reported a quadratic increase in tillering with increasing PD of 20 upland rice

TABLE 7.25
Relationship between P Rate and PD and TGW of Upland Rice

Variable	Regression Equation	R^2	PMV
PD			
P rate with SSP vs. PD	$Y = 2.38 + 0.07X - 0.00014X^2$	0.79**	250
P rate with PSSP vs. PD	$Y = 2.77 + 0.07X - 0.00015X^2$	0.55**	233
P rate with TSP vs. PD	$Y = 1.75 + 0.06X - 0.00012X^2$	0.83**	250
P rate with PTSP vs. PD	$Y = 2.08 + 0.06X - 0.00010X^2$	0.83**	300
P rate with MAP vs. PD	$Y = 41.71 + 0.07X - 0.00014X^2$	0.83**	250
P rate with PMAP vs. PD	$Y = 2.25 + 0.04X - 0.000096X^2$	0.67**	208
Average	$Y = 2.16 + 0.06X - 0.00013X^2$	0.81**	231
TGW			
P rate with SSP vs. TGW	$Y = 20.37 + 0.056X - 0.00013X^2$	0.26^{NS}	—
P rate with PSSP vs. TGW	$Y = 20.29 + 0.08X - 0.00019X^2$	0.50*	211
P rate with TSP vs. TGW	$Y = 20.54 + 0.07X - 0.00018X^2$	0.44*	194
P rate with PTSP vs. TGW	$Y = 19.41 + 0.018X - 0.000032X^2$	0.51**	—
P rate with MAP vs. TGW	$Y = 21.4 + 0.016X - 0.000027X^2$	0.04^{NS}	—
P rate with PMAP vs. TGW	$Y = 19.90 + 0.027X - 0.000055X^2$	0.12^{NS}	—
Average	$Y = 20.42 + 0.042X - 0.000093X^2$	0.41*	226

Source: Fageria, N.K. et al., *Commun. Soil Sci. Plant Anal.*, 45, 1399, 2014b.
Note: PMV, P rate for maximum panicle density, spikelet sterility (SS), and TGW.
SSP, simple superphosphate; PSSP, polymer-coated simple superphosphate; TSP, triple superphosphate; PTSP, polymer-coated triple superphosphate; MAP, monoammonium phosphate; PMAP, polymer-coated monoammonium phosphate.
$^{*,**, \text{and NS}}$Significant at the 5% and 1% probability levels and nonsignificant, respectively.

genotypes in Brazilian Oxisol soils when P was applied in the range of 0–175 mg kg^{-1}. Similarly, Fageria (1991) reported a quadratic response of PD in upland rice to increasing P rates (0–200 mg kg^{-1}) on Brazilian Oxisol soils. Fageria et al. (1982) reported a quadratic increase in tillering of upland rice under field conditions with increasing P rate (0–66 kg ha^{-1}). Tillering was measured during the crop growth cycle, starting from 20 days after seeding until maturity. Upland rice did not produce any tiller without the addition of P (Table 7.24 and Figures 7.22 through 7.27).

TGW varied from 19.5 to 29.3 g, with a mean value of 23.35 g. Fageria (2007b) reported that thousand grain weight of upland rice varied from 18.6 to 27.9 g, with a mean value of about 24 g. TGWs in some P sources increased quadratically with increasing P rates (0–400 mg kg^{-1}) (Table 7.25). P rates required for maximum 1000 weights were 211 mg kg^{-1} for PSSP and 194 mg P kg^{-1} for TSP, and mean value was 226 mg P kg^{-1}. Fageria (1991) studied the influence of P on thousand grain weights of three upland rice cultivars and concluded that thousand grain weights of two cultivars increased with the addition of P, and there was no increase in the third cultivar.

RDW, MRL, and PUE (mg grain produced per mg P applied) were significantly affected by P treatments (Table 7.26). RDW varied from 0.48 g plant produced by control treatment to 3.9 g plant^{-1} produced by 200 mg P kg^{-1} PSSP, with a mean value of 1.96 g plant^{-1}. Similarly, MRL varied from 18.7 cm produced by control treatment to 29 cm produced with 400 mg P kg^{-1} added by polymer-coated superphosphate, with a mean value of 23.5 cm. A quadratic increase in root weight occurred with increasing rates (0–400 mg P kg^{-1}) of SSP, PSSP, and TSP (Table 7.27). The highest values of RDW were obtained with the addition of 221–286 mg P kg^{-1} of three P sources, with a mean value of six P sources of 261 mg P kg^{-1}.

Root length increased quadratically with increasing P rates (0–400 mg kg^{-1}) (Table 7.27). MRL was obtained with the addition of 232 mg P kg^{-1} by SSP, 402 mg P kg^{-1} by PSSP, and 306 mg P kg^{-1}

FIGURE 7.22 Upland rice growth at different P levels applied by SSP. (From Fageria, N.K. et al., *Commun. Soil Sci. Plant Anal.*, 45, 1399, 2014b.)

FIGURE 7.23 Upland rice growth at different P levels applied by PSSP. (From Fageria, N.K. et al., *Commun. Soil Sci. Plant Anal.*, 45, 1399, 2014b.)

by MAP. The MRL that occurred with the addition of six P sources was obtained with the addition of 298 mg P kg^{-1}. Root length required more P as compared to RDW. Improvement in upland rice root growth with the addition of P in Brazilian Oxisol soil has been reported by Fageria (2009). Fageria and Moreira (2011) reported that P fertilization improved RDW and root length of upland rice genotypes grown on a Brazilian Oxisol soil. P level and upland rice genotype interactions for RDW and root length were significant, indicating different responses of genotypes to varying P levels. Root growth of upland rice under different P sources is presented (Figures 7.28 through 7.33). Improvement in root growth with the addition of P is clear, but the magnitude of growth varied among sources. A more extensive root system potentially has access to larger reserves of soil, water, and nutrients (Fageria, 1992).

FIGURE 7.24 Upland rice growth at different P levels applied by TSP. (From Fageria, N.K. et al., *Commun. Soil Sci. Plant Anal.*, 45, 1399, 2014b.)

FIGURE 7.25 Upland rice growth at different P levels applied by PTSP. (From Fageria, N.K. et al., *Commun. Soil Sci. Plant Anal.*, 45, 1399, 2014b.)

The authors of this book studied the influence of eight P sources applied at different rates to lowland rice (Figure 7.34). Minimum GY was produced by control treatment, and maximum GY was produced by TSP applied at the rate of 200 mg kg^{-1}. GY increased quadratically with increasing P rates (0–400 mg kg^{-1}) (Figure 7.34). Maximum GY was obtained with the addition of 179–325 mg P kg^{-1} depending on the P source, with a mean value of 273 mg P kg^{-1}. Variation in GY was 81%–96% due to P fertilization depending on P source. P is an important nutrient in soil when determining GY

FIGURE 7.26 Upland rice growth at different P levels applied by MAP. (From Fageria, N.K. et al., *Commun. Soil Sci. Plant Anal.*, 45, 1399, 2014b.)

FIGURE 7.27 Upland rice growth at different P levels applied by PMAP. (From Fageria, N.K. et al., *Commun. Soil Sci. Plant Anal.*, 45, 1399, 2014b.)

of rice. Response of lowland rice to P fertilization in Brazilian Inceptisol soil has been reported by Fageria (1980), Fageria and Barbosa Filho (2007), and Fageria et al. (2011b).

Data in Figure 7.34 show that the response curves were quadratic for all the P sources evaluated. However, the magnitude of response was different. For example, ammoniated simple superphosphate (ASSP) produced lowest GY, and PSSP, TSP, and PTSP produced maximum yield at mean P rate of 273 mg P kg^{-1}. Maximum GY occurred at 273 mg kg^{-1}. P sources were classified for PUE in the order of PSSP = TSP, > PTSP > polymer-coated ammoniated simple superphosphate (PASSP)

TABLE 7.26
Influence of P Treatments on RDW, MRL, GHI, and PUE of Upland Rice

Code of P Treatment	RDW (g Plant^{-1})	MRL (cm)	PUE (mg mg^{-1})
1. P0 (control)	0.48l	18.66b	—
2. SSP50	1.80d–j	23.00ab	75.24a
3. SPS100	3.03abc	26.00ab	29.71d
4. SPS200	2.46b–f	26.33ab	16.49ef
5. SPS400	1.91c–i	23.33ab	2.31l
6. PSSP50	3.65ab	21.00ab	68.28a
7. PSSP100	3.89a	24.33ab	32.42d
8. PSSP200	3.90a	25.66ab	15.19e–i
9. PSSP400	2.52b–f	29.00a	6.87f–l
10. TSP50	2.86a–d	23.33ab	27.62d
11. TSP100	2.32c–h	21.00ab	28.76d
12. TSP200	2.76a–e	21.33ab	15.24e–h
13. TSP400	2.36c–g	21.66ab	3.59jl
14. PTSP50	1.38f–l	23.00ab	57.33b
15. PTSP100	1.63e–l	21.66ab	32.28d
16. PTSP200	1.50f–l	26.00ab	13.36f–j
17. PTSP400	1.46f–l	23.33ab	4.63ijl
18. MAP50	0.62jl	23.33ab	43.14c
19. MAP100	0.70jl	24.33ab	29.71d
20. MAP200	1.00ijl	25.33ab	16.38ef
21. MAP400	0.96ijl	25.00ab	5.57g–l
22. PMAP50	2.14c–i	23.66ab	35.05cd
23. PMAP100	1.41f–l	20.00ab	24.86de
24. PMAP200	1.16h–l	23.66ab	16.09efg
25. PMAP400	1.19g–l	22.66ab	5.17efg
Average	1.96	23.46	25.22
F-test	**	*	**
CV (%)	19.34	12.33	13.35

Source: Fageria, N.K. et al., *Commun. Soil Sci. Plant Anal.*, 45, 1399, 2014b.

1 = 0 mg P kg^{-1} (control); 2 = 50 mg P kg^{-1} with SSP; 3 = 100 mg P kg^{-1} with SSP; 4 = 200 mg P kg^{-1} with SSP; 5 = 400 mg P kg^{-1} with SSP; 6 = 50 mg P kg^{-1} with PSSP; 7 = 100 mg P kg^{-1} with PSSP; 8 = 200 mg P kg^{-1} with PSSP; 9 = 400 mg P kg^{-1} with PSSP; 10 = 50 mg P kg^{-1} with TSP; 11 = 100 mg P kg^{-1} with TSP; 12 = 200 mg P kg^{-1} with TSP; 13 = 400 mg P kg^{-1} with TSP; 14 = 50 mg P kg^{-1} with PTSP; 15 = 100 mg P kg^{-1} with PTSP; 16 = 200 mg P kg^{-1} with PTSP; 17 = 400 mg P mg P kg^{-1} with PTSP; 18 = 50 mg P kg^{-1} with MAP; 19 = 100 mg P kg^{-1} with MAP; 20 = 200 mg P kg^{-1} with MAP; 21 = 400 mg P kg^{-1} with MAP; 22 = 50 mg P kg^{-1} with PMAP; 23 = 100 mg P kg^{-1} with PMAP; 24 = 200 mg P kg^{-1} with PMAP; and 25 = 400 mg P kg^{-1} with PMAP.

*,** and NS Significant at the 1% or 5% probability level and nonsignificant, respectively.

> SSP > MAP > ASSP. Engelstad and Terman (1980) reported that there was a quadratic increase in rice yield with increasing P rates (0–88 kg ha^{-1}) and response varied with P source.

The authors of this book evaluated two sources of P for upland and lowland rice production under field conditions. PH and GY of upland rice were significantly affected by P rate and sources (Table 7.28). Similarly, there was an improvement in the SY, PD, and GHI with the addition of P rates with two sources. PH varied from 100 cm produced by control treatment to 111 cm produced by polymer-coated ammonium sulfate applied at the rate of 100 kg P_2O_5 ha^{-1}, with a mean value of 108 cm (Table 7.28). Increases in PH with the addition of P in upland rice grown on Brazilian

TABLE 7.27
Relationship between P Rate and RDW and MRL of Upland Rice

Variable	Regression Equation	R^2	PMV
RDW			
P rate with SSP vs. RDW	$Y = 0.82 + 0.02X - 0.000043X^2$	0.66**	238
P rate with PSSP vs. RDW	$Y = 1.27 + 0.03X - 0.000068X^2$	0.72**	221
P rate with TSP vs. RDW	$Y = 1.09 + 0.02X - 0.000035X^2$	0.56**	286
P rate with PTSP vs. RDW	$Y = 0.73 + 0.009X - 0.000017X^2$	0.28^{NS}	—
P rate with MAP vs. RDW	$Y = 0.46 + 0.004X - 0.0000059X^2$	0.31^{NS}	—
P rate with PMAP vs. RDW	$Y = 1.06 + 0.005X - 0.0000011X^2$	0.11	—
Average	$Y = 0.91 + 0.01X - 0.000031X^2$	0.66**	261
MRL			
P rate with SSP vs. MRL	$Y = 19.38 + 0.07X - 0.000151X^2$	0.49*	232
P rate with PSSP vs. MRL	$Y = 18.91 + 0.05X - 0.000062X^2$	0.86**	402
P rate with TSP vs. MRL	$Y = 20.12 + 0.02X - 0.000032X^2$	0.05^{NS}	—
P rate with PTSP vs. MRL	$Y = 19.06 + 0.05X - 0.00011X^2$	0.27^{NS}	—
P rate with MAP vs. MRL	$Y =19.66 + 0.06X - 0.000098X^2$	0.71**	306
P rate with PMAP vs. MRL	$Y = 19.66 + 0.03X - 0.000053X^2$	0.32^{NS}	—
Average	$Y = 19.47 + 0.05X - 0.000084X^2$	0.89**	298

Source: Fageria, N.K. et al., *Commun. Soil Sci. Plant Anal.*, 45, 1399, 2014b.
Note: PMV = P rate for maximum RDW and MRL.
SSP, simple superphosphate; PSSP, polymer-coated simple superphosphate; TSP, triple superphosphate; PTSP, polymer-coated triple superphosphate; MAP, monoammonium phosphate; PMAP, polymer-coated monoammonium phosphate.
*,**, and NS Significant at the 5% and 1% probability levels and nonsignificant, respectively.

FIGURE 7.28 Upland rice growth root growth at different P levels applied by SSP. (From Fageria, N.K. et al., *Commun. Soil Sci. Plant Anal.*, 45, 1399, 2014b.)

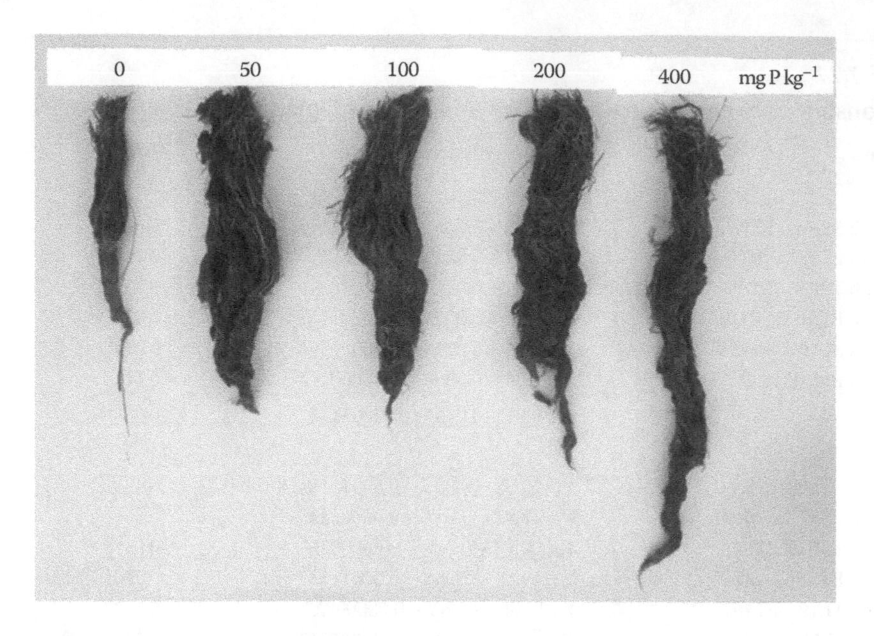

FIGURE 7.29 Upland rice root growth at different P levels applied by PSSP. (From Fageria, N.K. et al., *Commun. Soil Sci. Plant Anal.*, 45, 1399, 2014b.)

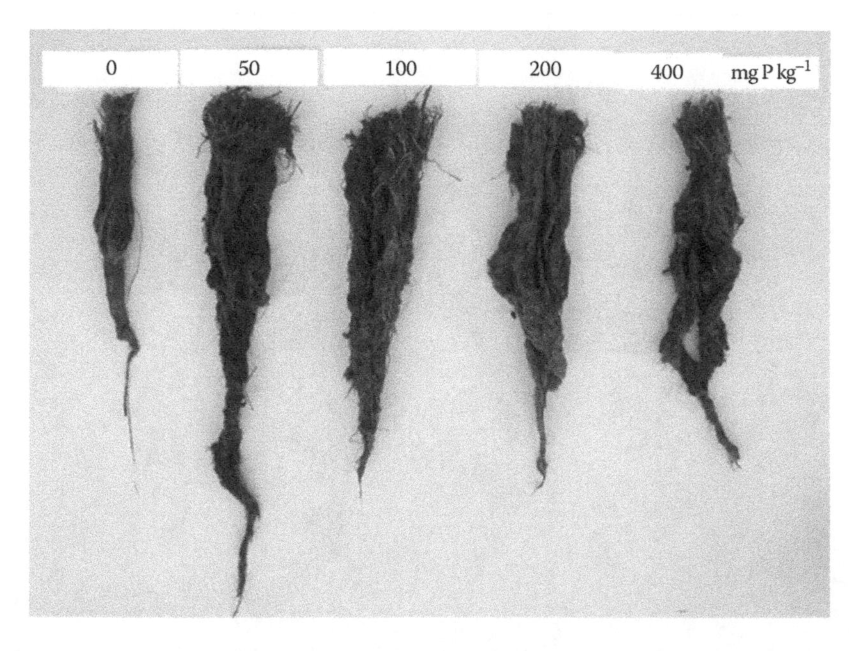

FIGURE 7.30 Upland rice growth root growth at different P levels applied by TSP. (From Fageria, N.K. et al., *Commun. Soil Sci. Plant Anal.*, 45, 1399, 2014b.)

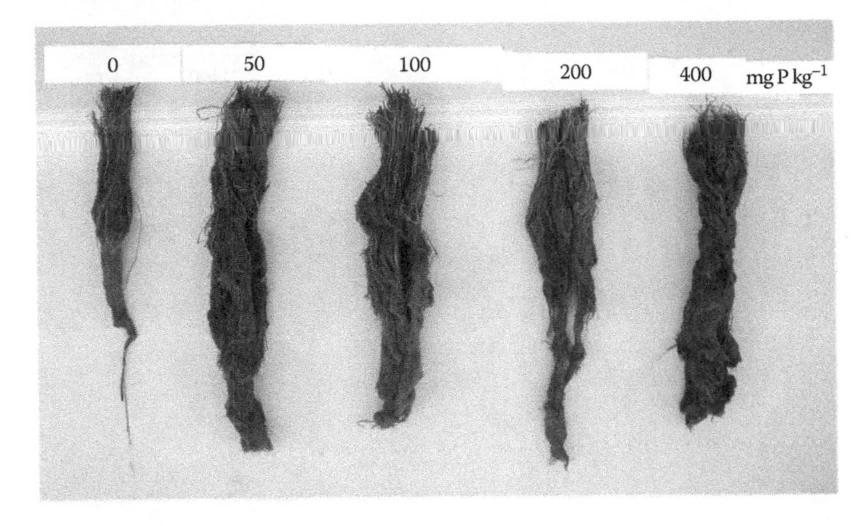

FIGURE 7.31 Upland rice root growth at different P levels applied by PTSP. (From Fageria, N.K. et al., *Commun. Soil Sci. Plant Anal.*, 45, 1399, 2014b.)

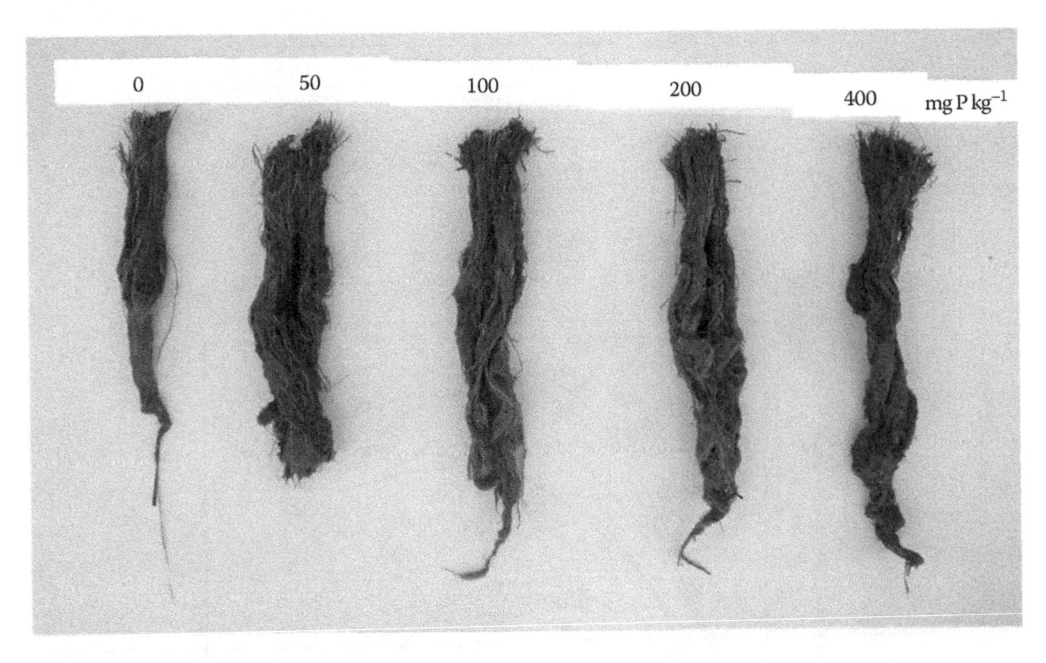

FIGURE 7.32 Upland rice root growth at different P levels applied by MAP. (From Fageria, N.K. et al., *Commun. Soil Sci. Plant Anal.*, 45, 1399, 2014b.)

Oxisol soils have been reported by Fageria and Baligar (1997) and Fageria et al. (1991a, 2013a). Fageria (2007b) reported that PH is influenced by environmental factors and mineral nutrition and is genetically controlled.

Two P sources did not differ significantly in PH in upland rice (Table 7.28). However, PMAP produced taller plants at 25, 50, and 100 kg P_2O_5 ha^{-1}. Based on regression equations, maximum PH was obtained with the addition of 148 kg P_2O_5 ha^{-1} by MAP and 113 kg P_2O_5 by PMAP (Table 7.29). PH is an important trait in determining crop response to fertilization. Taller plants lodge easily at high fertility levels (Fageria, 2007b). GY increases quadratically with increasing

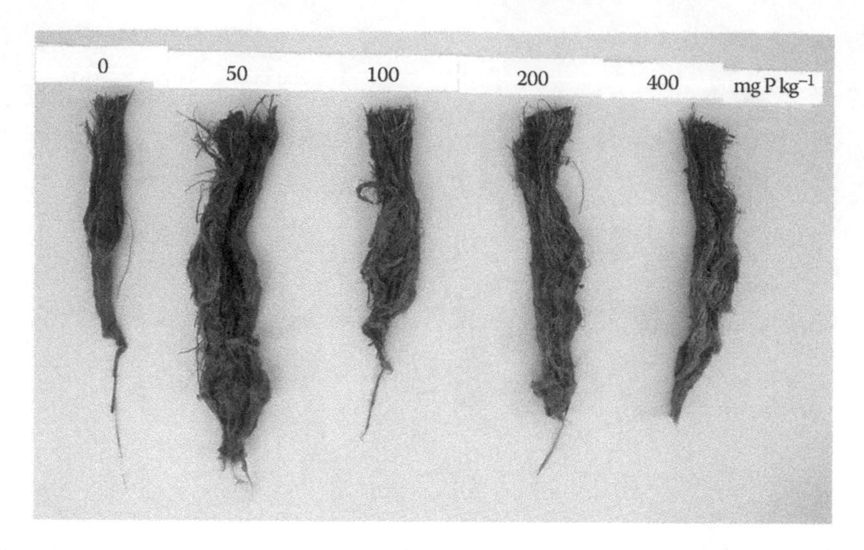

FIGURE 7.33 Upland rice root growth at different P levels applied by PMAP. (From Fageria, N.K. et al., *Commun. Soil Sci. Plant Anal.*, 45, 1399, 2014b.)

Variable	Regression Equation	R^2	PRMV (mg kg^{-1})
	Grain yield		
PR of SSP vs. GY	$Y = -0.41 + 0.07X - 0.00012X^2$	0.90**	292
PR of PSSP vs. GY	$Y = 0.83 + 0.08X - 0.00015X^2$	0.96**	267
PR of ASSP vs. GY	$Y = 0.38 + 0.03X - 0.000084X^2$	0.88**	179
PR of PASSP vs. GY	$Y = 1.48 + 0.05X - 0.000077X^2$	0.88**	325
PR of TSP vs. GY	$Y = 1.06 + 0.08X - 0.00015X^2$	0.94**	267
PR of PTSP vs. GY	$Y = 1.75 + 0.06X - 0.000098X^2$	0.81**	306
PR of MAP vs. GY	$Y = 0.34 + 0.04X - 0.000074X^2$	0.90**	270
PR of PMAP vs. GY	$Y = 1.21 + 0.05X - 0.00012X^2$	0.82**	208
Average	$Y = 0.83 + 0.06X - 0.00011X^2$	0.96**	273

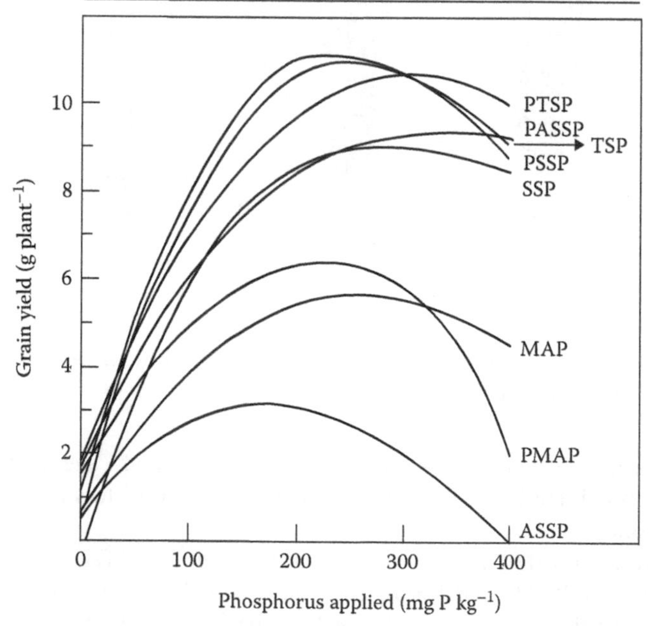

FIGURE 7.34 GY of lowland rice under different P levels and sources. (From Fageria, N.K. et al., *Commun. Soil Sci. Plant Anal.*, 45, 2067, 2014.)

TABLE 7.28
PH, SY, GY, PD, and GHI of Upland Rice as Influenced by Source and Rate of P Fertilization

P Rate (kg P_2O_5 ha^{-1}) and Source	PH (cm)	SY (kg ha^{-1})	GY (kg ha^{-1})	PD (m^{-2})	GHI
Control	100.25b	6416.66a	4152.08b	153.75a	0.39a
MAP25	105.00ab	7823.33a	5713.43a	184.16a	0.42a
PMAP25	108.75ab	7885.83a	5631.33a	206.66a	0.42a
MAP50	109.25ab	7286.66a	5695.41a	188.33a	0.44a
PMAP50	111.25a	7443.33a	5961.04a	224.16a	0.44a
MAP100	108.00ab	7501.66a	5345.83a	201.66a	0.41a
PMAP100	111.50a	7690.00a	5464.44a	215.00a	0.41a
MAP200	110.00ab	6945.83a	5011.31ab	214.16a	0.42a
PMAP200	106.25ab	7461.66a	5077.01ab	233.33a	0.40a
Average	107.80	7383.88	5339.10	202.36	0.42a
F-test	*	NS	**	NS	NS
CV (%)	4.07	12.10	7.85	16.71	9.08

Source: Fageria, N.K. et al., *Commun. Soil Sci. Plant Anal.*, 46, 1097, 2015b.

*,**, and NSSignificant at the 5% and 1% probability levels and nonsignificant, respectively. Means in the same column followed by the same letter are not significantly different at the 5% probability level by Tukey's test. MAP, monoammonium phosphate; PMAP, polymer-coated monoammonium phosphate.

TABLE 7.29
Relationship between P Rate and Growth, Yield, and Yield Components of Upland Rice under Two P Sources

Variable	Regression Equation	R^2	P_2O_5 Rate (kg ha^{-1}) for Maximum Growth and Yield (MGY)
P rate by MAP vs. PH	$Y = 101.49 + 0.13X - 0.00044X^2$	0.40*	148
P rate by PMAP vs. PH	$Y = 102.08 + 0.21X - 0.00093X^2$	0.49**	113
P rate by MAP vs. GY	$Y = 4635.22 + 21.54X - 0.10X^1$	0.38*	108
P rate by PMAP vs. GY	$Y = 4605.09 + 25.20X - 0.12X^2$	0.40*	105
P rate by MAP vs. PD	$Y = 159.90 + 0.65X - 0.0019X^2$	0.36NS	NS
P rate by PMAP vs. PD	$Y = 169.32 + 0.97X - 0.0033X^2$	0.49**	150

Source: Fageria, N.K. et al., *Commun. Soil Sci. Plant Anal.*, 46, 1097, 2015b.

PH, plant height; GY, grain yield; PD, panicle density; MGY, maximum growth and yield; MAP, monoammonium phosphate; PMAP, polymer-coated monoammonium phosphate.

*,**, and NSSignificant at the 5% and 1% probability levels and nonsignificant, respectively.

PH (Table 7.30). Fageria et al. (2010) reported that PH was significantly related to GY of upland rice genotypes and relationship was quadratic.

Addition of P did not improve SY among P rate and sources (Table 7.28). GY increased significantly with the addition of P fertilizers. Maximum GY occurred with the addition of 108 kg P_2O_5 ha^{-1} by the addition of MAP (Table 7.29). Similarly, maximum GY was obtained with the addition of 105 kg P_2O_5 ha^{-1} of PMAP. Improvement in the upland rice GY with the addition of P in the Brazilian Oxisol soils has been reported by Fageria and Barbosa Filho (1981), and Fageria et al. (1982, 2013a).

TABLE 7.30
Relationship between PH, GHI, and GY of Upland Rice

Variable	Regression Equation	R^2
PH vs. GY	$Y = -134{,}850.80 + 2{,}562.83X - 11.68X^2$	0.45**
GHI vs. GY	$Y = -10{,}799.19 + 64{,}672.19X - 62{,}029.62X^2$	0.41**

Source: Fageria, N.K. et al., *Commun. Soil Sci. Plant Anal.*, 46, 1097, 2015b.
Note: Values are averages of two P sources.
**Significant at the 1% probability level.

PD did not differ among treatments (Table 7.28). However, regression analysis showed that PD was significantly influenced by the P rates applied with PMAP (Table 7.29). Maximum panicle number occurred with the addition of 150 kg P_2O_5 ha^{-1} (Table 2.29). Improvement in the panicle number in upland rice with the addition of P has been reported by Fageria et al. (2011a,b) and Fageria (2014). Improvement in GHI occurred with the addition of P. However, GHI was not affected significantly by P treatments although it had significant quadratic association with GY (Table 7.30). Variation in GY was 40% and was due to GHI. Significant association between GY and GHI of rice has been reported by Fageria (2009) and Fageria et al. (2010).

In lowland rice, PH, SY, GY, PD, and panicle length (PL) were significantly increased with the addition of P by two fertilizer sources (Table 7.31). PH varied from 63 cm with control to 94 cm with 200 kg P_2O_5 added as MAP and PMAP, with a mean value of 83.6 cm. Maximum PH was obtained with the addition of 170 kg P_2O_5 by MAP and 120 kg P_2O_5 by PMAP (Table 7.32). Improvement in PH of upland and lowland rice with the addition of P in the Brazilian Inceptisol soil has been reported by Fageria and Barbosa Filho (2007), Fageria et al. (2011a,b), and Fageria (2014).

TABLE 7.31
PH, SY, GY, PD, and PL of Lowland Rice as Influenced by Source and Rate of P Fertilization

P Rate (kg P_2O_5 ha^{-1}) and Source	PH (cm)	SY (kg ha^{-1})	GY (kg ha^{-1})	PD (m^{-2})	PL (cm)
Control	63.00b	5,153.75b	1365.21e	242.50b	19.22c
MAP25	79.00ab	6,708.75ab	5874.38cd	407.50a	21.10bc
PMAP25	77.50ab	5,755.00b	5115.10d	405.00	20.65bc
MAP50	85.25ab	7,140.00ab	7700.52abc	420.00a	21.50abc
PMAP50	82.00ab	6,976.25ab	6732.50bcd	438.75a	21.50abc
MAP100	84.50ab	6,877.50ab	6946.77abcd	431.25a	21.92ab
PMAP100	93.25a	10,101.25a	8019.27abc	516.25a	22.60ab
MAP200	94.00a	8,222.50ab	8494.47ab	453.75a	22.67ab
PMAP200	94.00a	8,422.50ab	9040.52a	496.25a	23.52a
Average	83.61	7,261.94	6587.63	423.47	21.63
F-test	**	**	**	**	**
CV (%)	11.19	22.32	13.81	13.16	4.44

Source: Fageria, N.K. et al., *Commun. Soil Sci. Plant Anal.*, 46, 1097, 2015b.
MAP, monoammonium phosphate; PMAP, polymer-coated monoammonium phosphate.
**Significant at the 1% probability level. Means within the same column followed by the same letter are not significantly different at the 5% probability level by Tukey's test.

TABLE 7.32
Relationship between P Rate and Growth, Yield, and Yield Components of Lowland Rice under Two P Sources

Variable	Regression Equation	R^2	P_2O_5 Rate in kg ha^{-1} for MGY
P rate by MAP vs. PH	$Y = 67.43 + 0.32X - 0.00094X^2$	0.52**	170
P rate by PMAP vs. PH	$Y = -35.38 + 2.19X - 0.0091X^2$	0.69**	120
P rate by MAP vs. SY	$Y = 5655.09 + 23.36X - 0.055X^2$	0.35*	195
P rate by PMAP vs. SY	$Y = 4564.66 + 75.01X - 0.27X^2$	0.52**	139
P rate by MAP vs. GY	$Y = 2655.92 + 88.88X - 0.31X^2$	0.72**	143
P rate by PMAP vs. GY	$Y = 2063.92 + 99.37X - 0.33X^2$	0.85**	151
P rate by MAP vs. PD	$Y = 286 + 2.85X - 0.01X^2$	0.55**	143
P rate by PMAP vs. PD	$Y = 269.15 + 4.17X - 0.02X^2$	0.77**	104
P rate by MAP vs. PL	$Y = 19.66 + 0.04X - 0.00012X^2$	0.64**	167
P rate by PMAP vs. PL	$Y = 19.38 + 0.05X - 0.00013X^2$	0.71**	192

Source: Fageria, N.K. et al., *Commun. Soil Sci. Plant Anal.*, 46, 1097, 2015b.
MAP, monoammonium phosphate; PMAP, polymer-coated monoammonium phosphate; MGY, maximum growth and yield; PH, plant height; SY, straw yield; GY, grain yield; PD, panicle density; PL, panicle length.
*,**Significant at the 5% and 1% probability levels, respectively. Means within the same column followed by the same letter are not significantly different at the 5% probability level by Tukey's test.

SY varied from 5,154 kg ha^{-1} with the control treatment to 10,101 kg ha^{-1} with PMAP applied at the rate of 100 kg P_2O_5 ha^{-1}, with a mean value of 7,262 kg ha^{-1}. Similarly, GY varied from 1365 kg ha^{-1} with control to 9041 kg ha^{-1} with PMAP applied at the rate of 200 kg P_2O_5 ha^{-1}, with a mean value of 6588 kg ha^{-1}. Maximum SY occurred with the addition of 195 kg P_2O_5 ha^{-1} by MAP and 139 kg P_2O_5 by PMAP (Table 7.32). Similarly, maximum GY occurred with the addition of 143 kg P_2O_5 by MAP and 151 kg P_2O_5 by PMAP (Table 7.32).

SY and GY of lowland rice increase with the addition of P in the Brazilian Inceptisol soil has been reported by Fageria et al. (1997, 2000, 2003) and Fageria and Baligar (1999) reported that P is a major yield-limiting factor for crop production in Brazilian Inceptisol soil due to low natural levels and high immobilization capacity of P. Fageria et al. (1991b) characterized the chemical properties of Varzea soils in central part of Brazil and reported a 65-fold differences in Mehlich 1 extractable P among 23 sites of surface 0–20 cm soil covering different municipalities. Soil samples (40%) analyzed had low (<3 mg P kg^{-1}) to medium (3–7 mg P kg^{-1}) levels of extractable P, where crop response to P is expected. Fageria (1980) reported a significant increase in lowland rice yield with P fertilization, planted in Varzea soil.

PD varied from 242 to 516 m^{-2}, with a mean value of 423 panicles m^{-2} (Table 7.31). Similarly, PL varied from 19.2 to 23.5 cm, with a mean value of 21.6 cm. Improvement in PD and PL of lowland rice with the addition of P in the Brazilian Inceptisol soil has been reported by Fageria et al. (2003). Both the parameters increased quadratically with increasing P rates (0–200 kg P_2O_5 ha^{-1}) by both fertilizers (Table 7.32). PD and PL increased quadratically in lowland rice with increasing P rates (Fageria and Santos, 2008). They reported that control treatment (0 kg P ha^{-1}) produced minimum panicle number as compared to P-applied treatment as happened in the present study.

P rate for maximum plant density was 143 kg P_2O_5 ha^{-1} for MAP and 104 kg P_2O_5 ha^{-1} for PMAP. Similarly, in case of PL, maximum value was obtained with the addition of 167 kg P_2O_5 ha^{-1} by MAP and 192 kg P_2O_5 ha^{-1} by PMAP. Comparisons of different P sources for lowland rice production are limited.

TGW, SS, and GHI were significantly influenced by P fertilization (Table 7.33). TGWs varied from 23.0 to 28.7 g, with a mean value of 27.2 g, depending on P treatment. Similarly, SS varied from 52.4% to 7.66%, with a mean value of 17.2%. GHI varied from 0.21 to 52, with a mean value of 0.46. Improvement in thousand grain weight and GHI in lowland rice has been reported by Fageria

TABLE 7.33
TGW, SS, and GHI of Lowland Rice as Influenced by Source and Rate of P Fertilization

P Rate (kg P_2O_5 ha^{-1}) and Source	Thousand Grain Wt. (g)	Spike Sterility (%)	GHI
Control	23.05b	52.42a	0.21b
MAP25	28.04a	19.82b	0.46a
PMAP25	25.54ab	16.61b	0.46a
MAP50	27.89a	12.75b	0.52a
PMAP50	27.96a	15.56b	0.49a
MAP100	27.02ab	11.78b	0.50a
PMAP100	28.72a	7.66b	0.44a
MAP200	27.85a	10.03b	0.51a
PMAP200	28.25a	8.40b	0.51a
Average	27.15	17.22	0.46
F-test	**	**	**
CV (%)	7.24	12.55	10.32

Source: Fageria, N.K. et al., *Commun. Soil Sci. Plant Anal.*, 46, 1097, 2015b.
MAP, monoammonium phosphate; PMAP, polymer-coated monoammonium phosphate.
**Significant at the 1% probability level. Means within the same column followed by the same letter are not significantly different at the 5% probability level by Tukey's test.

(2014). Fageria (2014) reported that P fertilization decreased SS when applied at an adequate rate. These three plant traits are important in increasing rice yield since thousand grain weight and GHI are positively correlated with rice GY (Fageria, 2007a, 2009) and SS is negatively correlated with GY (Fageria, 2007a, 2009, 2014). P rate for maximum thousand grain weight was 130 kg P_2O_5 ha^{-1} by MAP and 133 kg P_2O_5 by PMAP (Table 7.34). Similarly, maximum GHI was obtained with the application of 139 kg P_2O_5 by MAP and 154 kg P_2O_5 by PMAP.

P concentration (content per unit dry matter) and uptake (concentration × dry matter) in the straw and in grain was significantly influenced by P treatments (Table 7.35). However, P concentration in straw or in grain did not differ significantly among P sources. P concentration increased linearly with increasing P rates (Table 7.36). Similarly, P uptake in the straw was linear. An increase in P concentration in straw and grain with increasing P rate has been reported by Fageria and Santos (2008).

TABLE 7.34
Relationship between P Rate and TGW, SS, and GHI of Lowland Rice under Two P Sources

Variable	Regression Equation	R^2	P_2O_5 Rate in kg ha^{-1} for MV
P rate by MAP vs. TGW	$Y = 24.65 + 0.06X - 0.00023X^2$	0.36*	130
P rate by PMAP vs. TGW	$Y = 23.38 + 0.10X - 0.00036X^2$	0.48**	133
P rate by MAP vs. SS	$Y = 44.14 - 0.65X + 0.0024X^2$	0.71**	—
P rate by PMAP vs. SS	$Y = 44.09 - 0.68X + 0.0025X^2$	0.77**	—
P rate by MAP vs. GHI	$Y = 0.28 + 0.005X - 0.000018X^2$	0.67**	139
P rate by PMAP vs. GHI	$Y = 0.29 + 0.004X - 0.000013X^2$	0.52**	154

Source: Fageria, N.K. et al., *Commun. Soil Sci. Plant Anal.*, 46, 1097, 2015b.
MAP, monoammonium phosphate; PMAP, polymer-coated monoammonium phosphate; MV, maximum value.
*,**Significant at the 5% and 1% probability levels, respectively. Means within the same column followed by the same letter are not significantly different at the 5% probability level by Tukey's test.

TABLE 7.35
Concentration and Uptake of P in Straw and Grain and P Harvest Index (PHI) of Lowland Rice as Influenced by Source and Rate of P Fertilization

P Rate (kg P_2O_5 ha^{-1}) and Source	Conc. of P in Straw (g kg^{-1})	Conc. of P in Grain (g kg^{-1})	Uptake of P in Straw (kg ha^{-1})	Uptake of P in Grain (kg ha^{-1})	PHI
Control	0.21b	0.77d	1.06b	1.05d	0.50b
MAP25	0.25b	0.91cd	1.63b	5.38c	0.76a
PMAP25	0.25b	0.94cd	1.46b	4.78cd	0.76a
MAP50	0.28b	0.99cd	2.05b	7.66bc	0.78a
PMAP50	0.32ab	1.05bcd	2.30ab	7.03bc	0.75a
MAP100	0.33ab	1.08bcd	2.29ab	7.53bc	0.77a
PMAP100	0.34ab	1.21abc	3.73ab	9.83ab	0.73a
MAP200	0.40ab	1.38ab	3.25ab	11.79a	0.77a
PMAP200	0.59a	1.49a	4.96a	13.37a	0.73a
Average	0.33	1.09	2.53	7.60	0.73
F-test	**	**	**	**	**
CV (%)	17.35	13.85	15.27	21.19	8.87

Source: Fageria, N.K. et al., *Commun. Soil Sci. Plant Anal.*, 46, 1097, 2015b.
MAP, monoammonium phosphate; PMAP, polymer-coated monoammonium phosphate.
**Significant at the 1% probability level. Means within the same column followed by the same letter are not significantly different at the 5% probability level by Tukey's test.

TABLE 7.36
Relationship between P Rate and P Concentration and Uptake in Straw and Grain and PHI of Lowland Rice under Two P Sources

Variable	Regression Equation	R^2
P rate by MAP vs. PCS	$Y = 0.22 + 0.001X$	0.59**
P rate by PMAP vs. PCS	$Y = 0.21 + 0.002X$	0.49**
P rate by MAP vs. PCG	$Y = 0.81 + 0.003X$	0.71**
P rate by PMAP vs. PCG	$Y = 0.84 + 0.003X$	0.77**
P rate by MAP vs. PUS	$Y = 1.29 + 0.01X$	0.67**
P rate by MAP vs. PUS	$Y = 1.19 + 0.02X$	0.53**
P rate by MAP vs. PUG	$Y = 2.25 + 0.09X - 0.00022X^2$	0.77**
P rate by MAP vs. PUG	$Y = 1.52 + 0.12X - 0.00029X^2$	0.90**
P rate by MAP vs. PHI	$Y = 0.58 + 0.0043X - 0.000017X^2$	0.56**
P rate by PMAP vs. PHI	$Y = 0.58 + 0.0037X - 0.000015X^2$	0.38*

Source: Fageria, N.K. et al., *Commun. Soil Sci. Plant Anal.*, 46, 1097, 2015b.
MAP, monoammonium phosphate; PMAP, polymer-coated monoammonium phosphate; MV, maximum value; PCS, P concentration in straw; PCG, P concentration in grain; PCG, P concentration in grain; PUS, P uptake in straw; PUG, P uptake in grain; PHI, P harvest index; MV, maximum value.
*,**Significant at the 5% and 1% probability levels, respectively. Means within the same column followed by the same letter are not significantly different at the 5% probability level by Tukey's test.

P uptake in grain increased quadratically with increasing P rates (0–200 kg P_2O_5 ha^{-1}). Quadratic increases in P uptake in lowland rice with increasing P rate have been reported by Fageria (2014). An increase in P concentration and uptake with increasing P rate was expected. P concentration and uptake were higher in grain as compared to straw. Fageria et al. (2003) reported that after flowering, filling rice grain became a strong sink for P and the straw P concentration declined.

Among inorganic P sources, ammonium phosphates, superphosphates, and nitric phosphate are generally equal as P sources for optimizing crop yields (Hedley et al., 1995; Armstrong, 1999; Mortvedt et al., 1999). Cihacek (1993) has compared nine P sources (58 kg P ha^{-1} annually) for alfalfa production in New Mexico and the United States. TSP and MAP produced the highest yields over the 3-year study. Ammonium N can stimulate P availability from fertilizers (Engelstad and Terman, 1980). This enhancement of plant P uptake by ammonium N may be due to several factors including (1) increasing root and shoot growth, (2) altering plant metabolisms, and (3) increasing the solubility and availability of P. Increased root mass is largely responsible for increasing P uptake (Havlin et al., 1999). These benefits may be less apparent in soils with adequate or high available P levels (Engelstad and Terman, 1980). Dual application of ammonium N and P with both nutrients placed in close proximity in subsurface bands has increased winter wheat yields in Great Plains region of the United States (Leikam, 1992).

7.3.2 Efficient Method of Application

Use of efficient method is important for P fertilization in crop plants since P movement in the soil is mainly by diffusion process. Major methods of P application are broadcast, band placement, seed coating, and foliar spraying. Broadcast and incorporation method involves uniformly broadcasting and incorporating P fertilizers onto the soil prior to seeding. At low and medium soil test levels for P, the broadcast and incorporation methods are less effective than seed placed or banded P. When soil test value of P is low, broadcast application requires two to four times more P fertilizers, as compared to band or seed placement, methods to have an equivalent crop response. Therefore, broadcast and incorporation of P may not be economical or practical if high rates are needed to increase yield.

High rates of broadcast and incorporated P on eroded soil, and high P-fixing soils are recommended to improve P status of these soils and to consolidate P level of these soils to an optimum level. These types of soils have high response of crops to applied P fertilizers. Broadcasting P at higher rates involves a high initial cost, but these costs can be recovered over several years. Availability of applied P (water-soluble P sources) to plants tends to decrease over a time period of years because of the reaction products formed in the soil. However, it is recommended that the one-time large application be followed by an annual application of seed-placed P fertilizer with annual crops.

Broadcast P fertilizer without incorporation is only recommended for established forage crops because it is the only practical application method currently available. Granular phosphate can be broadcast. Liquid phosphate fertilizer can be dribble banded onto forage stands. There are times where dribble-banded liquid P fertilizer may be superior to a broadcast application of granular P fertilizers. However, liquid fertilizer is generally more expensive per unit of P versus granular fertilizer. P fertilizer is immobile in soil; therefore, availability of surface-applied P may be low in the first year after application. However, alfalfa and grasses do have feeder roots near the soil surface and can absorb some broadcast P fertilizer when surface soil moisture conditions are adequate.

For forage crops, a 3–4-year supply of P can be either deep banded or broadcast and incorporated before establishment. Subsequent applications may be top-dressed. Maximum safe rate of P that can be applied with the seed for cereal crops is 50–70 kg ha^{-1} of P_2O_5 depending on soil moisture conditions and the opener used. For oilseed crops, seed-placed P rate should not exceed 15 kg ha^{-1} when a seedling implement that places the seed and fertilizer in a narrow band is used. For peas (*P. sativum*), seed-placed P_2O_5 rates should not exceed 30–35 kg ha^{-1}, especially when a seeding implement that places the seed and fertilizer in a narrow band (double disc drill) is used.

Side-banding P is the method that places P in a band near the seed row during the sowing operation. Fertilizer is commonly banded 2.5–5 cm below and/or to the side of the seed row for small seeded and row crops such as sugar beets (*Beta vulgaris* L.), potatoes, sunflowers (*Helianthus annuus*), corn, and beans. This method is the optimum placement option for crops that are sensitive to seed-placed P fertilizers. P at all recommended rates can be applied by side banding. Application of P both with the seed and deep banded is an effective method to achieve optimum benefits. For P, banded spacing of more than 30 cm is not recommended, and narrower spacing may be more effective, particularly where no phosphate is seed placed. Generally, phosphate should not be banded with N fertilizer if the N rate is higher than 70–80 kg ha^{-1} to avoid reduced uptake efficiency of the P fertilizer. The major reason for this effect is that plant roots cannot penetrate the concentrated N band and, therefore, cannot absorb P effectively.

A high concentration of MAP can generally be placed with the seed than with diammonium phosphate (Bundy et al., 2005). Crops vary in response to P sources and sensitivity to fertilizers placed directly with the seed. For example, oilseed crops such as sunflower or canola (*Brassica rapa* L.) are less tolerant of direct fertilizer contact with the seed than small grain crops such as wheat or barley (Armstrong, 1999; Grant et al., 2001). Halvorson and Havlin (1992) and Halvorson et al. (2002) reported that when sufficient P was applied to eliminate P deficiency, methods of P application had minimal influence on crop yields.

7.3.3 Adopting Appropriate Time of Application

P is an immobile nutrient in soil–plant system. The optimum time of application is just prior to sowing. If water-soluble sources are applied in advance, a major portion of the P may be fixed in the soil and not available immediately to plants. Studies conducted in the North Great Plains of the United States indicate that applying sufficient amount of P in a single one-time application to eliminate P deficiency is an effective method to optimize crop yields and long-term economic returns (Halvorson, 1989; Halvorson and Havlin, 1992). High rates of P application to wheat generally were not economically feasible during the first year of application. However, the residual effects of large P application paid dividends for several years (Bundy et al., 2005). Roberts and Stewart (1988) reported that in Manitoba and Saskatchewan, a single application of broadcast P (80 kg P ha^{-1}) incorporation resulted in higher wheat yields and economic returns over a 5-year period than annual applications of 20 kg P ha^{-1} with the seed. Lowenberg-Deboer and Reetz (2002) suggested that it is more profitable to build up soil P levels rapidly than to use a slow consolidating approach. P can be considered an investment in long-term soil fertility and probably should be treated as a capital investment in the land with costs amortized over several years (Lowenberg-Deboer and Reetz, 2002; Bundy et al., 2005).

7.3.4 Adequate Phosphorus Rate

Use of adequate P rate in crop production is important in optimizing P availability to crop plants, improving P efficiency and reducing P losses from soil–plant systems (Fageria, 2009). Under- as well as overfertilization has negative effects on crop yield and environmental pollution. Adequate rate is based on experimental results under field conditions that varied from soil to soil, environmental conditions, crop to crop, and genotypes within crops (Fageria, 2014). No amount of P fertilizer, however, would correct poor management decisions. Attention should be given to well-planned rotations, timely operations, cultivar selection, pests, soil moisture, and inputs. P supply is a key determinant of GY, whereas the quality of that grain is determined by the supply of other nutrients, especially N and K and water.

At low rates of applied P, water solubility may be more important than at high rates. This is due to the same factors influencing crop response under conditions of poor distribution, suggesting that when optimum application rates cannot be used, it is important that materials of high water solubility be used in order to obtain full benefits of limited amounts of fertilizers (Tisdale et al., 1985).

Adequate rate of P is determined by a soil test. Soil testing is the major diagnostic tool used for formulating P recommendations and includes soil sampling, chemical analysis using an appropriate method, interpreting test results in terms of nutrient sufficiency for crops, and making fertilizer recommendations (Mallarino and Schepers, 2005). Use of an adequate level of P for maximum economic yield is possible if P soil test calibration data are available for a specific crop and soil types. Fageria (1996) calibrated soil P test results for lowland rice grown on a Brazilian Inceptisol soil (Figure 7.35). GY increased with broadcast P rates, with a maximum yield obtained at about 290 kg P ha^{-1}. Relative GY of this experiment was plotted against soil-extractable P for calibration of soil test P (Figure 7.36). Four categories were established for the P soil test: very low (VL), low (L), medium (M), and high (H). Relative yield zone (0%–70%) is called VL, 70%–90% relative yield zone is called L, 90%–100%

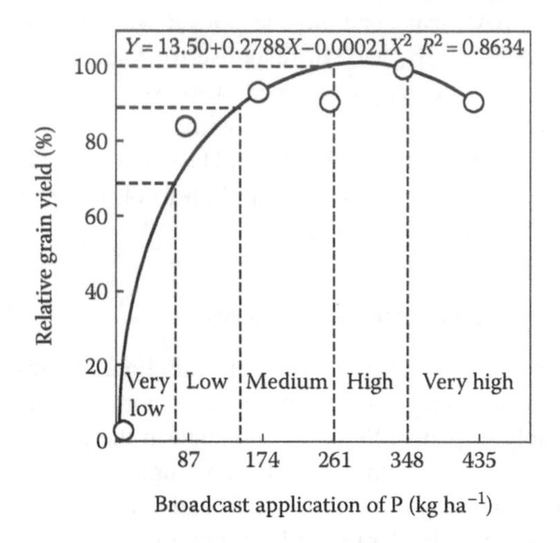

FIGURE 7.35 The relationship between P applied as broadcast and relative GY of lowland rice. (From Fageria, N.K., Annual report of the project, The study of liming and fertilization for rice and common bean in Cerrado region, National Rice and Bean Research Center, Goiania, Brazil, 1996.)

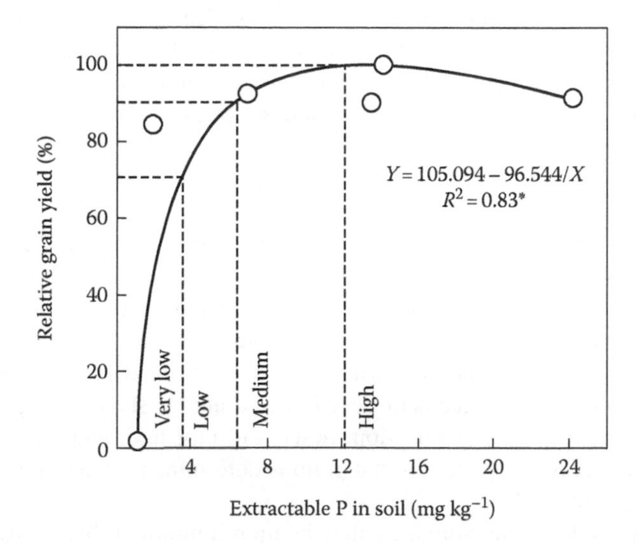

FIGURE 7.36 The relationship between Mehlich 1 extractable P and relative GY of lowland rice. (From Fageria, N.K., Annual report of the project, The study of liming and fertilization for rice and common bean in Cerrado region, National Rice and Bean Research Center, EMBRAPA Goiania, Brazil, 1996.)

TABLE 7.37
Soil P Test Availability Indices and P Fertilizer Recommendations for Lowland Rice in an Inceptisol Soil

Soil P Test (mg kg^{-1})	Interpretation	Relative Yield (%)	Broadcast P Application (kg ha^{-1})	Band P Application (kg ha^{-1})
0–3.6	Very low	0–70	100	66
3.6–6.4	Low	70–90	170	66
6.4–12.0	Medium	90–100	275	44
>12.0	High	>100	>275	22

Source: Fageria, N.K. et al., *Growth and Mineral Nutrition of Field Crops*, 3rd edn., CRC Press, Boca Raton, FL, 2011a.

relative yield zone is called M, and more than 100% relative yield is called high soil P test. These zones are selected arbitrarily, as suggested by Raij (1991), for a soil P calibration study under Brazilian soil conditions. Soil P test availability indices and P fertilizer recommendations for the soil under investigation calculated on the basis of Figures 7.35 and 7.36 are presented (Table 7.37).

In addition to soil test calibration study, P fertilizer recommendations can be made based on crop response data to P fertilization. The authors of this book conducted field experiment that relates P rate with GY for lowland (Figure 7.37). GY of 12 lowland rice genotypes increased quadratically with increasing P rates (0–200 kg P_2O_5). Mean GY of 12 genotypes versus P rates is presented (Figure 7.38). Maximum GY of 12 genotypes was obtained with the application of 120 kg P_2O_5 ha^{-1}.

In addition to soil test calibrations, the information used for making P recommendations may include P removal in harvest products, the fertilizer application method, economic considerations, and philosophies concerning nutrient management (Dahnke and Olson, 1990; Hergert et al., 1997; Mallarino and Schepers, 2005). Most concepts for soil testing and fertilizer recommendations in use today were developed in regions when responses to P fertilizer were large and highly probable and where most field areas were responsive (Mallarino and Schepers, 2005). Most soil testing for P and P recommendation systems are based on a combination of strict sufficiency levels and buildup and maintenance concepts (Dahnke and Olson, 1990).

Differences in the application of these concepts partly explain the discrepancies in P recommendations for similar conditions across regions (Cox, 1994; Mallarino and Schepers, 2005). Recommended fertilizer rates for low testing soils in the United States, for example, may be designed to maximize the economic response to fertilization of one crop but often are designed to apply large amounts so that soil test P is increased to optimum levels at a certain rate over time (Mallarino and Schepers, 2005). Maintenance fertilization sometimes is recommended for soil test P levels for which the probability of response is usually low (Mallarino and Schepers, 2005). Maintenance fertilization usually is based on P removal in harvest products, which requires the use of established yield potential or producer yield goals for different soil series or more detailed soil map units. Recommendations are producer preferences and vary among fertilizer application for one crop or a one-time application for two or more crops in a rotation. Majority of corn and soybean are produced in the United States. Corn Belt applies fertilizer once (usually before the corn crop) to meet the P needs of the second-year corn–soybean rotation (Mallarino and Schepers, 2005).

7.4 CROP MANAGEMENT

In addition to soil and fertilizer management practices, P availability can be improved with the adoption of adequate crop management practices. These practices include the use of crop rotation, conservation tillage, supply of adequate soil moisture and P-efficient crop species/genotypes, and control of diseases, insects, and weeds.

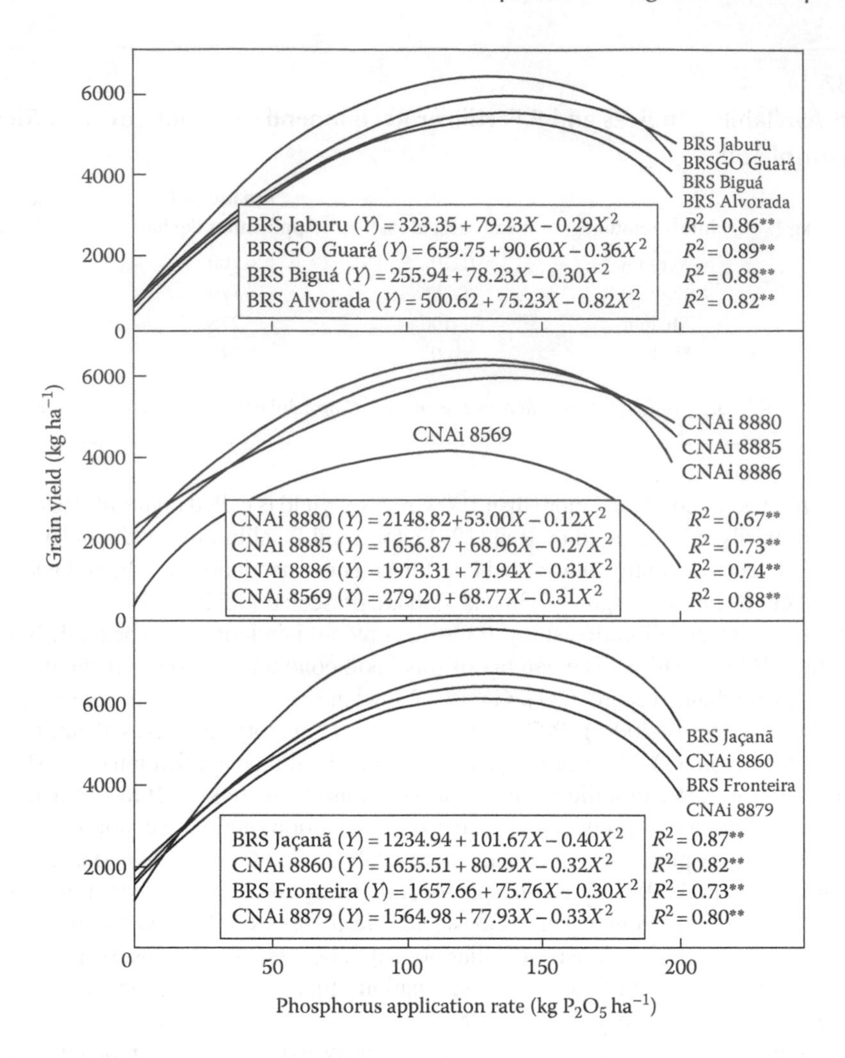

FIGURE 7.37 Response of lowland rice genotypes to P fertilization. (From Fageria, N.K. et al., *J. Plant Nutr.*, 34, 1087, 2011b.)

7.4.1 Crop Rotation

P availability can be optimized with the use of adequate crop rotation. Legumes are rotated with cereals to optimize resources due to different root systems of cereals and legumes, and legumes fix atmospheric N. Legumes contribute to greenhouse gas (N_2O and CO_2) emissions during nitrification and denitrification of fixed N. However, because less fertilizer N is used in legume-based cropping systems, overall greenhouse gas emissions are usually less than those in fertilized monoculture cereals (Lupwayi and Kennedy, 2007). Lupwayi and Kennedy (2007) reported that grain legumes in Northern Great Plains have positive effects on agriculture by adding and recycling biologically fixed N_2, enhancing nutrient uptake, reducing greenhouse gas emissions by reducing N fertilizer use, and breaking nonlegume crop pest cycles (Fageria and Stone, 2013).

Appropriate crop rotation reduces the risk of diseases, insects, and weeds in cropping systems. In central part of Brazil, locally known as Cerrado region, appropriate crop rotations include dry bean and soybean being rotated with upland rice and corn. Including cover crops in cropping systems is a viable strategy in improving soil fertility, reducing soil erosion, conserving soil moisture, controlling weeds, improving crop yields, and enhancing water and nutrient use efficiency

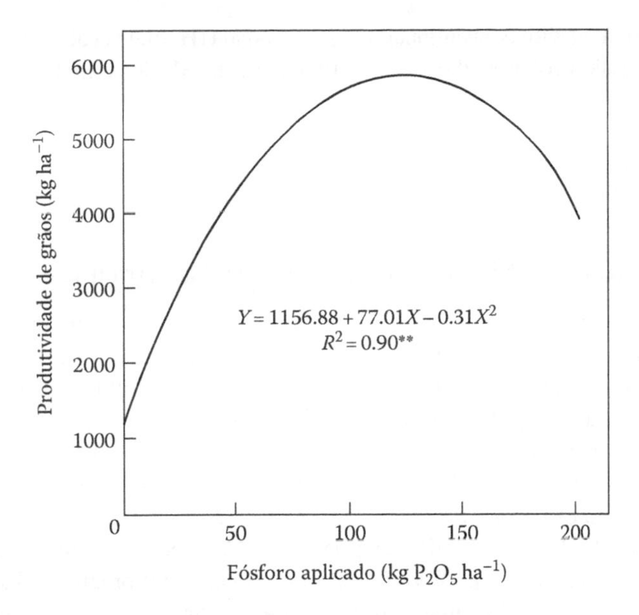

FIGURE 7.38 Response of lowland rice genotypes to P fertilization. Values are averages of 12 genotypes. (From Fageria, N.K. et al., *J. Plant Nutr.*, 34, 1087, 2011b.)

(Baligar and Fageria, 2007; Fageria, 2007a). Many tropical and temperate legume cover crops are available with a high potential of producing adequate amounts of dry matter to cover soil surface on low fertility acid soils (Baligar and Fageria, 2007; Fageria, 2007a).

Sogbedji et al. (2006) reported that the addition of Mucuna (*Mucuna pruriens* L. D. C.) and pigeon pea (*C. cajan* L.) in rotation with corn in sub-Saharan Africa cropping systems improved soil fertility and increased corn yield by about 38% and 32%, respectively. Hulugalle and Lal (1986) reported that in West Africa cropping systems involving grain legumes such as cowpea (*V. unguiculata* L. Walp.), pigeon pea, soybean, and peanut (*Arachis hypogaea*) in rotation with corn improved soil fertility and increased corn yields by approximately 50%. In South America, tropical legume cover crops can be useful in improving soil fertility and consequently crop yields (Fageria, 2009).

7.4.2 Conservation Tillage

Conservation tillage is defined as any tillage sequence that minimizes or reduces loss of soil and water. Operationally, it means a tillage or tillage and planting combination that facilitate a 30% or greater cover of crop residue on the soil surface (Soil Science Society of America, 2008). WUE and precipitation use efficiency increase with residue management practices, thereby increasing water storage efficiency and soil surface alterations that reduce runoff and soil temperature and consequently reduce evaporation (Hatfield et al., 2001; Nielsen et al., 2005). In addition, the increased soil–water storage is a result of reducing the number of times that moist soil is brought to the surface as tillage intensity is reduced. Crop residues increase precipitation infiltration by protecting the soil surface from raindrop impact and subsequent crusting, thus reducing runoff (Unger, 2001; Nielsen et al., 2002, 2005). Nielsen et al. (2005) reported that no-till systems increased crop productivity in the Northern Great Plains of the United States by harvesting a greater proportion of incident precipitation.

In Australia, Gibson et al. (1992) reported that retaining sorghum stubble on the soil increased the sorghum yield by 393 kg ha^{-1} due to increased WUE because of a greater amount of water stored in the soil profile as compared with conventional tillage. WUE for barley was increased in the dry year by 21%

with no tillage as compared with conventional tillage in Canada (Hatfield et al., 2001). In western Kansas, in a wheat–row crop–fallow rotation, the use of no tillage increased corn yield by 31% (Nodwood, 1999).

In Brazil, compared with conventional tillage, no tillage provided water saving in dry beans ranging from 14% to 30%, depending on the architecture of the plant (Fageria and Stone, 2013). Similarly, for upland rice, the water economy was improved by 15% in Brazilian Oxisol soils (Fageria and Stone, 2013).

7.4.3 Supply of Adequate Moisture/Improve Water Use Efficiency

Soil moisture is an important factor that affects availability of P to crop plants. In soils, water is the medium, support, and regulator of all chemical, physical, and biological reactions. Water deficit is a major yield-limiting factor in many parts of the world. Agricultural sustainability around the world depends on adequate supply of water or moisture to crop plants (Stone and Schlegel, 2006). Important soil management practices that can improve WUE are conservation tillage, increased SOM content, reduced length of fallow periods, contour farming, furrow dikes, control of plowpans, crop selection, and use of appropriate crop rotation (Nielsen et al., 2005; Stone and Schlegel, 2006). An increase in WUE by 25%–40% can be achieved through soil management practices that involve tillage and 15%–25% by modifying nutrient management practices (Hatfield et al., 2001). Precipitation use efficiency can be enhanced through adoption of more intensive cropping systems in semiarid environments and increased plant populations in more temperate and humid environments (Hatfield et al., 2001).

7.4.4 Efficient Crop Species/Genotypes

Significant variation exists among crop species and cultivars in nutrient uptake and utilization, including P (Gerloff and Gabelman, 1983; Baligar and Duncan, 1990; Baligar et al., 2001; Epstein and Bloom, 2005). Siliceous and calcareous sandy soils of South Australia are considered severely deficient in micronutrients for growth of wheat, oats, or barley, but growth and yield of rye was optimum (Graham, 1984). Native vegetation in this area is fully adapted to these soils mainly due to their slow growth rate (Loneragan, 1978).

Difference in nutrient uptake and utilization may be associated with appropriate root geometry; plant capacity to acquire sufficient nutrients from soils with low availability of these nutrients; improved transport, distribution, and utilization within plants; and balanced source and sink relationship (Graham, 1984; Baligar et al., 2001; Fageria and Stone, 2013). Antagonistic (uptake of one nutrient is restricted by another nutrient) and synergistic (uptake of one nutrient is enhanced by other nutrient) effects of nutrients on nutrient use efficiency among various plant species and cultivars have not been explored sufficiently (Fageria et al., 2008). Variation in PUE among 30 dry bean genotypes is presented (Table 7.37). There was a significant difference in PUE occurring among genotypes. Based on GY efficiency index (GYEI), dry bean genotypes could be classified as P efficient, moderately efficient, and inefficient (Table 7.38). P-efficient genotypes were Pérola, BRSMG Talisma, BRS Requinte, BRS Pontal, BRS 9435 Cometa, BRS Estilo, BRSMG Majestoso, CNFC 10429, CNFC 10408, CNFC 10467, CNFC 10470, Diamante Negro, Corrente, BRS 7762 Supremo, BRS Esplendor, BRS Marfim, BRS Agreste, and BRS Executivo. Aporé, BRS Valente, BRS Grafite, BRS Campeiro, CNFP 10104, Bambuí, BRS Pitamda, BRS Verede, EMPOPA Ouro, BRS Radiante, and Jalo Precoce were considered moderately efficient in PUE. Genotype BRS Embaixador was only P inefficient among the 30 dry bean genotypes. Differences in PUE among dry bean genotypes have been reported by Fageria (1992, 1998). Genotype differences in PUE may be related to their ability to alter rhizosphere conditions that are known to influence the bioavailability of soil P,

TABLE 7.38
GYEI of 30 Fry Bean Genotypes

Genotype	GYEI
1. Aporé	0.73im (ME)[a]
2. Pérola	1.24b–e (E)
3. BRSMG Talisma	1.04d–j (E)
4. BRS Requinte	1.10c–h (E)
5. BRS Pontal	1.79a (E)
6. BRS 9435 Cometa	1.20b–f (E)
7. BRS Estilo	1.27b–e (E)
8. BRSMG Majestoso	0.92e–m (E)
9. CNFC 10429	1.10c–h (E)
10. CNFC 10408	1.41bc (E)
11. CNFC 10467	1.03e–j (E)
12. CNFC 10470	1.53ab (E)
13. Diamante Negro	1.10c–h (E)
14. Corrente	1.39b–d (E)
15. BRS Valente	0.87f–m (ME)
16. BRS Grafite	0.66m (ME)
17. BRS Campeiro	0.76h–m (ME)
18. BRS 7762 Supremo	1.08c–I (E)
19. BRS Esplendor	1.06c–I (E)
20. CNFP 10104	0.83g–m (ME)
21. Bambuí	0.92e–m (ME)
22. BRS Marfim	1.01e–l (E)
23. BRS Agreste	1.09c–h (E)
24. BRS Pitamda	0.69m (ME)
25. BRS Verede	0.62n (ME)
26. EMGOPA Ouro	0.87f–m (ME)
27. BRS Radiante	0.85f–m (ME)
28. Jalo Precoce	0.80h–m (ME)
29. BRS Executivo	1.16c–g (E)
30. BRS Embaixador	0.28n (IE)
Average	1.01
F-test	
Genotype	**
CV (%)	10.98

Source: Fageria, N.K. et al., *Commun. Soil Sci. Plant Anal.*, 43, 2752, 2012.

[a] Values in the parenthesis represent E, efficient; ME, moderately efficient; IE, inefficient.

**Significant at the 1% probability level. Means followed by the same letter within the same column are not significantly different at the 5% probability level by Tukey's test.

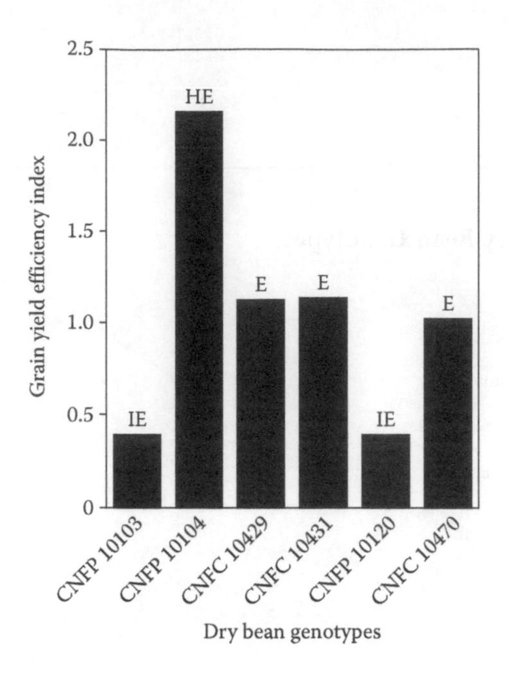

FIGURE 7.39 GYEI of six dry bean genotypes. E, efficient; IE, inefficient; HE, highly efficient. (From Fageria, N.K. et al., *Commun. Soil Sci. Plant Anal.*, 43, 2289, 2012.)

via the release of protons, organic anions, or phosphatase-like enzymes (Gaume et al., 2001; Hinsinger, 2001; Tang et al., 2004).

Fageria et al. (2012) evaluated PUE of six dry bean genotypes (Figure 7.39). GYEI was used as a parameter to classify genotypes into efficient, highly efficient, or inefficient groups. GYEI was calculated by using following equation:

$$GYEI = \frac{GY_1}{AGY_1} \times \frac{GY_2}{AGY_2}$$

where

GY_1 is GY at low P level
AGY_1 is mean GY of six genotypes at low P level
GY_2 is GY at high P level
AGY_2 is mean GY of six genotypes at high P level

Genotypes having GYEI values >1.5 were classified as highly efficient (HE), genotypes having GYEI values between 1.0 and 1.5 were classified as efficient (E), genotypes that were having GYEI between 0.5 and 1.0 were classified as moderately efficient (ME), and genotypes with GYEI values <0.5 were classified as inefficient (IE) in P use (Figure 7.39).

Based on the GYEI, genotypes were classified into different efficiency categories. The GYEI varied from 0.39 to 2.16, with a mean value of 1.03. Genotype CNFP 10104 was grouped into highly efficient, and genotypes CNFP and CNFP 10120 were inefficient. Remaining three genotypes, CNFC 10429, CNFC 10431, and CNFC 10470, were classified as efficient (Figure 7.39). Fageria et al. (2011a), Baligar et al. (2001), and Fageria (2009) have reported variability in PUE among dry bean genotypes. They reported that variability in P uptake among dry bean genotypes may be associated with their physiological or biochemical mechanisms.

7.4.5 Control of Diseases, Insects, and Weeds

Diseases, insects, and weeds are main constraints in crop production around the world. Plants infested with diseases, insects, and weeds have lower yield and lower resource use efficiency, including nutrients. To improve resource use efficiency, it is fundamental to control diseases, insects, and weeds. Diseases, insects, and weeds can be controlled by the use of herbicides. However, integrated control measures should be adopted to control diseases, inscets, and weeds. These include preparing soil adequately before or during the crop sowing, crop rotation, and using organic and inorganic fertilizers in adequate amounts, effective sources, and appropriate methods of application. Use of high-yielding and nutrient-efficient cultivars is important in reducing risks of diseases, insects, and weeds. Appropriate date of sowing of lowland rice significantly reduced brown spot (*Acidah'a trigeminata*) diseases in the state of Rio Grande do Sul of Brazil (Menezes et al., 2013). Brown spot infestation occurred linearly when rice was planted from the first of November to December. However, its severity was low when rice was planted before the first of November. This may be associated with favorable climatic and other conditions favorable for this disease in the late planting season.

7.5 CONCLUSIONS

P is important for growth and development of crop plants. There are several functions of P in the growth and development of plants, but its role in energy storage and transfer is the most important. Availability of P is lower in the acid and alkaline soils. Soil pH is a major chemical property in determining P availability to crop plants. Optimum soil pH for the growth of most crop plants is in the range of 6.0–7.0. Use of efficient source, effective methods of application, and adequate rate are fundamental to achieve higher P availability and maximize crop yields. Diffusion rate of P is low, and P should be applied in bands or furrows to obtain higher availability to plant roots. Soluble P sources improve their availability to plants as compared to water-insoluble sources for annual crops. As a general rule, those materials that have a high percentage of P in a water-soluble form are more accepted than those with no or only a small fraction.

Adequate rate of P is determined by field trials and or/ soil test. Soil test values are determined by extracting solution used in determining P levels in the soil. In addition to these considerations, crop management practices are important in improving P availability. These practices include appropriate crop rotation, use of conservation tillage, improved water availability to crops, and use of P-efficient crop species or genotypes within species. Use of liming and gypsum for acid soils is an important soil management practice for improving P availability to crop plants.

REFERENCES

Adriano, D. C. 1986. *Trace Elements in the Terrestrial Environment.* New York: Springer-Verlag.

Alazard, D. 1985. Stem and root nodulation in *Aeschynomene* spp. *Appl. Environ. Microbiol.* 50:732–734.

Alexander, M. 1977. Mineralization and immobilization of nitrogen. In: *Introduction to Soil Microbiology*, ed. M. Alexander, pp. 225–250. New York: Wiley.

Al-Khatib, K., C. Libbey, and R. Boydston. 1997. Weed suppression with *Brassica* green manure crops in green Pea. *Weed Sci.* 45:439–445.

Alvarez, M. A., S. Gagne, and H. Antoun. 1995. Effect of compost rhizosphere microflora of the tomato and on the incidence of plant growth-promoting rhizobacteria. *Appl. Environ. Microbiol.* 61:194–199.

Angle, J. S. 1998. Impact of biosolids and co-utilization wastes on rhizobia, nitrogen fixation and growth of legumes. In: *Beneficial Co-Utilization of Agricultural, Municipal and Industrial by-Products*, eds. S. Brown, J. S. Angle, and L. Jacobs, pp. 235–245. Dordrecht, the Netherlands: Kluwer Academic Publishers.

Anjos, J. T. and D. L. Rowell. 1987. The effect of lime on phosphorus adsorption and barley growth in three acid soils. *Plant Soil* 103:75–82.

Armstrong, D. L. 1999. Phosphorus for agriculture. *Better Crops Plant Food* 83:1–39.

Badruddin, M., M. P. Reynolds, and O. A. A. Ageeb. 1999. Wheat management in warm environment: Effect of organic and inorganic fertilizers, irrigation frequency, and mulching. *Agron. J.* 91:975–983.

Baligar, V. C. and R. R. Duncan. 1990. *Crops as Enhancers of Nutrient Use*. San Diego, CA: Academic Press.

Baligar, V. C. and N. K. Fageria. 2007. Agronomy and physiology of tropical cover crops. *J. Plant Nutr.* 30:1287–1339.

Baligar, V. C., N. K. Fageria, and Z. L. He. 2001. Nutrient use efficiency in plants. *Commun. Soil Sci. Plant Anal.* 32:921–950.

Barbarick, K. A., J. A. Ippolito, and D. G. Westfall. 1995. Biosolids effect on phosphorus, copper, zinc, nickel, and molybdenum concentrations in dryland wheat. *J. Environ. Qual.* 24:608–611.

Barker, A. V. 1997. Composition and uses of compost. In: *Agricultural Uses of By-Products and Wastes*, eds. J. E. Recheigl and H. C. Mackinnon, pp. 140–162. Washington, DC: American Chemical Society.

Basta, N. T. and M. A. Tabatabai. 1992. Effect of cropping systems on adsorption of metals by soils. II. Effect of pH. *Soil Sci.* 153:195–204.

Becker, M., M. Ali, J. K. Ladha, and J. C. G. Ottow. 1995. Agronomic and economic evaluation of *Sesbania rostrata* green manure establishment in irrigated rice. *Field Crops Res.* 40:135–141.

Becker, M., J. K. Ladha, I. C. Simpson, and J. C. G. Ottow. 1994. Parameters affecting residue N mineralization in flooded soils. *Soil Sci. Soc. Am. J.* 58:1666–1671.

Becker, M., J. K. Ladha, I. Watanable, and J. C. G. Ottow. 1988. Stem nodulating legumes as green manure for lowland rice. *Phil. J. Crop Sci.* 13:121–127.

Beckie, H. J. and S. A. Brandt. 1997. Nitrogen contribution of field pea in annual cropping systems: I. Nitrogen residual effect. *Can. J. Plant Sci.* 77:311–322.

Bending, G. D., M. K. Turner, and I. G. Burns. 1998. Fate of nitrogen from crop residues as affected by biochemical quality and the microbial biomass. *Soil Biol. Biochem.* 30:2055–2065.

Bhandhari, B. and D. J. D. Nicholas. 1985. Proton motive force in washed cells of *Rhizobium japonicum* and bacteroids from *Glycine max*. *J. Bacteriol.* 164:1383–1385.

Biederbeck, V. O., H. H. Janzen, C. A. Campbell, and R. P. Zentner. 1994. Labile soil organic matter as influenced by cropping practices in arid environment. *Soil Biol. Biochem.* 26:1647–1656.

Bolan, N. S., D. C. Adriano, and D. Curtin. 2003. Soil acidification and liming interactions with nutrient and heavy metal transformation and bioavailability. *Adv. Agron.* 78:215–272.

Bolan, N. S., R. Naidu, R. J. K. Syers, and R. W. Tillman. 1999. Surface charge and solute interactions in soils. *Adv. Agron.* 67:88–141.

Bouldin, D. R. 1988. Effects of green manure on soil organic matter content and nitrogen availability. In: *Sustainable Agriculture: Green Manure in Rice Farming*, ed. IRRI, pp. 152–163. Los Baños, Philippines: International Rice Research Institute.

Bouwer, H. and R. L. Chaney. 1974. Land treatment of wastewater. *Adv. Agron.* 26:133–176.

Brady, N. C. and R. R. Weil. 2002. *The Nature and Properties of Soils,* 13th edn. Upper Saddle River, NJ: Prentice Hall.

Brockwell, J., P. J. Bottomley, and E. T. Janice. 1995. Manipulation of rhizobia microflora for improving legume productivity and soil fertility: A critical assessment. *Plant Soil* 174:143–180.

Broder, M. W. and G. H. Wagner. 1988. Microbial colonization and decomposition of corn, wheat and soybean residue. *Soil Sci. Soc. Am. J.* 52:112–117.

Brosius, M. R., G. K. Evanylo, L. R. Bulluck, and J. B. Ristaino. 1998. Comparison of commercial fertilizer and organic by products on soil chemical and biological properties and vegetable yields. In: *Beneficial Co-Utilization of Agricultural, Municipal and Industrial by Products*, eds. S. Brown, J. S. Angle, and L. Jacobs, pp. 195–202. Dordrecht, the Netherlands: Kluwer Academic Publishers.

Brummer, G. W. and U. Herms. 1983. Influence of soil reaction and organic matter on solubility of heavy metals in soils. In: *Effects of Accumulation of Air Pollutants in Forest Ecosystems*, eds. B. Ulrich and J. Pankrath, pp. 233–243. Dordrecht, the Netherlands: Reidel.

Bundy, L. G., H. Tunney, and A. D. Halvorson. 2005. Agronomic aspects of phosphorus management. In: *Phosphorus: Agriculture and the Environment*, ed. L. K. Al-Amoodi, pp. 685–727. Madison, WI: ASA, CSSA, and SSSA.

Buresh, R. J. and S. K. De Datta. 1991. Nitrogen dynamics and management in rice-legume cropping systems. *Adv. Agron.* 45:1–59.

Burgess, M. S., G. R. Mehuys, and C. A. Madramootoo. 2002. Nitrogen dynamics of decomposing corn residue components under three tillage systems. *Soil Sci. Soc. Am. J.* 66:1350–1358.

Campbell, C. A., R. P. Zentner, F. Selles, V. O. Biederbeck, and A. J. Leyshon. 1992. Comparative effects of grain lentil-wheat and monoculture wheat on crop production, N economy and N fertility in a Brown Chernozem. *Can. J. Plant Sci.* 72:1091–1107.

Carter, M. R. 2002. Soil quality for sustainable land management: Organic matter and aggregation interactions that maintain soil functions. *Agron. J.* 94:38–47.

Carvalho, M. M., F. T. Oliveira, O. F. Saraiva, and C. E. Martins. 1985. Nutritional factors limiting growth of tropical forage species in two soils from "Zona da Mata" Minas Gerais (Brazil) I. Red-Yellow Latosol. *Pesq. Agropec. Bras.* 20:519–528.

Cassman, K. G. and D. N. Munns. 1980. Nitrogen mineralization as affected by soil moisture, temperature and depth. *Soil Sci. Soc. Am. J.* 44:1233–1237.

Chan, K. Y. and D. P. Heenan. 1998. Effect of lime ($CaCO_3$) application on soil structural stability of a red earth. *Aust. J. Soil Res.* 36:73–86.

Christie, P., D. L. Easson, J. R. Picton, and S. C. P. Love. 2001. Agronomic value of alkaline-stabilized biosolids for spring barley. *Agron. J.* 93:144–151.

Cihacek, L. J. 1993. Phosphorus source effects on alfalfa yield, total nitrogen content, and soil test phosphorus. *Commun. Soil Sci. Plant Anal.* 24:2043–2057.

Clement, A., J. K. Ladha, and F. P. Chalifour. 1998. Nitrogen dynamics of various green manure species and the relationship to lowland rice production. *Agron. J.* 90:149–154.

Clough, T. J., F. M. Kelliher, R. R. Sherlock, and C. D. Ford. 2004. Lime and soil moisture effects on nitrous oxide emissions from a urine patch. *Soil Sci. Soc. Am. J.* 68:1600–1609.

Clough, T. J., R. R. Sherlock, and F. M. Kelliher. 2003. Can liming mitigate N_2O fluxes from a urine-amended soil? *Aust. J. Soil Res.* 41:439–457.

Cole, C. V., J. Williams, M. Shaffer, and J. Hanson. 1987. Nutrient and organic matter dynamics as components of agricultural production systems models. In: *Soil Fertility and Organic Matter as Critical Agricultural Production Systems Models*, ed. R. F. Follett, pp. 147–166. Madison, WI: ASA, CSSA, and SSSA.

Cox, R. R. 1994. Current phosphorus availability indices: Characteristics and shortcomings. In: *Soil Testing: Prospects for Improving Nutrient Recommendations*, eds. J. L. Havlin and J. S. Jacobsen, pp. 101–113. Madison, WI: SSSA.

Cregan, D. D., J. R. Hirth, and M. K. Conyers. 1989. Amelioration of soil acidity by liming and other amendments. In: *Soil Acidity and Plant Growth*, ed. A. D. Robson, pp. 205–264. Sydney, New South Wales, Australia: Academic Press.

Crookston, R. K., J. E. Kurle, P. J. Copeland, J. H. Ford, and W. E. Lueschen. 1991. Rotational cropping sequence affects yield of corn and soybean. *Agron. J.* 83:108–113.

Curi, N. and O. A. Camargo. 1988. Phosphorus adsorption characteristics of Brazilian Oxisols. In: *Proceedings of the Eighth International Soil Classification Workshop: Part 1*, eds. F. H. Beinroth, M. N. Camargo, and H. Eswarn, pp. 56–63. Washington, DC: Soil Management Support Services, U.S. Department of Agriculture.

Curtin, D. and J. K. Syers. 2001. Lime-induced changes in indices of phosphate availability. *Soil Sci. Soc. Am. J.* 65:147–152.

Dahnke, W. C. and R. A. Olson. 1990. Soil test correlation, calibration, and recommendation. In: *Soil Testing and Plant Analysis*, 3rd edn., ed. R. L. Westerman, pp. 45–71. Madison, WI: SSSA.

Dao, T. H. 1996. Tillage system and crop residue effects on surface compaction of a Paleustoll. *Agron. J.* 88:141–148.

Dao, T. H. 1999. Coamendments to modify phosphorus extractability and nitrogen/phosphorus ratio in feedlot manure and composted manure. *J. Environ. Qual.* 28:1114–1121.

Davelouis, J. R., P. A. Sanchez, and J. C. Alegre. 1991. Green manure incorporation and soil acidity amelioration. In: TropSoils, Technical report 1988–1989, ed. T. P. McBridge, pp. 286–289. Raleigh, NC: North Carolina State University.

Devine, T. E., J. H. Bouton, and T. Mabrahtu. 1990. Legume genetics and breeding for stress tolerance and nutrient efficiency. In: *Crops as Enhancers of Nutrient Use*, eds. V. C. Baligar and R. R. Duncan, pp. 211–252. New York: Academic Press.

Dias-Zorita, M., D. E. Buschiazzo, and N. Peinemann. 1999. Soil organic matter and wheat productivity in the semiarid Argentine Pampas. *Agron. J.* 91:276–279.

Diekmann, K. H., S. K. De Datta, and J. C. G. Ottow. 1993. Nitrogen uptake and recovery from urea and green manure in lowland rice measured by 15N and non-isotopic techniques. *Plant Soil* 148:91–99.

Dinnes, D. L., D. L. Karlen, D. B. Jaynes, T. C. Kaspar, J. L. Hatfield, T. S. Colvin, and C. A. Cambardella. 2002. Nitrogen management strategies to reduce nitrate leaching in tile-drained midwestern soils. *Agron. J.* 94:153–171.

Doran, J. W. and M. S. Smith. 1991. Role of cover crops in nitrogen cycling. In: *Cover Crops for Clean Water*, ed. W. L. Hargrove, pp. 85–90. Ankeny, IA: Soil and Water Conservation Society.

Douglas, C. L. Jr. and R. W. Rickman. 1992. Estimating crop residue decomposition from air temperature, initial nitrogen content, and residue placement. *Soil Sci. Soc. Am. J.* 56:272–278.

Dreyfus, B. L. and Y. R. Dommergues. 1981. Nitrogen-fixing nodules induced by *Rhizobium* on the stem of the tropical legume *Sesbania rostrata*. *FEMS Microbial. Lett.* 10:313–317.
Eagle, A. J., J. A. Bird, J. E. Hill, W. R. Horwath, and C. V. Kessel. 2001. Nitrogen dynamics and fertilizer use efficiency in rice following straw incorporation and winter flooding. *Agron. J.* 93:1346–1354.
Edmeades, D. C. and K. W. Perrott. 2004. The calcium requirements of pastures in New Zealand: A review. *N. Z. J. Agric. Res.* 47:11–21.
Eghball, B. and J. F. Power. 1999. Phosphorus and nitrogen based manure and compost applications: Corn production and soil phosphorus. *Soil Sci. Soc. Am. J.* 63:895–901.
Eghball, B., J. F. Power, J. E. Gilley, and J. W. Doran. 1997. Nutrient, carbon, and mass loss during composting of beef cattle feedlot manure. *J. Environ. Qual.* 26:189–193.
Elliott, L. F. and N. P. Swanson. 1976. Land use of animal wastes. In: *Land Application of Waste Materials*, ed. Soil Conservation Society of America, pp. 80–90. Ankeny, IA: Soil Conservation Society of America.
EMBRAPA (Empresa Brasileira de Pesquisa Agropecuaria). 1995. Technical recommendations for soybean production in the central region of Brazil. Londrina, Brazil: EMBRAPA-National Soybean Research Center, Document 88.
Eneji, A. E., S. Yamamoto, T. Honna, and A. Ishiguro. 2001. Physico-chemical changes in livestock feces during composting. *Commun. Soil Sci. Plant Anal.* 32:477–489.
Engelhard, A. W. 1989. Historical highlights and prospects for the future. In: *Soil Born Plant Pathogens: Management of Diseases with Macro and Microelements*, ed. A. W. Engelhard, pp. 9–17. St. Paul, MN: The American Phytopathological Society.
Engelstad, O. P. and G. L. Terman. 1980. Agronomic effectiveness of phosphorus fertilizers. In: *The Role of Phosphorus in Agriculture*, ed. R. C. Dinauer, pp. 311–332. Madison, WI: ASA, CSSA and SSSA.
Entz, M. H., W. J. Bullied, and F. Katepa-Mupondwa. 1995. Rotational benefits of forage crops in Canadian Prairie cropping systems. *J. Prod. Agric.* 8:521–529.
Epstein, E. and A. J. Bloom. 2005. *Mineral Nutrition of Plants: Principles and Perspectives*, 2nd edn. Sunderland, MA: Sinauer Associates, Inc.
Fageria, N. K. 1980. Influence of phosphorus application on growth, yield and nutrient uptake by irrigated rice. *Revista brasileira de ciência do solo* 4:26–31.
Fageria, N. K. 1984. Response of rice cultivars to liming in cerrado soil. *Pesq. Agropec. Bras.* 19:883–889.
Fageria, N. K. 1989. *Tropical Soils and Physiological Aspects of Crops*. Goiânia, Brazil: EMBRAPA-CNPAF.
Fageria, N. K. 1991. Response of rice cultivars to phosphorus fertilization on a dark red latosol of central Brazil. *Revista brasileira de ciência do solo* 15:63–67.
Fageria, N. K. 1992. *Maximizing Crop Yields*. New York: Marcel Dekker.
Fageria, N. K. 1996. Annual report of the project. The study of liming and fertilization for rice and common bean in Cerrado region. Goiania, Brazil: National Rice and Bean Research Center.
Fageria, N. K. 1998. Phosphorus use efficiency by bean genotypes. *Revista Brasileira de Engenharia Agrícola e Ambiental* 2:128–131.
Fageria, N. K. 2001a. Effect of liming on upland rice, common bean, corn, and soybean production in cerrado soil. *Pesq. Agropec. Bras.* 36:1419–1424.
Fageria, N. K. 2001b. Response of upland rice, dry bean, corn and soybean to base saturation in cerrado soil. *Revista Brasileira de Engenharia Agrícola e Ambiental* 5:416–424.
Fageria, N. K. 2002. Soil quality vs. environmentally based agricultural management practices. *Commun. Soil Sci. Plant Anal.* 33:2301–2329.
Fageria, N. K. 2006. Liming and copper fertilization in dry bean production on an Oxisol in no-tillage system. *J. Plant Nutr.* 29:1–10.
Fageria, N. K. 2007a. Green manuring in crop production. *J. Plant Nutr.* 30:691–719.
Fageria, N. K. 2007b. Yield physiology of rice. *J. Plant Nutr.* 30:843–879.
Fageria, N. K. 2007c. Soil fertility and plant nutrition research under field conditions: Basic principles and methodology. *J. Plant Nutr.* 30:203–223.
Fageria, N. K. 2008. Optimum soil acidity indices for dry bean production on an Oxisol in no-tillage system. *Commun. Soil Sci. Plant Anal.* 39:845–857.
Fageria, N. K. 2009. *The Use of Nutrients in Crop Plants*. Boca Raton, FL: CRC Press.
Fageria, N. K. 2012. Role of soil organic matter in maintaining sustainability of cropping systems. *Commun. Soil Sci. Plant Anal.* 43:2063–2113.
Fageria, N. K. 2013. *The Role of Plant Roots in Crop Production*. Boca Raton, FL: CRC Press.
Fageria, N. K. 2014. *Mineral Nutrition of Rice*. Boca Raton, FL: CRC Press.
Fageria, N. K. and V. C. Baligar. 1996. Response of lowland rice and common bean grown in rotation to soil fertility levels on a Varzea soil. *Fert. Res.* 13:13–20.

Fageria, N. K. and V. C. Baligar. 1997. Upland rice genotypes evaluation for phosphorus use efficiency. *J. Plant Nutr.* 20:499–509.

Fageria, N. K. and V. C. Baligar. 1999. Growth and nutrient concentrations of common bean, lowland rice, corn, soybean, and wheat at different soil pH on an Inceptisol. *J. Plant Nutr.* 22:1495–1507.

Fageria, N. K. and V. C. Baligar. 2003. Fertility management of tropical acid soils for sustainable crop production. In: *Handbook of Soil Acidity*, ed. Z. Rengel, pp. 359–385. New York: Marcel Dekker.

Fageria, N. K. and V. C. Baligar. 2005. Enhancing nitrogen use efficiency in crop plants. *Adv. Agron.* 88:97–185.

Fageria, N. K. and V. C. Baligar. 2008. Ameliorating soil acidity of tropical Oxisols by liming for sustainable crop production. *Adv. Agron.* 99:345–399.

Fageria, N. K., V. C. Baligar, and B. A. Bailey. 2005. Role of cover crops in improving soil and row crop productivity. *Commun. Soil Sci. Plant Anal.* 40:1148–1160.

Fageria, N. K., V. C. Baligar, and R. B. Clark. 2002. Micronutrients in crop production. *Adv. Agron.* 77:185–268.

Fageria, N. K., V. C. Baligar, and C. A. Jones. 2011a. *Growth and Mineral Nutrition of Field Crops*, 3rd edn. Boca Raton, FL: CRC Press.

Fageria, N. K., V. C. Baligar, and Y. C. Li. 2008. The role of nutrient-efficient plants in improving crop yields in the twenty first century. *J. Plant Nutr.* 31:1121–1157.

Fageria, N. K., V. C. Baligar, and Y. C. Li. 2009. Differential soil acidity tolerance of tropical legume cover crops. *Commun. Soil Sci. Plant Anal.* 40:1148–1160.

Fageria, N. K., V. C. Baligar, and Y. C. Li. 2014a. Nutrient uptake and use efficiency by tropical legume cover crops at varying pH of an oxisol. *J. Plant Nutr.* 37:294–311.

Fageria, N. K., V. C. Baligar, and R. J. Wright. 1991a. Influence of phosphate rock sources and rates on rice and common bean production in an Oxisol. In: *Plant–Soil Interactions at Low pH*, eds. R. J. Wright, V. C. Baligar, and R. P. Murmann, pp. 539–546. Dordrecht, the Netherlands: Kluwer Academic Publishers.

Fageria, N. K. and M. P. Barbosa Filho. 1981. Screening rice cultivars for higher efficiency of phosphorus absorption. *Pesq. Agropec. Bras.* 16:777–782.

Fageria, N. K. and M. P. Barbosa Filho. 2007. Dry-matter and grain yield, nutrient uptake, and phosphorus use-efficiency of lowland rice as influenced by phosphorus fertilization. *Commun. Soil Sci. Plant Anal.* 38:1289–1297.

Fageria, N. K., M. P. Barbosa Filho, and J. R. P. Carvalho. 1982. Response of upland rice to phosphorus fertilization on an Oxisol of central Brazil. *Agron. J.* 74:51–56.

Fageria, N. K., E. M. Castro, and V. C. Baligar. 2004. Response of upland rice genotypes to soil acidity. In: *The Red Soils of China: Their Nature, Management and Utilization*, eds. M. J. Wilson, Z. He, and X. Yang, pp. 219–237. Dordrecht, the Netherlands: Kluwer Academic Publishers.

Fageria, N. K. and H. R. Gheyi. 1999. *Efficient Crop Production*. Campina Grande, Brazil: Federal University of Paraiba.

Fageria, N. K., A. B. Heinemann, and R. A. Reis Jr. 2014b. Comparative efficiency of phosphorus sources for upland rice production. *Commun. Soil Sci. Plant Anal.* 45:1399–1420.

Fageria, N. K., L. C. Melo, J. P. Oliveira, and A. M. Coelho. 2012. Yield and yield components of dry bean genotypes as influenced by phosphorus fertilization. *Commun. Soil Sci. Plant Anal.* 43:2752–2766.

Fageria, N. K. and O. P. Morais. 1987. Evaluation of rice cultivars for utilization of calcium and magnesium in the cerrado soil. *Pesq. Agropec. Bras.* 22:667–672.

Fageria, N. K., O. P. Morais, M. C. S. Carvalho, and J. M. Colombari Filho. 2015a. Upland rice genotypes evaluation for acidity tolerance. *Commun. Soil Sci. Plant Anal.* 46:1076–1096.

Fageria, N. K., O. P. Morais, and A. B. Santos. 2010. Nitrogen use efficiency in upland rice. *J. Plant Nutr.* 33:1696–1711.

Fageria, N. K., O. P. Morais, and M. J. Vasconcelos. 2013a. Upland rice genotypes evaluation for phosphorus use efficiency. *J. Plant Nutr.* 36:1868–1880.

Fageria, N. K. and A. Moreira. 2011. The role of mineral nutrition on root growth of crop plants. *Adv. Agron.* 110:251–331.

Fageria, N. K., A. Moreira, C. Castro, and M. F. Moraes. 2013b. Optimal acidity indices for soybean production in Brazilian Oxisols. *Commun. Soil Sci. Plant Anal.* 44:2941.

Fageria, N. K., A. Moreira, L. A. C. Moraes, and M. F. Moraes. 2014c. Influence of lime and gypsum on yield and yield components of soybean and changes in soil chemical properties. *Commun. Soil Sci. Plant Anal.* 45:271–283.

Fageria, N. K. and A. B. Santos. October 18–20, 2005. Influence of base saturation and micronutrient rates on their concentration in the soil and bean productivity in cerrado soil in no-tillage system. In: *Paper presented at the VIII National Bean Congress*, Goiânia, Brazil.

Fageria, N. K. and A. B. Santos. 2008. Lowland rice response to thermophosphate fertilization. *Commun. Soil Sci. Plant Anal.* 39:873–889.

Fageria, N. K., A. B. Santos, and V. C. Baligar. 1997. Phosphorus soil test calibration for lowland rice on an Inceptisol. *Agron. J.* 89:737–742.

Fageria, N. K., A. B. Santos, and M. C. S. Carvalho. 2015b. Agronomic evaluation of phosphorus sources applied to upland and lowland rice. *Commun. Soil Sci. Plant Anal.* 46:1097–1111.

Fageria, N. K., A. B. Santos, and A. B. Heinemann. 2011b. Lowland rice genotypes evaluation for phosphorus use efficiency in tropical lowland. *J. Plant Nutr.* 34:1087–1095.

Fageria, N. K. and J. M. Scriber. 2002. The role of essential nutrients and minerals in insect resistance in crop plants. In: *Insect and Plant Defense Dynamics*, ed. T. N. Ananthakrishnan, pp. 23–54. Enfield, NH: Science Publisher.

Fageria, N. K., N. A. Slaton, and V. C. Baligar. 2003. Nutrient management for improving lowland rice productivity and sustainability. *Adv. Agron.* 80:63–152.

Fageria, N. K. and N. P. Souza. 1995. Response of rice and common bean crops in succession to fertilization in Cerrado soil. *Pesq. Agropec. Bras.* 30:359–368.

Fageria, N. K. and L. F. Stone. 1999. *Managing Soil Acidity of Cerrado and Varzea of Brazil*. Document no. 42. Santo Antônio de Goiás, Brazil: National Rice and Bean Research Center of EMBRAPA.

Fageria, N. K. and L. F. Stone. 2004. Yield of dry bean in no-tillage system with application of lime and zinc. *Pesq. Agropec. Bras.* 39:73–78.

Fageria, N. K. and L. F. Stone. 2013. Water and nutrient use efficiency in food production in South America. In: *Improving Water and Nutrient Use Efficiency in Food Production Systems*, ed. Z. Rengel, pp. 275–296. New York: John Wiley & Sons.

Fageria, N. K., R. J. Wright, V. C. Baligar, and C. M. R. Sousa. 1991b. Characterization of physical and chemical properties of Varzea soils of Goias state of Brazil. *Commun. Soil Sci. Plant Anal.* 22:1631–1646.

FAO. 2013. *FAOSTAT Data Base-Agricultural Production*. Food and Agricultural Organization of the United Nations, Rome, Italy. http://faostat3.fao.org/faostat-gateway/go/to/home/E (accessed November 4, 2013).

Fauci, M. F. and R. P. Dick. 1994. Plant response to organic amendments and decreasing inorganic nitrogen rates in soil from a long-term experiment. *Soil Sci. Soc. Am. J.* 58:134–138.

Fischer, K. S. 1998. Toward increasing nutrient-use efficiency in rice cropping systems: The next generation of technology. *Field Crops Res.* 56:1–6.

Fox, R. H., R. J. Myers, and I. Vallis. 1990. The nitrogen mineralization rate of legume residues in soil as influenced by their polyphenol, lignin, and nitrogen contents. *Plant Soil* 129:251–259.

Foy, C. D. 1984. Physiological effects of hydrogen, aluminum and manganese toxicities in acid soil. In: *Soil Acidity and Liming*, 2nd edn., ed. F. Adams, pp. 57–97. Madison, WI: ASA, CSSA and SSSA.

Franco, A. A. and D. N. Munns. 1982. Acidity and aluminum restraints on nodulation, nitrogen fixation, and growth of *Phaseolus vulgaris* in nutrient solution. *Soil Sci. Soc. Am. J.* 46:296–301.

Friesen, D. K., A. S. R. Juo, and M. H. Miller. 1980a. Liming and lime phosphorus–zinc interactions in two Nigerian Ultisols. I. Interactions in the soil. *Soil Sci. Soc. Am. J.* 44:1221–1226.

Friesen, D. K., M. H. Miller, and A. S. R. Juo. 1980b. Lime and lime-phosphate-zinc interactions in two Nigerian Ultisols. II. Effects on maize root and shoot growth. *Soil Sci. Soc. Am. J.* 44:1227–1232.

Frossard, E., S. Sinaj, and P. Dufour. 1996. Phosphorus in urban sewage sludges as assessed by isotopic exchange. *Soil Sci. Soc. Am J.* 60:179–182.

Furoc, R. E., M. A. Dizon, R. A. Morris, and E. P. Marqueses. 1985. Effects of flooding regimes and planting dates to N accumulation of three Sesbania species and consequently to transplanted rice. In: Paper Presented at the 16th Annual Scientific Convention of the Crop Science Society of Philippines, Muñoz. Nueva Ecija, Philippines: Central Luzon State University, May 1985, pp. 8–10.

Gaume, A., F. Machler, C. Leon, L. Narro, and E. Frossard. 2001. Low P tolerance by maize (*Zea mays* L.) genotypes: Significance of root growth and organic acids and acid phosphatase root exudation. *Plant Soil* 228:253–264.

George, T., R. J. Buresh, J. K. Ladha, and G. Punzalan. 1998. Recycling in situ of legume-fixed and soil nitrogen in tropical lowland rice. *Agron. J.* 90:429–437.

Gerloff, G. C. and W. H. Gabelman. 1983. Genetic basis of inorganic plant nutrition. In: *Inorganic Plant Nutrition*, Encyclopaedia and Plant Physiology New Series, Vol. 15B, eds. A. Lauchli and R. L. Bieleski, pp. 453–480. New York: Springer Verlag.

Ghai, K., D. I. N. Rao, and I. Batra. 1985. Comparative study of the potential of *Sesbania* for green manuring. *Trop. Agric.* 62:52–56.

Gibson, G., B. J. Radford, and R. G. H. Nielsen. 1992. Fallow management, soil water, plant-available soil nitrogen and grain sorghum production in south west Queensland. *Aust. J. Exp. Agric.* 32:473–482.

Gil, J. L. and W. H. Fick. 2001. Soil nitrogen mineralization in mixtures of eastern gamagrass with alfalfa and red clover. *Agron. J.* 93:902–910.

Gill, H. S. and O. P. Meelu. 1982. Studies on the substitution of inorganic fertilizers with organic manure and their effect on soil fertility in rice–wheat rotation. *Fert. Res.* 3:303–314.

Gilmour, J. T., A. Mauromoustakos, P. M. Gale, and R. J. Norman. 1998. Kinetics of crop residue decomposition: Variability among crops and years. *Soil Sci. Soc. Am. J.* 62:750–755.

Graham, R. D. 1984. Breeding for nutritional characteristics in cereals. In: *Advances in Plant Nutrition*, Vol. 1, eds. P. B. Tinker and A. Lauchi, pp. 57–102. New York: Praeger Publisher.

Grant, C. A., D. N. Flaten, D. J. Tomasiewicz, and S. C. Sheppard. 2001. The importance of early season phosphorus nutrition. *Can. J. Plant Sci.* 81:211–224.

Grant, C. A., G. A. Peterson, and C. A. Campbell. 2002. Nutrient considerations for diversified cropping systems in the Northern Great Plains. *Agron. J.* 94:186–198.

Grant, R. F., N. G. Juma, and W. B. McGill. 1993. Simulation of carbon and nitrogen transformations in soil: Mineralization. *Soil Biol. Biochem.* 25:1317–1329.

Green, C. J. and A. M. Blackmer. 1995. Residue decomposition effects on nitrogen availability to corn following corn or soybean. *Soil Sci. Soc. Am. J.* 59:1065–1070.

Guldan, S. J., C. A. Martin, W. C. Lindemann, J. Cueto-Wong, and R. L. Steiner. 1997. Yield and green manure benefits of interseeded legumes in a high desert environment. *Agron. J.* 89:757–762.

Habte, M. 1995. Soil acidity as a constraint to the application of vesicular-arbuscular mycorrhizal technology. In: *Mycorrhiza*, eds. A. Varma and B. Hock, pp. 593–605. New York: Springer-Verlag.

Hall, J. L. 2002. Cellular mechanisms for heavy metal detoxification and tolerance. *J. Exp. Bot.* 53:1–11.

Halvorson, A. D. 1989. Multiple year response of winter wheat to a single application of phosphorus fertilizer. *Soil Sci. Soc. Am. J.* 53:1862–1868.

Halvorson, A. D. and J. L. Havlin. 1992. No-till winter wheat response to phosphorus placement and rate. *Soil Sci. Soc. Am. J.* 56:1635–1639.

Halvorson, A. D., J. L. Havlin, and C. A. Reule. 2002. Fertilizer management options for no-till dryland winter wheat. *Better Crops Plant Food* 86:4–7.

Hargrove, W. L. 1986. Winter legumes as nitrogen sources for no-tillage grain sorghum. *Agron. J.* 78:70–74.

Hatch, D. J., S. C. Jarvis, and S. E. Reynolds. 1991. An assessment of the contribution of net mineralization to N cycling in grass swards using a field incubation method. *Plant Soil* 138:23–32.

Hatfield, J. L., T. J. Sauer, and J. H. Prueger. 2001. Managing soils to achieve greater water use efficiency: A review. *Agron. J.* 93:271–280.

Havlin, J. L., J. D. Beaton, S. L. Tisdale, and W. L. Nelson. 1999. *Soil Fertility and Fertilizers*, 6th edn. Upper Saddle River, NJ: Prentice Hall.

Hayman, D. S. and M. Tavares. 1985. Plant growth responses to vesicular-arbuscular mycorrhiza. XV. Influence of soil pH on the symbiotic efficiency of different endophytes. *New Phytol.* 100:367–377.

Haynes, R. J. 1982. Effects of liming on phosphate availability in acid soils. A critical review. *Plant Soil* 68:289–308.

Haynes, R. J. 1984. Lime and phosphate in the soil-plant system. *Adv. Agron.* 37:249–315.

Haynes, R. J. and R. S. Swift. 1988. Effects of lime and phosphate additions on changes in enzyme-activities, microbial biomass and levels of extractable nitrogen, sulfur and phosphorus in an acid soil. *Biol. Fert. Soils* 6:153–158.

He, X. T., T. J. Logan, and S. J. Traina. 1995. Physical and chemical characteristics of selected U.S. municipal solid waste composts. *J. Environ. Qual.* 24:543–552.

He, Z., X. Yang, V. C. Baligar, and D. V. Calvert. 2003. Microbiological and biochemical indexing systems for assessing quality of acid soils. *Adv. Agron.* 78:89–138.

Hedley, M. J., J. J. Mortvedt, N. S. Bolan, and J. K. Syers. 1995. Phosphorus fertility management in agroecosystems. In: *Phosphorus in the Global Environment: Transfer, Cycles, and Management*, ed. H. Tiessen, pp. 59–92. New York: John Wiley & Sons.

Henriksen, T. M. and T. A. Breland. 1999. Evaluation of criteria for describing crop residue degradability in a model of carbon and nitrogen turnover in soil. *Soil Biol. Biochem.* 31:1135–1149.

Hergert, G. W., W. L. Pan, D. R. Huggins, G. H. Gwve, and T. R. Peck. 1997. Adequacy of current fertilizer recommendations for site-specific management. In: *The State of Site-Specific Management for Agriculture*, eds. F. J. pierce and E. J. Sadler, pp. 981–1020. Madison, WI: ASA, CSSA, and SSSA.

Hesterman, O. B., T. S. Griffin, P. T. Williams, G. H. Harris, and D. R. Christenson. 1992. Forage legume-small grain intercrops: Nitrogen production and response of subsequent corn. *J. Prod. Agric.* 5:340–348.

Hinsinger, P. 2001. Bioavailability of soil inorganic P in the rhizosphere as affected by root-induced chemical changes: A review. *Plant Soil* 237:173–195.

Holdings, A. J. and J. F. Lowe. 1971. Some effects of acidity and heavy metals on the rhizobium-leguminous plant association. *Plant Soil* 35:153–166.

Huang, J. W. and J. Chen. 2003. Role of pH in phytoremediation of contaminated soils. In: *Handbook of Soil Acidity*, ed. Z. Rengel, pp. 449–472. New York: Marcel Dekker.

Hue, N. V. 1988. Residual effects of sewage-sludge application on plant and soil profile chemical composition. *Commun. Soil Sci. Plant Anal.* 19:1633–1643.

Hue, N. V. 1995. Sewage sludge. In: *Soil Amendments and Environmental Quality*, ed. J. E. Rechcigl, pp. 199–247. Boca Raton, FL: CRC Press.

Hue, N. V., H. Ikawa, and J. A. Silva. 1994. Increasing plant available phosphorus in an Ultisol with a yard-waste compost. *Commun. Soil Sci. Plant Anal.* 25:3291–3303.

Hulugalle, N. R. and L. Lal. 1986. Root growth of maize in a compacted gravelly tropical Alfisol as affected by rotation with a woody perennial. *Field Crops Res.* 13:33–44.

Ibekwe, M. A., J. S. Angle, R. L. Chaney, and P. Van Berkum. 1995. Sewage sludge and heavy metal effects on nodulation and nitrogen fixation of legumes. *Environ. Sci. Technol.* 24:1199–1204.

IPCC. 2001. Climate change: The scientific basis. In: *Contribution of Workshop Group I to Third Assessment Report of the Intergovernmental Panel on Climate Change*. Cambridge, U.K.: Cambridge University Press.

Islam, M. M., F. Iyamuremye, and R. P. Dick. 1998. Effect of organic residue amendment on mineralization of nitrogen in flooded rice soils under laboratory conditions. *Commun. Soil Sci. Plant Anal.* 29:971–981.

Izaurralde, R. C., R. L. Lemke, T. W. Goddard, B. McConkey, and Z. Zhang. 2004. Nitrous oxide emissions from agriculture top sequences in Alberta and Saskatchewan. *Soil Sci. Soc. Am. J.* 68:1285–1294.

Jackson, L. E. 2000. Fates and losses of nitrogen from a nitrogen-15 labeled cover crop in an intensively managed vegetable system. *Soil Sci. Soc. Am. J.* 64:1404–1412.

Jannink, J. L., M. Liebman, and L. C. Merrick. 1996. Biomass production and nitrogen accumulation in pea, oat, and vetch green manure mixtures. *Agron. J.* 88:231–240.

Janzen, H. H. and R. M. N. Kucey. 1988. Carbon, nitrogen, and sulfur mineralization of crop residues as influenced by crop species and nutrient regime. *Plant Soil* 106:35–41.

Jensen, E. S. 1992. The release and fate of nitrogen from catch-crop materials decomposing under field conditions. *J. Soil Sci.* 43:335–345.

Jewell, W. J. 1994. Engineering and cost considerations: Sludge management and land application. In: *Sewage Sludge: Land Utilization and the Environment*, eds. C. E. Clapp, W. E. Larson, and R. H. Dowdy, pp. 41–54. Madison, WI: ASA, CSSA, and SSSA.

Johansen, A., I. Jakobsen, and E. S. Jensen. 1993. External hyphae of vesicular-arbuscular mycorrhizal fungi associated with *Trifolium subterraneum* L.: Hyphal transport of ^{32}P and ^{15}N. *New Phytol.* 124:61–68.

Joner, E. J. and I. Jakobsen. 1994. Contribution by two arbuscular mycorrhizal fungi to P uptake by cucumber (*Cucumis sativus* L.) from ^{32}P-labeled organic matter during mineralization in soil. *Plant Soil* 163:203–209.

Kafkafi, U., B. Bar-Yosef, R. Rosenberg, and G. Sposito. 1988. Phosphorus adsorption by kaolinite and montmorillonite: II. Organic anion competition. *Soil Sci. Soc. Am. J.* 52:1585–1589.

Kalidurai, M. and S. Kannaiyan. 1989. Effect of *Sesbania rostrata* on N uptake and yield of upland rice. *J. Agron. Crop Sci.* 163:284–288.

Kashket, E. 1985. The proton motive force in bacteria: A critical assessment of methods. *Annu. Rev. Microbiol.* 39:219–242.

Kelling, K. A., A. E. Peterson, L. M. Walsh, J. A. Ryan, and D. R. Keeney. 1977. A field study of the agricultural use of sewage sludge: I. Effect on crop yield and uptake of N and P. *J. Environ. Qual.* 6:339–345.

Kiraly, Z. 1976. Plant disease resistance as influenced by biochemical effects on nutrients in fertilizers. In: *Fertilizer Use and Plant Health*, ed. International Potash Institute, pp. 33–46. Bern, Switzerland: International Potash Institute.

Kirchmann, H. 1988. Shoot and root growth and nitrogen uptake by six green manure legumes. *Acta. Agric. Scand.* 38:25–31.

Kirkham, M. B. 1974. Disposal of sludge on land: Effect on soils, plants and ground water. *Compost Sci.* 15:6–10.

Kirkham, M. B. 1982. Agricultural use of phosphorus in sewage sludge. *Adv. Agron.* 35:129–163.

Koelsch, R. K. 2007. Estimating manure nutrient excretion. https://articles.extension.org/pages/10997/estimating-manure-nutrient-excretion (accessed October 23, 2015).

Krishnan, G., D. L. Holshauser, and S. J. Nissen. 1998. Weed control in soybean (*Glycine max*) with green manure crops. *Weed Technol.* 12:97–102.

Kumar, V., S. Singh, S. Singh, and H. D. Yadav. 1999. Performance of sugarcane grown under sodic soil and water conditions. *Agric. Water Manage.* 41:1–9.

Kuo, S. and E. J. Jellum. 2002. Influence of winter cover crop and residue management on soil nitrogen availability and corn. *Agron. J.* 94:501–508.

Kuo, S. and U. M. Sainju. 1998. Nitrogen mineralization and availability of mixed leguminous and non-leguminous cover crop residues in soil. *Biol. Fert. Soils* 22:310–317.

Ladd, J. N., M. Amato, R. B. Jackson, and J. H. Butler. 1983. Utilization by wheat crops of nitrogen from legume residues decomposing in soils in the field. *Soil Biol. Biochem.* 15:231–238.

Ladha, J. K., D. K. Kundu, M. G. Angelo-Van Coppenolle, M. B. Peoples, V. R. Carangal, and P. J. Dart. 1996. Legume productivity and soil nitrogen dynamics in lowland rice-based cropping systems. *Soil Sci. Soc. Am. J.* 60:183–191.

Ladha, J. K., R. P. Pareek, and M. Becker. 1992. Stem-nodulating legume-*Rhizobium* symbiosis and its agronomic use in lowland rice. *Adv. Soil Sci.* 20:147–192.

Ladha, J. K., I. Watanabe, and S. Saono. 1988. Nitrogen fixation by leguminous green manure and practices for its enhancement in tropical lowland rice. In: *Sustainable Agriculture: Green Manure in Rice Farming*, ed. IRRI, pp. 165–183. Los Baños, Philippines: The International Rice Research Institute.

Larney, F. and H. H. Janzen. 1996. Restoration of productivity to a desurfaced soil with livestock manure, crop residue, and fertilizer amendments. *Agron. J.* 88:921–927.

Lathwell, D. J. 1990. Legume green manures. TropSoils Bulletin Number 90-01, Soil Management Collaborative Research Support Program. Raleigh, NC: North Carolina State University.

Leikam, D. F. 1992. Summary of P fertilizer use effects on soil test phosphorus. In: *Proceedings North Central Extension Industry Soil Fertility Workshop*, Vol. 8, ed. R. Lamond, pp. 108–117. Manhattan, KS: Potash and Phosphate Institute.

Linderman, R. G. 1992. Vesicular-arbuscular mycorrhizae and soil microbial interactions. In: *Mycorrhizae in Sustainable Agriculture*, eds. G. J. Bethlenfalvay and R. G. Linderman, pp. 45–70. Madison, WI: ASA, CSSSA, and SSSA.

Lindsay, W. L. 1979. *Chemical Equilibria in Soils*. New York: John Willey & Sons.

Logan, T. J. and B. J. Harrison. 1995. Physical characteristics of alkaline stabilized sewage sludge and their effects on soil physical properties. *J. Environ. Qual.* 24:153–164.

Loneragan, J. F. 1978. The physiology of plant tolerance to low phosphorus availability. In: *Crop Tolerance to Suboptimal Land Conditions*, ed. G. A. Jung, pp. 329–343. Madison, WI: ASA.

Lowenberg-Deboer, J. and H. F. Reetz Jr. 2002. Phosphorus and potassium economics for the 21st Century. *Better Crops Plant Food* 86:4–8.

Lupwayi, N. Z. and C. Kennedy. 2007. Grain legumes in northern grain plains: Impacts on selected biological soil processes. *Agron. J.* 99:1700–1709.

Lynch, D. L. and L. J. Cotnoir. 1956. The influence of clay minerals on the breakdown of certain organic substrates. *Soil Biol. Biochem.* 20:367–370.

MacRae, R. J. and G. R. Mehuys. 1985. The effect of green manuring on the physical properties of temperate-area soils. *Adv. Soil Sci.* 3:71–94.

Mahler, R. L. and R. E. McDole. 1985. The influence of lime and phosphorus on crop production in northern Idaho. *Commun. Soil Sci. Plant Anal.* 16:485–499.

Mallarino, A. P. and J. S. Schepers. 2005. Role of precision agriculture in phosphorus management practices. In: *Phosphorus: Agriculture and Environment*, ed. L. K. Al-Amoodi, pp. 881–908. Madison, WI: ASA, CSSA, and SSSA.

Mansell, G. P., R. M. Pringle, D. C. Edmeades, and P. W. Shannon. 1984. Effects of lime on pasture production on soils in the North Island of New Zealand. III. Interaction of lime with phosphorus. *N. Z. J. Agric. Res.* 27:363–369.

Marschner, H. and B. Dell. 1994. Nutrient uptake in mycorrhizal symbiosis. *Plant Soil* 159:89–102.

Martens, D. A. and J. Frankenberger. 1992. Modification of infiltration rates in an organic amended irrigated soil. *Agron. J.* 84:707–717.

Mary, B. and S. Recous. 1994. Measurement of nitrogen mineralization and immobilization fluxes in soil as a means of predicting net mineralization. *Eur. J. Agron.* 3:291–300.

McCracken, D. V., M. S. Smith, J. H. Grove, C. T. Ackown, and R. L. Blevins. 1994. Nitrate leaching as influenced by cover cropping and nitrogen source. *Soil Sci. Soc. Am. J.* 58:1476–1483.

Meelu, O. P., R. E. Furoc, M. A. Dizon, R. A. Morris, and F. P. Marqueses. 1985. Evaluation of different green manures on rice yield and soil fertility. In: Paper Presented at the *16th Annual Scientific Convention of the Crop Science Society of Philippines*. Muñoz, Nueva Ecija, Philippines: Central Luzon State University, May 1985, pp. 8–10.

Menezes, V. G., I. Anghinoni, P. R. F. Silva, V. R. M. C. C. Petry, D. S. Grohs, T. F. S. Freitas, and L. A. L. Valente. 2013. Project 10-Management strategies to increase productivity and sustainability of irrigated rice field growth in the state of Rio Grande do Sul, Brazil: Developments and new challenges. Porto Alegre, Brazil: IRGA/Estação Experimental do Arroz.

Mian, M. I. A., M. A. Rouf, M. A. Rashid, M. A. Mazid, and M. Eaqubal. 1985. Residual effects of triple super phosphate (TSP) and farmyard manure (FYM) under renewed application of urea on the yield of crops and some chemical properties of soil. Bangladesh. *J. Agric. Sci.* 10:99–109.

Midwest Plan Service. 2000. Manure characteristics. MWPS-18, Sect. 1. Ames, IA: Iowa State University.

Mills, H. A. and J. B. Jones Jr. 1996. *Plant Analysis Handbook II*. Athens, GA: Micromacro Publishing, Inc.

Moreira, A. and N. K. Fageria. 2010. Liming influence on soil chemical properties, nutritional status and yield of alfalfa grown in acid soil. *Revista brasileira de ciência do solo* 34:1231–1239.

Morel, C. and C. Plenchette. 1994. Is the isotopically exchangeable phosphate of a loamy soil the plant available P? *Plant Soil* 158:287–297.

Morris, R. A., R. F. Furoc, and M. A. Dizon. 1986. Rice response to a short duration green manure. II. Nitrogen recovery and utilization. *Agron. J.* 78:413–416.

Morris, R. A., R. E. Furoc, N. K. Rajbhandari, E. P. Marqueses, and M. A. Dizon. 1989. Rice response to water tolerant green manures. *Agron. J.* 81:803–809.

Mortvedt, J. J. 2000. Bioavailability of micronutrients. In: *Handbook of Soil Science*, ed. M. E. Sumner, pp. 71–87. Boca Raton, FL: CRC Press.

Mortvedt, J. J., L. S. Murphy, and R. H. Follett. 1999. *Fertilizer Technology and Application*. Willoughby, OH: Meister Publishing Company.

Muchovej, R. M. C., A. C. Borges, R. F. Novias, and J. T. L. Thiebaut. 1986. Effect of liming levels and Ca-Mg ratios on yield, nitrogen content and nodulation of soybean grown in acid Cerrado. *J. Soil Sci.* 37:235–240.

Mulder, E. G., T. A. Lie, and A. Houwers. 1977. The importance of legumes under temperate conditions. In: *Treatise on Dinitrogen Fixation. IV. Agronomy and Ecology*, eds. R. W. F. Hardy and H. A. Gibson, pp. 221–242. New York: John Wiley & Sons.

Muller, M. M., V. Sundman, and A. Merilainen. 1988. Effect of chemical composition on the release of nitrogen from agricultural plant materials decomposing in soil under field conditions. *Biol. Fert. Soils* 6:78–83.

Mullins, G., B. Joern, and P. Moore. 2005. By-product phosphorus: Sources, characteristics, and management. In: *Phosphorus: Agriculture and Environment*, ed. L. K. Al-Amoodi, pp. 829–879. Madison, WI: ASA, CSSA, and SSSA.

Naidu, R., J. K. Syers, R. W. Tillman, and J. H. Kirkman. 1990. Effect of liming and added phosphate on charge characteristics of acid soils. *J. Soil Sci.* 41:157–164.

Nielsen, D. C., P. W. Unger, and P. R. Miller. 2005. Efficient water use in dryland cropping systems in the Great Plains. *Agron. J.* 97:364–372.

Nielsen, D. C., M. F. Vigil, and R. L. Anderson. 2002. Cropping system influence on planting water content and yield of winter wheat. *Agron. J.* 94:962–967.

Norwood, C. A. 1999. Water use and yield of dryland row crops as affected by tillage. *Agron. J.* 91:108–115.

Nurlaeny, N., H. Marschner, and E. George. 1996. Effects of liming and mycorrhizal colonization on soil phosphate depletion and phosphate uptake by maize (*Zea mays* L.) and soybean (*Glycine max* L.) grown in two tropical acid soils. *Plant Soil* 181:275–285.

Pang, X. P. and J. Letey. 2000. Organic farming: Challenge of timing nitrogen availability to crop nitrogen requirements. *Soil Sci. Soc. Am. J.* 64:247–253.

Parr, J. F. and S. B. Hornick. 1992. Utilization of municipal wastes. In: *Soil Microbial Ecology: Applications in Agricultural and Environmental Management*, ed. F. B. Meeting, pp. 549–559. New York: Marcel Dekker.

Pikul, J. L. Jr., J. K. Aase, and V. L. Cochran. 1997. Lentil green manure as fallow replacement in the semiarid northern Great Plains. *Agron. J.* 89:867–874.

Radcliffe, D. E., E. W. Tollnec, W. L. Hargrove, R. L. Clark, and M. H. Golabi. 1988. Effect of tillage practices on infiltration and soil strength of a typic Hapludult soil after 10 years. *Soil Sci. Soc. Am. J.* 52:798–804.

Raij, B. V. 1991. *Soil Fertility and Fertilization*. São Paulo, Brazil: Editora Agronomica Ceres.

Rannells, N. N. and M. J. Wagger. 1996. Nitrogen release from grass and legume cover crop monocultures and biocultures. *Agron. J.* 88:777–782.

Raun, W. R. and G. V. Johnson. 1999. Improving nitrogen use efficiency for cereal production. *Agron. J.* 91:357–363.

Reeve, W. G., R. P. Tiwari, M. J. Dilworth, and A. R. Glenn. 1993. Calcium affects the growth and survival of *Rhizobium meliloti*. *Soil Biol. Biochem.* 25:581–586.

Reinertsen, S. A., F. F. Elliott, V. L. Cochran, and G. S. Campbell. 1984. Role of available carbon and nitrogen in determining the rate of wheat straw decomposition. *Soil Biol. Biochem.* 16:459–464.

Ribeiro, M. R., J. O. Siqueira, N. Curi, and J. R. R. Simão. 2001. Fractioning and bioavailability of heavy metals in contaminated soil incubated with organic and inorganic materials. *Revista brasileira de ciência do solo* 25:495–507.

Rice, W. A., P. E. Olsen, L. D. Bailey, V. O. Biederbeck, and A. E. Slinkard. 1993. The use of annual legume green-manure crops as a substitute for summer fallow in the Peace River region. *Can. J. Soil Sci.* 73:243–252.

Roberts, T. L. and J. W. B. Stewart. 1988. Residual phosphorus application effective for cereal crops. *Better Crops Plant Food* 72:18–19.

Robertson, G. P., E. A. Paul, and R. R. Harwood. 2000. Greenhouse gases in intensive agriculture: Contributions of individual gases to the radiative forcing of the atmosphere. *Science* 289:1922–1925.

Robertson, G. P. and J. M. Tiedje. 1987. Nitrous oxide sources in aerobic soils: Nitrification, denitrification and other biological processes. *Soil Biol. Biochem.* 19:187–193.

Roe, N. E., P. J. Stoeffela, and D. Greetz. 1997. Composts from various municipal solid waste feed-stocks affect vegetable crops. II. Growth, yields, and fruit quality. *J. Am. Soc. Hortic. Sci.* 123:433–437.

Ruffo, M. L. and G. A. Bollero. 2003. Residue decomposition and prediction of carbon and nitrogen release rates based on biochemical fractions using principal component regression. *Agron. J.* 95:1034–1040.

Rynk, R., M. V. Kamp, G. B. Wilson, M. E. Singley, T. L. Rihard, J. J. Kolega, F. R. Gouin, L. Laliberty Jr., D. Kay, and D. W. Murphy. 1992. *On Farm Composting Handbook*. Ithaca, NY: NRAES.

Sainju, U. M. and B. P. Singh. 1997. Winter cover crops for sustainable agricultural systems: Influence on soil properties, water quality and crop yields. *HortScience* 2:21–28.

Sainju, U. M., B. P. Singh, and W. Whitehead. 1998. Cover crop root distribution and its effects on soil nitrogen cycling. *Agron. J.* 90:511–518.

Sainju, U., B. P. Singh, and S. Yaffa. 2002. Soil organic matter and tomato yield following tillage, cover cropping, and nitrogen fertilization. *Agron. J.* 94:594–602.

Sanchez, P. A. and J. G. Salinas. 1981. Low-input technology for managing Oxisols and Ultisols in tropical America. *Adv. Agron.* 34:280–406.

Sattar, M. A. and A. C. Gaur. 1989. Effect of VA-mycorrhiza and phosphate dissolving microorganism on the yield and phosphorus uptake of wheat in Bangladesh. *Bangladesh J. Agric. Res.* 14:233–239.

Sauve, S., C. E. Martinez, M. McBridge, and W. Hendershot. 2000. Adsorption of free lead (Pb^{2+}) by pedogenic oxides, ferrihydrite, and leaf compost. *Soil Sci. Soc. Am. J.* 64:595–599.

Schlegel, A. J. 1992. Effect of composted manure on soil chemical properties and nitrogen use by grain sorghum. *J. Prod. Agric.* 5:153–157.

Sharpley, A. N., T. C. Daniel, J. T. Sims, and D. H. Pote. 1996. Determining environmentally sound soil phosphorus levels. *J. Soil Water Conserv.* 51:160–166.

Shipley, P. R., J. J. Meisinger, and A. M. Decker. 1992. Conserving residual corn fertilizer nitrogen with winter cover crops. *Agron. J.* 84:869–876.

Sikora, L. J. and R. A. K. Szmidt. 2001. Nitrogen sources, mineralization rates, and nitrogen nutrition benefits to plants from composts. In: *Compost Utilization in Horticultural Cropping Systems*, eds. P. J. Stoffella and B. A. Hahn, pp. 287–305. Boca Raton, FL: CRC Press.

Sims, J. T. 1998. Phosphorus soil testing: Innovations for water quality protection. *Commun. Soil Sci. Plant Anal.* 29:1471–1489.

Sims, J. T., A. C. Edwards, O. F. Schoumans, and R. R. Simard. 2000. Integrating soil phosphorus testing into environmental based agricultural management practices. *J. Environ. Qual.* 29:60–72.

Singh, Y., C. S. Khind, and B. Singh. 1991. Efficient management of leguminous green manures in wetland rice. *Adv. Agron.* 45:135–189.

Singh, Y., B. Singh, and C. S. Khind. 1992. Nutrient transformation in soils amended with green manures. *Adv. Agron.* 20:237–309.

Smith, M. S., W. W. Frye, and J. J. Varco. 1987. Legume winter cover crops. *Adv. Soil Sci.* 7:95–139.

Smyth, T. J. and M. S. Cravo. 1992. Aluminum and calcium constraints to continuous crop production in a Brazilian Oxisol. *Agron. J.* 84:843–850.

Sogbedji, J. M., H. M. van Es, and K. L. Agbeko. 2006 Cover cropping and nutrient management strategies for maize production in Western Africa. *Agron. J.* 98:883–889.

Soil Science Society of America. 2008. *Glossary of Soil Science Terms*. Madison, WI: Soil Science Society of America.

Sommers, J. E., D. W. Nelson, and K. J. Yost. 1976. Variable nature of chemical composition of sewage sludge. *J. Environ. Qual.* 5:303–306.
Sommers, L. E., D. W. Nelson, J. E. Yahner, and J. V. Mannering. 1972. Chemical composition of sewage sludge from selected Indiana cities. *Ind. Acad. Sci.* 82:424–432.
Sorensen, L. H. 1975. The influence of clay on the rate of decay of amino acid metabolites synthesized in soil during the decomposition of cellulose. *Z. Pflanzenernaehr. Bodenkd.* 7:171–177.
Springett, J. A. and J. K. Syers. 1984. Effect of pH and calcium content of soil on earthworm cast production in the laboratory. *Soil Biol. Biochem.* 16:185–189.
Steen, I. 1996. Putting the concept of environmentally balanced fertilizer recommendations into practice on the farm. *Fert. Res.* 43:235–240.
Stevens, R. J., R. J. Laughlin, and J. P. Malone. 1998. Soil pH affects process reducing nitrate to nitrous oxide and DI-nitrogen. *Soil Biol. Biochem.* 30:1119–1126.
Stone, L. R. and A. J. Schlegel. 2006. Yield-water supply relationships of grain sorghum and winter wheat. *Agron. J.* 98:1359–1366.
Stratton, M. L., A. V. Barker, and J. E. Rechcigl. 1995. Compost. In: *Soil Amendments and Environmental Quality*, ed. J. E. Rechcigl, pp. 249–310. Boca Raton, FL: CRC Press.
Stratton, M. L. and J. E. Rechcigl. 1998. Agronomic benefits of agricultural, municipal, and industrial by products and their co-utilization: An overview. In: *Beneficial Co-Utilization of Agricultural, Municipal and Industrial by Products*, eds. S. Brown, J. S. Angle, and L. Jacobs, pp. 9–34. Dordrecht, the Netherlands: Kluwer Academic Publishers.
Tanaka, D. L., J. M. Krupinsky, M. A. Liebig, S. D. Merrill, R. E. Ries, J. R. Hendrickson, H. A. Johnson, and J. D. Hanson. 2002. Dynamic cropping systems: An adaptable approach to crop production in the Great Plains. *Agron. J.* 94:957–961.
Tang, C., J. J. Drevon, B. Jaillard, G. Souche, and P. Hinsinger. 2004. Proton release of two genotypes of bean (*Phaseolus vulgaris* L.) as affected by N nutrition and P deficiency. *Plant Soil* 26:59–68.
Tarafdar, J. C. and H. Marschner. 1995. Dual inoculation with *Aspergillus fumigatus* and *Glomus mosseae* enhances biomass production and nutrient uptake in wheat supplied with organic phosphorus as Na-phytate. *Plant Soil* 173:97–102.
Teasdale, J. R. and C. L. Mohler. 1993. Light transmittance, soil temperature, and soil moisture under residue of hairy vetch and rye. *Agron. J.* 85:673–680.
Thomann, R. V. and J. A. Mueller. 1987. *Principles of Surface Water Quality Modeling and Control*. New York: Harper Collins Publications.
Thomas, G. A., G. Gibson, R. G. H. Nielsen, W. D. Martin, and B. J. Radford. 1995. Effects of tillage, stubble, gypsum, and nitrogen fertilizer on cereal cropping on a red brown earth in southwest Queensland. *Aust. J. Exp. Agric.* 35:997–1008.
Thonnissen, C., D. J. Midmore, J. K. Ladha, R. J. Holmer, and U. Schmidhalter. 2000a. Tomato crop response to short-duration legume green manures in tropical vegetable systems. *Agron. J.* 92:245–253.
Thonnissen, C., D. J. Midmore, J. K. Ladha, D. C. Olk, and U. Schmidhalter. 2000b. Legume decomposition and nitrogen release when applied as green manures to tropical vegetable production systems. *Agron. J.* 92:253–260.
Tisdale, S. L., W. L. Nelson, and J. D. Beaton. 1985. *Soil Fertility and Fertilizers*, 4th edn. New York: Macmillan Publishing Company.
Tissen, H., E. Cuevas, and P. Chacon. 1994. The role of organic matter in sustaining soil fertility. *Nature* 371:783–785.
Toma, M., M. E. Sumner, G. Weeks, and M. Saigusa. 1999. Long-term effects of gypsum on crop yield and subsoil chemical properties. *Soil Sci. Soc. Am.* 39:891–895.
Tortoso, A. C. and G. L. Hutchinson. 1990. Contributions of autotrophic and heterotrophic nitrifiers to soil NO and N_2O emissions. *Appl. Environ. Microbiol.* 56:1799–1805.
Treder, W. and G. Cieslinski. 2005. Effect of silicon application on cadmium uptake and distribution in strawberry plants grown on contaminated soils. *J. Plant Nutr.* 28:917–929.
Unger, P. W. 2001. Alternative and opportunity dryland crops and related soil conditions in the southern Great Plains. *Agron. J.* 93:216–226.
Varco, J. J., W. W. Frye, M. S. Smith, and C. T. Mackown. 1989. Tillage effects on nitrogen recovery by corn from a nitrogen-15 labeled legume cover crop. *Soil Sci. Soc. Am. J.* 53:822–827.
Verbene, E. L., J. Hassink, P. Willingen, J. J. Goot, and J. A. Van Veen. 1990. Modeling organic matter dynamics in different soils. *Neth. J. Agric. Sci.* 38:221–238.
Viator, R. P., J. L. Kovar, and W. B. Hallmark. 2002. Gypsum and compost effects on sugarcane root growth, yield, and plant nutrients. *Agron. J.* 94:1332–1336.

Vigil, M. F. and D. E. Kissel. 1995. Rate of nitrogen mineralization from incorporated crop residues as influenced by temperature. *Soil Sci. Soc. Am. J.* 59:1636–1644.

Vogtmann, H. K. F. and T. Turk. 1993. Quality, physical characteristics, nutrient content, heavy metals and organic chemical in biogenic waste compost. *Compost Sci. Util.* 1:69–87.

Vyn, T. J., K. J. Janovicek, M. H. Miller, and E. G. Beauchamp. 1999. Spring soil nitrate accumulation and corn response to preceding small-grain N fertilization and cover crops. *Agron. J.* 91:17–24.

Walker, J. M. 1975. Sewage sludge: Management aspects for land application. *Compost Sci.* 16:12–21.

Wani, S. P., W. B. McGill, K. L. Haugen-Kozyra, J. A. Robertson, and J. J. Thurston. 1994. Improved soil quality and barley yields with faba beans, manure, forages and crop rotation on a Gray Luvisol. *Can. J. Soil Sci.* 74:75–84.

Watanabe, I. 1984. Use of green manure in northeast Asia. In: *Organic Matter and Rice*, ed. International Rice Research Institute, pp. 229–234. Los Baños, Philippines: IRRI.

Withers, P. J. A., D. M. Nash, and C. A. M. Laboski. 2005. Environmental management of phosphorus fertilizers. In: *Phosphorus: Agriculture and the Environment*, ed. L. K. Al-Amoodi, pp. 781–827. Madison, WI: ASA, CSSA, and SSSA.

Yakovchenko, V., L. J. Sikora, and D. D. Kaufman. 1996. A biological based indicator of soil quality. *Biol. Fert. Soils* 21:245–251.

Yang, J. L., S. J. Zheng, Y. F. He, C. X. Tang, and G. D. Zhou. 2005. Genotypic differences among plant species in response to aluminum stress. *J. Plant Nutr.* 28:949–961.

Yang, Z. M., M. Sivaguru, W. J. Horst, H. Matsumoto, and Z. M. Yang. 2000. Aluminum tolerance is achieved by exudation of citric acid from roots of soybean (*Glycine max*). *Physiol. Plant.* 110:72–77.

Vigil, M. F., and D. E. Kissel. 1995. Rate of nitrogen mineralized from incorporated crop residues as influenced by temperature. *Soil Sci. Soc. Am. J.* 59:1636–1644.

W[illegible]. [illegible] physical characteristics, nutrient content, [illegible] and [illegible] compost. *Compost Sci.* [illegible]

W[illegible], [illegible] Miller, and P. [illegible] [illegible] *Agron. J.* [illegible]:17–23.

W[illegible]. [illegible] Management aspects for [illegible] application. [illegible] 164.

Wong, S. [illegible] [illegible] and [illegible] Thornton. 1994. Improved [illegible]

W[illegible]. [illegible] International Rice Research [illegible] Los Baños, Philippines [illegible]

[illegible]

8 Phosphorus Nutrition Research in Crop Plants

Basic Principles and Methodology

8.1 INTRODUCTION

Mineral nutrition, including phosphorus (P), is the process of nutrient application to soil, movement of nutrients to plant roots, absorption by roots, translocation, and utilization in plants. Numerous soil, plant, microbial, and environmental factors affect nutrient availability to crop plants. These factors vary from region to region and sometimes even from field to field in the same region. Research is needed for each crop species under different agroecological regions and socioeconomic conditions of the growers. Experiments need to be conducted under field and controlled conditions to generate basic and applied information. Research into optimizing essential nutrients that are related to soil fertility and plant nutrition is dynamic, complex, and challenging for agricultural scientists.

Research is the foundation of technological improvements. The standard of living in a country is in correlation with the use of technology. In agricultural science, soil fertility and plant nutrition are important to increasing crop yields. Borlaug and Dowswell (1994) reported that up to 50% of the increase in crop yields worldwide during the twentieth century was due to the use of chemical fertilizers. In the twenty-first century, the importance of chemical fertilizers in improving crop yields will continue and is expected to be still higher due to the necessity of increase in yields per unit land area rather than increasing land areas. Further, judicious use of chemical fertilizers along with other complementary methods such as use of organic manures, and exploiting genetic potential of crop species and cultivars in nutrient utilization will be extremely useful and necessary (Fageria, 2005, 2007).

Low yields of crops in some parts of the world are the result of actions and interactions of several factors, and there are no simple, easily implementable solutions. Improved understanding of biological, climatic, edaphic, and management factors through research and the development of production technologies that are in the appropriate socio-political-economic climate can assist in increasing crop production in such regions. In twenty-first-century research, one of the key guiding factors will be the need to develop low-cost technology components that do not require significant inputs and result in minimum degradation of natural resources. The objective of this chapter is to present basic principles and research methodologies for soil fertility and plant nutrition, especially in relations to P under controlled and field conditions. The information provided in this chapter may assist agricultural scientists, professors, and students of agronomy in planning and conducting analysis and interpretation of their research activities in the field of soil fertility and plant nutrition. In addition, the information may be useful for the research needs to meet the challenges of soil fertility and plant nutrition problems in the twenty-first century that will improve crop production and reduce environmental degradation (Fageria, 2005, 2007).

8.2 RESEARCH UNDER CONTROLLED CONDITIONS

In modern agriculture, use of essential plant nutrients in adequate amounts and proper balance is one of the key components in increasing crop yields (Figure 8.1). In developing crop production technologies, research under field and controlled conditions is necessary to generate

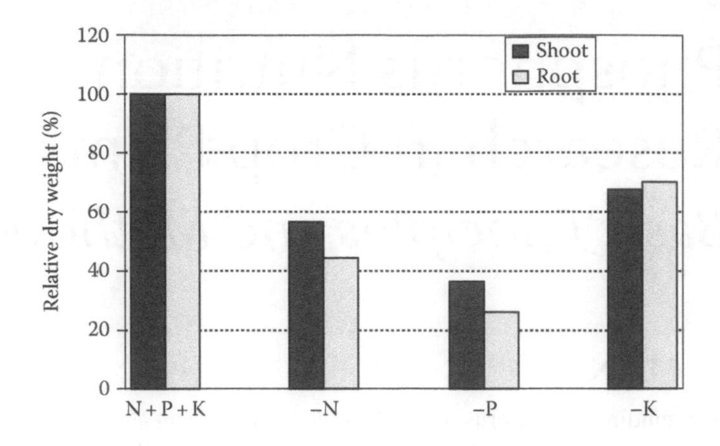

FIGURE 8.1 Relative dry weight of shoot and root of dry bean at different N, P, and K treatments. The N rate was 400 mg kg^{-1} soil (half applied at sowing + remaining half at 35 days after sowing), 200 mg P kg^{-1}, and 200 mg K kg^{-1}. (From Fageria, N.K. and Moreira, A., The role of mineral nutrition on root growth of crop plants, in: *Advances in Agronomy*, Sparks, D.L., ed., Vol. 110, Academic Press, Burlington, NJ, 2011, pp. 251–331.)

basic and applied information. In addition, agriculture research is dynamic and complex due to variation in climatic, soil, and plant factors and their interactions. This demands that basic research information can only be obtained under controlled conditions to avoid or reduce effects of environmental factors on treatments. The objective of this section is to discuss basic principles and methodologies of research in soil fertility and mineral nutrition (P as a case study) under controlled conditions.

The main objective of controlled conditions experiments are to understand basic principles. In the case of soil fertility and plant nutrition, such experiments are mainly conducted to understand nutrient movements and absorption and utilization processes in soil plant systems. Nutrient/elemental deficiency/toxicity symptoms and adequate and toxic concentrations in plant tissues are determined under controlled conditions. For example pot experiments with different types of soils can be used to measure the degree of response that may be anticipated at different soil test levels. Since such tests provide no measure of the cumulative effects of treatments on yield or soil buildup or depletion, they have limited value in determining rates of fertilizer that should be recommended for sustained productivity. Greenhouse pot studies, in which plants are used for estimating the relative availability of nutrients, can provide useful indexes of the relative availability of a standard fertilizer source in different soils and indices of different fertilizer sources. In this section, the methodology aspects of controlled conditions experiment will be discussed. This information will assist those who are involved in soil fertility and plant nutrition research to improve understanding of the principles of experimentation under controlled conditions.

8.2.1 Experimental Procedure and Techniques

Experimental plan and procedure are crucial to the success of research. In a research project, well-formulated hypothesis and clearly defined objectives are essential parts of the experimental techniques. Most of the controlled experiments are conducted in pots using soil, solution culture, or sand as a growth medium. Polyethylene pots are commonly used in controlled conditions experiments. Polyethylene pots are suitable for all soil fertility and plant nutrition experiments. A wide variety of sizes and colors are available in the markets. Most suppliers offer pots with drainage holes, as well as polyethylene saucers for bottom watering or collection of leachate in case of overwatering. Pots without drainage holes are available. In the opinion of the authors, holes are not necessary, if irrigation water is applied carefully in the soil fertility and plant nutrition experiments. Pots with holes may leach nutrients and it may affect the treatments adversely. Soils containing montmorillonite clays shrink upon drying, thus permitting loss of water and nutrients during routine watering. Pots without holes solve the problems; however, they require attention to avoid overwatering. Some crops are sensible to overwatering, such as common bean (*Phaseolus vulgaris* L.). Unglazed or glazed earthen pots are no longer used in greenhouse experiments because of excess weight, water loss, and possible absorption of salts. However, they may be satisfactory for some experiments, provided polyethylene liners are used.

Most of experimental designs under controlled conditions are completely randomized or randomized complete block design with three or four replications of each treatment. It is convenient to group pots of each replication together on a bench and treatments randomized within each replicate. Pots should be rotated twice a week to eliminate any environmental effects, especially solar radiation. In plant nutrition study under controlled conditions, it is better to have separate small experiments with various levels of a nutrient rather than a factorial experiment with two or three levels of each nutrient.

8.2.1.1 Soil Culture

Soil selected for greenhouse experiments for soil fertility and plant nutrition should be low in fertility in order to obtain yield response to applied nutrients. The effectiveness of fertilizers may be interpreted in terms of plant growth yield and uptake of nutrients. Pot experiment can be used as a reference as to whether a given site of field experiment is appropriate for a fertility trial. Soils which have yield response to a given nutrient in the field may fail to express a response under greenhouse conditions. Probable explanation for this is that high temperatures and moisture levels that usually prevail in the greenhouse result in more rapid decomposition of organic materials and greater release of nutrients than in the field (Allen et al., 1976).

Cultivated soils having a history of no fertilization with a given nutrient for several years are preferred over virgin or fertilized soils for obtaining yield responses. Usefulness of results on a particular soil type may be an important consideration if studies are being made of fertilizer problems related largely to that soil type. However, if broader principles are being studied, then physical and chemical properties of a nutrient-responsive soil may be more important than employing a particular soil type (Allen et al., 1976). Distance from greenhouse to soil collecting site should be considered in order to minimize the cost of transportation. Selected soil should be as free from soil-borne diseases, nematodes, insects, and weed seeds as possible. Fumigation by dry heat or steam tends to change a number of soil properties and may render a soil unfit for some soil fertility studies. Methyl bromide or some other organic fumigant may be more satisfactory. Soils for problem solving must be selected from the site where a specific problem is known to exist, regardless of their suitability by other standards.

After selecting a site, soil should be transported near the greenhouse. While collecting a soil, soil depth should be considered from which it is collected. Generally, it is recommended that soils for greenhouse studies should represent the arable soil depth, that is, 0–20 cm. However, in practice it is always higher soil depth is covered in collecting the soil for greenhouse experiments. After drying the soil, it should be screened through a 0.5–1 cm screen. A screen lower than 0.5 cm mesh can change the soil physical properties and may create problems during experimentation, especially compaction in the pots. Soil prepared in this method, if not used immediately for experimentation, can be stored in polyethylene bags or polyethylene drums.

8.2.1.1.1 Fertilizer Application and Sowing

After soil preparation, the next step in experimental technique is application of fertilizer treatments and sowing the crop seeds under investigation. Each pot should be filled with prepared soil and weight should be recorded on a portable balance. To determine the optimum level of a nutrient in greenhouse for a particular crop, a simple experiment with several rates of a single nutrient and nonlimiting levels of other nutrients should usually be supplied. The quantity of fertilizer applied should be equivalent to the quantity that is used under field conditions. However, experiments conducted by the author of this book at the National Rice and Bean Research Center, Goiânia, Goiás, Brazil, indicated that under greenhouse conditions the quantity of fertilizer required is much higher for rice (*Oryza glaberrima*), common bean, corn (*Zea mays*), cowpea (*Vigna unguiculata* L. Walp.), and wheat (*Triticum aestivum* L.) crops on an Oxisol soil (Fageria et al., 1982, 2010, 2012, 2014; Fageria, 1989; Fageria and Santos, 2014).

Greenhouse experiment in which upland and lowland rice shoot and grain yield under different levels of soil fertility and plant density were evaluated is presented in Table 8.1. Nutrient levels were zero (F0), normal level (F1) recommended under field conditions, and four (F4) and eight (F8) times the normal level. These levels were evaluated at four plant densities, that is, one, two, three, and four plants per pot of 6 kg soil. Grain yield and dry matter production were significantly affected by nutrient level and plant density. Adequate nutrient levels for upland and lowland rice cultivated in 6 kg pots were approximately eight times those recommended for field conditions (Fageria et al., 1982). Optimum plant density was obtained with two to three plants per pot (Fageria et al., 1982). Terman and Mortvedt (1978) and Mortvedt and Terman (1978) reported that nutrient rates adequate for small greenhouse pots (4.5 kg soil pot^{-1}) are higher than those equivalent to normal rates recommended for crops grown under field conditions. Length of growth period and other growth-limiting factors are equally important. Adequate plant density determined by authors for common bean are two plants per pot of 6 kg soil (Fageria, 1989), for wheat four plants per pot of 6 kg (Fageria, 1990), and for cowpea two plants per 6 kg soil (Fageria, 1991) until maturity. However, if plants are grown for a short duration, higher plant density can be used. But in authors' opinion this density should not be more than double in any case to get meaningful results related to soil fertility and plant nutrition problem.

TABLE 8.1
Shoot Dry Weight and Grain Yield of Upland and Lowland Rice under Different Soil Fertility and Plant Density under Greenhouse Conditions

	Shoot Dry Weight (g Pot^{-1})		Grain Yield (g Pot^{-1})	
Fertility Level[a]	Upland	Lowland	Upland	Lowland
F0	3.36	10.95	2.10	12.55
F1	16.82	20.04	12.24	23.27
F4	30.80	36.41	26.11	40.75
F8	44.42	42.45	34.73	51.87
F-test	**	**	**	**
D1	22.00	29.10	17.18	31.15
D2	24.59	25.84	19.24	31.19
D3	24.70	29.64	19.46	34.87
D4	24.11	25.29	19.30	31.23
F-test	*	*	*	*

Source: Fageria, N.K. et al., *Pesq. Agropec. Bras.*, 17, 1279, 1982.

[a] F0 = zero fertility level; F1 = 35 kg N ha^{-1}, 50 kg P_2O_5 ha^{-1}, 40 kg K_2O ha^{-1}; 5 kg Zn ha^{-1}; and 2 Mg ha^{-1} dolomitic lime for upland rice and 60 kg N ha^{-1}; 80 kg P_2O_5 ha^{-1}, 60 kg K_2O ha^{-1}; 5 kg Zn ha^{-1}; and 2 Mg ha^{-1} dolomitic lime for lowland rice. These levels correspond to recommend under field conditions in Brazil during 1980–1990 for upland and lowland rice. The F4 and F8 are four and eight times nutrient levels those recommended under field conditions. D1, D2, D3, and D4 correspond to one, two, three, and four plants per pot.

*,**Significant at the 5% and 1% levels, respectively.

Granular or powder fertilizers are usually used in greenhouse studies. Analytical-grade reagents should not be used for soil fertility experiment in the greenhouse when soil is used as a growth medium. Due to differences in quality, water solubility, and composition, reactions with soil will be different as compared to commercial fertilizers. Due to a small quantity, it is sometimes difficult to mix fertilizers with soil for each individual pot (Fageria, 2005). This problem can be solved by weighing fertilizers for each treatment for all replications. For example, if there are four replications of defined treatments, the quantity required for all the four replications can be weighed together and mixed with soil of four replications and then separated into different pots. When polyethylene bags are used as pot liners, weigh the soil into bags and fit a bag into each pot. If there are holes in the bottom of a pot, put a filter paper in the bottom before filling each pot. All the pots for an experiment should be filled simultaneously to reduce errors in dry soil weighing attributed to drying of the initial soil supply. After applying the fertilizer, treatment through mixing the soil is important. Mixing can be accomplished through a simple soil mixer or by hand-stirring or by rolling on a heavy polyethylene sheet.

8.2.1.1.2 Liming Acid Soils

Liming acid soils is one essential practice either in field or greenhouse experimentation. A satisfactory comparison among fertilizer treatments cannot be made if soil acidity is a limiting factor. To solve the issue, dolomitic lime should be applied and soil under investigation should be incubated for several weeks before the experiment is planted. The quantity of lime required for a soil should be determined through a lime calibration curve. Fageria (1984) developed a lime requirement curve for Oxisol soil of central Brazil and reported that after 30 days of incubation period, soil pH was nearly stable (raised from 5 to 7 in water) (Figure 8.2) and Ca + Mg levels increased from 0.5 cmol kg^{-1} to more than 5 cmol kg^{-1} (Figure 8.3). In Oxisol soils, about 4-week incubation period is sufficient to obtain soil pH and Ca + Mg level at desired levels. All the chemical reactions in the soils are dynamic processes and never reach equilibrium. After determining the lime requirement through calibration curves, lime may be added to individual pots or bulk lots of soil may be limed to the desired pH prior to establishing an experiment.

8.2.1.1.3 Care, Duration, and Observations

During the experimentation, watering, control of insects and diseases, and rotation of pots should be carefully conducted to minimize environmental effects among the replications. Watering should occur approximately at near field capacity of a soil. Weighing is the most widely used method in watering experimental pots. Polyethylene pots currently available are uniform in weight and tarring is not necessary. Deionized water is preferred in irrigating pots. However, in many developing

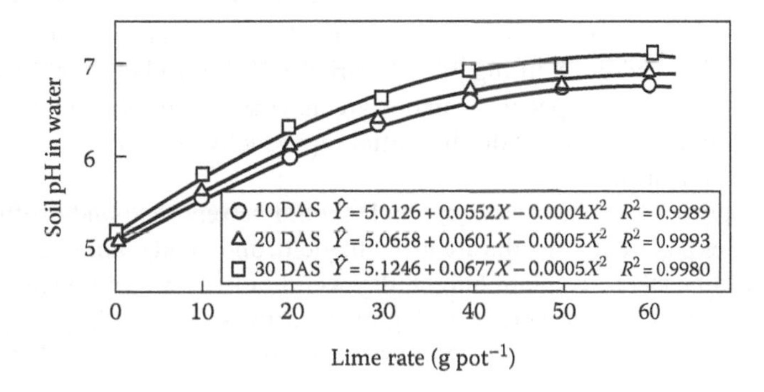

FIGURE 8.2 Relationship between lime rate and soil pH at three incubation period of an Oxisol. (From Fageria, N.K., *Pesq. Agropec. Bras.*, 19, 883, 1984.)

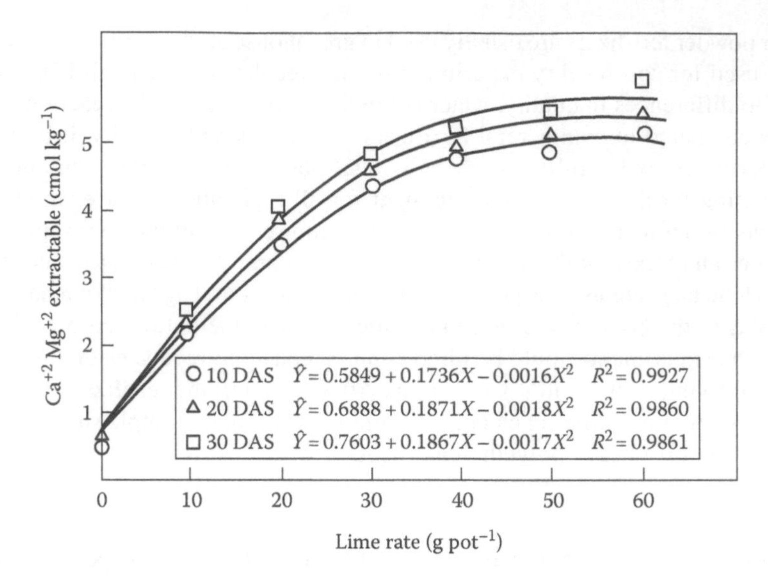

FIGURE 8.3 Relationship between lime rate and soil extractable Ca + Mg at three incubation period of an Oxisol. (From Fageria, N.K., *Pesq. Agropec. Bras*., 19, 883, 1984.)

countries, these facilities are either not available or too expensive. In such situation use of tap water is the only solution. If the laboratory is not equipped to determine soil moisture retention curves, an alternative simple method is to add varying amounts of water to the surface of soil in pots lined with polyethylene bags and subsequently let stand overnight. A satisfactory level of water is the amount which just wets the entire soil volume. This may be observed visually by carefully lifting the bag of soil from the pot. During initial watering, protect the soil surface from washing by a filter paper. Most of the water should be added along the rim of the pot. Depending on the climatic conditions, in the beginning of the experiment, generally watering twice a week is sufficient. On later stages, depending on growth of the crop plants, watering is necessary every day.

Dry matter yield or grain yield is a parameter for evaluating crop response to fertilization. Fageria et al. (1999) studied rice crop response to applied P in two greenhouse experiments conditions on an Oxisol soil. In one experiment harvesting was conducted in 60 days after sowing and in another until maturity. The authors concluded that when relative dry matter and grain yields are plotted against P levels, results were almost identical. Dry matter can be used as a criterion in greenhouse experiment to evaluate crop responses to fertilization provided that all other factors are at optimum level. Factors that can alter greenhouse experimental results, including dry matter and grain yields, are infestation of insects and diseases, and low temperature. If these adverse factors occur during reproductive and grain filling period, variation can exist between dry matter and grain yield variables. In soil fertility experiment under greenhouse conditions plants can be harvested during flowering for legume crops and during initial reproductive growth stage for cereal crops to evaluate soil fertility treatments.

In greenhouse experiments, observations which should be recorded and useful in the analysis and interpretation of experimental results are growth, yield, and yield components. In cereal crops, measurements should be taken for plant height, dry weight of straw, grain yield, panicle number, grain sterility, and thousand grain weight. In legumes crops, straw yield, grain yield, number of pods, seeds per pod, and hundred grain weights are generally recorded. Grain harvest index (GHI) is calculated based on straw and grain yield data to determine how photosynthetic products are translocated to economic parts (grain or seeds) of the plant. Greenhouse experimental data should be transformed to a per-plant basis (generally three to four plants per pot are used for annual crops) for statistical analysis. Following equations are used for statistical analysis of greenhouse experiments (Fageria, 2015):

$$\text{Straw yield}\left(\text{g plant}^{-1}\right) = \frac{\text{Straw yield in g pot}^{-1}}{\text{Number of plants per pot}}$$

$$\text{Grain yield}\left(\text{g plant}^{-1}\right) = \frac{\text{Grain yield in g pot}^{-1}}{\text{Number of plants per pot}}$$

$$\text{Number of panicles}\left(\text{plant}^{-1}\right) = \frac{\text{Number of panicles per pot}}{\text{Number of plants per pot}}$$

$$\text{Spikelet sterility}\left(\%\right) = \frac{\text{Unfilled spikelets}}{\text{Unfilled and filled spikelets}} \times 100$$

$$\text{Thousand grain weight}\left(\text{g}\right) = \frac{\text{Grain weight per pot in g}}{\text{Number of grain per pot}} \times 1000$$

$$\text{Number of pods}\left(\text{plant}^{-1}\right) = \frac{\text{Number of pods per pot}}{\text{Number of plants per pot}}$$

$$\text{Grain harvest index}\left(\text{GHI}\right) = \frac{\text{Grain yield in g}}{\text{Grain} + \text{straw yield in g}}$$

$$\text{Hundred grain weight}\left(\text{g}\right) = \frac{\text{Grain weight per pot in g}}{\text{Number of grain per pot}} \times 100$$

$$\text{Grain per pod} = \frac{\text{Number of grain per pot}}{\text{Pods per pot}}$$

8.2.1.1.4 Soil Test Calibration for Phosphorus

Soil test calibration is defined as the process of determining the crop nutrient requirement at different soil test values (Soil Science Society of America, 2008). Soil test calibration studies are used to make fertilizer recommendations at different soil test levels for a crop. Fageria (1990) conducted a greenhouse experiment to establish correlation between soil P test and productivity utilizing two upland rice cultivars. Dry matter yield of two rice cultivars is presented in Table 8.2. Relationship between soil P extracted by Mehlich 1 extracting solution and relative dry matter yield at different growth stages is presented in Figures 8.4 and 8.5. Soil test calibration results and their classification are presented in Table 8.3. Critical P levels in the soil varied between cultivars and with stage of the crop growth. These results can serve as a guide for greenhouse experiments to make P recommendations based on soil test results for rice.

8.2.1.2 Solution Culture

Growing plants in solution culture is an established technique in mineral nutrition studies. Some important discoveries in mineral nutrition have been made using solution culture techniques such as discovery of essentiality of nutrients. Solution culture studies are useful for developing deficiency symptoms of nutrients essential for plant growth. These symptoms can be used as a guide to

TABLE 8.2
Dry Matter Yield (g Plants^{-1}) of Two Upland Rice Cultivars at Different Growth Stages

	Plant Age in Days (Cultivar IAC 47)						
P Rate (mg kg^{-1})	**28**	**43**	**57**	**70**	**84**	**98**	**139**
0	0.20	0.54	1.27	1.34	6.02	8.56	11.59
25	0.43	1.77	5.50	9.73	17.14	37.11	53.74
50	0.49	2.66	6.49	12.52	18.95	32.78	57.95
75	0.50	2.53	6.52	13.23	20.05	34.59	60.54
100	0.53	2.86	8.90	11.97	23.40	35.67	69.05
125	0.62	3.00	9.26	14.53	20.53	38.60	66.71
150	0.67	3.20	8.71	15.26	21.70	40.54	67.41
175	0.53	2.53	7.87	16.59	21.34	40.65	67.90
200	0.55	3.74	8.59	15.30	23.70	41.12	72.66
	Plant Age in Days (Cultivar IR 43)						
P Rate (mg kg^{-1})	**28**	**43**	**57**	**70**	**84**	**98**	**149**
0	0.18	0.34	1.04	1.86	2.43	5.70	16.14
25	0.34	1.45	5.09	12.27	18.77	34.62	62.01
50	0.40	2.82	9.17	11.13	25.16	42.19	72.64
75	0.40	2.32	9.08	16.63	24.80	46.88	73.95
100	0.33	2.47	8.42	17.27	29.89	46.56	78.68
125	0.38	2.77	10.74	17.17	31.49	44.04	70.34
150	0.50	3.20	10.86	16.71	31.21	55.12	78.86
175	0.38	1.92	9.70	18.06	30.73	50.35	87.70
200	0.40	2.45	11.11	20.26	29.36	50.31	88.64

Source: Fageria, N.K., *Pesq. Agropec. Bras.*, 25, 579, 1999.

identify nutritional disorders in crop plants under field conditions. In addition to deficiency symptoms, toxicity symptoms of elements should be identified, and possible correction measures be adopted. For example, aluminum (Al) toxicity in acidic soils, iron (Fe) toxicity in flooded rice, and soil salinity problems in saline soils can be determined by solution culture.

Critical tissue concentrations for the diagnosis of nutrient deficiencies and toxicities are frequently established from water culture or sand culture experiments. Plant and environmental factors can affect measured critical concentrations (Bates, 1971). Critical tissue concentrations are a comparatively stable plant characteristic unlikely to be affected by temporal variation in the external supply of the element concerned (Asher and Edwards, 1983). However, consideration should be taken when the greenhouse results are extrapolated to field conditions. Under field conditions, variability in environmental factors may influence nutrient concentrations in plant tissues. Composition of nutrient solutions commonly used in hydroponic techniques is presented in Table 8.4. In preparing nutrient solutions, all the chemicals should be of reagent grade. Commonly used chemicals for preparing nutrient solutions are presented in Table 8.5. Iron is generally chelated with ethylene diamine di-[*o*-hydroxyphenylacetic] acid (EDDHA), hydroxyethylene diaminetriacetic acid (HEDTA) or ethylene diaminetetraacetic acid (EDTA).

According to Chaney and Bell (1987), the Fe-chelate of choice for solution culture of dicots is FeEDDHA and for gramineae the recommended chelate is FeHEDTA. Commercial FeEDDHA or FeHEDTA is not pure enough for use in controlled solution experiments, and it is not easily available especially in developing countries. Therefore, one should purchase pure chelates and make their own solutions. Solutions of these chelates are difficult to prepare. Therefore, methodology of their preparation is given here.

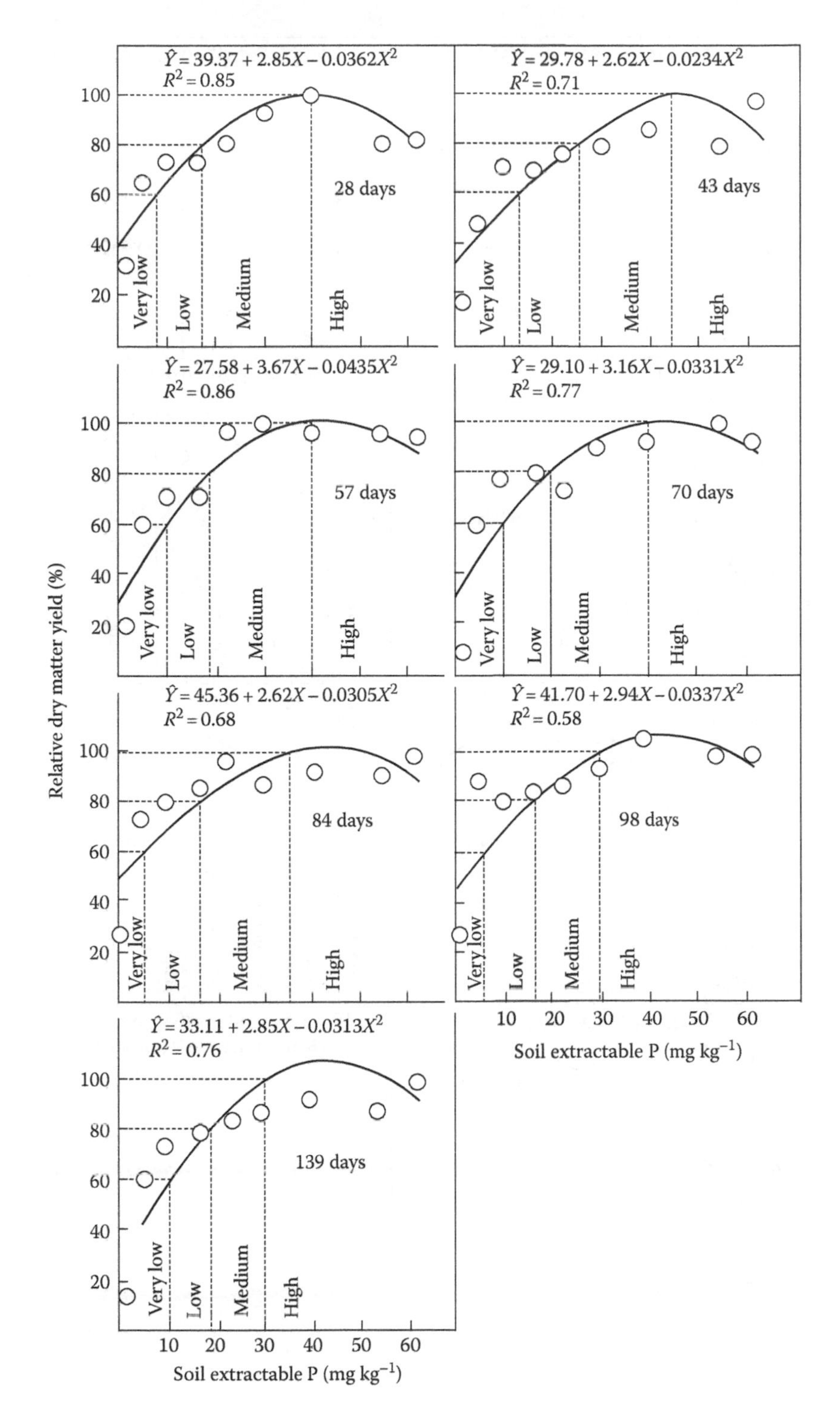

FIGURE 8.4 Relationship between soil extractable P by Mehlich 1 extracting solution and relative dry matter yield of shoot of upland rice cultivar IAC 47. (From Fageria, N.K., *Pesq. Agropec. Bras.*, 25, 579, 1999.)

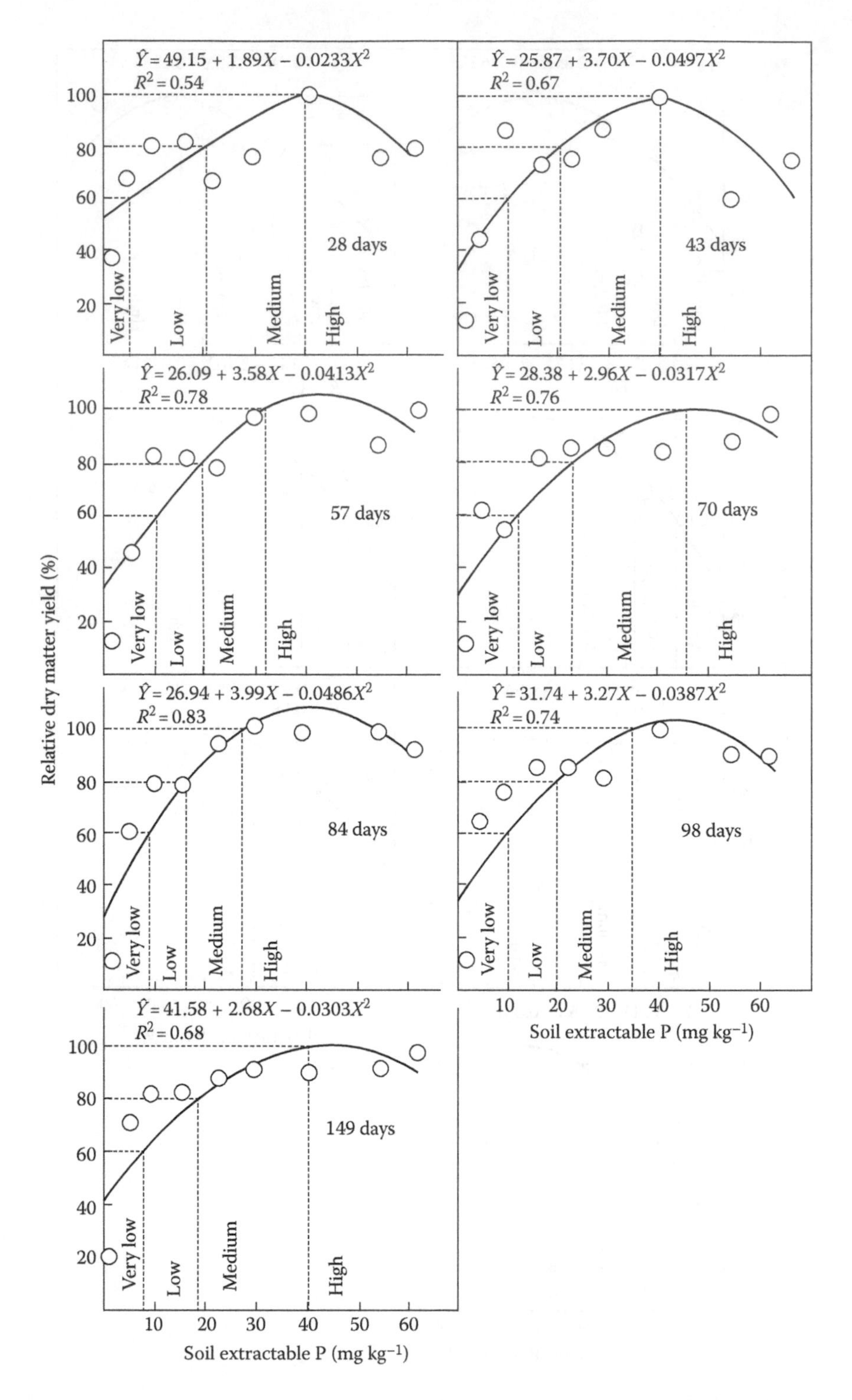

FIGURE 8.5 Relationship between soil extractable P by Mehlich 1 extracting solution and relative dry matter yield of shoot of upland rice cultivar IR 43. (From Fageria, N.K., *Pesq. Agropec. Bras.*, 25, 579, 1999.)

TABLE 8.3
Classification of P Soil Test in Relation to Relative Dry Matter Yield of Two Upland Rice Cultivars Grown on an Oxisol Soil

	Soil P Level (mg kg^{-1})			
Plant Age (Days)	Cultivar IAC 47	Cultivar IR 43	Average	Soil Test Classification
28	0–8	0–5	0–6	Very low
	8–18	5–20	6–18	Low
	18–40	20–40	18–37	Medium
	>40	>40	>37	High
43	0–13	0–10	0–11	Very low
	13–26	10–20	11–26	Low
	26–45	20–40	26–42	Medium
	>45	>40	>42	High
57	0–10	0–10	0–10	Very low
	10–18	10–20	10–20	Low
	20–40	20–33	20–38	Medium
	>40	>33	>38	High
70	0–10	0–12	0–11	Very low
	10–20	12–22	11–21	Low
	20–40	22–45	21–40	Medium
	>40	>45	>40	High
84	0–5	0–9	0–8	Very low
	5–17	9–16	8–18	Low
	17–35	16–28	18–34	Medium
	>35	>28	>34	High
98	0–5	0–10	0–8	Very low
	5–17	10–20	8–19	Low
	17–35	20–35	19–35	Medium
	>35	>35	>35	High
At harvest	0–10	8–8	0–9	Very low
	10–19	8–19	9–19	Low
	19–30	19–40	19–37	Medium
	>30	>40	>37	High

Source: Fageria, N.K., *Pesq. Agropec. Bras.*, 25, 579, 1999.
Cultivar IAC was harvested at 139 days and IR 43 149 days.

8.2.1.2.1 Preparation of Fe-EDDHA Solution

To prepare FeEDDHA solution for solution culture experiments, EDDHA salt, ferrous sulfate or ferric nitrate, and KOH or NaOH salts are required. All these salts should be of reagent grades.

1. Prepare 8 mM solution of EDDHA by dissolving 2.97 g L^{-1} in deionized water. The currently available EDDHA salt is about 97% pure with formula weight of 360.4 g mol^{-1}.
2. Prepare 8 mM solution of ferrous sulfate ($FeSO_4 \cdot 7H_2O$) or ferric nitrate ($Fe(NO_3)_2 \cdot 9H_2O$). In the case of $FeSO_4 \cdot 7H_2O$ (FW 278.02 g mol^{-1}, 99% pure), 2.25 g L^{-1} reagent is required. In the case of $Fe(NO_3)_2 \cdot 9H_2O$ (FW 404 g mol^{-1}, 100% pure), 3.23 g L^{-1} of salt is required.

TABLE 8.4
Nutrient Solution Composition Used in the Solution Culture

Nutrient	Hoagland and Arnon (1950)	Johnson et al. (1957)	Andrew et al. (1973)	Clark (1982)	Yoshida et al. (1976)
Macronutrients (mM)					
Nitrogen					
NO_3^-	14.0	14.0	2.00	7.26	2.21
NH_4^+	1.0	2.0	—	0.90	0.64
Phosphorus	1.0	2.0	0.07	0.07	0.29
Potassium	6.0	6.0	1.10	1.80	1.02
Calcium	4.0	4.0	1.00	2.60	1.00
Magnesium	2.0	1.0	0.50	0.60	1.64
Sulfur	2.0	1.0	1.50	0.50	—
Micronutrients (μM)					
Manganese	9.1	5.0	4.60	7.00	9.00
Zinc	0.8	2.0	0.80	2.00	0.15
Copper	0.3	0.5	0.30	0.50	0.16
Boron	46.3	25.0	46.30	19.00	18.50
Molybdenum	0.1	0.1	0.10	0.60	0.50
Iron	32.0	40.0	17.90	38.0	36.0
Chlorine	—	50	—	—	—

Source: Fageria, N.K., *J. Plant Nutr.*, 28, 1975, 2005.
In the earlier literature nutrient concentration is generally expressed in ppm. Values of ppm can be converted into mM with the help of equation: mM = ppp/atomic wt. of the element.

3. Prepare 32 mM solution of KOH or NaOH. In the case of KOH formula weight 66 g mol^{-1} and 100% pure, 2.11 g L^{-1} salt is required. In the case of NaOH with formula weight 40 g mol^{-1}, 1.28 g L^{-1} of reagent is required for a solution of 32 mM.
4. Add 32 mM KOH or NaOH solution to 8 mM EDDHA solution and stir for 30–60 min.
5. Add ferrous or ferric 8 mM solution to mixed EDDHA and KOH or NaOH solution.
6. Adjust pH 6–7 by KOH or HCl. Stir overnight at >50°C. Then filter with Whatman filter paper No. 42 to remove $Fe(OH)_3$.
7. Store in a brown bottle in a refrigerator.
8. Two liters of this solution contain 4 mM or 224 ppm of Fe^{3+}. If Fe^{2+} salt is used to prepare the Fe^{3+} chelates, one should recognize that the ligand will catalyze oxidation of the Fe^{2+} if oxygen is present.

8.2.1.2.2 Preparation of Fe-HEDTA Solution

To prepare 1 L of Fe-chelate solution, 30 mM solution of HEDTA and 30 mM of ferrous sulfate or ferric nitrate solutions are required.

1. Prepare 30 mM solution of HEDTA by dissolving 11.52 g L^{-1} in deionized water. Commercially available HEDTA has a formula weight of 380.24 g mol^{-1} with 99% purity.
2. To prepare ferrous sulfate solution, dissolve 8.4 g $FeSO_4 \cdot 7H_2O$ (FW 278.02 g mol^{-1} with 99% purity) in 1 L deionized water to obtain 30 mM solution. If ferric nitrate is used, it generally has 404 g mol^{-1} formula weight and requires 12.12 g L^{-1} of deionized water to obtain 30 mM solution.

TABLE 8.5
Commonly Used Reagent-Grade Chemicals for Nutrient Solution

Nutrients	Reagents	1 Molar Solution (g L^{-1})
Nitrogen	$(NH_4)_2SO_4$	132.0
Nitrogen	NH_4NO_3	80.0
Phosphorus	$NaH_2PO_4 \cdot 2H_2O$	156.0
Phosphorus	KH_2PO_4	136.0
Potassium	KCl	74.6
Potassium	K_2SO_4	174.2
Potassium	KNO_3	101.1
Calcium	$CaCl_2 \cdot 2H_2O$	147.0
Calcium	$Ca(NO_3)_2 \cdot 4H_2O$	236.2
Calcium	$CaSO_4 \cdot 2H_2O$	172.1
Magnesium	$MgSO_4 \cdot 7H_2O$	246.3
Manganese	$MnCl_2 \cdot 4H_2O$	197.9
Manganese	$MnSO_4 \cdot 4H_2O$	169.0
Zinc	$ZnSO_4 \cdot 7H_2O$	287.4
Zinc	$ZnCl_2$	136.3
Copper	$CuSO_4 \cdot 5H_2O$	249.5
Copper	$CuCl_2 \cdot 2H_2O$	170.4
Boron	H_3BO_3	61.8
Molybdenum	$(NH4)_6Mo_7O_{24} \cdot 4H_2O$	1235.6
Molybdenum	$Na_2MoO_4 \cdot 2H_2O$	241.9
Iron	$FeCl_3 \cdot 6H_2O$	270.0
Iron	$FeSO_4 \cdot 7H_2O$	278.0
Iron	$Fe(NO_3) \cdot 9H_2O$	404.0

Source: Fageria, N.K., *J. Plant Nutr.*, 28, 1975, 2005.
Iron is generally chelated with EDDHA, HEDTA, and EDTA.

3. Add ferrous sulfate or ferric nitrate solutions to HEDTA solution. Stir for 30–60 min at >50°C. Then filter the solution through Whatman filter paper No. 42. Store in a brown bottle and keep it in a refrigerator. This 2 L solution has 15 mM or 840 ppm Fe.

8.2.1.2.3 Preparation of FeEDTA Solution

To prepare FeEDTA solution, sodium salt of EDTA and ferrous sulfate along with NaOH solution are used. To prepare 1 L of FeEDTA solution, the following composition and procedure should be adopted.

1. Dissolve 33.2 g NaEDTA salt ($C_{10}H_{14}N_2Na_2O_8 \cdot 2H_2O$ FW 372.24 g mol^{-1}) in about 200 mL of deionized water.
2. Make an 89.2 mL solution of 1 N NaOH.
3. Dissolve 29.4 g of $FeSO_4 \cdot 7H_2O$ in about 100 mL of deionized water.
4. Mix NaEDTA and NaOH solution slowly and stir it. After mixing this solution with $FeSO_4 \cdot 7H_2O$ solution, leave this solution overnight in a dark ambient place. Next day complete the volume to 1 L. This solution will contain 106 mM or 5936 ppm Fe.

According to Novais et al. (1991), FeEDTA solution can be prepared by mixing 14.1 g of Na_2 EDTA and 10.3 g of $FeCl_3 \cdot 6H_2O$, diluted separately in about 300–400 mL of deionized water, and then mixed together to make a volume of 1 L. This solution contains 38 mM Fe. Care should be taken while chelating Fe is used in nutrient solution.

Chelators added at sufficiently high concentrations to hydroponic solution can induce micronutrient deficiencies by chelating Cu, Zn, Mn, and Fe, making the metals unavailable to plants (Fageria and Gheyi, 1999). This loss in metal bioavailability can be counteracted by increasing the amount of metal in solution. Hydroponic solutions that have a chelator concentration greater than the sum of the concentrations of Fe, Cu, Zn, Mn, Co, and Ni are called chelator buffered solutions (these solutions are usually designed to prevent Fe precipitation) (Chaney et al., 1989). Chelator-buffered solutions offer more precise control of micronutrients phytoavailability than do conventional hydroponic solutions because, in the buffered solutions, (1) there is a greater range in the total-metal concentration from deficiency to toxicity, and (2) the free metal ion activity decreases only a small amount during plant uptake of metals because metal activities are buffered (Bell et al., 1991).

8.2.1.2.4 pH of Solution Culture

In solution culture experiments, special attention should be given to control of pH and maintaining of stable supply of nutrients. In general, pH shifts in nutrient solutions are likely to be greater than in soils because of the lack of an exchange complex to adsorb or desorb hydrogen ions. Principal factor which leads to change in nutrient solution is unequal absorption of cations and anions. Nitrogen (N) is absorbed in large quantity and the form in which N is supplied exerts a great influence on pH change. Adsorption of more anions such as NO_3^- can liberate the OH ions in the rhizosphere by growing plant roots and pH is generally increased. If more cations such as NH_4^+ are absorbed, pH is decreased due to liberation of H^+ ions in the growth medium. Trelease and Trelease (1935) reported that in water culture experiments with wheat, varying the NO_3^-/NH_4^+ ratio causes the pH to increase, decrease, or remain about constant. Crop species are important in changing nutrient solution pH due to their different nutrient absorption capacities. Even cultivars within a crop species are important in modifying rhizosphere pH in solution culture. It has been consistently observed with non-nodulated jack beans (*Canavalia ensiformis*) that the pH of the nutrient solution decreases markedly with time even when all the N is supplied in the NO_3^- form (Asher and Edwards, 1983). Appropriate pH in nutrient solution is certainly different than in soil. Range reported in literature for conducting solution culture experiments varies from 5 to 7. For example Yoshida et al. (1976) reported that for rice growth in nutrient solution pH should be maintained around 5.0. Rice plants can grow well in nutrient solution even at pH 4.0 provided all essential nutrients are maintained at adequate levels (Fageria, 1989). Romero et al. (1981) used a modified Hoagland's solution with a pH of 6.0–6.4 to compare sand, soil, and solution culture systems as methods of assessing K fertilization effects on alfalfa (*Medicago sativa*) yield and K uptake. Bell et al. (1991) maintained the nutrient solution pH 5.9 with the addition of 1 M HCl in a study to determine copper (Cu^{2+}) activity required by maize (*Zea mays*) using chelator-buffered nutrient solutions. Ben-Asher et al. (1982) working with tomato (*Solanum lycopersicum*) plants in nutrient solution to study nutrient uptake reported that solution pH in their study was maintained between 5.5 and 6.5. Hohenberg and Munns (1984) studied the effect of pH on nodulation of cowpea in nutrient solution and concluded that pH 5.3 ± 0.3 was superior in modulation as compared to pH of 4.4 ± 0.2. Although there were differences among cultivars in relation to pH tolerance. McElhannon and Mills (1978) studied the influence of various percentages of NO_3^- to NH_4^+ and N concentration on N absorption, assimilation, growth, and yield of lima beans (*Phaseolus limensis* L.) in nutrient solution. They adjusted initial solution pH to 6.6, and change in pH due to nutrient absorption was not adjusted during experimentation so the influence of pH on N uptake at weekly intervals could be evaluated. Peaslee et al. (1981) studied absorption and accumulation of Zn by two corn cultivars in nutrient solution and they maintained the culture solution pH about 6.0. Teyker and Hobbs (1992) studied the effects of N forms on growth and root morphology of corn in hydroponic study. They maintained the pH 5.5 in the NH_4^+ treatment and pH 5.0 in NO_3^- treatments during the experimentation. Elliot and Lauchli (1985) compared rates of P absorption, P accumulation, and P utilization in inbred maize genotypes under varying conditions of P supply in a solution culture under controlled conditions. For the first 6 days following transplanting, pH adjusted to between 5.0 and 5.5 and thereafter to 4.5–5.0. Brown and Jones

(1976) studied the Fe uptake efficiency in corn, sorghum (*Sorghum bicolor* L. Moench), tomato, and soybeans (*Glycine max*) plants in nutrient solution. The initial pH of the solution was adjusted to 4.5 and during the experimentation pH was not adjusted. Itoh and Barber (1983) in a controlled climate chamber studied P uptake by six plant species in solution culture. Species tested were onion (*Allium cepa*), corn, wheat, lettuce (*Lactuca sativa*), carrot (*Daucus carota*), and Russian thistle (*Salsola kali* L.). Solution pH in this study was adjusted to 5.5 during the experimentation. Miyasaka and Grunes (1990) studied the effects of increased root temperature and Ca levels on shoot and root growth of winter wheat (*Triticum aestivum*) grown in solution culture, and the pH of the solutions was maintained to 6.0 by adding HCl or NaOH as required.

To maintain a desired pH, adjust the pH of the culture solution at the desired level every other day by either 1 N NaOH or 1 N HCl. Islam et al. (1980) using flowing nutrient solutions reported greatest growth of six species: ginger (*Zingiber officinale*), cassava (*Manihot esculenta* Crantz), tomato, common bean, wheat, and maize at about 5.5. Yields of ginger and tomato were not significantly changed at higher pH values, whereas yields of the other four species were depressed. Breeze et al. (1987) reported that the increase of dry weight of white clover (*Trifolium repens*) over a 20-day period, whether fixing atmospheric N or dependent on NO_3^- in solution, was not significantly lower at pH 4.0 than at pH 5.0, 6.0, or 7.0. Edmeades et al. (1991) reported that increasing the nutrient solution pH from 4.7 to 6.0 had no significant effect on the yield of the temperate grasses examined but significantly decreased yields of paspalum (*Paspalum dilatatum*) and veld grass (*Ehrharta longiflora* Sm.).

There is no consistent recommendation regarding pH of the nutrient solutions in the experimental studies. In the authors' opinion, with a pH value of around 5.5, all crops can be grown in nutrient solution satisfactorily. At pH values higher than 5.5, there are always the possibility of precipitation of many nutrients, especially micronutrients, thereby affecting their availability. Only in studies of Al toxicity, solution culture pH should not be more than 4.0 to avoid precipitation of Al. To maintain a desired pH, adjust the pH of the culture solution at the desired level every other day by either 1 N NaOH or 1 N HCl. However, pH control that recently has been reported to show promise are ion-exchange resins (Checkai et al., 1987) and the organic buffer, 2-(*N*-morpholino) ethanesulfonic acid (MES) (Wehr et al., 1986). Miyasaka et al. (1988) compared an unbuffered nutrient solution titrated once or twice a day, with solution buffered either by the organic buffer, MES, or by an ion-exchange system using a weakly acidic cation exchange resin loaded with Ca, Mg, K, and H. They reported that among the pH buffer methods studied, 1 mM MES method is recommended as a pH buffer for the hydroponic culture for winter wheat. Five millimolar MES proved the most consistent control of solution pH, but it inhibited Zn accumulation by wheat. Imsande and Ralston (1981) reported that 1–2 mM MES had excellent buffering capacity, did not interfere nutritionally with soybean growth, and did not impede N_2 fixation.

However, Rys and Phung (1985) reported that MES at 9 and 12 mM concentrations resulted in reduced growth of *Trifolium repens* L., which is dependent on symbiosis to provide N, and they suggested that the N-fixing ability of nodules was impaired by high levels of MES. Wehr et al. (1986) considered MES to be the most useful buffer in the pH range of 5.0–6.5 for growth of algae because of its biological inertness, high buffering capacity, and minimal metal-complexing ability. However, Clark (1982) stated that buffered solutions often induce more complications than original solutions. Research is needed to clearly understand the effects of buffering reagent interactions with nutrient uptake and plant growth.

8.2.1.2.5 Stable Supply of Nutrients

In soil culture mineralization of organic matter, weathering of primary minerals, biological activities, and chemical reaction provide replenishment of mineral nutrients. As roots elongate, they come in contacts with more soil volume and more nutrients are available for absorption. In soil environment, depletion of nutrients occurs over a longer time and soil provides a buffering capacity. In water culture, the composition of the nutrient solution is essentially unbuffered and large changes in nutrients concentrations occur within a relatively short period of time. This may affect nutrient absorption pattern of a crop and consequently growth and yield. Depletion of nutrients in nutrient solution by plants depends on original concentration, crop species or cultivar's rate of absorption, temperature of root rhizosphere, and volume of solution in which plants are grown. Measures can be adopted to minimize the rapid depletion of the nutrients in a nutrient culture experiment. These measures are use of high concentration, planting in large containers, maintaining adequate temperature, and frequent renewal of culture solution. Yoshida et al. (1976) suggested that change of culture solution should occur once a week at early growth stages and twice a week from active tillering until flowering. Asher and Edwards (1983) suggested a mathematical equation for the calculation of time interval between solution renewals.

$$T = DVC/100RU$$

where

T is the time interval in hours
D is the maximum acceptable depletion (%)
V is the volume of solution per pot or per plant (L)
C is the initial concentration of an ion in the solution (μM)
R is the root weight per pot or per plant (g fresh wt.)
U is the uptake rate per unit root weight (μmol g^{-1} fresh wt. h^{-1} at concentration C)

Disadvantage of this equation is that the maximum acceptable depletion may depend on yield reduction of a crop due to particular depletion value and rate of nutrient uptake. Information may not be available for a particular crop species or cultivars. Use of continuous flow technique is another way to maintain stable nutrient concentration in solution culture studies. A consideration of maintaining stable concentrations of nutrients, and renewing a large series of solutions, leads obviously to the suggestion that the solution may be controlled to flow continuously through the culture vessel, the inflow being of known composition and the outflow being discarded or reutilized. The continuous flow system has many advantages:

1. Concentration of the dilute solution can be kept constant at a given value.
2. Suitable for experiments where pH is to be maintained at a given value.
3. Keeps a constant flow rate of the nutrient solution at a given temperature and humidity.
4. Technique is well suited for comparative studies in the nutrition of plant species.
5. There is no risk of injury to plant material on renewal or replenishment of the solution during the experiment.
6. Ideally suited for studies of nutrient interactions, since the concentration of all the nutrients can be controlled throughout the period of experimentation.
7. An important technique in screening crop genotypes for nutrients use efficiency.

Basic principle of the continuous flow system is that the rate of nutrient uptake (U) is equal to the product of the flow rate (F) and the differences between the concentration of the solution entering the system (Co) and of the outgoing solution (Cs). A mathematical equation can be written as follows:

$$U = F\left(Co - Cs\right)$$

Rate of ion uptake expressed in μg h^{-1} g^{-1} root weight (may be fresh or dry) is calculated from the following formula (Hai and Laudelout, 1966; Fageria, 1974, 1976):

$$\text{Rate of ion uptake} = \left[\left(1 - \frac{Cs}{Co}\right) \times \frac{(F \times C)}{(\text{Root weight})}\right]$$

where

Cs is the concentration of the outgoing solution
Co is the concentration of the ingoing solution
C is the concentration of the stable ion in nutrient solution (ppm or μmol L^{-1})

TABLE 8.6
Composition of Flowing Culture Solution

Nutrient	Islam et al. (1980)	Fageria (1976)
Macronutrients (μM)		
Nitrogen (NO_3^-)	250	193
Nitrogen (NO_4^+)	—	227
Phosphorus	15	31
Potassium	250	250
Calcium	250	125
Magnesium	10	41
Sulfur	261	26.8
Micronutrients (μM)		
Manganese	0.25	2.0
Zinc	0.50	0.17
Copper	0.10	0.2
Boron	3.0	9.7
Iron	20.0	9.5
Molybdenum	0.02	0.004
Chlorine	5.0	46

Source: Fageria, N.K., *J. Plant Nutr.*, 28,1975, 2005.

In the continuous-flow culture technique, flow rate through the system is an important parameter to be considered in ion uptake studies. Technical and methodological aspects of this technique have been reported (Fageria, 1974; Asher and Edwards, 1983; Edwards and Asher, 1983; Callahan and Engel, 1986; Wild et al., 1987). Concentrations of nutrients used in flowing culture solution are lower than nonflowing solutions (Table 8.6).

8.2.1.2.6 Duration of the Solution Culture Experiment and Observations

Solution culture experiments are generally short-duration experiments. The objective of solution culture experiments is to gain understanding of fundamental factors which govern plant growth and nutrient uptake. In solution culture experiments, seeds are generally germinated in quartz sand or moistened paper towels. Paper towels are generally soaked in dilute solutions (1/10 of strength) or simply in 0.1 mM $CaCl_2$ solution during germination. Seeds should be surface sterilized for a few minutes in an appropriate solution to avoid fungus development during experimentation. For rice seed germination Yoshida et al. (1976) recommended that seeds should be sterilized for 4 min with 0.1% mercuric chloride solution or soaked in a formalin solution for 15 min. Then wash thoroughly with several changes of demineralized water. Grain legume seedlings should be transferred in nutrient solution treatments after 2–3 days of germination and cereals seedlings may be transferred 4–5 days after germination. This timetable is arbitrary and it may change according to needs or objective of an experiment. Duration of growing plants in treated solution depends on objective of the experiments, but 3–4 weeks' growth duration is sufficient for mineral nutrition studies. Some researchers use few hours or few days in nutrient uptake studies. In the authors' opinion, such short-duration experiments cannot produce any meaningful results of the subject under investigation. Plants should grow at least a few weeks to express their response to applied treatments. Some researchers argue that longer-duration experiments may change the nutrient uptake behavior of the plants and exact mechanisms are not understood. However, plants would not produce meaningful dry matter accumulation in a few hours or even days.

Change in pH and, depletion of nutrient concentration, plant shoots and roots (fresh and dry weight) and root lengths may be recorded in solution culture experiments. These variables can provide information for analysis and interpretation of experimental results. Plant variables and their

TABLE 8.7
Soil and Plant Parameters and Their Unit Commonly Used in Soil Fertility and Plant Nutrition Research

Parameter	Unit	Symbol or Preferred SI Unit
Land area	Square meter, hectare	m^2, ha
Grain or dry matter yield	Gram per square meter, kilogram per hectare, megagram per hectare, tons per hectare	g m^{-2}, kg ha^{-1}, Mg ha^{-1}, t ha^{-1}
Ion uptake	Mole per kilogram per second dry plant tissue, mole of charge per kilogram per second dry plant tissue	mol kg^{-1} S^{-1}, mol_c S^{-1}
Nutrient conc. in plant tissue	Millimole per kilogram, gram per kilogram, milligram per kilogram	mmol kg^{-1}, g kg^{-1}, mg kg^{-1}
Nutrient conc. in solution	Milligram per liter, centimol per liter	mg L^{-1}, cmol L^{-1}
Soil extractable ion (mass basis)	Centimole per kilogram, milligram per kilogram	cmol kg^{-1}, mg kg^{-1}
Fertilizer application rate to soil	Gram per square meter, kilogram per hectare	g m^{-2}, kg ha^{-1}
Lime or gypsum application rate to soil	Ton per hectare, megagram per hectare	t ha^{-1}, Mg ha^{-}
Soil bulk density	Megagram per cubic meter, gram per cubic centimeter	Mg m^{-3}, g cm^{-3}
Electrical conductivity	Siemen per meter, decisiemen per meter	S m^{-1}, dS m^{-1}
Cation exchange capacity	Cation exchange capacity per kilogram	cmol kg^{-1}
Absolute growth rate	Milligram per day	mg day^{-1}
Crop growth rate	Milligram per square meter per day	mg m^{-2} day^{-1}
Relative growth rate	Milligram per gram per day	mg g^{-1} day^{-1}
Leaf area index	Square meter per square meter	m^2 m^{-2}
Leaf area ratio	Square meter per kilogram	m^2 kg^{-1}
Leaf weight ratio	Gram per gram	g g^{-1}
Net assimilation rate	Gram per square meter per day	g m^{-2} day^{-1}
Specific leaf area	Square meter per kilogram	m^{-2} kg^{-1}

Source: Fageria, N.K., *J. Plant Nutr.*, 28, 1975, 2005.

unit commonly used in plant nutrition research are presented in Table 8.7. Conversion factors from nonmetric units to metric unit of soil fertility and plant nutrition research are presented in Table 8.8.

8.2.1.2.7 Screening Crop Genotypes for Phosphorus Efficiency under Controlled Conditions

Screening of crop species or cultivars for P use efficiency can be accomplished under controlled conditions using soil or solution culture as a growth medium. In solution culture, it is possible to manipulate nutrient concentration as desired, but it requires pH control and maintainance of stable concentrations of nutrients. Soil can be easily used as a growth medium as compared with solution culture. Nutrient levels can be manipulated by mixing upper and subsoil. Generally, immobile nutrients (P and K) are concentrated in the upper layer, and subsoils have lower concentrations of these nutrients. Levels of nutrients required to grow normal plants in a greenhouse are quite high, as compared with field conditions. Therefore, careful consideration should be given in selecting nutrient levels under greenhouse conditions.

In screening for a particular nutrient efficiency, a response curve should be developed before starting the mass screening of genotypes. In developing a response curve a large range of concentrations should be covered to obtain a quadratic response with more than one cultivar. From such a curve, two or three P levels can be selected for P-screening purposes such as low, medium, and high. Fageria and Baligar (1999) reported screening of 15 wheat genotypes for P-efficiency growth variable and P uptake responses to increasing soil P levels (Table 8.9). Plant growth and

TABLE 8.8
Conversion Factors for Non-SI Units to SI Units Most Commonly Used in Soil Fertility and Plant Nutrition Research

Conversion Unit	Multiplied By	Converted Unit
Acre to hectare	0.405	Hectare
Acre to square kilometer	4.05×10^{-3}	Square kilometer
Square mile to square kilometer	2.590	Square kilometer
Square foot to square meter	9.29×10^{-2}	Square meter
Square inch to square millimeter	645	Square millimeter
Pound per acre to kilogram per hectare	1.12	Kilogram per hectare
Pound per acre to tones per hectare	1.12×10^{-3}	Tones per hectare
Pound per acre to megagram per hectare	1.12×10^{-3}	Megagram per hectare
Millimhos per centimeter to siemen per meter	0.10	Siemen per meter
Millimhos per centimeter to decisiemen per meter	1	Decisiemen per meter
Milliequivalent per 100 grams to centimol per kilogram	1	Centimol per kilogram
Percent to gram per kilogram	10	Gram per kilogram
Parts per million to milligram per kilogram	1	Milligram per kilogram
Milliequivalent per liter to milligram per liter	Equivalent weight	Milligram per liter
Millimole to mol per cubic meter	1	mol m^{-3}
P_2O_5 to P	0.4365	P
K_2O to K	0.8301	K
CaO to Ca	0.7147	Ca
MgO to Mg	0.6032	Mg
SO_4 to S	0.3339	S
NH_3 to N	0.8225	N
NO_3^-N to N	0.23	N

Source: Fageria, N. K., *J. Plant Nutr.*, 28, 1975, 2005.
To convert converted unit into conversion unit divide by multiple factor.

TABLE 8.9
Significance of *F* Values and Orthogonal Contrasts Derived from Analysis of Variance for Variables Measured on 15 Wheat Genotypes with Three P Levels

Variable	P Levels	Genotypes	P × G	P Linear	P Quadratic	CV (%)
Tillers	**	**	**	**	NS	9
Plant height	**	**	**	**	**	8
Root length	**	**	NS	**	NS	13
Shoot dry wt.	**	**	**	**	**	22
Root dry wt.	**	*	NS	**	NS	19
Shoot/root ratio	**	**	**	**	**	18
P conc. in shoot	**	NS	NS	*8	**	18
P conc. in root	**	**	*	**	**	19
P uptake in shoot	**	**	**	**	**	28
P uptake in root	**	**	**	**	**	25
P use efficiency	NS	**	NS	—	—	16

Source: Fageria, N.K. and Baligar, V.C., *J. Plant Nutr.*, 22, 331, 1999.
*,**, and NS Significant at the 5% and 1% probability levels and not significant, respectively.

TABLE 8.10
Phosphorus Use Efficiency in Wheat Genotypes

Genotype	P Use Efficiency (mg mg^{-1})
Anahuac	152ab
BR 10	188a
BR 26	152ab
BR 33	138ab
PF 87949	185a
PF 87950	150ab
PF 89481	168ab
PF 89490	182a
CPAC 8909	183a
CPAC 8947	143ab
CPAC 89128	182a
CPAC 89194	125b
CPAC 89321	145ab
IPAR 8745	152ab
NL 459	162ab

Source: Fageria, N.K. and Baligar, V.C., *J. Plant Nutr.*, 22, 331, 1999.
Means within the same column followed by the same letter are not significantly different at the 5% probability level by Tukey's test. P use efficiency (mg mg^{-1}) = (Dry weight of shoot + root in mg)/(P uptake in shoot plus root in mg).

P uptake variables increased significantly with increasing levels of soil P. Soil used in the experiment was appropriate for screening purposes. One of the prerequisites of cultivars screening for mineral stress is that the growth medium should be deficient and/or toxic with respect to the nutrient under study.

Nutrient uptake efficiencies of crop genotypes can be evaluated in solution culture at various concentrations. Genotypes can be classified on the basis of dry matter produced per unit nutrient absorbed. Higher dry matter production per unit nutrient absorbed means more efficiency and vice versa. Variation among 15 wheat genotypes in P use efficiency is presented in Table 8.10. Wheat genotypes differ significantly in P use efficiency.

8.2.1.2.8 Presentation of Results and Discussion

Research in agriculture is a complex process and demands constant efforts due to changes in weather conditions, difference in soil properties, difference in adaptation of crop species, and different socioeconomic conditions of growers. Soil fertility and plant nutrition research, like any agricultural research, involves laboratory, greenhouse or growth chamber, and field experimentation. Laboratory and greenhouse experiments are generally short-duration experiments conducted to develop and understand basic principles. For example, pot experiment with different types of soils can express the degree of response that may be anticipated at different soil test levels and serve as excellent checks on ratings being used. Since such tests provide no measure of the cumulative effects of treatments on yield or soil buildup or depletion, they have limited value in determining rates of fertilizer that should be recommended for sustained productivity (Fageria, 2005).

Due to large variations in environmental factors, results of controlled condition experiments can hardly be extrapolated to field conditions and vice versa. However, these two types of experiments should serve as complementary components in developing a crop production technology. In controlled condition experiments soil and solution culture are generally used as medium of plant growth to test for treatment effects. Although the use of nutrient solutions allows precise control of experimental variables, it eliminates entirely the soil–root aspect (Fageria, 2005). The pattern of exploration and activity in root systems subjected to zonal salinization as well as the significance of ionic motilities in determining quantities of a given element absorbed from soil suggests the importance of testing hypothesis in a soil system, especially a system similar to that reported in the field (Fageria, 2005).

Many of the successful conditions and details involved for successful growth of plants in soil and solution cultures are not fully explained in literature. Information about conducting controlled condition experiments is taken for granted and left to the ingenuity and experience of investigators (Fageria, 2005). Many useful ideas and practices are derived only from experience. Some of the concerns, issues, and precautions required to conduct controlled condition experiments have been discussed. Hopefully these comments and suggestions are useful in conducting controlled condition experiments in the field of soil fertility and plant nutrition (Fageria, 2005).

8.3 RESEARCH UNDER FIELD CONDITIONS

In the agricultural science, soil fertility and plant nutrition had played an important role during the twentieth century in increasing crop yields. In the twenty-first century, importance of this field is still expected to be of greater importance due to limited natural resources (land and water), sustainable agriculture, and concern of environmental pollution. In this context, increasing crop yields will be associated with rational use of chemical fertilizers, increasing use of organic sources of nutrients, recycling of plant available nutrients, and exploiting genetic potential of crop species or cultivars. In the future, increasing crop yields will be a major challenge for agricultural scientists in general and soil scientists in particular. Conducting fertilizer field trials for adequate sources, methods, rates, and timing of application along with crop species or genotypes within species under different agroecological regions are necessary to generate data and their use for achieving maximum economic crop yields.

The *era* of field experimentation, which began in 1834 when J.B. Boussingault, a French Chemist established the first field experiments at Bechelbonn, Alsace (France), was placed on a modern scientific basis by Liebig's report of 1840 (Collis-George and Davey, 1960). First field experiments in the form used today was established by Lawes and Gilbert at Rothamsted in 1843. Since then, the field experiments have sought for and have confirmed the importance of the essential elements in influencing the production of field crops. However, an evidence for discovering the essentiality of nutrients has been found in laboratory experiments using nutrient solution and not from field experiments (Collis-George and Davey, 1960).

Application of field trial results led to a large increase in agricultural production around the world. Research in agriculture is a complex process and demands constant efforts and experimentation due to changes in weather conditions, soil heterogeneity, and release of new cultivars (Barley, 1964). These changes are often so significant that all management practices in use to produce optimum yields of crops need reevaluation and adjustments. For example, when a new cultivar is released, its nutritional requirements are different to those under cultivation due to difference in yield potential, diseases, and insect resistance and change in architecture. Field experiments are the basic need in modern agriculture to evaluate nutritional requirements under different agroecological regions. Transferring experimental results of one region to another is difficult due to differences in soil properties, climatic differences, and socioeconomic conditions of growers. These factors determine the technological development and its adaptation by the growers. In conducting field experimentation, basic principles should be followed to obtain meaningful conclusions. Principles discussed will assist agricultural scientists in planning and execution of research trials. Discussion is mainly concerned with the field of soil fertility and plant nutrition, but basic principles are also applicable to other disciplines of agricultural science. These principles are applicable everywhere with slight modification according to the circumstances of a particular situation. Most of the points discussed are the outcome of the authors' practical experience of over 45 years in the field of agriculture, in general soil fertility, and plant nutrition, in particular.

8.3.1 Basic Considerations for Conducting Field Experimentation

For conducting field experiments, there are basic considerations that should be followed to achieving desired results. These considerations are (1) hypothesis and objectives should be well defined; (2) selection of appropriate site(s); (3) adequate land preparation; (4) appropriate plot size, shape, and orientation; (5) selection of appropriate experimental design; (6) selection of adequate nutrient levels or treatments; (7) use of adequate seed rate, row and plant spacing; (8) timely conduct of required cultural practices such as control of insects, diseases, and weeds, topdressing N, and irrigation; (9) collection of yield and yield components data; (10) harvesting at physiological maturity, (11) repeating field experiments at least for 2 years; (12) use of adequate statistical methods for data analysis; and (13) publishing experimental results in scientific journals, book chapters, or technical bulletins.

8.3.1.1 Hypothesis and Objectives

A scientific experiment is needed to answer questions or to solve some issues. In agriculture, field experiments are designed on the basis of improving crop production. In the field of soil fertility, it may be necessary to determine optimum levels of nutrients for a crop in a particular soil. A hypothesis is a statement about the parameter (s) in one or more populations. A hypothesis typically arises in the form of speculation concerning an observed phenomenon. Examples of hypothesis are new cultivars require more nutrients, tropical Oxisols are deficient in P, and micronutrients are not yield-limiting factors in newly cleared forestlands. Statistical hypothesis is often referred to as the null hypothesis. Once a hypothesis is formed, the next step is to design a procedure for its verification. This is referred to as the experimental procedure. When experimental procedures are outlined, the objectives should be clearly defined. What answers are expected from the study under investigation? For example, optimum levels of N and P, optimum method of application. A review of pertinent literature is a valuable aid in evaluating hypothesis and achieving the objectives of an experiment. A review of literature can provide an idea of what type of experiments had been conducted in the past related to the issues, how it was performed, and what results were obtained. Review of literature will assist all from planning of experiments to interpretation of results.

8.3.1.2 Selection of Appropriate Site

Test plots are the foundation of most modern agricultural research programs. Sites of agronomic field research should ideally represent extensive areas of similar cropland. Where similar areas identified and mapped, research results can be extrapolated across a large region. Often sites have been selected on the basis of the availability of land and other socioeconomic considerations with limited attention to climate, pedologic, and geographic characteristics. When such selection occurs, research may be conducted on the sites having environmental characteristics of limited importance, and the information will only be useful for a small portion of the surrounding area (Ford and Nielsen, 1982). Therefore, a first step in site selection is to consider key soil, climatic, and socioeconomic factors. These factors have to be measured or determined at potential sites and evaluated for transferability of agrotechnology for recommendation domains.

Sites where field experiment is established should be uniform in physical and chemical properties. If the objective of the field trial is to calibrate for soil test P, then the soil or experimental site should be deficient in P. To verify soil fertility, at least 25–30 subsamples should be taken for a composite sample at the depth of 0–20 cm. If an experiment is conducted during the rainy season, supplementary irrigation during drought may be an appropriate strategy to avoid moisture influence on crop response to fertilization. One should try to avoid areas that have been previously used for experiments involving treatments that may have different effects on soil conditions. Treatments involving fertilizers, depth of plowing, different cultivars, and plant densities may have such an effect. In such areas, one or more uniform planting should precede an experimental planting.

Avoid areas in which alleys were left unplanted between plots in the previous crop. In such areas, one or more uniform planting should precede an experimental planting. Choose areas where the history of the site use is available. Such information may assist to identify possible causes of added heterogeneity from previous cropping so that appropriate remedies can be made. Sometimes, experiments should be conducted on whatever land is available and the scientist has no options.

8.3.1.3 Land Preparation

Experimental area should be well prepared to break hard pan or incorporate crop or weed residues for sowing a test crop under investigation. Soil preparation affects seed germination, plant growth, water infiltration, erosion, and weeds. Soils prepared adequately will favor higher yield and desired experimental results. Land should be plowed, harrowed, followed by leveling with spike-tooth or drag harrow. All these operations should be conducted at the ideal time and with adequate moisture level in the soil. If a large amount of crop or weed residues exist, incorporation should be performed in advance of planting. Generally, one plowing with a moldboard plow and harrowing once or twice, followed by leveling with spike-tooth or drag harrow, will be adequate.

8.3.1.4 Plot Size, Shape, and Orientation

Experimental plots refer to the unit areas on which treatments are tested. Plot size is the whole unit receiving the treatment. Shape of the plot refers to the ratio of its length to its width. Orientation of plots refers to the choice of direction along which the lengths of the plots will be placed. Orientation of plots naturally is not defined for square plots (Gomez, 1972). Plot size, shape, and orientation can affect the magnitude of experimental error in a field trial as well as soil preparation, planting, and cultural operations including harvesting. In general, experimental error decreases as plot size increases, but the reduction is not proportional. Gomez (1972) reported that there was a significant relationship between plot size and coefficient of variation for lowland rice yield. There was a sharp decrease in coefficient of variation up to 10 m^2 plot size. After this size, the decrease in coefficient of variation was small or almost constant. Coefficient of variation indicates the precision of experimental data. Under field conditions, a value of coefficient of variation $<10\%$ is considered low, 10%–20% medium, 20%–30% high, and more than 30% high (Gomes, 1984). According to Gomez (1972) in rice field experiments, the plot size commonly varies from 8 to 25 m^2. He suggested that whatever the size and shape of plots chosen, it is essential to make sure that an area is not smaller than 5 m^2, free from all types of competition and border effects, and available for harvesting and yield determination. Fertilizer trials require larger plots than cultivar trials. If a fertilizer trial is of longer duration, possibilities of contamination of adjacent plots of different fertilizer treatments exist. In these situations, larger plots with ample border area are advisable. MacDonald and Peck (1976) determined that horizontal movement of soil across plot boundary results from normal tillage operations. Horizontal movement of P and K was determined by soil tests at the completion of a 10-year fertility experiment, during which nine applications of differential rates of broadcast concentrated superphosphate and potassium chloride had been applied. Soil tests revealed that soil P was depleted to a distance of 2.7 m in the high concentrated superphosphate residual plots and moved into the low residual phosphate plots 1.8 m. Soil K was depleted to a distance of 2.7 m in the high KCl residual plots and moved into low KCl residual plots of 2.7 m. Normal tillage operations moved soil in both directions across plot boundaries. Effective length of plots was reduced from 15 to 10 m for P plots and 9 m for KCl plots in the direction of tillage operations after a period of 10 years. They concluded that dimension of long-term fertilizer plots should be adjusted according to the number of years the experiment will be continued.

Harvest area is influenced by the type of study and experimental material available. Laird (1968) indicated that fertility trial plots in Mexico should be 50 m^2. Younis and Tamimi (1970) reported that a study involving saline soil conditions required a plot size of 10 m^2. In general optimum harvest area estimated for relative yield comparison ranges from 5 to 10 m^2 (Gomez and Alicbusan, 1969; Kavitkar et al., 1971; Johnston and Miller, 1973). Harvest area for yield determination appears to be in excess of 30 m^2 (Singh et al., 1975). Harvest areas smaller than 30 m^2 overestimate the true yield, though the variance may be small; as the area decreased below 30 m^2, the percent overestimation increased. Areas as small as 1 m^2 overestimated yields by over 42% (Gomez and De Datta, 1971). According to Nelson and Rawlings (1983), formerly use of large plots in field experiments was recommended because the variance of large plots is small. But now, the recommendation is the use of small plots with a compensating increase in number of replications to use the available resources.

According to the authors' experience, a minimum plot size for fertilizer and liming experiments should be 6 × 5 m for an experimental duration of 3–5 years, if soil preparation operations are performed mechanically. In some developing countries where animal-drawn implements are used for soil preparation, a smaller plot size can be used. When soil heterogeneity is large, a large plot should be used in fertilizer trials to reduce the effects of soil properties on yield of evaluated crop.

Choice of field plot shape is not critical when the experimental area is uniform in soil physical and chemical properties. Plots may be square or rectangular in shape. If a gradient is present, plots should have their largest dimension in the direction of the greatest variation. When the fertility gradient of the experimental field is known, a rectangular plot with appropriate orientation will provide higher precision. Normally, one or two composite soil samples are taken prior to starting the experiment, to determine the soil fertility status of the experimental area. In this case, it is not possible to determine the gradient of the soil fertility. Under these situations square plots should be used to obtain a mean precision in experiments where border effects and square plots are desirable since they have a minimum perimeter for a given plot size (Gomez, 1972).

Cost of conducting field trials using expensive labeled ^{15}N fertilizer is minimized by limiting the application of ^{15}N to small areas (microplots) within larger experimental plots. Microplot and the surrounding macroplots are usually treated with the same fertilizer type and fertilizer management practices, with the exception that fertilizer applied to microplot is labeled with ^{15}N. Stumpe et al. (1989) using corn as a test crop suggested that microplots with 3.0 × 2.0 m border dimensions, containing four and eight plants/row, resulted in the desired number of plants (eight) unaffected by proximity to the border from a minimum microplot, and thus minimized the requirement for ^{15}N. Follett et al. (1991) recommended that the minimum microplot size for studies of fall applied ^{15}N on winter wheat grown in the great plains of the United States is 1.5 × 1.5 m.

8.3.1.5 Experimental Design

Experimental design refers to the method of arranging the experimental units (plots) and the method of assigning treatments to the units, usually with replications and randomization. The objective of replication in an experiment is to provide a measure of experimental error. One of the simplest means of increasing precision in an experiment is increasing number of replications. However, beyond a certain number of replications, the improvement in precision is too small to be worth the addition cost. The magnitude of the experimental error in an experiment is measured by the coefficient of variation. Generally, if an experimental site has uniform soil, using four replications and adequate sampling of character under study can provide a coefficient of variation about 8%–10%, a value considered quite low for field experiments. Tables are available to determine the number of replications necessary for a prescribed degree of precision under the magnitude of experimental error likely to be encountered in an experiment. Cochran and Cox (1957) reported an estimate the number of replications required for the stated precision of each main effect and interaction. A standard rule in experimental design is that each estimated experimental error term should have a minimum 10 degrees of freedom. For example, with four treatments in a randomized complete block design, four replications are not sufficient because they provide only nine degrees of freedom for error. Minimum of five replications are necessary to provide 12 degree of freedom for error. In soil fertility trials field experiments are conducted at several locations to obtain a mean value of change in soil and climatic conditions of the region. In such cases, if resources are limited, it is desirable to have more experiments at different locations rather than having more replications at few locations.

For the same degree of precision, more replications are required with heterogeneous soil than with homogeneous soil. Scientific value of an experiment increases with the number of treatments, whereas the cost of the experiment increases with the number of plots. Therefore, for a given cost, there is a trade-off between the number of treatments and the number of replications.

Randomization is essential to avoid anomalies in interpretation of results from systematic assignment of treatments to field plots. Randomization is a procedure for allocating treatments so that each experimental plot has identical chance of receiving any treatment. Randomization is fundamental to the validity of statistical significance tests and confidence interval estimation. In brief, the design of field plots should include a randomization scheme (Cady, 1991). Randomization of treatments can be performed with a table of random numbers or by drawing lots.

Choice of an experimental design has an important influence on the precision of the experimental results. Optimum treatment designs provide the greatest precision with a given number of replications or, alternatively, provide a given level of precision with the smallest number of replications. Common experimental designs used in soil fertility and plant nutrition experiments are complete randomized design, randomized complete block design, and split-plot design.

8.3.1.5.1 Completely Randomized Design

This is a simple type of design in which the whole experimental area is divided into plots depending upon the number of treatments and number of replications. Treatments are allotted to the plots entirely by chance. Completely randomized design is most appropriate for experiments with homogeneous experimental units. The design is commonly used in greenhouse experiments where environmental effects are easily controlled. This design is rarely used in the field experiments.

8.3.1.5.2 Randomized Complete Block Design

Randomized complete block design is one of the most widely used experimental designs in agricultural research. Blocks of equal size contain a complete set of all treatments. Major advantage of this design is that it reduces experimental error through proper blocking while retaining much of the flexibility and simplicity of the completely randomized design (Gomez and Gomez, 1976). Major objective of blocking is to reduce heterogeneity among plots within each block. In soil fertility experiment blocking has an advantage if the gradient of fertility in the experimental area is known. For an area with a unidirectional fertility gradient, rectangular blocks should be used. Blocks should be oriented in a way that their length is perpendicular to the direction of the fertility gradient. For example, for a field with a gradient along the length of the field, blocks should be established across the width of the field, cutting across the gradient. When fertility gradient occurs in two directions, square blocks are recommended. Similarly, when a fertility gradient does not exist or is not known, blocks should be square in shape.

8.3.1.5.3 Split-Plot Design

Split-plot design is especially used in factorial experiments when large plots are needed for one treatment and smaller plots for another treatment. For example, evaluating more than one cultivar with various levels of soil fertility. Similarly, evaluating irrigation treatments along with various levels of N or P. Larger plot size is called the main plot, and each main plot is subdivided into smaller plots to accommodate a second set of treatments. In this design, the precision for main plot treatment is less than the subplot treatments. Smaller differences can be detected among subplot treatment than among the main plot treatments. Consideration of the relative importance of the factors involved should be made before using the design.

8.3.1.6 Selection of the Treatments

A treatment is a single entity in an experiment. Treatment design refers to the selection of the treatments to be included in an experiment and is one of the components of statistical design (Federer, 1979). Selection of appropriate treatment is an important step in evaluating the hypothesis formulated. If a hypothesis, for example, is that P is one of the most yield-limiting factors in tropical Oxisol soils for crop production. To test this hypothesis, it is necessary to apply different levels of P to a crop to determine the response of this crop to applied P levels. In this case, P levels (treatments) should cover a wide application rates and response curve could be determined. All other nutrients and factors should be at optimum levels to evaluate the P treatments effects. When defining the optimum level of a nutrient for a crop in a particular soil, there must be a minimum of five rates that should be included to achieve a satisfactory response curve. A control treatment should always be included to compare the results of fertilized plots.

8.3.1.7 Sowing

Adequate seed rate, row spacing, plant spacing, and sowing depth should be used. In addition, seeds should be treated with appropriate fungicides and insecticides to minimize disease and insect problems.

8.3.1.8 Cultural Practices

After proper installation, proper management of experiment in the field is as important as proper planning. If it is desired that in order for the experiment to provide valid results, all other variable should be at an optimum level. If an experiment is designed to test levels of soil fertility for a particular crop on a particular soil, all other factors such as weeds, diseases, and insects should be controlled. Similarly, soil moisture should be at an adequate level. All management practices should be conducted on block basis to control any variation that may occur in the management and operation processes. For instance, when an operation such as weeding cannot be completed for the entire experiment in one day, at least weeding should be completed in one block. Therefore, blocking, if any, from day to day can control the difference. Similarly, when more than one person is performing any operation, each one should be assigned one block for maintaining uniformity for that operation. If herbicides are used for weed control, they should be sprayed in preemergence, which is at 1–2 days after seeding time in moist soil. If the soil is dry, generally herbicides are not effective in weed control.

8.3.1.9 Data Collection

In a soil fertility experiment, data should be collected which are related to plant, soil, and climate. Grain yield is the most important plant variable to measure the effects of an applied treatment under field condition. Grain yield refers to the weight of cleaned and dried grains harvested from a unit area. After discarding border areas on all four sides of a plot, harvest as large an area as possible. Dry matter yield should be determined to obtain data regarding nutrient accumulation by a crop during the season. In cereals this can be performed at harvesting. In the case of legumes, the appropriate time of plant sampling for dry matter determination is at flowering, when dry matter accumulation is at the maximum. During harvest time, most of the legume leaves fall down and therefore do not provide an accurate measure of dry matter accumulation.

To determine dry matter yield for nutrient uptake studies, there are numerous techniques such as random selection of a number of plants and the multiplication of the per plant value by a population estimate or sampling of a section of row and multiplication by a factor for conversion to a unit area basis. Statistical aspects of subsampling are discussed under the topic of two stages sampling by Steel et al. (1980). Hunt et al. (1987) concluded that adequate accuracy and precision were obtained by the use of the 1 or 2 m sampling technique, but not by the 0.3 m or 4 plants plot^{-1} technique in soybean dry matter determination per unit area. Yield components should be collected since grain yield in field crops is determined by various ratios of yield components. Yield components in cereals are panicles or ear per unit area, the number of spikelets per panicle or ear, and spikelet weight. Similarly, in legumes, the number of pods per unit area, grains per pod, and weight of grain determine the yield.

In field experiments, at the time of harvest, plant height and panicle density from 1 m row should be determined. One-meter row should be harvested from each plot to determine straw dry weight. In addition, 10–20 panicles should be harvested from each plot to determine spikelet sterility and 1000 grain weight in the case of cereals such as rice. Grain and straw yield ha^{-1}, panicle density m^{-2}, spikelet sterility, 1000 grain weight, grain harvest index (GHI), and grain yield efficiency index (GYEI) can be determined by using the following equations:

$$\text{Straw yield}\left(\text{kg ha}^{-1}\right) = \frac{\text{Dry weight of straw per m row in g}}{\text{Spacing between row in m}} \times 10$$

$$\text{Grain yield}\left(\text{kg ha}^{-1}\right) = \frac{\text{Grain yield per plot in g}}{\text{Area harvested per plot in m}^2} \times 10$$

$$\text{Panicle density}\left(\text{m}^{-2}\right) = \frac{\text{Panicle number per meter row}}{\text{Row spacing in m}}$$

$$\text{Spikelet sterility}(\%) = \frac{\text{Number of unfilled spikeltes in 10 panicles}}{\text{Number of filled and unfilled spikeltes in 10 panicles}} \times 100$$

$$\text{Thousand grain wt.}(\text{g}) = \frac{\text{Filled grain weight in 10 panicles in g}}{\text{Number of filled grain in 10 panicles}}$$

$$\text{Grain harvest index}(\text{GHI}) = \frac{\text{Grain yield}}{\text{Grain plus straw yield}}$$

In addition to plant variable determination, soil samples should be taken from each plots or treatments, and soil chemical indices should be determined to evaluate fertilizer treatment effects and correlate these soil indices with grain yield. These are

$$\text{CEC}\left(\text{cmol}_c\ \text{kg}^{-1}\right) = \sum(\text{Ca, Mg, K, H, Al})$$

where Ca, Mg, K, H, and Al are in $cmol_c\ kg^{-1}$, H + Al determined at pH 7

$$\text{Base saturation}(\%) = \frac{\sum(\text{Ca, Mg, K})}{\text{CEC at pH 7}} \times 100$$

$$\text{Acidity saturation}(\%) = \frac{\text{H} + \text{Al}}{\text{CEC}} \times 100$$

$$\text{Saturation of Ca, Mg, or K}(\%) = \frac{\text{Ca}}{\text{CEC}} \times 100,\ \frac{\text{Mg}}{\text{CEC}} \times 100,\ \text{or}\ \frac{\text{K}}{\text{CEC}} \times 100$$

$$\text{Ca, Mg, or K ratios} = \frac{\text{Ca}}{\text{Mg}},\ \frac{\text{Ca}}{\text{K}},\ \text{or}\ \frac{\text{Mg}}{\text{K}}$$

In cereals, the relationship between yield and its components can be expressed in the form of the following equation (Fageria et al., 2011):

$$\text{Grain yield}\left(\text{Mg ha}^{-1}\right) = \text{Number of panicles or ears m}^{-2}\ \text{or number of spikelets per panicle or ear} \times 1000\ \text{grain weight}(\text{g}) \times \text{filled spikelet}(\%) \times 10^{-5}$$

For example, if the target of a rice crop is to produce 6 Mg ha^{-1} grain yield, it is necessary to have a combination of the following yield-attributing characters or components: (1) 400 panicles m^{-2}, (2) 80 spikelets per panicle, (3) 85% filled spikelets, and (4) 22 g weight of 1000 grains. By incorporating these values in the previous equation, we get

$$\text{Grain yield}\left(\text{t ha}^{-1}\right) = 400 \times 80 \times 0.85 \times 22 \times 10^{-5} = 6$$

Similarly, the yield equation for grain legumes can be written as follows:

$$\text{Grain yield}\left(\text{Mg ha}^{-1}\right) = \text{Number of pods m}^{-2} \times \text{number of grains per pod}$$

$$\times \text{ weight of 1000 grains}\left(\text{g}\right) \times 10^{-5}$$

or

$$\text{Number of grains m}^{-2} \times \text{weight of 1000 grains}\left(\text{g}\right) \times 10^{-5}$$

To obtain a yield of 1.5 Mg ha^{-1} in cowpea, the following combination of yield components is required: (1) 155 pods m^{-2}, (2) grains per pod, (3) 140 g weight of 1000 grains

While putting these values in the previous equation, the following results will be obtained:

$$\text{Grain yield}\left(\text{Mg ha}^{-1}\right) = 155 \times 7 \times 140 \times 10^{-5} = 1.5$$

Each of the yield components differs not only in the time of determination, but in the contribution to the grain yield. In cereals, the number of panicles or ears is determined during vegetative growth stage, the number of grains per panicle or ear is determined during reproductive stage, and grain weight is determined during ripening or grain-filling stage. Reproductive growth stage is most sensitive to environmental stresses in cereals as well as legumes. If there is drought, deficiency of N, low/high temperature, or low solar radiation during the reproductive stage, yield will be reduced more than if these stresses occurred during the vegetative and ripening growth stages. Within the reproductive stage, there is a differential sensitivity to stresses. In cereals, the growth period 14–10 days prior to flowering (reduction-division stage) is considered to be the most sensitive period to stress.

Yield components are not independent and an increase in one component at a certain level often leads to a decrease in another component. In general, the number of panicles or ears per plant declines as the number of plants per unit area increases. Similarly, the weight per grain decreases as the number of grains per panicle or ear increases. To achieve an optimum yield, all these yield components should be in an appropriate balance.

Fageria (1992) studied correlations between grain yield of upland rice and yield components in a field experiment conducted at the National Rice and Bean Research Center of Embrapa, Santo Antônio de Goiás, GO, Brazil. Grain yield was significantly correlated with dry matter and grains per panicle. Panicles per square meter had the highest correlation, which means that this parameter was responsible for a higher contribution to grain yield. Leaf canopy of barley (*Hordeum vulgare* L.) is related to yield and yield components. Grafius and Barnard (1976) concluded that components of yield are not passive participants in the determination of grain yield, but on the contrary they exert active influence on yield through the source–sink and transport relationships. Fageria (1992) provides a detailed discussion of physiology of yield in cereals and legumes.

After harvesting a fertility experiment, soil sampling should be performed at 0.20 m depth for annual crops to evaluate the fertility status of the experimental plots. This analysis will provide information on the accumulation and depletion level of each nutrient, organic matter, and soil pH. This information will assist in fertilizing the succeeding crop in the cropping system. Air temperature, precipitation, and radiation data should be recorded during the experimental period.

These data can furnish information in interpretation of experimental results. Data should be recorded on diseases, insects, and other environmental stresses, which may be prevalent during the conduct of the experiment.

Electronic data collection is used in agricultural sciences as a means of improving the accuracy and efficiency of data collection. Data loggers are commonly used for the automated collection of data produced by environmental monitors or laboratory equipment. Portable keypad-entry data loggers, which allow entry and the temporary storage field or laboratory observations, have been available for several years (Cooney, 1985). Utilization of portable notebook-size microcomputers for experimental data collection is considered a common and efficient method (Kidger and McNicol, 1986; Jackson and Stone, 1987).

8.3.1.10 Harvesting

Harvesting at appropriate time of a crop is an important step in a field experiment. Plants should be harvested at physiological maturity. For cereals and grain legumes, it is related to the grains full-stage development. If a crop is harvested earlier than physiological maturity, yield will be reduced, owing to low grain weights. If harvesting is delayed after physiological maturity, yield may be reduced owing to shattering of grains or a large percentage of grains lost during harvesting. Harvesting at inappropriate times may reduce the quality of grains. Physiological maturity of a crop can be determined by a systematic sampling and dry weight determination of grains. When no further increase in grain dry weight is observed, the plant is said to have reached physiological maturity. Data regarding appropriate grain moisture content for harvest of cereals and legume crops are available as a reference point (Fageria, 1992).

Research plots are generally harvested manually and threshed with a stationary thresher. This is an effective method for breeding research where samples must be kept pure. However, in soil fertility experiments, small plot combines can be used for harvesting where absolute seed purity is not necessary.

Mechanical harvesting equipment has been developed to reduce labor requirements in field research. Alleys must be maintained between plots to improve the efficiency of mechanized harvesting and to reduce the risk of contamination between plots (Wolkowski et al., 1988). Wolkowski et al. (1988) reported that mechanized harvesting methods could perform as equally well as hand harvesting, assuming that plants are not lodged. The plot lengths were 9, 12, and 15 m.

8.3.1.11 Statistical Analysis of the Data

Use of adequate statistical methods in data analyses is an essential experimental technique. Experimental and treatment designs dictate the proper method of statistical analysis and the basis for assessing the precision of treatment means. The aim of a statistical analysis of data from an agronomic experiment is to provide information on experimental response to the applied treatments. Data are subjected to an analysis of variance to determine whether or not significant differences exist among treatment means. The data are then analyzed in an attempt to explain the nature of the response in more detail. A number of statistical procedures may be used for this purpose (Petersen, 1977): (1) fitting response functions using regression techniques, (2) planned sets of contrasts among means, or groups of means, and (3) pairwise multiple comparison procedures. These procedures may not be applicable to all situations. Of these procedures, the most often used are the multiple comparison tests.

8.3.1.11.1 Analysis of Variance

Analysis of variance is the first step in data analysis to determine treatment effects. Several statistical programs are available for analysis of variance. When treatment effect is significant for a determined plant variable under a given fertility or nutrient treatment, further analysis like regression technique is used to determine the optimum nutrient rate for maximum yield.

8.3.1.11.2 Regression Technique

Regression technique or curve fitting is the most appropriate method for quantitative factors or treatments. Quantitative factors can be cited such as levels of fertilizers, pH, nutrient concentrations, temperature, pressure, plant density, and seeding rates. Regression equations fitted to the data depend on the response curve. Linear (first-degree), quadratic (second-degree), and cubic (third-degree) are typical equations. In fertilizer experiments generally a second-degree polynomial is sufficient to describe the response surface. The linear and quadratic relationships can be expressed mathematically as follows:

$$Y = a + bX$$

$$Y = a + b_1X - b_2X^2$$

where

Y is the estimated crop yield
a is the x intersect of crop yield at zero fertilizer rate
b is the slope of the line or regression coefficient
b_1 is the linear regression coefficient
b_2 is the quadratic regression coefficient
X is the fertilizer application rate

What type of regression equation fits to a given set of data can be determined by visual observation of the trend of the response curve or more scientifically by calculating R^2 value for the fitted function. Higher R^2 value indicates an improved fit of the response function. Analysis of the fitted function may permit the estimation of the factor levels at which the response is a maximum or minimum within the range of the factors. From a quadratic regression equation, the fertilizer rate that produces maximum yield can be calculated with the following formula:

$$\text{Fertilizer rate} = \frac{b_1}{2b_2}$$

where

b_1 is linear regression coefficient
b_2 is quadratic regression coefficient

There is no other statistical method by which this information (fertilizer rates that produce maximum yield) is effectively obtained. If the linear and/or quadratic regression is significant, then a multiple comparison procedure is not necessary. All treatments are significantly different in their effects in such cases.

8.3.1.11.3 Orthogonal Contrasts

When the treatments consist of qualitative factors such as forms of fertilizers, type of soils, and method of fertilizer application, orthogonal contrasts can be used to compare treatments. A linear combination is called a comparison or a contrast if the coefficients add up to zero (Chew, 1976). Many agronomic trials are conducted as factorial experiments in which each level of every factor occurs in combination with each level of every other factor. Experiments in which the treatments are made up of all possible combinations of the levels of two or more factors (qualitative or quantitative) are called a set of factorial treatments. Factorial describes the nature of the treatments and not the design of the experiment (Chew, 1976). In factorial experiments the first step in the analysis should be to compute and test the main effects and interactions of the several factors. Proper interpretation of the data depends on the outcome of these tests. If the interactions are not significant, then all of the information in the experiment is contained in the main effect means. Response curves, orthogonal comparisons, or multiple comparison tests should be performed on these means, not the individual treatment means. If the interactions are significant, then the response to changes in one factor depends on the level of the interacting factor. In this case, response curves, orthogonal contrast, or multiple comparison tests for one factor may be performed separately at each level of the interacting factor (Petersen, 1977).

8.3.1.11.4 Pairwise Multiple Comparison

Pairwise multiple comparison tests are applied for qualitative types of treatments such as cultivars, fungicides, locations, methods of soil analysis, and sources of fertilizers. Multiple comparison tests may be useful for grouping means from experiments involving unstructured qualitative treatments. Such tests may be included as least significant difference test (LSD), Duncan's multiple range tests (DMRT), and Tukey's honestly significant difference test (THSDT). Multiple comparison tests were developed to permit data snooping among the sample means after the experiment has been concluded. Their most common use is to make comparison of each mean with every other mean. Their purpose is to detect possible groups among a set of unstructured treatments. They are not intended to be used for quantitative treatments for which response surface methodology is more appropriate (Petersen, 1977). In soil fertility and plant nutrition research, experimental data, especially yield, are transformed into relative yield form. Soil test correlation data are generally characterized by considerable scatter. This is particularly true when soil test values are plotted against actual yields. To eliminate some of the scatter, most soil test correlation work uses relative or percent yields. Relative yield is defined as adequate but not excessive amounts of all nutrients other than the one being correlated, divided by the yield of a treatment which is the same except that it includes the nutrient under study (Cate and Nelson, 1971). Relative yield can be calculated using the following formula:

$$\text{Relative yield}\,(\%) = \left(\frac{\text{Yield of control plot}}{\text{Maximum yield of treated plot}}\right) \times 100$$

According to Evans (1987), a wide scattering of absolute yields may occur as a result of factors other than soil fertility. Scattering of absolute yields does not necessarily indicate that there is a poor correlation, but an improved relationship may be obtained by using a relative yield to eliminate some of the climate and site influence.

8.3.1.12 Duration of the Experiment

It is difficult to define an optimum duration of a soil fertility field experiment. Field experiments conducted for one to several years have been reported. An experiment is normally conducted to test a hypothesis. When the hypothesis is tested and objectives are achieved, the experiment is then terminated. A soil fertility field experiment needs to be conducted for several years due to variability in environmental factors from year to year and within a season of the same year. By repeating the experiment several years, mean values of the applied treatment are altered under different environmental conditions. This value can serve as a basis for making fertilizer recommendations for a particular crop under a given agroecological region. In addition, long-term experiments permit a measurement of the effect of treatment on buildup or depletion of nutrients in the soil and change in soil pH and organic matter content of the soil. Further, long-term fertility experiments can be useful in providing information on sustainability of a farming system under use. Sustainability of an agricultural system is influenced by soil physical, chemical, and biological properties in addition to climatic components. Environmental change pattern is not well established in short-duration experiments. When long-term fertility experiments are planned, it is important to decide the level of soil fertility that should be tested, crop rotation, soil preparation methods, cultural practices adapted in the management of the experiment, and observations to be recorded. In addition to appropriate crop rotations, use of some form of organic manures is an important component of sustainable agricultural systems. Further, fallowing may be an important component of a farming system to give rest to the soil for one season. This can provide favorable changes in soil physical, chemical, and biological properties and can stabilize or assist in sustainability of a farming system.

Long-term experiments have been reported (Li and Barber, 1988; Barber, 1995). The Rothamsted long-term experiments were started nearly 150 years ago (Jenkinson, 1991). These experiments were originally designed to study the N, P, K, Na, Mg, and Si needs of the field crops then grown in the United Kingdom. Inorganic nutrients in various combinations were compared with farmyard manure (the traditional source of fertility at that time). Experiment results indicated that grain yields can be sustained for almost 150 years in monocultures of wheat and barley given annual application of organic or inorganic fertilizers. Long-term N balances indicate that there are considerable inputs of N to the soil–plant system, amounting to some 30 kg N ha^{-1} $year^{-1}$ in unfertilized wheat and up to 65 kg ha^{-1} $year^{-1}$ in an arable soil reverting to woodland (Jenkinson, 1991). Mitchell et al. (1991) reviewed long-term agronomic research in the United States. Research plots that have been monitored continuously since the late nineteenth century exist in several states. According to these authors, 25 experiments have been identified that have been monitored for over 25 years. Twelve of these are more than 50 years old, and yield and treatment records provide valuable information on the effects of cropping systems, tillage, manuring, and fertilization practices on yields and on soil physical and chemical properties. Most of these early experiments were nonreplicated studies using large plots and crop rotation systems. Four of the U.S. oldest, continuous agronomic research tests were reviewed in more detail: (1) Illinois, Morrow plots (C.1876), (2) Missouri's, Sanborn Field (C.1888), (3) Oklahoma's Magruder plots (C.1892), and (4) Alabama's old rotation (C.1896). These long-term experiments are listed on the National Register of Historical places. These experiments indicate that long-term crop production can be sustained and improved in different regions and on different soils of the United States. Long-term studies have indicated that crop rotations are useful to consolidate soil fertility. Practices which may or may not include legumes and manuring are essential to maintaining high, sustained crop production (Army and Kemper, 1991).

A fertility experiment well planned and conducted for a period of 3 years can provide adequate information in understanding the complex interactions of plants, soils, climate, pests, and management, and their effects on sustainable crop production. If an experiment is conducted for more than 3 years, it is important to include new technologies in the experiments such as cultivars and other management practices. If treatments are changed to make it practical, then the initial objectives of the experiments will be changed. Under these situations, it is a short-term experiment conducted on the same site for a long period of time. Further, rapid change in priorities makes planning and funding of long-term yield studies extremely difficult. A major priority research need today may not be a major priority in 2, 5, or 10 years in the future (Army and Kemper, 1991). A field experiment of longer duration is quite expensive, especially for developing countries. Therefore, it is important to determine the duration of the experiment within which maximum necessary information can be generated to improve crop yield of the region. A field experiment was conducted to determine effects of broadcast P application on lowland rice for 3 years (Figure 8.6). Response curve was quadratic in shape, and 87% variation in yield was due to P fertilization. Soil of this experimental site had about 2 mg P kg^{-1} by Mehlich 1 extracting solution in the initial stage or before the application of P treatment.

8.3.1.13 Dissemination of Results

Dissemination of results of field experimentation for the benefit of the society is the last step in the process of improving agricultural production of a country or region. Study indicates that unless disseminated to those who need them most research data would become meaningless and a waste of time and resources. Best instrument of divulgation of research results is through demonstration

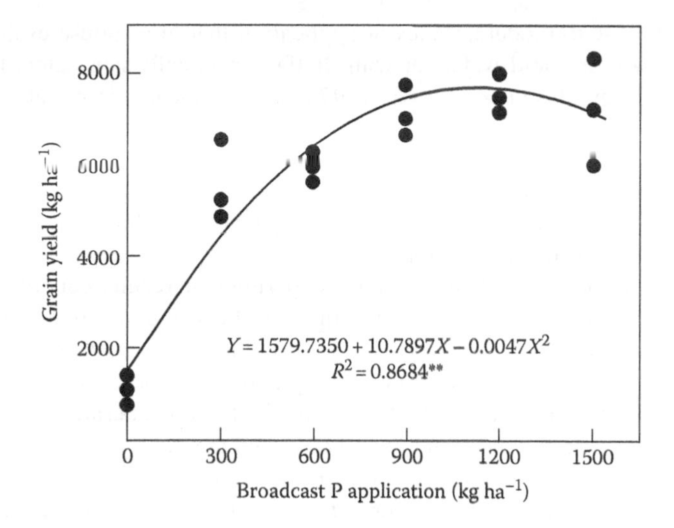

FIGURE 8.6 Response of lowland rice to phosphorus application. (From Fageria, N.K., *The Use of Nutrients in Crop Plants*, CRC Press, Boca Raton, FL, 2008.)

in the grower's fields along with extension service. In the final stage of a technology, it can be evaluated in the farmer's field. Package of recommendations can be tested side by side with the farmer's conventional methods of production. New technology can be compared with those of growers and if differences are significant, growers can be convinced to adopt the new technology. In this evaluation process, extension specialists should have involvement while scientists should also participate. Another important function of the extension service is to provide feedback to researcher on how the research results are being accepted by growers, where problems and bottlenecks have occurred and how they may be solved. Other ways of divulgation of research results are through publications. Field research is generally applied research and its results can be summarized in the form of technical bulletins. Such publications can be distributed to farmers through extension service or through credit-providing agencies. Private enterprise can assist in technology transfer, as has been accomplished in many developed countries.

In authors' opinion the results of research generated by public funds belong to public property and are available to everyone with no attempt to channel results and recommendations to specific users are justified.

8.3.1.14 Comparison of Controlled Conditions and Field Experiments

Controlled conditions experiments are of short duration conducted to develop and understand basic principles of soil fertility and plant nutrition. Field experiments are relatively of longer duration and are conducted to understand the applied portion of crop production. Because of limited soil volume in greenhouse pots, crops are usually harvested at a stage prior to full maturity. Small grains and grasses are harvested when heads begin to appear, legumes at early bloom stage, and corn when plants are 50–100 cm tall. Harvests commonly occurred at a time comparable to early growth responses commonly observed in field experiments (Allen et al., 1976). Conflicting reports about extrapolation of controlled conditions experimental results to field conditions are presented. According to Collis-George and Davy (1960) transference of greenhouse results to field conditions is possible if the environments of the greenhouse and field situation were known quantitatively. Cook and Millar (1946) reported techniques which assist to make greenhouse results comparable to those in field experiments. Higher nutrient levels in the greenhouse, adequate size of pots, and other factors were reported to be important. Significant correlation occurred, particularly with soil from the same site, used for both the greenhouse and field experiments. Hausenbuiller and Weaver

(1960) reported with low to moderate rates of application that greenhouse evaluation with ladino clover (*Trifolium repens*) and field evaluation with alfalfa were equally satisfactory for determining the P supply capacity of several P sources. Terman (1974) stressed the need for high nutrient application in pot experiments.

Duncan et al. (1983) reported that a major problem of acid soil stress research has been the lack of correlation between laboratory and field data, especially since germplasm, which has been screened under controlled conditions (greenhouse and growth chamber), has not consistently performed well under field stress conditions.

In the authors' opinion, controlled condition experimental results cannot be extrapolated to field conditions due to large variation in environmental factors under field conditions. Properly conducted greenhouse and laboratory studies often provide useful evidence as to the factors that merit investigation under field conditions. For example, if greenhouse experiments indicate rather marked differences in response among the fertilizers, further consideration of field experiment is justified. Pot experiments are valuable in relating observed laboratory measurement to actual plant response and thereby are important in establishing basic principles on soil–fertilizer–plant relationships. Both types of experimentation are important and should be complimentary in solving the problem of crop production in general.

8.4 CONCLUSIONS

Research in agriculture is a complex process and demands constant efforts and experimentation due to change in weather conditions, difference in soil properties, difference in adaptation of crop species, and different socioeconomic conditions of growers. Soil fertility and plant nutrition research involves laboratory, greenhouse or growth chamber, and field experimentation. Laboratory and greenhouse experiments are generally short-duration experiments conducted to develop and understand some basic principles of a subject. For example, pot experiment with different types of soils can indicate the degree of response that may be anticipated at different soil test levels and serve as excellent checks on ratings being used. Since such tests provide no measure of the cumulative effects of treatments on yield, soil accumulation, or depletion, they have limited value in determining rates of fertilizer that should be recommended for sustained productivity.

Due to large variation in environmental factors, results of controlled conditions experiments can hardly be extrapolated to field conditions and vice versa. However, these two types of experiments should serve as complementary components in developing a crop production technology. In controlled condition experiments, soil and solution culture are generally used as medium of plant growth to evaluate treatment effects. Although use of nutrient solutions facilitates precise control of experimental variables, it eliminates entirely the soil–root aspect, an important part of soil–plant system. The pattern of exploration and activity in root systems subjected to zonal salinization as well as the significance of ionic mobility in determining quantities of a given element absorbed from soil suggests the importance of testing hypothesis in a soil system, especially a system similar to that occurred in the field.

Technological process from generation to verification and to dissemination must be viewed as a continuum providing a strong mechanism for forward and backward linkages to monitor and provide feedback of needs and problems of both technology source and ultimate users. Local adaption of the package of recommendations to growers is the responsibility of extension. This does not imply that researcher cannot participate in the process of technology transfer. Low yields of crops in some parts of the world or countries are the result of actions and interactions of many factors, and there are no simple, easily implementable solutions. Improved understanding of biological, climatic, edaphic, and management factors through research and development of production technologies that occur in the appropriate socio-political-economic climate can assist in increasing crop production in such regions.

Soil fertility is one of the important factors in determining crop yields. Maintaining soil fertility at an appropriate level is vital for sustainable agriculture and in reducing environmental pollution.

To achieve these objectives, research data are required for different agroecological regions for different crops and cropping systems. Projects involving experimentation should have appropriate planning to get meaningful results. Planning includes well-defined objectives based on priority of problems, and to achieve the objectives experimental methodology should be adequate. Statistical analysis and interpretation of experimental data are as important as planning and execution of the experiments. In this review chapter, basic principles and methodology of conducting field experimentation in soil fertility and plant nutrition were discussed. This information may be useful for those who are involved in developing technology in the field of agriculture in general and soil fertility and plant nutrition in particular.

REFERENCES

Allen, S. E., G. L. Terman, and L. B. Clements. 1976. Greenhouse techniques for soil-plant-fertilizer research, Bulletin 4-104. Muscle Shoals, AL: National Fertilizer Development Center.

Andrew, C. S., A. D. Johnson, and R. L. Sandland. 1973. Effect of aluminum on the growth and chemical composition of some tropical and temperate pasture legumes. *Aust. J. Agric. Res.* 24:325–339.

Army, T. J. and W. D. Kemper. 1991. Support for long-term agriculture research. *Agron. J.* 83:62–65.

Asher, C. J. and D. G. Edwards. 1983. Modern solution culture techniques. In: *Encyclopedia of Plant Physiology*, eds. A. Lauchi and R. L. Bieleski, pp. 94–119. Berlin, Germany: Springer Verlag.

Barber, S. A. 1995. *Soil Nutrient Bioavailability: A Mechanistic Approach*. New York: Wiley.

Barley, K. P. 1964. The utility of field experiments. *Soils Fert.* 27:267–269.

Bates, T. E. 1971. Factor affecting critical nutrient concentrations in plants and their evaluation: A review. *Soil Sci.* 112:116–130.

Bell, P. F., R. L. Chaney, and J. S. Angle. 1991. Determination of copper2+ activity required by maize using chelator-buffered nutrient solutions. *Soil Sci. Soc. Am. J.* 55:1366–1374.

Ben-Asher, J., J. M. Gordon, A. Linear, and Y. Zarmi. 1982. Nutrient uptake and supply to tomato plants in a water culture system. *Agron. J.* 74:640–644.

Borlaug, N. E. and C. R. Dowswell. 1994. Feeding a human population that increasingly crowds a fragile planet. In: Paper Presented at the *15th World Congress of Soil Science*, Acapulco, Mexico, July 10–16, 1994.

Breeze, V. G., D. G. Edwardds, and M. J. Hopper. 1987. Effect of pH in flowing nutrient solution on the growth and phosphate uptake of white clover supplied with nitrate, or dependent upon symbiotically fixed nitrogen. *New Phytol.* 106:101–114.

Brown, J. C. and W. E. Jones. 1976. A technique to determine iron efficiency in plants. *Soil Sci. Soc. Am. J.* 40:398–404.

Cady, F. B. 1991. Experimental design and data management of rotation experiments. *Agron. J.* 83:50–56.

Callahan, L. M. and R. E. Engel. 1986. A compact continuous flow and constant level solution culture renewal system. *Agron. J.* 78:547–549.

Cate, R. B. Jr. and L. A. Nelson. 1971. A simple statistical procedure for partitioning soil test correlation data into two classes. *Soil Sci. Soc. Am. Proc.* 35:658–660.

Chaney, R. L. and P. F. Bell. 1987. Complexity of iron nutrition: Lessons for plant-soil interaction research. *J. Plant Nutr.* 10:963–994.

Chaney, R. L., P. F. Bell, and A. Coulombe. 1989. Screening strategies for improved nutrient uptake and use by plants. *Hort. Sci.* 24:565–572.

Checkai, R. T., R. L. Hendrickson, R. B. Corey, and P. A. Helmke. 1987. A method for controlling the activities of free metal, hydrogen, and phosphate ions in hydroponic solutions using ion exchange and chelating resins. *Plant Soil* 9:321–334.

Chew, V. 1976. Comparing treatment means: A compendium. *HortScience* 11:348–357.

Clark, R. B. 1982. Nutrient solution growth of sorghum and corn in mineral nutrition studies. *J. Plant. Nutr.* 5:1039–1057.

Cochran, W. G. and G. M. Cox. 1957. *Experimental Design*, 2nd edn. New York: John Wiley & Sons.

Collis-George, N. and B. G. Davey. 1960. The doubtful utility of present day field experimentation and other determinations involving soil–plant interactions. *Soil Fert.* 23:307–310.

Cook, R. L. and C. E. Miller. 1946. Some techniques which help to make greenhouse investigations comparable with field plot experiments. *Soil Sci. Soc. Am. Proc.* 11:298–304.

Cooney, T. M. 1985. Portable data collectors, and how they are becoming useful. *J. For.* 83:18–23.

Duncan, R. R., R. B. Clark, and P. R. Furlani. 1983. Laboratory and field evaluation of sorghum for response to aluminum and acid soil. *Agron. J.* 75:1023–1026.

Edmeades, D. C., D. M. Wheeler, and R. A. Christie. 1991. The effects of aluminum and pH on the growth of a range of temperate grass species and cultivars. In: *Plant–Soil Interactions at Low pH*, eds. R. J. Wright, V. C. Baligar, and R. P. Murrmann, pp. 913–924. Dordrecht, the Netherlands: Kluwer Academic Publishers.

Edwards, D. G. and C. J. Asher. 1983. The significance of solution flow rate in flowing culture experiments. Ionic interactions in rice plants from dilute solutions. *Plant Soil* 70:309–316.

Elliott, G. C. and A. Lauchli. 1985. Phosphorus efficiency and phosphate–iron interaction in maize. *Agron. J.* 77:399–403.

Evans, C. E. 1987. Soil test calibration. In: *Soil Testing: Sampling, Correlation, Calibration, and Interpretation*, ed. J. R. Brown, pp. 23–29. Madison, WI: SSSA.

Fageria, N. K. 1974. Continuous-nutrient-flow method: A new approach to determine nutrient uptake. *Indian J. Agric. Sci.* 44:262–266.

Fageria, N. K. 1976. Effect of P, Ca, and Mg concentrations in solution culture on growth and uptake of these ions by rice. *Agron. J.* 68:726–732.

Fageria, N. K. 1984. Response of rice cultivars to liming in cerrado soil. *Pesq. Agropec. Bras.* 19:883–889.

Fageria, N. K. 1989. Effects of phosphorus on growth, yield and nutrient accumulation in common bean. *Trop. Agric.* 66:249–255.

Fageria, N. K. 1990. Response of wheat to phosphorus fertilization on an Oxisol. *Pesq. Agropec. Bras.* 25:530–537.

Fageria, N. K. 1991. Response of cowpea to phosphorus on an Oxisol. *Trop. Agric.* 68:384–388.

Fageria, N. K. 1992. *Maximizing Crop Yields*. New York: Marcel Dekker.

Fageria, N. K. 1999. Phosphorus calibration analysis for rice crop in greenhouse. *Pesq. Agropec. Bras.* 25:579–586.

Fageria, N. K. 2005. Soil fertility and plant nutrition research under controlled conditions: Basic principles and methodology. *J. Plant Nutr.* 28:1975–1999.

Fageria, N. K. 2007. Soil fertility and plant nutrition research under field conditions: Basic principles and methodology. *J. Plant Nutr.* 30:203–223.

Fageria, N. K. 2008. *The Use of Nutrients in Crop Plants*. Boca Raton, FL: CRC Press.

Fageria, N. K. 2015. *Nitrogen Management in Crop Production*. Boca Raton, FL: CRC Press.

Fageria, N. K. and V. C. Baligar. 1999. Phosphorus use efficiency in wheat genotypes. *J. Plant Nutr.* 22:331–340.

Fageria, N. K., V. C. Baligar, and C. A. Jones. 2011. *Growth and Mineral Nutrition of Field Crops*, 3rd edn. Boca Raton, FL: CRC Press.

Fageria, N. K., V. C. Baligar, A. Moreira, and T. A. Portes. 2010. Dry bean genotypes evaluation for growth, yield components and phosphorus use efficiency. *J. Plant Nutr.* 33:2167–2181.

Fageria, N. K., M. P. Barbosa Filho, and J. J. Garber. 1982. Adequate nutrient levels and plant density in grenhouse experiments for rice. *Pesq. Agropec. Bras.* 17:1279–1284.

Fageria, N. K. and J. R. P. Carvalho. 1982. Influence of aluminum in nutrient solution on chemical composition in upland rice cultivars. *Plant Soil* 69:31–44.

Fageria, N. K. and H. R. Gheyi. 1999. *Efficient Crop Production*. Campina Grande, Brazil: Federal University of Paraiba.

Fageria, N. K., L. C. Melo, J. P. Oliveira, and A. M. Coelho. 2012. Yield and yield components of dry bean genotypes as influenced by phosphorus fertilization. *Commun. Soil Sci. Plant Anal.* 43:2752–2766.

Fageria, N. K. and A. Moreira. 2011. The role of mineral nutrition on root growth of crop plants. In: *Advances in Agronomy*, ed. D. L. Sparks, Vol. 110, pp. 251–331. Burlington, NJ: Academic Press.

Fageria, N. K., A. Moreira, L. C. A. Moraes, and M. F. Moraes. 2014. Influence of lime and gypsum on yield and yield components of soybean and changes in soil chemical properties. *Commun. Soil Sci. Plant Anal.* 45:271–283.

Fageria, N. K. and A. B. Santos. 2014. Requirement of micronutrients by lowland rice. *Commun. Soil Sci. Plant Anal.* 45:844–863.

Fageria, N. K., L. F. Stone, and A. B. Santos. 1999. *Maximizing Crop Yields*. Brasilia, Brazil: Embrapa Department of Communication and Transfer of Technology.

Federer, W. T. 1979. Statistical designs and response models for mixtures of cultivars. *Agron. J.* 71:701–706.

Follett, R. F., L. K. Porter, and A. D. Halvorson. 1991. Border effects on nitrogen-15 fertilized winter wheat microplots grown in the Great Plains. *Agron. J.* 83:608–612.

Ford, G. L. and G. A. Nielsen. 1982. Selecting environmentally analogous areas for agronomic research using computer graphs. *Agron. J.* 74:261–265.

Gomes, F. P. 1984. *Modern Statistics in Agricultural Research*. São Paulo, Brazil: Brazilian Potassium and Phosphate Association.

Gomez, K. A. 1972. *Techniques for Field Experiments with Rice*. Los Baños, Philippines: International Rice Research Institute.

Gomez, K. A. and R. C. Alicbusan. 1969. Estimation of optimum plot size from rice uniformity data. *Phil. Agric.* 52:586–601.

Gomez, K. A. and S. K. De Datta. 1971. Border effects in rice experimental plots. I. Unplanted borders. *Exp. Agric.* 7:87–92.

Gomez, K. A. and A. A. Gomez. 1976. *Statistical Procedures for Agricultural Research, with Emphasis on Rice*. Los Baños, Philippines: International Rice Research Institute.

Grafius, J. E. and J. Barnard. 1976. The leaf canopy as related to yield in barley. *Agron. J.* 78:398–402.

Hai, T. V. and H. Laudelout. 1966. Phosphate uptake by intact rice plants by the continuous flow method at low phosphate concentrations. *Soil Sci.* 101:408–417.

Hausenbuiller, R. L. and W. H. A. Weaver. 1960. Comparison between greenhouse and field procedures in phosphate fertilizer testing. *Soil Sci.* 90:298–301.

Hoagland, D. R. and D. I. Arnon. 1950. *The Water Culture for Growing Plants without Soil.* California Agricultural Experimental Station Circular 347, 32 pp. Berkeley, CA: College of Agriculture, University of California.

Hohenberg, J. S. and D. N. Munns. 1984. Effect of soil acidity factors on nodulation and growth of *Vigna unguiculata* in solution culture. *Agron. J.* 76:477–481.

Hunt, P. G., K. P. Burnham, and T. A. Matheny. 1987. Precision and bias of various soybeans dry matter sampling techniques. *Agron. J.* 79:425–428.

Imsande, J. and E. J. Ralston. 1981. Hydrophonic growth and the nondestructive assay for dinitrogen fixation. *Plant Physiol.* 68:1380–1384.

Islam, A. K. M. S., D. G. Edwards, and C. J. Asher. 1980. pH optima for plant growth: Results of a flowing solution culture experiment with six species. *Plant Soil* 54:339–357.

Itoh, S. and S. A. Barber. 1983. Phosphorus uptake by six plant species as related to root hairs. *Agron. J.* 75:457–461.

Jackson, H. O. and J. A. Stone. 1987. Improved data collection program for a notebook-size microcomputer. *Agron. J.* 79:1087–1089.

Jenkinson, D. S. 1991. The Rothamsted long-term experiments: Are they still of use? *Agron. J.* 83:2–10.

Johnson, C. M., P. R. Stout, T. C. Broyer, and A. B. Carlton. 1957. Comparative chlorine requirement of different plant species. *Plant Soil* 8:337–354.

Johnston, J. H. and M. D. Miller. 1973. Culture. In: *Rice in the United States: Varieties and Production*, eds. A. L. Merrill, and B. K. Watt, Agriculture Handbook 289, pp. 88–134. Washington, DC: USDA-Agriculture Research Service.

Kavitkar, A. G., P. N. Saxena, and M. Prashad. 1971. Sampling method for estimating plot yield of wheat. *Indian J. Agron.* 16:510–511.

Kidger, R. and J. W. McNicol. 1986. A program for data recording on a portable microcomputer. *Comput. Electron. Agric.* 1:213–217.

Laird, R. J. 1968. Field plot technique for fertilizer experiments. Research Bulletin 9. Mexico City, Mexico: International Maize and Wheat Improvement Center.

Li, R. G. and S. A. Barber. 1988. Effect of phosphorus and potassium fertilizer on crop response and soil fertility in a long-term experiment. *Fert. Res.* 15:123–136.

MacDonald, G. E. and N. H. Peck. 1976. Border effects in a long-term fertility experiment. *Agron. J.* 68:530–534.

McElhannon, W. S. and H. A. Mills. 1978. Influence of NO_3^-/NH_4^+ on growth, N absorption, and assimilation by lima beans in solution culture. *Agron. J.* 70:1027–1032.

Mitchell, C. C., R. L. Westerman, J. R. Brown, and T. R. Peck. 1991. Overview of long-term agronomic research. *Agron. J.* 83:24–29.

Miyasaka, S. C., R. T. Checkai, D. L. Grunes, and W. A. Norvell. 1988. Methods for controlling pH in hydrophonic culture of winter wheat forage. *Agron. J.* 80:213–220.

Miyasaka, S. C. and D. L. Grunes. 1990. Root temperature and calcium level effects on winter wheat forage: I. Shoot and root growth. *Agron. J.* 82:236–242.

Mortvedt, J. J. and G. L. Terman. 1978. Nutrient effectiveness in relation to rates applied for pot experiments: II. Phosphorus sources. *Soil Sci. Soc. Am. J.* 42:302–306.

Nelson, L. A. and J. O. Rawlings. 1983. Ten common misuses of statistics in agronomic research and reporting. *J. Agron. Educ.* 12:100–105.

Novais, R. F., J. C. L. Neves, and N. F. Barros. 1991. Controlled conditions experiment. In: *Research Methods in Soil Fertility*, ed. EMBRAPA, pp. 189–253. Brasilia, Brazil: EMBRAPA.

Peaslee, D. E., R. Isarangkura, and J. E. Leggett. 1981. Accumulation and translocation of zinc by two corn cultivars. *Agron. J.* 73:729–732.

Petersen, R. G. 1977. Use and misuse of multiple comparison procedures. *Agron. J.* 69:205–208.

Romero, N. A., C. C. Sheaffer, and G. L. Malzer. 1981. Potassium response of alfalfa in solution, sand and soil culture. *Agron. J.* 73:25–28.

Rys, G. J. and T. Phung. 1985. Nutrient solution pH control using dipolar buffers in studies of *Trifolium repens* L.: Nitrogen nutrition. *J. Exp. Bot.* 36:426–431.

Singh, D., K. S. Krishnan, and P. N. Bhargava. 1975. Plot size in crop cutting surveys on paddy rice. *J. Indian Soc. Agric. Stat.* 27:71–80.

Soil Science Society of America. 2008. *Glossary of Soil Science Terms*. Madison, WI: SSSA.

Steel, R. G., J. H. Dad, and D. J. H. Torrie. 1980. *Principles and Procedures of Statistics*, 2nd edn. New York: McGraw-Hill Book Company.

Stumpe, J. M., P. L. G. Vlek, S. K. Mughogho, and F. Gamy. 1989. Microplot size requirements for measuring balances of fertilizer nitrogen-15 applied to maize. *Soil Sci. Soc. Am. J.* 53:797–800.

Terman, G. L. 1974. Amounts of nutrients supplies for crops grown in pot experiments. *Commun. Soil Sci. Plant. Anal.* 5:115–121.

Terman, G. L. and J. J. Mortvedt. 1978. Nutrient effectiveness in relation to rates applied for pot experiments: I. Nitrogen and potassium. *Soil Sci. Soc. Am. J.* 42:297–302.

Teyker, R. H. and D. C. Hobbs. 1992. Growth and root morphology of corn as influenced by nitrogen form. *Agron. J.* 84:694–700.

Trelease, S. F. and H. M. Trelease. 1935. Physiologically balanced culture solution with stable hydrogen ion concentration. *Science* 78:438–439.

Wehr, J. D., L. M. Brown, and I. E. Vanderelst. 1986. Hydrogen ion buffering of culture media for algae from moderately acidic oligotrophic waters. *J. Phycol.* 22:88–94.

Wild, A., L. H. Jones, and J. H. Macduff. 1987. Uptake of mineral nutrients and crop growth: The use of flowing nutrient solutions. *Adv. Agron.* 41:171–219.

Wolkowski, R. P., T. A. Reisdore, and L. G. Bundy. 1988. Field plot technique comparison for estimating corn grain and dry matter yield. *Agron. J.* 80:278–280.

Yoshida, S., D. Forno, J. H. Cock, and K. A. Gomez. 1976. *Laboratory Manual for Physiological Studies of Rice*. Los Banos, Philippines: International Rice Research Institute.

Younis, M. A. and S. A. Tamimi. 1970. Optimum plot size for irrigated wheat grown in saline soils. *World Crops* 22:236.

Index

N

O

P

Q

R

S